nature

The Living Record of Science

《自然》百年科学经典

英汉对照版（平装本）

第九卷（下）

总顾问：李政道（Tsung-Dao Lee）

英方主编：Sir John Maddox
Sir Philip Campbell

中方主编：路甬祥

IX

1998-2001

U0220974

外语教学与研究出版社 · 麦克米伦教育 · 自然科研

FOREIGN LANGUAGE TEACHING AND RESEARCH PRESS · MACMILLAN EDUCATION · NATURE RESEARCH

北京 BEIJING

图书在版编目 (CIP) 数据

《自然》百年科学经典. 第九卷. 下, 1998—2001 : 英汉对照 ／（英）约翰·马多克斯 (John Maddox)，（英）菲利普·坎贝尔 (Philip Campbell)，路甬祥主编. —— 北京：外语教学与研究出版社，2019.12
ISBN 978-7-5213-1473-1

Ⅰ. ①自… Ⅱ. ①约… ②菲… ③路… Ⅲ. ①自然科学－文集－英、汉 Ⅳ. ①N53

中国版本图书馆 CIP 数据核字 (2020) 第 014446 号

地图审图号：GS (2018) 5219 号

出 版 人　徐建忠
项目统筹　章思英
项目负责　刘晓楠　黄小斌
责任编辑　王丽霞
责任校对　黄小斌
封面设计　孙莉明　曹志远
版式设计　孙莉明
出版发行　外语教学与研究出版社
社　　址　北京市西三环北路 19 号（100089）
网　　址　http://www.fltrp.com
印　　刷　北京华联印刷有限公司
开　　本　787×1092　1/16
印　　张　26.5
版　　次　2020 年 4 月第 1 版　2020 年 4 月第 1 次印刷
书　　号　ISBN 978-7-5213-1473-1
定　　价　168.00 元

购书咨询：（010）88819926　电子邮箱: club@fltrp.com
外研书店: https://waiyants.tmall.com
凡印刷、装订质量问题，请联系我社印制部
联系电话：（010）61207896　电子邮箱: zhijian@fltrp.com
凡侵权、盗版书籍线索，请联系我社法律事务部
举报电话：（010）88817519　电子邮箱: banquan@fltrp.com
物料号：314730001

记载人类文明
沟通世界文化
www.fltrp.com

《自然》百年科学经典（英汉对照版）

总顾问：李政道（Tsung-Dao Lee）

英方主编：Sir John Maddox　　中方主编：路甬祥

Sir Philip Campbell

编审委员会

英方编委　　　　　　　**中方编委**（以姓氏笔画为序）

Philip Ball　　　　　　　许智宏

Vikram Savkar　　　　　赵忠贤

David Swinbanks　　　　滕吉文

本卷审稿专家（以姓氏笔画为序）

于　军	万渝生	石锦卫	卢　磊	田立德	巩克瑞	刘守偈
许家喜	杜江峰	李　然	李军刚	李崇银	肖景发	狄增如
汪筱林	宋心琦	张颖奇	陈　阳	陈尔强	陈捷胤	周天军
周礼勇	赵凌霞	胡松年	俞永强	郭建栋	姬书安	曹　俊
彭小忠	韩汝珊	曾长青	解彬彬	黎　卓		

编译委员会

本卷翻译工作组稿人 （以姓氏笔画为序）

| 王丽霞 | 王晓蕾 | 王耀杨 | 刘 明 | 刘晓楠 | 关秀清 | 何 铭 |
| 沈乃澂 | 周家斌 | 郭红锋 | 黄小斌 | 蔡 迪 | 蔡则怡 | |

本卷翻译人员 （以姓氏笔画为序）

王海纳	王耀杨	毛晨晖	卢 皓	吕 静	任 奕	刘项琨
刘振明	刘皓芳	齐红艳	李 梅	杨 晶	肖 莉	余 恒
沈乃澂	张玉光	金世超	周家斌	姜 克	高如丽	崔 宁
董培智	蔡则怡					

本卷校对人员 （以姓氏笔画为序）

马 昊	王晓蕾	王德孚	牛慧冲	龙娉娉	卢 皓	田晓阳
吉 祥	任 奕	刘本琼	刘立云	刘项琨	刘琰璐	许静静
阮玉辉	李 龙	李 平	李 景	李 婷	李志军	李若男
李照涛	李霄霞	杨 晶	杨学良	吴 茜	邱珍琳	邱彩玉
何 敏	张亚盟	张茜楠	张美月	张瑶楠	陈 秀	陈贝贝
陈思原	周 晔	周少贞	郑旭峰	郑婧澜	赵凤轩	胡海霞
娄 研	洪雅强	贺舒雅	顾海成	徐 玲	黄小斌	第文龙
蒋世仰	韩少卿	焦晓林	蔡则怡	裴 琳	潘卫东	薛 陕
Eric Leher (澳)						

Contents
目录

Volume IX

(1998-2001)

Climate and Atmospheric History of the Past 420,000 Years from the Vostok Ice Core, Antarctica

J. R. Petit *et al.*

Editor's Note

This paper by a team of glaciologists and palaeoclimatologists reports the longest record of climate change ever to have been obtained by that time from cores of ice drilled from polar ice sheets. The composition of isotopes in the ice, and the presence of trapped gas bubbles and dust, enables changes in temperature, ice volume, and levels of atmospheric greenhouse gases and dust to be tracked over the hundreds of thousands of years during which the ice was deposited. The cores described here nearly double the previous timespan covered, allowing climate to be reconstructed over four ice-age cycles. Such records test our understanding of how climate change happened, in particular supporting the link between warming and greenhouse-gas concentrations.

The recent completion of drilling at Vostok station in East Antarctica has allowed the extension of the ice record of atmospheric composition and climate to the past four glacial–interglacial cycles. The succession of changes through each climate cycle and termination was similar, and atmospheric and climate properties oscillated between stable bounds. Interglacial periods differed in temporal evolution and duration. Atmospheric concentrations of carbon dioxide and methane correlate well with Antarctic air-temperature throughout the record. Present-day atmospheric burdens of these two important greenhouse gases seem to have been unprecedented during the past 420,000 years.

THE late Quaternary period (the past one million years) is punctuated by a series of large glacial–interglacial changes with cycles that last about 100,000 years (ref. 1). Glacial–interglacial climate changes are documented by complementary climate records[1,2] largely derived from deep sea sediments, continental deposits of flora, fauna and loess, and ice cores. These studies have documented the wide range of climate variability on Earth. They have shown that much of the variability occurs with periodicities corresponding to that of the precession, obliquity and eccentricity of the Earth's orbit[1,3]. But understanding how the climate system responds to this initial orbital forcing is still an important issue in palaeoclimatology, in particular for the generally strong ~100,000-year (100-kyr) cycle.

Ice cores give access to palaeoclimate series that includes local temperature and precipitation rate, moisture source conditions, wind strength and aerosol fluxes of marine,

南极洲东方站过去 42 万年冰芯气候和大气历史记录

珀蒂等

编者按

这篇文章由冰川学家和古气候学家组成的研究团队完成，它报道了当时从极地冰盖钻出的冰芯所能获得的最长气候变化记录。冰芯中同位素的组成以及气体包裹体和粉尘的存在，可用于捕捉过去几十万年冰沉积以来的温度、冰储量以及大气温室气体和粉尘水平的变化。这里描述的冰芯所包含的时间尺度几乎是前一次的两倍，可以在四个冰期–间冰期旋回内重建气候。这些记录检验了我们对气候变化如何发生的理解，特别是为气候变暖与温室气体浓度之间的联系提供了支持证据。

最近完成的东南极洲东方站（注：东方站也称沃斯托克站）的冰芯钻取工作，使大气成分与气候的冰芯记录延伸到了过去 4 个冰期–间冰期旋回。每一个气候旋回和冰期结束期连续变化都是相似的，而且大气和气候特性总是在稳定的范围内波动。间冰期在时间变化和持续时间方面存在差异。整个记录中，大气中二氧化碳和甲烷的浓度与南极的气温记录有很好的相关性。而现在这两种重要的温室气体在大气中的荷载量与过去 42 万年以来相比是史无前例的。

第四纪晚期（过去的 100 万年）被一系列持续时间约 10 万年的大的冰期–间冰期的周期性变化而分段（参考文献 1）。大量来自深海沉积物、大陆动植物、黄土沉积物及冰芯的记录[1,2]互为补充地说明了冰期–间冰期的气候变化。这些研究证实了地球气候变化有很大的幅度，表明许多周期性发生的变化对应于地球轨道参数岁差、地轴倾角及偏心率[1,3]的变化。但是，古气候学中，了解气候系统如何响应这些初始轨道驱动，特别对周期约 10 万年的强旋回来说，这一直是一个重要的问题。

通过冰芯记录可以了解一系列古气候变化序列，包括局地温度与降水率，水汽来源地条件，风的强度及源自海洋、火山、陆地、宇宙和人类活动的气溶胶通量。

volcanic, terrestrial, cosmogenic and anthropogenic origin. They are also unique with their entrapped air inclusions in providing direct records of past changes in atmospheric trace-gas composition. The ice-drilling project undertaken in the framework of a long-term collaboration between Russia, the United States and France at the Russian Vostok station in East Antarctica (78° S, 106° E, elevation 3,488 m, mean temperature −55°C) has already provided a wealth of such information for the past two glacial–interglacial cycles[4-13]. Glacial periods in Antarctica are characterized by much colder temperatures, reduced precipitation and more vigorous large-scale atmospheric circulation. There is a close correlation between Antarctic temperature and atmospheric concentrations of CO_2 and CH_4 (refs 5, 9). This discovery suggests that greenhouse gases are important as amplifiers of the initial orbital forcing and may have significantly contributed to the glacial–interglacial changes[14-16]. The Vostok ice cores were also used to infer an empirical estimate of the sensitivity of global climate to future anthropogenic increases of greenhouse-gas concentrations[15].

The recent completion of the ice-core drilling at Vostok allows us to considerably extend the ice-core record of climate properties at this site. In January 1998, the Vostok project yielded the deepest ice core ever recovered, reaching a depth of 3,623 m (ref. 17). Drilling then stopped ~120 m above the surface of the Vostok lake, a deep subglacial lake which extends below the ice sheet over a large area[18], in order to avoid any risk that drilling fluid would contaminate the lake water. Preliminary data[17] indicated that the Vostok ice-core record extended through four climate cycles, with ice slightly older than 400 kyr at a depth of 3,310 m, thus spanning a period comparable to that covered by numerous oceanic[1] and continental[2] records.

Here we present a series of detailed Vostok records covering this ~400-kyr period. We show that the main features of the more recent Vostok climate cycle resemble those observed in earlier cycles. In particular, we confirm the strong correlation between atmospheric greenhouse-gas concentrations and Antarctic temperature, as well as the strong imprint of obliquity and precession in most of the climate time series. Our records reveal both similarities and differences between the successive interglacial periods. They suggest the lead of Antarctic air temperature, and of atmospheric greenhouse-gas concentrations, with respect to global ice volume and Greenland air-temperature changes during glacial terminations.

The Ice Record

The data are shown in Figs 1, 2 and 3 (see Supplementary Information for the numerical data). They include the deuterium content of the ice (δD_{ice}, a proxy of local temperature change), the dust content (desert aerosols), the concentration of sodium (marine aerosol), and from the entrapped air the greenhouse gases CO_2 and CH_4, and the $\delta^{18}O$ of O_2 (hereafter $\delta^{18}O_{atm}$) which reflects changes in global ice volume and in the hydrological cycle[19]. (δD and $\delta^{18}O$ are defined in the legends to Figs 1 and 2, respectively.) All these measurements have been performed using methods previously described except for slight modifications (see figure legends).

冰芯独到之处在于其捕获的空气包裹体直接记录了过去大气中痕量气体成分的变化。这项在俄罗斯、美国、法国长期合作框架下开展的东南极洲俄罗斯东方站(78°S，106°E，海拔 3,488 米，平均气温 −55℃)冰芯钻取计划，已经为过去两个冰期–间冰期旋回[4-13]提供了大量的以上信息。冰期期间，南极地区温度更低，降水更少，大尺度大气环流更加剧烈。南极的温度与大气中 CO_2 和 CH_4 的浓度有密切的关系(参考文献 5 和 9)。这一发现表明这些温室气体是重要的原始轨道驱动作用的放大器，并可能对冰期–间冰期的变化[14-16]有显著的影响。未来人类活动导致的温室气体浓度升高会加速全球升温，东方站冰芯记录也可以被用来对这一敏感性做出经验性估算[15]。

最近东方站冰芯钻探的完成，使得我们可以大幅度地延伸在该站的冰芯关于气候属性的记录。1998 年 1 月，东方站计划到达了有史以来最长的冰芯深度，达到 3,623 米(参考文献 17)。为了避免钻孔液污染东方湖——一个存在于冰盖底部面积广大的深水冰下湖[18]，钻探停留在东方湖湖面以上约 120 米处。初步的资料[17]表明东方站冰芯记录覆盖了四个气候旋回，在 3,310 米深处冰的年龄稍老于 40 万年，因此，其覆盖的时间范围与很多海洋[1]或大陆[2]记录相当。

本文中我们展示了东方站冰芯包含这一段约 40 万年的一系列详细记录。研究表明东方站较新的气候旋回与早期的气候旋回在主要特征方面是相似的。特别是，我们证实了大气温室气体浓度与南极温度之间有强烈的相关性，而且在大多数气候时间序列里地轴倾角(即黄赤交角)和岁差也留下了强烈的印记。我们的记录揭示了相继的间冰期之间既有相似性也有差别。这些记录表明，在冰期结束期，南极温度和大气温室气体浓度的变化先于全球冰储量和格陵兰气温的变化。

冰 芯 记 录

图 1、2 和 3 展示了相关的资料(数值数据请见补充信息)，包括冰中氘的含量(δD_{ice}，一个代表局地气温变化的指标)，粉尘量(沙漠气溶胶)，钠的浓度(海洋气溶胶)，包裹气体的温室气体 CO_2 和 CH_4 以及 O_2 中的 $\delta^{18}O$ 含量(下文用 $\delta^{18}O_{atm}$ 表示)等资料，$\delta^{18}O_{atm}$ 的含量反映了全球冰储量和水循环[19]的变化。(δD 和 $\delta^{18}O$ 分别在图 1 和图 2 的图例说明中给出了定义)。所有这些测量都使用上述方法完成，只是稍有修改(参见图例说明)。

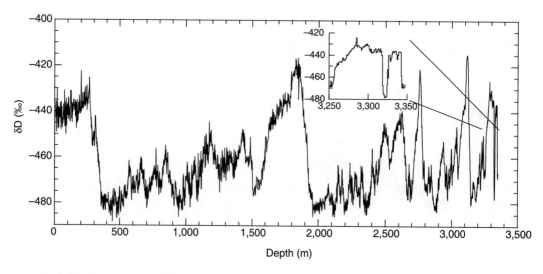

Fig. 1. The deuterium record. Deuterium content as a function of depth, expressed as δD (in ‰ with respect to Standard Mean Ocean Water, SMOW). This record combines data available down to 2,755 m (ref. 13) and new measurements performed on core 5G (continuous 1-m ice increments) from 2,755 m to 3,350 m. Measurement accuracy (1σ) is better than 1‰. Inset, the detailed deuterium profile for the lowest part of the record showing a δD excursion between 3,320 and 3,330 m. δD_{ice}(in ‰) = $[(D/H)_{sample}/(D/H)_{SMOW} - 1] \times 1,000$.

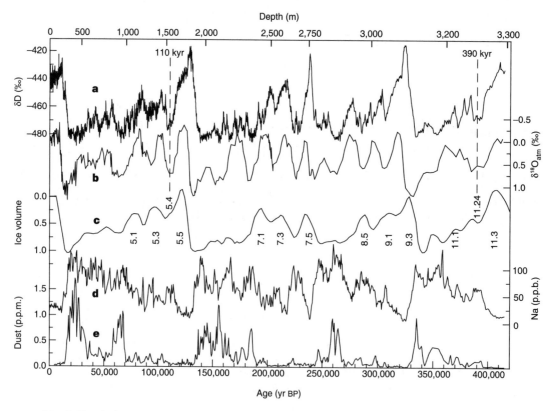

Fig. 2. Vostok time series and ice volume. Time series (GT4 timescale for ice on the lower axis, with indication of corresponding depths on the top axis and indication of the two fixed points at 110 and 390 kyr) of: **a**, deuterium profile (from Fig. 1); **b**, $\delta^{18}O_{atm}$ profile obtained combining published data[11,13,30] and

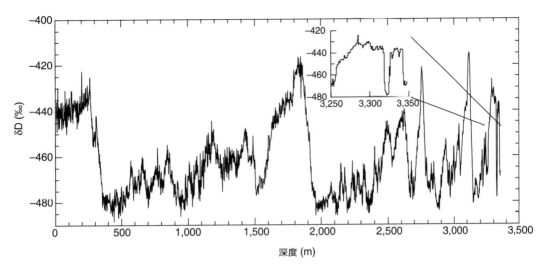

图 1. 氘的记录。氘含量随深度的变化，用 δD 表示（相对于标准平均大洋水（SMOW），用‰表示）。该记录结合了以前深达 2,755 米的资料（文献 13）以及从 5G 冰芯 2,755 米到 3,350 米段得到的新的测量结果（以 1 米间隔连续采样）。测量精度（1σ）优于 1‰。插图：记录中最低部分的详细氘剖面，展示了 δD 在 3,320 与 3,330 米之间的短期变化。δD$_{ice}$(‰) = [(D/H)$_{样本}$/(D/H)$_{SMOW}$−1] × 1,000。

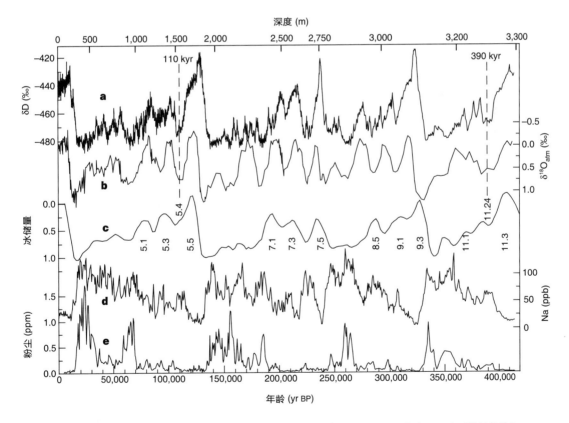

图 2. 东方站的时间序列和冰储量。各种时间序列（下面的横轴是冰的 GT4 时间标尺，上面的横轴是相应的深度，并标示出了 11 万年与 39 万年这两个固定点）：a，氘的剖面（来自图 1）；b，δ^{18}O$_{atm}$ 剖面结

81 new measurements performed below 2,760 m. The age of the gas is calculated as described in ref. 20; **c**, seawater $\delta^{18}O$ (ice volume proxy) and marine isotope stages adapted from Bassinot *et al.*[26]; **d**, sodium profile obtained by combination of published and new measurements (performed both at LGGE and RSMAS) with a mean sampling interval of 3–4 m (ng g^{-1} or p.p.b); and **e**, dust profile (volume of particles measured using a Coulter counter) combining published data[10,13] and extended below 2,760 m, every 4 m on the average (concentrations are expressed in µg g^{-1} or p.p.m. assuming that Antarctic dust has a density of 2,500 kg m^{-3}). $\delta^{18}O_{atm}$(in ‰) $= [(^{18}O/^{16}O)_{sample}/(^{18}O/^{16}O)_{standard} - 1] \times 1,000$; standard is modern air composition.

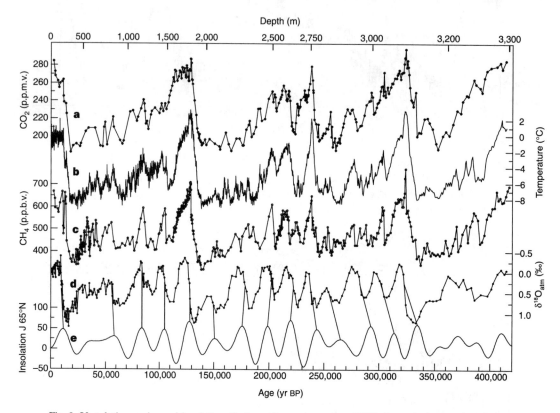

Fig. 3. Vostok time series and insolation. Series with respect to time (GT4 timescale for ice on the lower axis, with indication of corresponding depths on the top axis) of: **a**, CO_2; **b**, isotopic temperature of the atmosphere (see text); **c**, CH_4; **d**, $\delta^{18}O_{atm}$; and **e**, mid-June insolation at 65° N (in W m^{-2}) (ref. 3). CO_2 and CH_4 measurements have been performed using the methods and analytical procedures previously described[5,9]. However, the CO_2 measuring system has been slightly modified in order to increase the sensitivity of the CO_2 detection. The thermal conductivity chromatographic detector has been replaced by a flame ionization detector which measures CO_2 after its transformation into CH_4. The mean resolution of the CO_2 (CH_4) profile is about 1,500 (950) years. It goes up to about 6,000 years for CO_2 in the fractured zones and in the bottom part of the record, whereas the CH_4 time resolution ranges between a few tens of years to 4,500 years. The overall accuracy for CH_4 and CO_2 measurements are ± 20 p.p.b.v. and 2–3 p.p.m.v., respectively. No gravitational correction has been applied.

The detailed record of δD_{ice} (Fig. 1) confirms the main features of the third and fourth climate cycles previously illustrated by the coarse-resolution record[17]. However, a sudden decrease from interglacial-like to glacial-like values, rapidly followed by an abrupt return to interglacial-like values, occurs between 3,320 and 3,330 m. In addition, a transition from low to high CO_2 and CH_4 values (not shown) occurs at exactly the same depth. In undisturbed ice, the transition

合了已发表的资料[11,13,30]和 2,760 米以下的 81 个新测量值。气体的年龄根据参考文献 20 所述计算得到；c，海水 $\delta^{18}O$（代表冰储量的指标）值与海洋氧同位素阶段，来自于巴西诺等人的文章[26]；d，钠的剖面由已发表的测量值和平均每 3~4 米进行采样获得的新测量值（$ng \cdot g^{-1}$ 或 ppb）（在法国冰川与环境地球物理学实验室（LGGE）和美国迈阿密大学罗森斯蒂尔海洋和大气科学学院（RSMAS）都进行了相关的测量）相结合得到；e，粉尘剖面（用库尔特粒子仪测量粒子体积），结合了已发表的资料[10,13]和 2,760 米以下的延伸数据（平均每 4 米测量一个数据点）（粉尘浓度用 $\mu g \cdot g^{-1}$ 或 ppm 表示，假定南极粉尘的密度为 2,500 $kg \cdot m^{-3}$）。$\delta^{18}O_{atm}(‰) = [(^{18}O/^{16}O)_{样本}/(^{18}O/^{16}O)_{标准} - 1] \times 1,000$；标准是现代大气的成分。

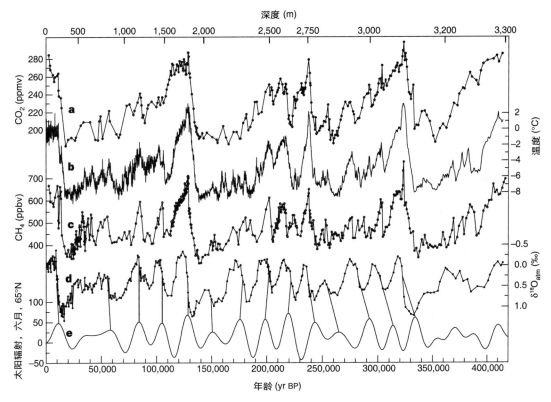

图 3. 东方站时间序列和太阳辐射。时间序列（下面的横轴是冰的 GT4 时间标尺，上面的横轴是相应的深度）a，CO_2；b，大气的同位素温度（参见正文）；c，CH_4；d，$\delta^{18}O_{atm}$；e，6 月中旬在 65° N 的太阳辐射（单位 $W \cdot m^{-2}$）（文献 3）。CO_2 与 CH_4 的测量按前述方法和分析流程[5,9]进行。然而，为了增加 CO_2 检测的敏感性，对测量方法稍有修改。用火焰离子化检测器替代了热导色谱检测器，去测量转化为 CH_4 后的 CO_2。$CO_2(CH_4)$ 剖面的平均时间分辨率约为 1,500(950) 年。在碎冰带和冰芯记录的底部，CO_2 的时间分辨率下降至约 6,000 年，而 CH_4 的时间分辨率范围跨度从几十年到 4,500 年。CH_4 和 CO_2 测量的总体精度分别为 ±20 ppbv 和 2~3 ppmv，这里没有经过重力校正。

　　详细的 δD_{ice} 记录（图 1）确认了以前用低分辨率记录[17]表示的第三和第四个气候旋回的主要特征。然而，在 3,320 米至 3,330 米之间出现了从类间冰期数值到类冰期数值的突然下降，紧接着迅速回到类间冰期数值。此外，在完全相同的深度，也发生了 CO_2 和 CH_4 从低值到高值的转变（图中未显示）。在未受扰动的冰内，出现大气

in atmospheric composition would be found a few metres lower (due to the difference between the age of the ice and the age of the gas[20]). Also, three volcanic ash layers, just a few centimetres apart but inclined in opposite directions, have been observed—10 m above this δD excursion (3,311 m). Similar inclined layers were observed in the deepest part of the GRIP and GISP2 ice cores from central Greenland, where they are believed to be associated with ice flow disturbances. Vostok climate records are thus probably disturbed below these ash layers, whereas none of the six records show any indication of disturbances above this level. We therefore limit the discussion of our new data sets to the upper 3,310 m of the ice core, that is, down to the interglacial corresponding to marine stage 11.3.

Lorius et al.[4] established a glaciological timescale for the first climate cycle of Vostok by combining an ice-flow model and an ice-accumulation model. This model was extended and modified in several studies[12,13]. The glaciological timescale provides a chronology based on physics, which makes no assumption about climate forcings or climate correlation except for one or two adopted control ages. Here, we further extend the Extended Glaciological Timescale (EGT) of Jouzel et al.[12] to derive GT4, which we adopt as our primary chronology (see Box 1). GT4 provides an age of 423 kyr at a depth of 3,310 m.

Box 1. The Vostok glaciological timescale

We use three basic assumptions[12] to derive our glaciological timescale (GT4); (1) the accumulation rate has in the past varied in proportion to the derivative of the water vapour saturation pressure with respect to temperature at the level where precipitation forms (see section on the isotope temperature record), (2) at any given time the accumulation between Vostok and Dome B (upstream of Vostok) varies linearly with distance along the line connecting those two sites, and (3) the Vostok ice at 1,534 m corresponds to marine stage 5.4 (110 kyr) and ice at 3,254 m corresponds to stage 11.2.4 (390 kyr).

Calculation of the strain-induced thinning of annual layers is now performed accounting for the existence of the subglacial Vostok lake. Indeed, running the ice-flow model[48] with no melting and no basal sliding as done for EGT[12] leads to an age > 1,000 kyr for the deepest level we consider here (3,310 m), which is much too old. Instead, we now allow for moderate melting and sliding. These processes diminish thinning for the lower part of the core and provide younger chronologies. We ran this age model[48] over a large range of values of the model parameters (present-day accumulation at Vostok, A, melting rate, M, and fraction of horizontal velocity due to base sliding, S) with this aim of matching the assumed ages at 1,534 and 3,254 m. This goal was first achieved (ages of 110 and 392 kyr) with $A = 1.96$ g cm^{-2} yr^{-1}, and M and S equal respectively to 0.4 mm yr^{-1} and 0.7 for the region 60 km around Vostok where the base is supposed to reach the melting point (we set $M = 0$ and $S = 0$ elsewhere). These values are in good agreement with observations for A (2.00 ± 0.04 g cm^{-2} yr^{-1} over the past 200 yr) and correspond to a reasonable set of parameters for M and S. We adopt this glaciological timescale (GT4), which gives an age of 423 kyr at 3,310 m, without further tuning (Fig. 2). GT4 never differs by more than 2 kyr from EGT over the last climate cycle and, in qualitative agreement with recent results[49], makes termination I slightly older (by ~700 yr). We

成分转折点的深度比同位素温度的相应转折点低几米(由于同层冰年龄和气体年龄不同[20])。同时,在 δD 发生短期流动的上方 10 米处(3,311 米)观测到相隔只几厘米、但倾斜方向相反的三个火山灰层。在格陵兰中部 GRIP 冰芯和 GISP2 冰芯的最底部也观测到了相似的倾斜层,人们相信这些倾斜层与冰流扰动有关。因此,东方站的气候记录中,火山灰层下面的部分可能受到扰动,然而这一层位之上,这六个记录中都没有表明存在有扰动的迹象。为此,我们把对这些新的数据的讨论限定在冰芯深度 3,310 米以上的部分,即只讨论到对应海洋氧同位素 11.3 阶的间冰期。

洛里于斯等人[4],用一个冰流模型结合积累模型建立了东方站的第一气候旋回的时间标尺。这一模式在数个研究中[12,13]得到拓展和修改。该冰川时间标尺基于物理学,除了采用一两个年龄控制点外,没有考虑任何有关气候强迫或气候相关性的假定。这里,我们进一步拓展了茹泽尔等人[12]建立的延长的冰川时标(EGT)以获得GT4,用作我们研究的基础年表(参见框 1)。按照 GT4 时标,3,310 米深处对应的年龄为 42.3 万年。

框 1. 东方站冰川的时间标尺

我们用 3 个基本假定[12]取得冰川的时间标尺(GT4):(1)过去的积累速率是与(在降水形成的高度处)饱和水汽压对温度的导数成比例的(参见同位素温度记录这一节);(2)在任何给定的时间,东方站与冰穹 B(在东方站的上游)的积累速率变化与两站连线的距离呈线性关系;(3)在东方站,1,534 米深度处的冰对应于海洋氧同位素 5.4 阶(距今 11 万年),3,254 米深度处的冰对应于海洋氧同位素 11.2.4 阶(距今 39 万年)。

计算考虑了冰下湖(东方湖)存在的情况下,冰川应力导致的年层减薄。的确,假设没有融化和没有基底滑动的冰流模型[48](如在 EGT[12]),运行模型得到的结果是冰川最底层(3,310 米)年龄大于 100 万年,这实在太古老了。不过,现在我们考虑有中等强度的融化和滑动。这些过程减少了深部冰芯的减薄,使冰川底部年龄更年轻。我们把模型的输入参数(东方站的现今积累量 A,融化率 M 和由于基底滑动引起的水平流速分量 S)的范围取大来运行这一模型[48],以使得运行结果与在 1,534 米与 3,254 米的设定年龄一致。当设定在东方站周围 60 km 的区域(假定基底达到熔点)$A = 1.96 \, g \cdot cm^{-2} \cdot yr^{-1}$, M 和 S 分别等于 0.4 mm·yr^{-1} 和 0.7(设其他地方的 $M = 0$ 且 $S = 0$)时,我们得到了预期的结果(年龄为距今 11 万年和距今 39.2 万年)。这些值与 A 的观测值(在过去的 200 年为 $2.00 \pm 0.04 \, g \cdot cm^{-2} \cdot yr^{-1}$)非常一致,并符合一组合理的参数 M 和 S 值。我们采用了这一冰川时间标尺(GT4),并未做进一步的调整,它在 3,310 米深度处给出了 42.3 万年的年龄值(图 2)。在最后一个气候旋回,GT4 从不会与 EGT 相差 2,000 年以上。与最近的研究结果[49]定性地一致,冰期结束期 I 稍微更早(提前了约 700 年)。我们注意到它对海洋

note that it provides a reasonable age for stage 7.5 (238 kyr) whereas Jouzel *et al.*[13] had to modify EGT for the second climate cycle by increasing the accumulation by 12% for ages older than 110 kyr. GT4 never differs by more than 4 kyr from the orbitally tuned timescale of Waelbroeck *et al.*[50] (defined back to 225 kyr), which is within the estimated uncertainty of this latter timescale. Overall, we have good arguments[11,50-52] to claim that the accuracy of GT4 should be better than ± 5 kyr for the past 110 kyr.

The strong relationship between $\delta^{18}O_{atm}$ and mid-June 65° N insolation changes (see text and Fig. 3) enables us to further evaluate the overall quality of GT4. We can use each well-marked transition from high to low $\delta^{18}O_{atm}$ to define a "control point" giving an orbitally tuned age. The mid-point of the last $\delta^{18}O_{atm}$ transition (~10 kyr ago) has nearly the same age as the insolation maximum (11 kyr). We assume that this correspondence also holds for earlier insolation maxima. The resulting control points (Fig. 3 and Table 1) are easy to define for the period over which the precessional cycle is well imprinted in 65° N insolation (approximately between 60 and 340 kyr) but not during stages 2 and 10 where insolation changes are small. The agreement between the $\delta^{18}O_{atm}$ control points and GT4 is remarkably good given the simple assumptions of both approaches. This conclusion stands despite the fact that we do not understand controls on $\delta^{18}O_{atm}$ sufficiently well enough to know about the stability of its phase with respect to insolation. We assume that the change in phase does not exceed ± 6 kyr (1/4 of a precessional period).

We conclude that accuracy of GT4 is always better than ± 15 kyr, better than ± 10 kyr for most of the record, and better than ± 5 kyr for the last 110 kyr. This timescale is quite adequate for the discussions here which focus on the climatic information contained in the Vostok records themselves.

Climate and Atmospheric Trends

Temperature. As a result of fractionation processes, the isotopic content of snow in East Antarctica (δD or $\delta^{18}O$) is linearly related to the temperature above the inversion level, T_I, where precipitation forms, and also to the surface temperature of the precipitation site, T_S (with $\Delta T_I = 0.67 \Delta T_S$, see ref. 6). We calculate temperature changes from the present temperature at the atmospheric level as $\Delta T_I = (\Delta \delta D_{ice} - 8\Delta\delta^{18}O_{sw})/9$, where $\Delta\delta^{18}O_{sw}$ is the globally averaged change from today's value of seawater $\delta^{18}O$, and 9‰ per °C is the spatial isotope/temperature gradient derived from deuterium data in this sector of East Antarctica[21]. We applied the above relationship to calculate ΔT_S. This approach underestimates ΔT_S by a factor of ~2 in Greenland[22] and, possibly, by up to 50% in Antarctica[23]. However, recent model results suggest that any underestimation of temperature changes from this equation is small for Antarctica[24,25].

To calculate ΔT_I from δD, we need to adopt a curve for the change in the isotopic composition of sea water versus time and correlate it with Vostok. We use the stacked $\delta^{18}O_{sw}$ record of Bassinot *et al.*[26], scaled with respect to the V19-30 marine sediment record over their common part that covers the past 340 kyr (ref. 27) (Fig. 2). To avoid distortions in the

同位素 7.5 阶（距今 23.8 万年）给出了一个合理的定年结果。而为了达到这一结果，茹泽尔等人[13]必须在 EGT 模型中调整第二气候旋回的输入参数，把 11 万年以前的积累率增加 12%。GT4 与韦尔布罗克等人[50]用轨道参数调整过的时间标尺（定义至距今 22.5 万年）相差从来没有超过 4,000 年，这一差值处在后者估计的不确定性范围之内。总之，我们有很好的论据[11,50-52]支撑，认为对过去的 11 万年，GT4 的定年精度应该优于 ±5,000 年。

从 $\delta^{18}O_{atm}$ 与 6 月中旬 65° N 的太阳辐射变化（见本文和图 3）之间的密切关系，我们可进一步评估 GT4 的总体效果。我们可以用每一个 $\delta^{18}O_{atm}$ 从高到低的显著转折点，定义一个能给出轨道修正时间的"控制点"。最后一个 $\delta^{18}O_{atm}$ 转折（约 1 万年前）的中点几乎与太阳辐射最大的年龄（距今 1.1 万年）一致。我们假定这种对应也适用于更早的太阳辐射最大值。对于 65° N 太阳辐射曲线中岁差周期显著的阶段（大约距今 6 万年至距今 34 万年之间），控制点（图 3 和表 1）是容易确定的，但 2 阶和 10 阶除外，这两个阶段太阳辐射变化很小。在给定两种方法的一些简单假设后，$\delta^{18}O_{atm}$ 控制点与 GT4 之间的一致性非常好。尽管我们对 $\delta^{18}O_{atm}$ 受到的控制还不太了解，因而不十分清楚 $\delta^{18}O_{atm}$ 与太阳辐射的相位变化的稳定性情况，但是结论仍然站得住脚。我们假定相位的变化不超过 ±6,000 年（岁差周期的 1/4）。

我们总结得出 GT4 的定年精确度总是优于 ±1.5 万年，对于大多数记录精度优于 ±1 万年，对最近的 11 万年精度优于 ±5,000 年。这一时间序列对于这里讨论东方站记录中包含的气候信息本身来说已经足够了。

气候与大气趋势

温度 由于同位素分馏效应，东南极洲雪的同位素值（δD 或 $\delta^{18}O$）与逆温层之上（降水形成处）的温度 T_I 呈线性相关，也与降水点的地面温度 T_s 呈线性相关（$\Delta T_I = 0.67\Delta T_s$，见参考文献 6）。我们从大气层的现今温度来计算温度的变化，$\Delta T_I = (\Delta\delta D_{冰} - 8\Delta\delta^{18}O_{sw})/9$，其中 $\Delta\delta^{18}O_{sw}$ 是相对于现今海水 $\delta^{18}O$ 值的全球平均变化，9‰/℃ 是从东南极洲区域内的氘数据得到的同位素与温度的空间梯度值[21]。我们应用上述关系计算 ΔT_s。这方法计算出来的格陵兰的 ΔT_s 被低估为约实际值的二分之一[22]，计算出来的在南极洲的值可能被低估了（最多）50%[23]。然而从现在的模式结果来看，从此等式得到的任何对南极洲温度变化的低估，都偏小了[24,25]。

为了通过 δD 计算 ΔT_I，我们需要采用海水同位素成分对时间的变化曲线，并将它与东方站相关联。我们将巴西诺等人[26]积累的 $\delta^{18}O_{sw}$ 记录，与 V19-30 海洋沉积物记录中的共同部分（覆盖了过去 34 万年）进行标定（文献 27）（图 2）。为避免在计

calculation of ΔT_{I} linked with dating uncertainties, we correlate the records by performing a peak to peak adjustment between the ice and ocean isotopic records. The $\delta^{18}O_{sw}$ correction corresponds to a maximum ΔT_{I} correction of $\sim 1°C$ and associated uncertainties are therefore small. We do not attempt to correct ΔT_{I} either for the change of the altitude of the ice sheet or for the origin of the ice upstream of Vostok[13]; these terms are very poorly known and, in any case, are also small ($< 1°C$).

The overall amplitude of the glacial–interglacial temperature change is $\sim 8°C$ for ΔT_{I} (inversion level) and $\sim 12°C$ for ΔT_{S}, the temperature at the surface (Fig. 3). Broad features of this record are thought to be of large geographical significance (Antarctica and part of the Southern Hemisphere), at least qualitatively. When examined in detail, however, the Vostok record may differ from coastal[28] sites in East Antarctica and perhaps from West Antarctica as well.

Jouzel *et al.*[13] noted that temperature variations estimated from deuterium were similar for the last two glacial periods. The third and fourth climate cycles are of shorter duration than the first two cycles in the Vostok record. The same is true in the deep-sea record, where the third and fourth cycles span four precessional cycles rather than five as for the last two cycles (Fig. 3). Despite this difference, one observes, for all four climate cycles, the same "sawtooth" sequence of a warm interglacial (stages 11.3, 9.3, 7.5 and 5.5), followed by increasingly colder interstadial events, and ending with a rapid return towards the following interglacial. The coolest part of each glacial period occurs just before the glacial termination, except for the third cycle. This may reflect the fact that the June 65° N insolation minimum preceding this transition (255 kyr ago) has higher insolation than the previous one (280 kyr ago), unlike the three other glacial periods. Nonetheless, minimum temperatures are remarkably similar, within 1°C, for the four climate cycles. The new data confirm that the warmest temperature at stage 7.5 was slightly warmer than the Holocene[13], and show that stage 9.3 (where the highest deuterium value, −414.8‰, is found) was at least as warm as stage 5.5. That part of stage 11.3, which is present in Vostok, does not correspond to a particularly warm climate as suggested for this period by deep-sea sediment records[29]. As noted above, however, the Vostok records are probably disturbed below 3,310 m, and we may not have sampled the warmest ice of this interglacial. In general, climate cycles are more uniform at Vostok than in deep-sea core records[1]. The climate record makes it unlikely that the West Antarctic ice sheet collapsed during the past 420 kyr (or at least shows a marked insensitivity of the central part of East Antarctica and its climate to such a disintegration).

The power spectrum of ΔT_{I} (Fig. 4) shows a large concentration of variance (37%) in the 100-kyr band along with a significant concentration (23%) in the obliquity band (peak at 41 kyr). This strong obliquity component is roughly in phase with the annual insolation at the Vostok site[4,6,15]. The variability of annual insolation at 78° S is relatively large, 7% (ref. 3). This supports the notion that annual insolation changes in high southern latitudes influence Vostok temperature[15]. These changes may, in particular, contribute to the initiation of Antarctic warming during major terminations, which (as we show below) herald the start of deglaciation.

算 ΔT_I 时与定年不确定度有关的失真，我们通过对冰与海洋之间同位素记录作峰值对峰值的调准。$\delta^{18}O_{sw}$ 的校正引起的 ΔT_I 最大校正值不会超过约 1℃，因此导致的不确定度是很小的。我们不打算从冰盖的高度变化或是东方站上游冰流的源地[13]对 ΔT_I 进行校正，我们对这些方面知之甚少，并且不管怎样，这些项对校正的温度影响也很小（<1℃）。

冰期-间冰期温度变化的总体幅度如下：逆温层的 ΔT_I 约为 8℃，地面的 ΔT_s 约为 12℃（图 3）。该记录具有更大地理范围（南极洲和南半球部分区域）上的意义，至少是定性的意义。然而当仔细检查时会发现，东方站的记录与南极洲东部海岸的记录存在差别[28]，也许同南极洲西部的也不同。

茹泽尔等人[13]指出从氘估计出的最后两个冰期的温度变化是相似的。在东方站的记录中，第三和第四气候旋回的持续时间比前两个周期的持续时间要短。深海记录也显示同样的特征，其中第三和第四旋回跨越的岁差周期是四个而不是像上两个气候旋回那样跨越了五个岁差周期（图 3）。尽管存在这一差异，对于这所有四个气候旋回，可以观测到相同的温度变化"锯齿形"序列温暖间冰阶（11.3 阶，9.3 阶，7.5 阶和 5.5 阶），其后紧随着逐渐转冷的间冰阶，最后以快速返回到间冰阶结束。除第三个旋回外，每个冰期最冷部分都出现在冰期结束期前。这也许反映了一个事实：出现在过渡期（25.5 万年前）之前的 65°N 6 月太阳辐射最小值比前一个（28 万年前）最小值要高，这与其他三个冰期阶段不同。无论如何，对于这四个气候旋回来说，最低温度明显相似，变化范围都在 1℃以内。新的资料证实 7.5 阶的最暖温度比全新世[13]还稍暖一点，9.3 阶（那里出现了氘的最高值 −414.8‰）至少和 5.5 阶一样温暖。东方站冰芯记录的 11.3 阶部分并不与深海沉积物记录的这一时期特别温暖的气候相符[29]。然而，如上所述，在 3,310 米以下东方站冰芯记录可能受到干扰，我们可能没有取到这一间冰期最温暖的冰样本。一般来说，东方站的气候旋回比深海岩芯记录[1]更均匀。此气候记录表明在过去 42 万年西南极冰盖没有可能垮塌过（或者至少表明东南极中部及其气候对这种垮塌表现不敏感）。

ΔT_I 的功率谱（图 4）在 10 万年带内出现一个很大的谱密度（37%），在地轴倾角带（峰值在 4.1 万年）还有一明显的谱密度（23%）。这个强的地轴倾角分量大致与东方站的年太阳辐射同相[4,6,15]。在 78°S，年日照量的变化相对较大，为 7%（文献 3）。这支持了南半球高纬度地区年日照量变化对东方站温度是有影响的观点[15]。特别是，这些变化促使主要冰期结束期南极变暖的开始，它也预示着冰川消退的开始（请见我们下面的说明）。

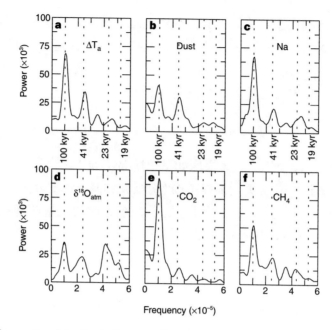

Fig. 4. Spectral properties of the Vostok time series. Frequency distribution (in cycles yr^{-1}) of the normalized variance power spectrum (arbitrary units). Spectral analysis was done using the Blackman-Tukey method (calculations were performed with the Analyseries software[47]): **a**, isotopic temperature; **b**, dust; **c**, sodium; **d**, $\delta^{18}O_{atm}$; **e**, CO_2; and **f**, CH_4. Vertical lines correspond to periodicities of 100, 41, 23 and 19 kyr.

Table 1. Comparison of the glaciological timescale and orbitally derived information

Depth (m)	Insolation maximum (kyr)	Age GT4 (kyr)	Difference (kyr)
305	11	10	1
900	58	57	1
1,213	84	83	1
1,528	105	105	0
1,863	128	128	0
2,110	151	150	1
2,350	176	179	−3
2,530	199	203	−4
2,683	220	222	−2
2,788	244	239	5
2,863	265	255	10
2,972	293	282	11
3,042	314	301	13
3,119	335	322	13

Control points were derived assuming a correspondence between maximum 65° N mid-June insolation and $\delta^{18}O_{atm}$ mid-transitions. Age GT4 refers to the age of the gas obtained after correction for the gas-age/ice-age differences[20].

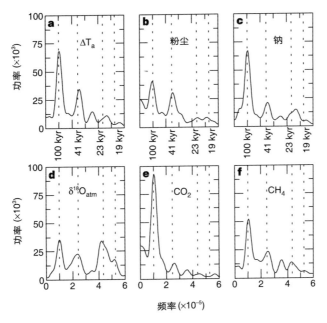

图 4. 东方站时间序列谱特征。标准方差功率谱（任意单位）的频率分布（周期：年）。用布莱克曼–图基方法（利用 Analyseries 软件[47] 计算的）完成了谱分析。a，同位素温度 ΔT_a；b，粉尘；c，钠；d，$\delta^{18}O_{atm}$；e，CO_2；f，CH_4；垂线相应于距今 10 万年，4.1 万年，2.3 万年和 1.9 万年的周期。

表 1. 冰期时间标尺和轨道推导信息的对比

深度（m）	最大太阳辐射（千年）	GT4 年龄（千年）	差值（千年）
305	11	10	1
900	58	57	1
1,213	84	83	1
1,528	105	105	0
1,863	128	128	0
2,110	151	150	1
2,350	176	179	−3
2,530	199	203	−4
2,683	220	222	−2
2,788	244	239	5
2,863	265	255	10
2,972	293	282	11
3,042	314	301	13
3,119	335	322	13

假定 65° N 的 6 月中旬太阳辐射的最大值对应 $\delta^{18}O_{atm}$ 的转折点的中点，从而可得到控制点，GT4 年龄参考经气体年龄/冰年龄差值校正后得来的气体年龄[20]。

There is little variance (11%) in ΔT_{I} around precessional periodicities (23 and 19 kyr). In this band, the position of the spectral peaks is affected by uncertainties in the timescale. To illustrate this point, we carried out, as a sensitivity test, a spectral analysis using the control points provided by the $\delta^{18}O_{atm}$ record (see Table 1). The position and strength of the 100- and 40-kyr-spectral peaks are unaffected, whereas the power spectrum is significantly modified for periodicities lower than 30 kyr.

Insolation. $\delta^{18}O_{atm}$ strongly depends on climate and related properties, which reflect the direct or indirect influence of insolation[19]. As a result, there is a striking resemblance between $\delta^{18}O_{atm}$ and mid-June insolation at 65° N for the entire Vostok record (Fig. 3). This provides information on the validity of our glaciological timescale (see Box 1). The precessional frequencies, which do not account for much variance in ΔT_{I}, are strongly imprinted in the $\delta^{18}O_{atm}$ record (36% of the variance in this band, Fig. 4). In addition, the remarkable agreement observed back to stage 7.5 between the amplitude of the filtered components of the mid-June insolation at 65° N and $\delta^{18}O_{atm}$ in the precessional band[13] holds true over the last four climate cycles (not shown). As suggested by the high variance of $\delta^{18}O_{atm}$ in the precessional band, this orbital frequency is also reflected in the Dole effect, the difference between $\delta^{18}O_{atm}$ and $\delta^{18}O_{sw}$, confirming results obtained on the last two climate cycles[19,30].

Aerosols. Figure 2 shows records of aerosols of different origins. The sodium record represents mainly sea-salt aerosol entrained from the ocean surface, whereas the dust record corresponds to the small size fraction ($\sim$2 μm) of the aerosol produced by the continent. The extension of the Vostok record confirms much higher fallout during cold glacial periods than during interglacials. Concentrations range up to 120 ng g^{-1}, that is, 3 to 4 times the Holocene value, for sea-salt. For dust, they rise from about 50 ng g^{-1} during interglacials to 1,000–2,000 ng g^{-1} during cold stages 2, 4, 6, 8 and 10.

The sodium concentration is closely anti-correlated with isotopic temperature ($r^2 = -0.70$ over the past 420 kyr). The power spectrum of the sodium concentration, like that of ΔT_{I}, shows periodicities around 100, 40 and 20 kyr (Fig. 4). Conditions prevailing during the present-day austral winter could help explain the observed glacial/interglacial changes in sodium. The seasonal increase of marine aerosol observed in the atmosphere and snow at the South Pole in September[31] corresponds to the maximum extent of sea ice; this is because the more distant source effect is compensated by the greater cyclonic activity, and by the more efficient zonal and meridional atmospheric circulation probably driven by the steeper meridional (ocean–Antarctica) temperature gradient. These modern winter conditions may be an analogue for glacial climates, supporting the apparent close anti-correlation between sodium concentration and temperature at Vostok.

The extension of the Vostok dust record confirms that continental aridity, dust mobilization and transport are more prevalent during glacial climates, as also reflected globally in many dust records (see ref. 10 and references therein). The presence of larger particles in the Vostok record, at least during the Last Glacial Maximum[10], indicates that the atmospheric circulation at high southern latitudes was more turbulent at that time. Lower atmospheric

ΔT_1 在岁差周期附近（距今 2.3 万年与距今 1.9 万年）的方差很小（11%）。在此带内，谱峰的位置受到时间标尺不确定性的影响。为了说明这一点，本文用 $\delta^{18}O_{atm}$ 记录（表 1）确定的控制点进行了谱分析，作为敏感性试验。它对距今 10 万年和 4 万年的谱峰的位置和强度没有影响，然而功率谱对低于 3 万年的周期有明显的影响。

太阳辐射　$\delta^{18}O_{atm}$ 强烈地依赖气候及其有关的特征，它们反映了太阳辐射的直接或间接影响[19]。因此，在东方站的全部记录中（图 3），$\delta^{18}O_{atm}$ 与 6 月中旬 65°N 太阳辐射之间极为相似。这为我们检验冰川时间标尺提供了信息（见框 1）。岁差频率在 ΔT_1 记录中并未占有很大的方差，但 $\delta^{18}O_{atm}$ 记录中留下了强烈的印记（在带中有 36% 的方差，图 4）。另外，65°N 的 6 月中旬的滤波分量的振幅与 $\delta^{18}O_{atm}$ 之间存在岁差带上显著的一致性，可以一直追踪到 7.5 阶[13]，这点在最后四个气候旋回中同样成立（文中未展示）。$\delta^{18}O_{atm}$ 在岁差带内大的方差，表明轨道的频率也反映在多尔效应（$\delta^{18}O_{atm}$ 与 $\delta^{18}O_{sw}$ 之间存在差异）内，证实了最后两个气候旋回中得到的结果[19,30]。

气溶胶　图 2 表示不同来源的气溶胶记录。钠的记录主要代表从海洋表面带来的海盐气溶胶，而粉尘记录则对应于陆地产生的气溶胶的细粒部分（约 2 μm）。东方站记录的延伸，证实了寒冷的冰期比间冰期沉降率要高得多。对于海盐，浓度范围可达 120 ng·g⁻¹，是全新世的 3~4 倍。对于粉尘，它们从间冰期 50 ng·g⁻¹ 左右上升到寒冷海洋同位素阶段 2、4、6、8 和 10 的 1,000~2,000 ng·g⁻¹。

钠的浓度与同位素温度几乎是反相关的（在过去的 42 万年 $r^2 = -0.70$）。钠浓度的功率谱，像 ΔT_1 的功率谱一样，在 10 万年、4 万年和 2 万年左右呈周期性（图 4）。当前南方冬季的主要条件能帮助说明钠在冰期和间冰期之间的变化。在南极点 9 月观察到的大气和雪中海洋气溶胶的季节性增长[31]与海冰的最大范围对应，这是因为更远源地的影响被以下因素所补偿：更大的气旋活动，以及可能由更陡的经向（海洋−南极）温度梯度导致的更有效的纬向和经向大气环流。这些现代冬季的条件可以比拟成冰期气候，来说明东方站钠浓度与温度的显著反相关关系。

这里延长的东方站冰芯粉尘记录证实，大陆干旱和粉尘的运移与输送在冰期气候时更为活跃，这也反映在全球许多其他粉尘记录中（参考文献 10 及其中的参考文献）。东方站记录中（至少在末次冰盛期[10]）更大颗粒的出现，指示那时南半球高纬地区的大气环流更为湍急。大气含水量和水循环通量的减少可能同样是整个冰期粉

moisture content and reduced hydrological fluxes may also have contributed significantly (that is, one order of magnitude[32]) to the very large increases of dust fallout during full glacial periods because of a lower aerosol-removal efficiency.

Unlike sodium concentration, the dust record is not well correlated with temperature (see below) and shows large concentrations of variance in the 100- and 41-kyr spectral bands (Fig. 4). The Vostok dust record is, in this respect, similar to the tropical Atlantic dust record of de Menocal[33] who attributes these spectral characteristics to the progressive glaciation of the Northern Hemisphere and the greater involvement of the deep ocean circulation. We suggest that there also may be some link between the Vostok dust record and deep ocean circulation through the extension of sea ice in the South Atlantic Ocean, itself thought to be coeval with a reduced deep ocean circulation[34]. Our suggestion is based on the fact that the dust source for the East Antarctic plateau appears to be South America, most likely the Patagonian plain, during all climate states of the past 420 kyr (refs 35, 36). The extension of sea ice in the South Atlantic during glacial times greatly affects South American climate, with a more northerly position of the polar front and the belt of Westerlies pushed northward over the Andes. This should lead, in these mountainous areas, to intense glacial and fluvial erosion, and to colder and drier climate with extensive dust mobilization (as evidenced by glacial loess deposits in Patagonia). Northward extension of sea ice also leads to a steeper meridional temperature gradient and to more efficient poleward transport. Therefore, Vostok dust peaks would correspond to periods of increased sea-ice extent in the South Atlantic Ocean, probably associated with reduced deep ocean circulation (thus explaining observed similarities with the tropical ocean dust record[33]).

Greenhouse gases. The extension of the greenhouse-gas record shows that the main trends of CO_2 and CH_4 concentration changes are similar for each glacial cycle (Fig. 3). Major transitions from the lowest to the highest values are associated with glacial–interglacial transitions. At these times, the atmospheric concentrations of CO_2 rises from 180 to 280–300 p.p.m.v. and that of CH_4 rises from 320–350 to 650–770 p.p.b.v. There are significant differences between the CH_4 concentration change associated with deglaciations. Termination III shows the smallest CH_4 increase, whereas termination IV shows the largest (Fig. 5). Differences in the changes over deglaciations are less significant for CO_2. The decrease of CO_2 to the minimum values of glacial times is slower than its increase towards interglacial levels, confirming the sawtooth record of this property. CH_4 also decreases slowly to its background level, but with a series of superimposed peaks whose amplitude decreases during the course of each glaciation. Each CH_4 peak is itself characterized by rapid increases and slower decreases, but our resolution is currently inadequate to capture the detail of millennial-scale CH_4 variations. During glacial inception, Antarctic temperature and CH_4 concentrations decrease in phase. The CO_2 decrease lags the temperature decrease by several kyr and may be either steep (as at the end of interglacials 5.5 and 7.5) or more regular (at the end of interglacials 9.3 and 11.3). The differences in concentration–time profiles of CO_2 and CH_4 are reflected in the power spectra (Fig. 4). The 100-kyr component dominates both CO_2 and CH_4 records. However, the obliquity and precession components are much stronger for CH_4 than for CO_2.

410

尘沉降大幅增长的重要原因之一 (一个数量级[32]), 因为气溶胶清除效率较低了。

与钠浓度不同, 粉尘记录与温度没有很好的相关性 (参见下文), 并在 10 万年和 4.1 万年的谱带内有很大的方差变化 (图 4)。在这方面, 东方站的粉尘记录与德梅诺卡尔[33]的热带大西洋粉尘记录相似, 德梅诺卡尔将这些 (粉尘记录的) 谱特征归因于北半球渐进的冰川活动以及深海环流进一步参与。我们提出在东方站粉尘记录和深海环流之间, 也可能通过南大西洋海冰的扩展产生一些联系, 这一扩展被认为与深海环流减弱[34]是同步的。我们的设想基于以下事实: 南极东部高原的粉尘在过去 42 万年的全部气候阶段 (文献 35 和 36) 都来源于南美洲, 其中最有可能是巴塔哥尼亚平原。南大西洋海冰在冰期内的扩展, 对南美的气候有很大影响, 导致极锋位置更加靠北, 西风带向北推进越过安第斯山。这将导致这些山区遭受更强烈的冰川侵蚀和河流侵蚀并使气候变得愈加干冷, 粉尘运移范围增大 (如巴塔哥尼亚冰川黄土的沉积所证实的那样)。海冰向北扩张也使经向温度梯度变得更陡, 并使极向输送效率更高。因此东方站的粉尘峰值对应于南大西洋海冰扩张的时期, 后者可能与深海洋流的减弱有关 (因而解释了为何与观测的热带海洋粉尘记录相似[33])。

温室气体 延展的冰芯温室气体记录表明 CO_2 和 CH_4 浓度变化的主要趋势在每个冰期旋回里都相似 (图 3)。从最低到最高值的主要转变阶段与冰期到间冰期的转变密切相关。这些时期大气中 CO_2 浓度从 180 ppmv 上升到 280~300 ppmv, CH_4 浓度从 320~350 ppbv 上升到 650~770 ppbv。但不同的冰消期 CH_4 浓度变化之间有明显的不同。冰消期 III CH_4 增长最小, 而冰消期 IV CH_4 增长最大 (图 5)。在冰消期, CO_2 变化的差异则并不明显。CO_2 减少到冰期最小值的过程比向间冰期增加的过程要慢, 证实了其锯齿形的特征。在每一个冰期过程中, CH_4 也缓慢减小到背景值, 但存在一系列叠加峰, 并且它们的幅度逐渐减小。每一个 CH_4 峰均以快速增长和缓慢减小为特征, 但是, 我们的分辨率不足以捕获 CH_4 的千年尺度变化的细节。在冰期开始时, 南极温度和 CH_4 浓度同步减小。CO_2 浓度下降比温度减小要滞后几千年, 并且可能很陡 (如在间冰期 5.5 和 7.5 的末期) 或比较规则 (如在间冰期 9.3 和 11.3 的末期)。CO_2 和 CH_4 的浓度随时间变化的差异也反映在功率谱内 (图 4)。10 万年的分量在 CO_2 和 CH_4 记录占主要地位, 然而 CH_4 地轴倾角和岁差分量强于 CO_2 的相应值。

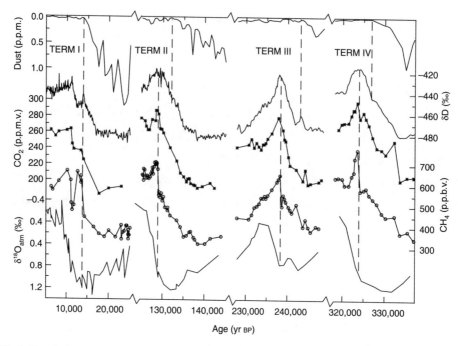

Fig 5. Vostok time series during glacial terminations. Variations with respect to time (GT4) of: **a**, dust; **b**, δD of ice (temperature proxy); **c**, CO₂; **d**, CH₄; and **e**, δ¹⁸O_atm for glacial terminations I to IV and the subsequent interglacial periods (Holocene, stage 5.5, stage 7.5 and stage 9.3).

The extension of the greenhouse-gas record shows that present-day levels of CO_2 and CH_4 (~360 p.p.m.v. and ~1,700 p.p.b.v., respectively) are unprecedented during the past 420 kyr. Pre-industrial Holocene levels (~280 p.p.m.v. and ~650 p.p.b.v., respectively) are found during all interglacials, while values higher than these are found in stages 5.5, 9.3 and 11.3 (this last stage is probably incomplete), with the highest values during stage 9.3 (300 p.p.m.v. and 780 p.p.b.v., respectively).

The overall correlation between our CO_2 and CH_4 records and the Antarctic isotopic temperature[5,9,16] is remarkable ($r^2 = 0.71$ and 0.73 for CO_2 and CH_4, respectively). This high correlation indicates that CO_2 and CH_4 may have contributed to the glacial–interglacial changes over this entire period by amplifying the orbital forcing along with albedo, and possibly other changes[15,16]. We have calculated the direct radiative forcing corresponding to the CO_2, CH_4 and N_2O changes[16]. The largest CO_2 change, which occurs between stages 10 and 9, implies a direct radiative warming of $\Delta T_{rad} = 0.75°C$. Adding the effects of CH_4 and N_2O at this termination increases the forcing to 0.95°C (here we assume that N_2O varies with climate as during termination I[37]). This initial forcing is amplified by positive feedbacks associated with water vapour, sea ice, and possibly clouds (although in a different way for a "doubled CO_2" situation than for a glacial climate[38]). The total glacial–interglacial forcing is important (~3 W m⁻²), representing 80% of that corresponding to the difference between a "doubled CO_2" world and modern CO_2 climate. Results from various climate simulations[39,40] make it reasonable to assume that greenhouse gases have, at a

412

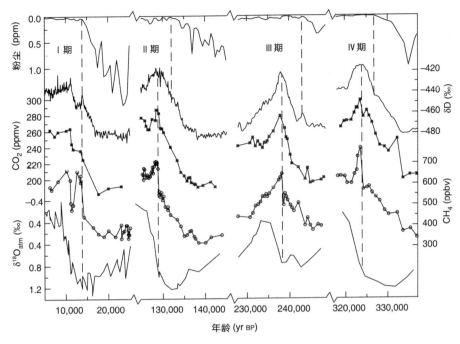

图 5. 东方站在冰消期时间序列。对应 (GT4) 时间尺度变化序列：a，粉尘；b，冰的 δD (温度指标)；c，CO_2；d，CH_4；e，冰消期 I 到 IV 及其后的间冰期 (全新世，5.5 阶，7.5 阶和 9.3 阶) 的 $\delta^{18}O_{atm}$。

延伸的温室气体记录显示目前的 CO_2 和 CH_4 浓度水平 (分别约为 360 ppmv 和 1,700 ppbv) 在过去 42 万年是史无前例的。工业化前的全新世浓度水平 (分别为约 280 ppmv 和约 650 ppbv) 出现于所有的间冰期，更高的数值出现在 5.5 阶，9.3 阶和 11.3 阶 (最后这一阶可能不完整)，最高值出现在 9.3 阶 (CO_2 和 CH_4 浓度水平分别为 300 ppmv 和 780 ppbv)。

总体上，CO_2 和 CH_4 记录与南极同位素温度[5,9,16]之间的相关性是显著的 (对 CO_2 和 CH_4 的相关系数分别为 $r^2 = 0.71$ 和 $r^2 = 0.73$)。这种高度相关性说明 CO_2 和 CH_4 在整个阶段都推动了冰期–间冰期的变化，这种作用是通过放大轨道强迫与反照率以及其他可能的变化[15,16]来实现的。我们计算了对应 CO_2，CH_4 和 N_2O 变化[16]的直接辐射强迫。最大的一次 CO_2 变化发生在 10 阶和 9 阶之间，其直接辐射强迫对应升温幅度为 $\Delta T_{rad} = 0.75℃$。再考虑到 CH_4 和 N_2O 的影响，在这次冰消期的辐射强迫导致升温增加到 0.95℃ (这里我们假定 N_2O 随气候变化与冰期 I[37]的情况一样)。初始强迫被与水汽、海冰，可能还与云有关 (虽然对 "2 倍 CO_2" 情景和冰期气候情景，反馈的方式不同[38]) 的正反馈放大。整个冰期–间冰期的辐射强迫是重要的 (约 3W·m^{-2})，相当于 "2 倍 CO_2" 情景和现代 CO_2 浓度情景二者之间气候变化的 80%。从各种气候模拟的结果[39,40]可以合理推定：在全球规模上，温室气体对冰期–间冰期

global scale, contributed significantly (possibly about half, that is, 2–3°C) to the globally averaged glacial–interglacial temperature change.

Glacial Terminations and Interglacials

Our complete Vostok data set allows us to examine all glacial commencements and terminations of the past 420 kyr. We can examine the time course of the following five properties during these events: δD_{ice}, dust content, $\delta^{18}O_{atm}$, CO_2 and CH_4 (Fig. 5). We consider that, during the terminations, $\delta^{18}O_{atm}$ tracks $\delta^{18}O_{sw}$ with a lag of ~2 kyr (ref. 11), the response time of the atmosphere to changes in $\delta^{18}O_{sw}$. $\delta^{18}O_{atm}$ can thus be taken as an indicator of the large ice-volume changes associated with the deglaciations. Broecker and Henderson[41] recently supported this interpretation for the last two terminations and discussed its limitations. Our extended $\delta^{18}O_{atm}$ record indeed reinforces such an interpretation, as it shows that the amplitudes of $\delta^{18}O_{atm}$ changes parallel $\delta^{18}O_{sw}$ changes for all four terminations. $\delta^{18}O_{sw}$ changes are similar for terminations I, II and IV (1.1–1.2‰) but much smaller for termination III (~0.6‰). The same is true for $\delta^{18}O_{atm}$ (1.4–1.5‰ for I, II and IV, and 0.8‰ for III).

A striking feature of the Vostok deuterium record is that the Holocene, which has already lasted 11 kyr, is, by far, the longest stable warm period recorded in Antarctica during the past 420 kyr (Fig. 5). Interglacials 5.5 and 9.3 are different from the Holocene, but similar to each other in duration, shape and amplitude. During each of these two events, there is a warm period of ~4 kyr followed by a relatively rapid cooling and then a slower temperature decrease (Fig. 5), rather like some North Atlantic deep-sea core records[42]. Stage 7.5 is different in all respects, with a slightly colder maximum, a more spiky shape, and a much shorter duration (7 kyr at mid-transition compared with 17 and 20 kyr for stages 5.5 and 9.3, respectively). This difference between stage 7.5 and stages 5.5 or 9.3 may result from the different configuration of the Earth's orbit (in particular concerning the phase of precession with respect to obliquity[3]). Termination III is also peculiar as far as terrestrial aerosol fallout is concerned. Terminations I, II and IV are marked by a large decrease in dust; high glacial values drop to low interglacial values by the mid-point of the δD_{ice} increase. But for termination III, the dust concentration decreases much earlier, with low interglacial values obtained just before a slight cooling event, as for termination I (for which interglacial values are reached just before the "Antarctic Cold Reversal").

Unlike termination I, other terminations show, with our present resolution, no clear temperature anomalies equivalent to the Antarctic Cold Reversal[6] (except possibly at the very beginning of termination III). There are also no older counterparts to the Younger Dryas CH_4 minimum[43] during terminations II, III and IV given the present resolution of the CH_4 record (which is no better than 1,000–2,000 yr before stage 5).

The sequence of events during terminations III and IV is the same as that previously observed for terminations I and II. Vostok temperature, CO_2 and CH_4 increase in phase

414

的全球平均温度变化有显著贡献（可能贡献了一半的变化，即 2~3℃）。

冰期结束和间冰期

东方站的完整资料可以使我们分析过去 42 万年的全部冰期的开始期和结束期。我们能分析以下 5 个参数在这两个阶段的时间变化：δD_{ice}，粉尘量，$\delta^{18}O_{atm}$，CO_2 和 CH_4（图 5）。本文认为在结束期，$\delta^{18}O_{atm}$ 随 $\delta^{18}O_{sw}$ 而变化，但滞后约 2,000 年（文献 11），也就是大气对 $\delta^{18}O_{sw}$ 变化的响应时间。因此 $\delta^{18}O_{atm}$ 可以作为冰消期冰量大幅度变化的一个指标。最近布勒克和亨德森[41]支持了对最后两个结束期的解释，并讨论了它的局限性。延伸了的 $\delta^{18}O_{atm}$ 记录也进一步支持了这一解释，发现在全部 4 个结束期，$\delta^{18}O_{atm}$ 与 $\delta^{18}O_{sw}$ 的变化幅度是平行的。对结束期 I、II 和 IV，$\delta^{18}O_{sw}$ 的变化是相似的（1.1‰ ~ 1.2‰）。但是，对结束期 III 来说该变化则小得多（约 0.6‰）。对 $\delta^{18}O_{atm}$ 而言也是同样的情况（I、II 和 IV 为 1.4‰ ~ 1.5‰，而 III 为 0.8‰）。

东方站氘记录的一个突出特征是，已经持续了 1.1 万年的全新世，是过去 42 万年间南极记录中最长的稳定暖期（图 5）。间冰期 5.5 和 9.3 与全新世是不同的，但他们之间彼此的持续时间、形态和幅度是相似的。这两个事件的每一个，都有约 4,000 年的暖期，随之是相对快速的变冷，接着温度较慢地降低（图 5），这很像北大西洋深海岩芯的记录[42]。7.5 阶在各方面都是不同的，有稍冷的极大值，更长而尖的形状，以及短得多的持续时间（以转折点中点计算的持续时间为 7,000 年，而 5.5 阶和 9.3 阶分别为 1.7 万年和 2 万年）。7.5 阶和 5.5 阶之间或 7.5 阶和 9.3 阶之间的差异可能是由地球轨道参数的配置不同（特别是相对于地轴倾角的岁差的相位[3]）导致的。考虑到陆地气溶胶沉降的话，结束期 III 也是特殊的。结束期 I、II 和 IV 以粉尘含量大幅度降低为标志；尘埃含量从冰期的高值向间冰期的低值降低，这种趋势一直持续到 δD_{ice} 增长阶段的中点。但对结束期 III 而言，粉尘浓度下降得更早，就在一个轻微的冷事件之前就达到了间冰期的最低值，这一点与结束期 I 有点相似（间冰期的最低值正好在"南极回冷"事件之前）。

与结束期 I 不同，在目前的分辨率下，其他结束期没有表现出与"南极回冷"事件相当的温度异常[6]（可能除了在结束期 III 的最开始阶段）。在给定的目前的 CH_4 记录的分辨率下（在 5 阶之前不会好于 1,000 ~ 2,000 年），结束期 II、III 和 IV 期间没有更老的新仙女木 CH_4 最低值[43]对应的事件。

结束期 III 和 IV 内的事件序列与之前的结束期 I 和 II 内观测到的顺序是相似的。在各个结束期内，东方站的温度、CO_2 和 CH_4 在冰期结束期同相增长。位相的不确

during terminations. Uncertainty in the phasing comes mainly from the sampling frequency and the ubiquitous uncertainty in gas-age/ice-age differences (which are well over ± 1 kyr during glaciations and terminations). In a recent paper, Fischer et al.[44] present a CO_2 record, from Vostok core, spanning the past three glacial terminations. They conclude that CO_2 concentration increases lagged Antarctic warmings by 600 ± 400 years. However, considering the large gas-age/ice-age uncertainty (1,000 years, or even more if we consider the accumulation-rate uncertainty), we feel that it is premature to infer the sign of the phase relationship between CO_2 and temperature at the start of terminations. We also note that their discussion relates to early deglacial changes, not the entire transitions.

An intriguing aspect of the deglacial CH_4 curves is that the atmospheric concentration of CH_4 rises slowly, then jumps to a maximum value during the last half of the deglacial temperature rise. For termination I, the CH_4 jump corresponds to a rapid Northern Hemisphere warming (Bölling/Allerød) and an increase in the rate of Northern Hemisphere deglaciation (meltwater pulse IA)[43]. We speculate that the same is true for terminations II, III and IV. Supportive evidence comes from the $\delta^{18}O_{atm}$ curves. During each termination, $\delta^{18}O_{atm}$ begins falling rapidly, signalling intense deglaciation, within 1 kyr of the CH_4 jump. The lag of deglaciation and Northern Hemisphere warming with respect to Vostok temperature warming is apparently greater during terminations II and IV ($\sim$9 kyr) than during terminations I and III ($\sim$4–6 kyr). The changes in northern summer insolation maxima are higher during terminations II and IV, whereas the preceding southern summer insolation maxima are higher during terminations I and III. We speculate that variability in phasing from one termination to the next reflects differences in insolation curves[41] or patterns of abyssal circulation during glacial maximum. Our results suggest that the same sequence of climate forcings occurred during each termination: orbital forcing (possibly through local insolation changes, but this is speculative as we have poor absolute dating) followed by two strong amplifiers, with greenhouse gases acting first, and then deglaciation enhancement via ice-albedo feedback. The end of the deglaciation is then characterized by a clear CO_2 maximum for terminations II, III and IV, while this feature is less marked for the Holocene.

Comparison of CO_2 atmospheric concentration changes with variations of other properties illuminates oceanic processes influencing glacial–interglacial CO_2 changes. As already noted for terminations I and II[41], the sequence of climate events described above rules out the possibility that rising sea level induces the CO_2 increase at the beginning of terminations. On the other hand, the small CO_2 variations associated with Heinrich events[45] suggest that the formation of North Atlantic Deep Water does not have a large effect on CO_2 concentrations. Our record shows similar relative amplitudes of atmospheric CO_2 and Vostok temperature changes for the four terminations. Also, values of both CO_2 and temperature are significantly higher during stage 7.5 than during stages 7.1 and 7.3, whereas the deep-sea core ice volume record exhibits similar levels for these three stages. These similarities between changes in atmospheric CO_2 and Antarctic temperature suggest that the oceanic area around Antarctica plays a role in the long-term CO_2 change. An influence of high southern latitudes is also suggested by the comparison with the dust profile, which exhibits a maximum during the periods of lowest CO_2. The link between dust and CO_2

416

定性主要来自取样频率，以及气体年龄与冰年龄比值之间普遍存在的差异的不确定性（在冰期和结束期超过 ±1,000 年）。菲舍尔等人[44]在最近的一篇论文中给出了来自东方站冰芯的 CO_2 记录，跨越了过去 3 个冰期的结束期。他们推断 CO_2 浓度的增长滞后于南极变暖 600±400 年。然而考虑到气体年龄与冰年龄比值之间巨大的不确定度（有 1,000 年，假如我们考虑积累率的不确定性的话则更多）。我们感觉在结束期开始时就推测 CO_2 与温度之间的位相关系的迹象还为时过早。我们还注意到，他们讨论的是冰消期的早期阶段，而不是整个转折阶段。

与冰川消退有关的 CH_4 曲线的一个有意思之处是，大气中的 CH_4 浓度上升很慢，然后在冰消期温度上升到最后一半时跳到最大值。就结束期 I 而言，CH_4 的跃升对应于北半球的快速变暖（波令/阿勒罗德）以及北半球冰川消退速率的增加（融水脉冲 IA）[43]。我们推测对结束期 II、III 和 IV 也是相同的情形。支持证据来自 $\delta^{18}O_{atm}$ 曲线。在每一个结束期，在 CH_4 跃升的 1,000 年内 $\delta^{18}O_{atm}$ 开始迅速下降，标志着强烈冰川消退。与东方站温度变暖相比，冰川消退和北半球变暖相对滞后于东方站冰芯气温转暖，而且这种滞后时间在结束期 II 和 IV（约 9,000 年）明显大于结束期 I 和 III（约 4,000 ~ 6,000 年）。在北半球夏季太阳辐射最大值阶段，CH_4 在结束期 II 和 IV 变化比较大，而在南半球夏季太阳辐射最大值阶段，CH_4 在结束期 I 和 III 阶段变化比较大。我们推测从一个结束期到下一个结束期 CH_4 相位上的变化反映了冰期最盛期太阳辐射曲线[41]或者深海环流模式的差异。我们的结果表明在每个结束期发生了同样的气候强迫序列：轨道强迫（可能是通过局地太阳辐射的变化，但是这点还没有把握，因为我们的绝对定年结果不够）之后跟随着两个放大器，首先是温室气体活动，然后是冰的反射率反馈导致的冰川消退增强。冰消期结束时，CO_2 出现一个清楚的最大值，尤其是对结束期 II、III、IV 而言；而在全新世，这一特征不明显。

大气 CO_2 浓度的变化与其他属性变化的对比阐明了海洋过程影响着冰期-间冰期 CO_2 变化。正如对结束期 I 和 II[41]已经指出的那样，上述一系列气候事件排除了在结束期一开始时海平面上升诱发 CO_2 增加的可能性。另一方面，与海因里希事件[45]对应时段 CO_2 变化非常小，说明北大西洋深水的形成对 CO_2 浓度没有太大的影响。我们的记录显示，四个结束期中的大气 CO_2 和东方站温度的相对变化幅度很相似。另外，CO_2 和温度两者的值在 7.5 阶都明显高于 7.1 阶和 7.3 阶，然而，深海岩芯的冰量记录显示在这三个阶其水平是相近的。在大气 CO_2 变化和南极温度变化之间的相似性表明南极附近的洋域对 CO_2 的长期变化是起作用的。南半球高纬的影响也可以从粉尘剖面的对比看出，在 CO_2 最低时粉尘记录是最大值。粉尘与 CO_2 变化

variations could be through the atmospheric input of iron[46]. Alternatively, we suggest a link through deep ocean circulation and sea ice extent in the Southern Ocean, both of which play a role in ocean CO_2 ventilation and, as suggested above, in the dust input over East Antarctica.

New Constraints on Past Climate Change

As judged from Vostok records, climate has almost always been in a state of change during the past 420 kyr but within stable bounds (that is, there are maximum and minimum values of climate properties between which climate oscillates). Significant features of the most recent glacial–interglacial cycle are observed in earlier cycles. Spectral analysis emphasises the dominance of the 100-kyr cycle for all six data series except $\delta^{18}O_{atm}$ and a strong imprint of 40- and/or 20-kyr periodicities despite the fact that the glaciological dating is tuned by fitting only two control points in the 100-kyr band.

Properties change in the following sequence during each of the last four glacial terminations, as recorded in Vostok. First, the temperature and atmospheric concentrations of CO_2 and CH_4 rise steadily, whereas the dust input decreases. During the last half of the temperature rise, there is a rapid increase in CH_4. This event coincides with the start of the $\delta^{18}O_{atm}$ decrease. We believe that the rapid CH_4 rise also signifies warming in Greenland, and that the deglacial $\delta^{18}O_{atm}$ decrease records rapid melting of the Northern Hemisphere ice sheets. These results suggest that the same sequence of climate forcing operated during each termination: orbital forcing (with a possible contribution of local insolation changes) followed by two strong amplifiers, greenhouse gases acting first, then deglaciation and ice-albedo feedback. Our data suggest a significant role of the Southern Ocean in regulating the long-term changes of atmospheric CO_2.

The Antarctic temperature was warmer, and atmospheric CO_2 and CH_4 concentrations were higher, during interglacials 5.5 and 9.3 than during the Holocene and interglacial 7.5. The temporal evolution and duration of stages 5.5 and 9.3 are indeed remarkably similar for all properties recorded in Vostok ice and entrapped gases. As judged from the Vostok record, the long, stable Holocene is a unique feature of climate during the past 420 kyr, with possibly profound implications for evolution and the development of civilizations. Finally, CO_2 and CH_4 concentrations are strongly correlated with Antarctic temperatures; this is because, overall, our results support the idea that greenhouse gases have contributed significantly to the glacial–interglacial change. This correlation, together with the uniquely elevated concentrations of these gases today, is of relevance with respect to the continuing debate on the future of Earth's climate.

(**399**, 429-436; 1999)

之间的关系，可以通过大气中铁[46]的输入来联系。或者，我们提出通过深海环流与南大洋海冰扩展的联系，二者对海洋 CO_2 流通以及如上所述的东南极洲粉尘输入都起着作用。

过去气候变化中的新约束

从东方站记录判断，气候在过去的 42 万年几乎总是处在变化状态中，但还在稳定的边界内（也就是说，气候特性在最大和最小值之间振荡）。最近的冰期–间冰期旋回中的重要特征在较早的旋回中就有观察到。谱分析强调了全部 6 条数据序列（除了 $\delta^{18}O_{atm}$）中 10 万年周期的主导性，以及 4 万年和/或 2 万年周期的强印记，尽管事实上冰芯 10 万年尺度上的定年是仅仅通过两个控制点进行拟合的。

在东方站冰芯记录中的 4 个冰期结束期中，一些主要的参数在每一个结束期都按以下的顺序发生变化。首先，温度与大气中的 CO_2 和 CH_4 浓度稳步上升，而粉尘的输入量却在减少。在温度上升的后半段，CH_4 浓度快速上升。这一事件与 $\delta^{18}O_{atm}$ 开始降低同步。我们相信 CH_4 迅速上升也意味着格陵兰变暖，同时冰消期 $\delta^{18}O_{atm}$ 减少记录了北半球冰盖的迅速融化。这些结果表明每一个结束期驱动气候变化的气候强迫序列是相同的：轨道强迫（也可能有局地变化的影响），之后跟随着 2 个强烈的放大器，首先是温室气体的作用，然后是冰川消退和冰反射率的反馈。我们的数据表明南大洋在调节大气 CO_2 的长期变化中起着重要的作用。

和全新世、间冰期 7.5 阶相比，间冰期 5.5 阶和 9.3 阶的南极温度更暖，大气 CO_2 和 CH_4 的浓度也更高。对东方站冰芯冰及其包裹气体中记录的所有指标，在 5.5 阶和 9.3 阶的时间演化和持续时长确实都极为相似。从东方站记录判断，持续时间长而稳定的全新世在过去 42 万年都是独一无二的，也许对人类文明的演化和发展具有深远的意义。最后，CO_2 和 CH_4 的浓度与南极温度有密切的相关性。这是因为，总的来说，我们的结果支持了温室气体对冰期–间冰期转变有重要作用的想法。考虑到关于地球未来气候变化的持续辩论，这种关系和今天这些温室气体浓度的异常升高不无联系。

（蔡则怡 翻译；田立德 审稿）

J. R. Petit[*], J. Jouzel[†], D. Raynaud[*], N. I. Barkov[‡], J.-M. Barnola[*], I. Basile[*], M. Bender[§], J. Chappellaz[*], M. Davis[‖], G. Delaygue[†], M. Delmotte[*], V. M. Kotlyakov[¶], M. Legrand[*], V. Y. Lipenkov[‡], C. Lorius[*], L. Pépin[*], C. Ritz[*], E. Saltzman[‖] & M. Stievenard[†]

[*] Laboratoire de Glaciologie et Géophysique de l'Environnement, CNRS, BP96, 38402, Saint Martin d'Hères Cedex, France

[†] Laboratoire des Sciences du Climat et de l'Environnement (UMR CEA/CNRS 1572), L'Orme des Merisiers, Bât. 709, CEA Saclay, 91191 Gif-sur-Yvette Cedex, France

[‡] Arctic and Antarctic Research Institute, Beringa Street 38, 199397, St Petersburg, Russia

[§] Department of Geosciences, Princeton University, Princeton, New Jersey 08544-1003, USA

[‖] Rosenstiel School of Marine and Atmospheric Science, University of Miami, 4600 Rickenbacker Causeway, Miami, Florida 33149, USA

[¶] Institute of Geography, Staromonetny, per 29, 109017, Moscow, Russia

Received 20 January; accepted 14 April 1999.

References:

1. Imbrie, J. *et al.* On the structure and origin of major glaciation cycles. 1. Linear responses to Milankovich forcing. *Paleoceanography* **7**, 701-738 (1992).

2. Tzedakis, P. C. *et al.* Comparison of terrestrial and marine records of changing climate of the last 500,000 years. *Earth Planet. Sci. Lett.* **150**, 171-176 (1997).

3. Berger, A. L. Long-term variations of daily insolation and Quaternary climatic change. *J. Atmos. Sci.* **35**, 2362-2367 (1978).

4. Lorius, C. *et al.* A 150,000-year climatic record from Antarctic ice. *Nature* **316**, 591-596 (1985).

5. Barnola, J. M., Raynaud, D., Korotkevich, Y. S. & Lorius, C. Vostok ice cores provides 160,000-year record of atmospheric CO_2. *Nature* **329**, 408-414 (1987).

6. Jouzel, J. *et al.* Vostok ice core: a continuous isotope temperature record over the last climatic cycle (160,000 years). *Nature* **329**, 402-408 (1987).

7. Raisbeck, G. M. *et al.* Evidence for two intervals of enhanced ^{10}Be deposition in Antarctic ice during the last glacial period. *Nature* **326**, 273-277 (1987).

8. Legrand, M., Lorius, C., Barkov, N. I. & Petrov, V. N. Vostok (Antarctic ice core): atmospheric chemistry changes over the last climatic cycle (160,000 years). *Atmos. Environ.* **22**, 317-331 (1988).

9. Chappellaz, J., Barnola, J.-M., Raynaud, D., Korotkevich, Y. S. & Lorius, C. Ice-core record of atmospheric methane over the past 160,000 years. *Nature* **345**, 127-131 (1990).

10. Petit, J. R. *et al.* Paleoclimatological implications of the Vostok core dust record. *Nature* **343**, 56-58(1990).

11. Sowers, T. *et al.* 135,000 year Vostok—SPECMAP common temporal framework. *Paleoceanography* **8**, 737-766 (1993).

12. Jouzel, J. *et al.* Extending the Vostok ice-core record of palaeoclimate to the penultimate glacial period. *Nature* **364**, 407-412 (1993).

13. Jouzel, J. *et al.* Climatic interpretation of the recently extended Vostok ice records. *Clim. Dyn.* **12**, 513- 521 (1996).

14. Genthon, C. *et al.* Vostok ice core: climatic response to CO_2 and orbital forcing changes over the last climatic cycle. *Nature* **329**, 414-418 (1987).

15. Lorius, C., Jouzel, J., Raynaud, D., Hansen, J. & Le Treut, H. Greenhouse warming, climate sensitivity and ice core data. *Nature* **347**, 139-145 (1990).

16. Raynaud, D. *et al.* The ice record of greenhouse gases. *Science* **259**, 926-934 (1993).

17. Petit, J. R. *et al.* Four climatic cycles in Vostok ice core. *Nature* **387**, 359-360 (1997).

18. Kapitza, A. P., Ridley, J. K., Robin, G. de Q., Siegert, M. J. & Zotikov, I. A. A large deep freshwater lake beneath the ice of central East Antarctica. *Nature* **381**, 684-686 (1996).

19. Bender, M., Sowers, T. & Labeyrie, L. D. The Dole effect and its variation during the last 130,000 years as measured in the Vostok core. *Glob. Biogeochem. Cycles* **8**, 363-376 (1994).

20. Barnola, J. M., Pimienta, P., Raynaud, D. & Korotkevich, Y. S. CO_2 climate relationship as deduced from the Vostok ice core: a re-examination based on new measurements and on a re-evaluation of the air dating. *Tellus B* **43**, 83-91 (1991).

21. Lorius, C. & Merlivat, L. in *Isotopes and Impurities in Snow and Ice. Proc. the Grenoble Symp. Aug./Sept. 1975* 127-137 (Publ. 118, IAHS, 1977).

22. Dahl-Jensen, D. *et al.* Past temperatures directly from the Greenland ice sheet. *Science* **282**, 268-271 (1998).

23. Salamatin, A. N. *et al.* Ice core age dating and paleothermometer calibration on the basis of isotopes and temperature profiles from deep boreholes at Vostok station (East Antarctica). *J. Geophys. Res.* **103**, 8963-8977 (1998).

24. Krinner, G., Genthon, C. & Jouzel, J. GCM analysis of local influences on ice core δ signals. *Geophys. Res. Lett.* **24**, 2825-2828 (1997).

25. Hoffmann, G., Masson, V. & Jouzel, J. Stable water isotopes in atmospheric general circulation models. *Hydrol. Processes* (in the press).

26. Bassinot, F. C. *et al.* The astronomical theory of climate and the age of the Brunhes-Matuyama magnetic reversal. *Earth Planet. Sci. Lett.* **126**, 91-108 (1994).

27. Shackleton, N. J., Imbrie, J. & Hall, M. A. Oxygen and carbon isotope record of East Pacific core V19-30: implications for the formation of deep water in the late Pleistocene North Atlantic. *Earth Planet. Sci. Lett.* **65**, 233-244 (1983).

28. Steig, E. *et al.* Synchronous climate changes in Antarctica and the North Atlantic. *Science* **282**, 92-95 (1998).

29. Howard, W. A warm future in the past. *Nature* **388**, 418-419 (1997).

30. Malaizé, B., Paillard, D., Jouzel, J. & Raynaud, D. The Dole effect over the last two glacial-interglacial cycles. *J. Geophys. Res.* (in the press).

31. Legrand, M. & Delmas, R. J. Formation of HCl in the Antarctic atmosphere. *J. Geophys. Res.* **93**, 7153- 7168 (1987).

32. Yung, Y. K., Lee, T., Chung-Ho & Shieh, Y. T. Dust: diagnostic of the hydrological cycle during the last glacial maximum. *Science* **271**, 962-963 (1996).

33. de Menocal, P. Plio-Pleistocene African climate. *Science* **270**, 53-59 (1995).

34. CLIMAP. *Seasonal Reconstructions of the Earth's Surface at the Last Glacial Maximum* (Geol. Soc. Am., Boulder, Colorado, 1981).

35. Basile, I. *et al.* Patagonian origin dust deposited in East Antarctica (Vostok and Dome C) during glacial stages 2, 4 and 6. *Earth Planet. Sci. Lett.* **146**, 573-589 (1997).

36. Basile, I. *Origine des Aérosols Volcaniques et Continentaux de la Carotte de Glace de Vostok (Antarctique)*. Thesis, Univ. Joseph Fourier, Grenoble (1997).

37. Leuenberger, M. & Siegenthaler, U. Ice-age atmospheric concentration of nitrous oxide from an Antarctic ice core. *Nature* **360**, 449-451 (1992).

38. Ramstein, G., Serafini-Le Treut, Y., Le Treut, H., Forichon, M. & Joussaume, S. Cloud processes associated with past and future climate changes. *Clim. Dyn.* **14**, 233-247 (1998).

39. Berger, A., Loutre, M. F. & Gallée, H. Sensitivity of the LLN climate model to the astronomical and CO_2 forcings over the last 200 ky. *Clim. Dyn.* **14**, 615-629 (1998).

40. Weaver, A. J., Eby, M., Fanning, A. F. & Wilbe, E. C. Simulated influence of carbon dioxide, orbital forcing and ice sheets on the climate of the Last Glacial Maximum. *Nature* **394**, 847-853 (1998).

41. Broecker, W. S. & Henderson, G. M. The sequence of events surrounding termination II and their implications for the causes of glacial-interglacial CO_2 changes. *Paleoceanography* **13**, 352- 364 (1998).

42. Cortijo, E. *et al.* Eemian cooling in the Norwegian Sea and North Atlantic ocean preceding ice-sheet growth. *Nature* **372**, 446-449 (1994).

43. Chappellaz, J. *et al.* Synchronous changes in atmospheric CH_4 and Greenland climate between 40 and 8 kyr BP. *Nature* **366**, 443-445 (1993).

44. Fischer, H., Wahlen, M., Smith, J., Mastroianni, D. & Deck, B. Ice core records of atmospheric CO_2 around the last three glacial terminations. *Science* **283**, 1712-1714 (1999).

45. Stauffer, B. *et al.* Atmospheric CO_2 concentration and millenial-scale climate change during the last glacial period. *Nature* **392**, 59-61 (1998).

46. Martin, J. H. Glacial-interglacial CO_2 change: The iron hypothesis. *Paleoceanography* **5**, 1-13 (1990).

47. Paillard, D., Labeyrie, L. & Yiou, P. Macintosh program performs time-series analysis. *Eos* **77**, 379 (1996).

48. Ritz, C. *Un Modele Thermo-mécanique d'évolution Pour le Bassin Glaciaire Antarctique Vostok-Glacier Byrd: Sensibilité aux Valeurs de Paramètres Mal Connus* Thesis, Univ. Grenoble (1992).

49. Blunier, T. *et al.* Timing of the Antarctic Cold Reversal and the atmospheric CO_2 increase with respect to the Younger Dryas event. *Geophys. Res. Lett.* **24**, 2683-2686 (1997).

50. Waelbroeck, C. *et al.* A comparison of the Vostok ice deuterium record and series from Southern Ocean core MD 88-770 over the last two glacial-interglacial cycles. *Clim. Dyn.* **12**, 113-123 (1995).

51. Blunier, T. *et al.* Asynchrony of Antarctic and Greenland climate change during the last glacial period. *Nature* **394**, 739-743 (1998).

52. Bender, M., Malaizé, B., Orchado, J., Sowers, T. & Jouzel, J. High precision correlations of Greenland and Antarctic ice core records over the last 100 kyr. in *The Role of High and Low Latitudes in Millennial Scale Global Change* (eds Clark, P. & Webb, R.) (AGU Monogr., Am. Geophys. Union, in the press).

Supplementary information is available on *Nature*'s World-Wide Web site (http://www.nature.com) or as paper copy from the London editorial office of *Nature*.

Acknowledgements. This work is part of a joint project between Russia, France and USA. We thank the drillers from the St Petersburg Mining Institute; the Russian, French and US participants for field work and ice sampling; and the Russian Antarctic Expeditions (RAE), the Institut Français de Recherches et Technologies Polaires (IFRTP) and the Division of Polar Programs (NSF) for the logistic support. The project is supported in Russia by the Russian Ministry of Sciences, in France by PNEDC (Programme National d'Études de la Dynamique du Climat), by Fondation de France and by the CEC (Commission of European Communities) Environment Programme, and in the US by the NSF Science Foundation.

Correspondence and requests for materials should be addressed to J.R.P. (e-mail: petit@glaciog.ujf-grenoble.fr).

Causes of Twentieth-century Temperature Change Near the Earth's Surface

S. F. B. Tett *et al.*

Editor's Note

Despite mounting evidence that human activities are causing significant and potentially catastrophic warming of the Earth's climate via the release of greenhouse gases, some sceptics—including major government leaders—continued to suggest that the observed climate changes might have predominantly natural causes, for example change in the radiation output of the Sun. Here British climatologists Simon Tett and colleagues supply an unequivocal response to that notion. Their extensive computer simulation of twentieth-century climate shows that, while natural processes might conceivably have explained warming early in the century, only a dominant anthropogenic influence can explain the temperature records since 1946. Such findings led to much stronger wording of later assessments of the human role by the Intergovernmental Panel on Climate Change.

Observations of the Earth's near-surface temperature show a global-mean temperature increase of approximately 0.6 K since 1900 (ref.1), occurring from 1910 to 1940 and from 1970 to the present. The temperature change over the past 30–50 years is unlikely to be entirely due to internal climate variability[2-4] and has been attributed to changes in the concentrations of greenhouse gases and sulphate aerosols[5] due to human activity. Attribution of the warming early in the century has proved more elusive. Here we present a quantification of the possible contributions throughout the century from the four components most likely to be responsible for the large-scale temperature changes, of which two vary naturally (solar irradiance and stratospheric volcanic aerosols) and two have changed decisively due to anthropogenic influence (greenhouse gases and sulphate aerosols). The patterns of time/space changes in near-surface temperature due to the separate forcing components are simulated with a coupled atmosphere–ocean general circulation model, and a linear combination of these is fitted to observations. Thus our analysis is insensitive to errors in the simulated amplitude of these responses. We find that solar forcing may have contributed to the temperature changes early in the century, but anthropogenic causes combined with natural variability would also present a possible explanation. For the warming from 1946 to 1996 regardless of any possible amplification of solar or volcanic influence, we exclude purely natural forcing, and attribute it largely to the anthropogenic components.

THE coupled model we use is HadCM2 (refs 6, 7), which has a horizontal resolution of 2.5° in latitude by 3.75° in longitude, 19 atmospheric and 20 oceanic levels, and a

二十世纪近地表温度变化的原因

尽管越来越多的证据表明，通过排放温室气体，人类活动导致了全球气候显著而具有灾变可能的变暖，某些怀疑论者（包括一些主要的政府领导人）仍不断提出观测到的气候变化主要源于自然因素，比如太阳辐射输出的变化。本文中英国气候学家西蒙·邰蒂和他的同事们为这种说法提供了明确的回复。他们对二十世纪气候所做的大量的计算机模拟结果显示，虽然自然过程很可能可以解释二十世纪初的气候变暖，但只有把人类活动作为主要的影响因素才有可能解释自 1946 年以来的气温的变化情况。这个研究结果为未来联合国政府间气候变化专门委员会评估人类活动在气候变暖中的作用提供了更强有力的支持。

近地表温度观测结果表明，自 1900 年以来，全球平均气温大约升高了 0.6 K（参考文献 1），气温升高分别发生在 1910～1940 年和 1970 年至今这两个阶段。过去 30～50 年间的气温变化并不完全是由气候系统内部变率引起的[2-4]，而是由人类活动引起的温室气体浓度以及硫酸盐气溶胶浓度的变化[5]所致。但造成二十世纪初气候变暖的因素是比较难以确定的。本文对二十世纪以来可能引起大尺度温度变化的四种因素作了量化。其中有两个因素是自然变化（太阳辐照度和平流层火山气溶胶），而另外两个由于人类的影响发生了显著的变化（温室气体和硫酸盐气溶胶）。利用全球海洋-大气耦合环流模式对各个强迫分量引起的近地表温度的时空变化模态进行了模拟，并将这些模拟值的线性组合与实测值进行拟合。因此，本文分析结果对上述响应的模拟幅值中存在的误差并不敏感。研究发现，二十世纪初期温度的变化可能是由太阳辐射变化造成的，不过，也有可能是人为因素与自然变率相叠加的结果。关于 1946～1996 年间的气候变暖，不管太阳辐射或火山爆发对温度变化是否有放大效应，我们排除了纯粹是自然强迫的可能，并认为人类活动的影响是其主要原因。

我们所采用的耦合模式为 HadCM2（参考文献 6 和 7），水平分辨率为 2.5°×3.75°（纬度 × 经度），大气和海洋在垂直方向上分别分为 19 层和 20 层，并进行了通量校

flux correction. Its climate sensitivity to a doubling of the atmospheric concentration of CO_2 is estimated to be 3.3 K (C. A. Senior, personal communication).

The main radiative forcings of climate since 1850 are likely to be anthropogenic changes in well-mixed greenhouse gases and tropospheric aerosols (mainly sulphate), and natural changes in solar irradiance and in stratospheric aerosol due to volcanic activity[8]. We compare observations[1] of 10-year mean near-surface temperature changes over five 50-year periods (1906–56, 1916–66, ..., 1946–96) with simulations of HadCM2 forced by the following factors (see also section 1 of Supplementary Information):

"G". Changes in well-mixed greenhouse gases from 1860 to 1996[8,9,10] expressed as equivalent CO_2.

"GS". As G but also with changes in surface albedo[11] representing the effects of anthropogenic sulphate aerosols from 1860 to 1996[6,9,10] derived from an atmospheric chemistry model[12]. We assume that this albedo represents both the direct and indirect[11,13,14] effects of sulphate aerosols. When we consider both G and GS we define a further signal S, the pure sulphate signal, as GS − G (see section 9 of Supplementary Information).

"Vol". Changes in stratospheric volcanic aerosols[15] from 1850 to 1996.

"Sol". Changes in total solar irradiance from 1700 to 1996 based on proxy data[16] for 1700–1991 and extended to 1996 using satellite observations[17].

The climate response to each of the above factors was computed from the ensemble average of four simulations started from different initial conditions. Internal climate variability is estimated from a 1,700-year control simulation (Control). 50-year segments (of 10-year mean near-surface temperatures) from the responses of the four ensembles (signals), Control and observations were processed identically to allow for the effect of changing observational coverage, to filter out variability on scales less than 5,000 km (ref. 18) and to remove the 50-year time-mean (see section 2 of Supplementary Information).

We assume that the observations (**y**) may be represented as the sum of simulated responses or signals ($\mathbf{x}_i$) and internal climatic variability (**u**), which we assume to be normally distributed:

$$\mathbf{y} = \sum_i \beta_i \mathbf{X}_i + \mathbf{u} \qquad (1)$$

This assumption of linearity has been shown to be a good approximation in at least one general circulation model[19,20]. The amplitude, β_i, represents the amount by which we have to scale the ith signal to give the best fit to the observations. We use an "optimal fingerprinting" algorithm[21-24], a form of multivariate regression, to estimate these amplitudes and uncertainty ranges (see section 3 of Supplementary Information). Here we use 5–95% uncertainty ranges and if the uncertainty range for β_i includes unity, then the amplitude of the simulated signal ($\mathbf{x}_i$) could be correct. The first millennium of Control is used to define a

正。当大气中 CO_2 的浓度增加一倍时，其气候敏感度约为 3.3 K(西尼尔，个人交流)。

1850 年以来，气候的主要辐射强迫来自两方面，一是人为因素引起的均匀混合的温室气体与对流层气溶胶(主要为硫酸盐)的变化；另一个是太阳辐照度和由火山活动引起的平流层气溶胶的自然变化[8]。我们以 50 年为一个时间段(1906~1956，1916~1966，…，1946~1996)，将近地表温度变化观测结果[1]的 10 年平均值与 HadCM2 的模拟结果作了比较，其中 HadCM2 模式受以下因素(亦可见于补充信息第 1 节)驱动：

"G"：1860~1996 年间均匀混合的温室气体的变化[8,9,10]，以 CO_2 当量表示。

"GS"：同 G，但还包含地表反照率[11]的变化，该变化代表了 1860~1996[6,9,10] 年间人类活动引起的硫酸盐气溶胶的影响，是根据大气化学模式[12]得到的。我们假定该反照率代表硫酸盐气溶胶的直接与间接[11,13,14]的效应。综合考虑 G 和 GS，我们进一步定义了信号 S，即 S = GS-G，仅代表硫酸盐的信号(见补充信息第 9 节)。

"Vol"：1850~1996 年间平流层火山气溶胶的变化[15]。

"Sol"：1700~1996 年间的太阳总辐照度，其中，1700~1991 年的数据是根据代用资料[16]得到的，之后直到 1996 年的数据为卫星观测结果[17]。

气候对上述因子的响应是通过计算不同初始条件的四组模拟结果的集合平均结果得到的。气候内部变率是利用 1,700 年的控制试验模拟得到的(控制试验)。将(十年平均近地表温度的)四组集合结果(信号)、控制试验以及观测值以 50 年为一段作分割，并保证对各资料的处理相同，以便考虑观测资料覆盖度变化的影响，同时还过滤掉尺度小于 5,000 km 的变率(参考文献 18)，并减去 50 年均值(见补充信息第 2 节)。

我们假设观测结果(y)代表模拟响应或信号(x_i)与气候内部变率(u)的总和，其中，假定它们都是正态分布的，则：

$$y = \sum_i \beta_i X_i + u \tag{1}$$

已有学者证明，至少在一个大气环流模式中，上述线性拟合是较好的近似[19,20]。为了与观测结果拟合得更好，我们对第 i 个信号进行了调整，幅值 β_i 就代表调整的比例。我们利用"最佳指纹"法[21-24](一种多元回归的形式)来估算这些幅值以及不确定区间(见补充信息第 3 节)。在这里我们取 5%~95% 作为不确定范围，并且如果 β_i 的不确定性范围包含 1 时，模拟信号(x_i)的幅值就是正确的。用控制试验的前

weighting function which minimizes the influence of patterns of high internal variability on the estimated signal amplitudes (giving "optimal fingerprints"). Observations and all simulated data are then further filtered by projection on to the leading ten modes of spatio-temporal variability from this part of Control (see section 7 of Supplementary Information). The latter seven centuries, which are statistically independent, are used to estimate the uncertainty ranges.

A minimum requirement for a signal or signal-combination to explain temperature changes over a 50-year period is that the amplitude of the estimated residual $(\Sigma_i \beta_i \mathbf{x}_i - \mathbf{y})$ should be consistent, at the 5% level, with internal climate variability[24] estimated from Control. For a signal-combination to explain change over the century we require that the estimated residual be consistent in all five 50-year periods. We test this using a F-test (see sections 4 and 6 of Supplementary Information) which rules out simulated internal variability as an explanation of temperature changes during 1906–56, 1916–66 and 1946–96 (Table 1), despite being a relatively weak test in that it makes no use of the shape of the simulated signals. Furthermore, no combination of our simulated natural signals alone can explain the warming since 1946—the period when the observational coverage is best. However, every case that includes an anthropogenic signal passes the consistency test. These results hold even if the global-mean model sensitivity or the amplitude of the forcing is wrong as we place no restriction on the amplitudes of the simulated signals.

Table 1. Consistency tests

	F-test*				
	1906–56	1916–66	1926–76	1936–86	1946–96
Int. var.†	**0.01**	**0.04**	0.69	0.76	**0.01**
G	0.18^{G}	0.11	0.75	0.80	0.10^{G}
GS	0.06^{GS}	0.23	0.76	0.85	0.30^{GS}
Sol†	0.11^{Sol}	0.06	0.75	0.87	**0.03**
Vol†	**0.03**	0.08	0.74	0.80	**0.04†**
G&S	0.21^{G}	0.17^{G}	0.68	0.83	0.31^{G,S}
G&Sol	0.23^{G}	0.17^{G,Sol}	0.68	0.81	0.07^{G}
G&Vol	0.24^{G}	0.09	0.67	0.73	0.08^{G}
GS&Sol	0.13^{Sol}	0.19^{GS}	0.68	0.82	0.25^{GS}
GS&Vol	0.05^{GS}	0.22^{GS}	0.68	0.80	0.25^{GS}
Sol&Vol†	0.08^{Sol}	0.09‡	0.67	0.82	**0.04‡**

* These columns show the probability (P) that the best-fit signal combination is consistent with observations using an F-test with 21 degrees of freedom. Values in bold are inconsistent with the observations. When a signal is detected, its name is shown as a superscript next to the P-value. Actual P-values are shown, although 0.03 is the lowest value that can be robustly estimated given the available length of Control.

† Signal combination is an inadequate explanation of twentieth-century temperature change as the F-test fails at least once.

‡ Volcanic amplitude in this signal-combination is significantly negative and thus unphysical. No other signal ever has a significant negative amplitude.

1,000 年来定义一个权重函数,以使高频内部变率模态对所估算信号幅值的影响降到最小(给出"最佳指纹图谱")。然后通过映射到控制试验中前 10 个时空变率主要模态上,对观测结果与所有模拟数据做进一步滤波(见补充信息第 7 节)。后 700 年的模拟数据在统计意义上是独立的,用以估算不确定范围。

要解释 50 年期间发生的温度变化,对于信号或信号组合的最低要求为,所估算的残差 $(\Sigma_i\beta_i\mathbf{x}_i-\mathbf{y})$ 的幅值应当与根据控制试验结果估算出的气候内部变率[24]在 5% 的显著性水平上保持一致。在用信号组合来解释百年尺度上的变化时,我们要求所估算的残差与所有五个 50 年时段上的值一致。我们对此进行了 F 检验(见补充信息第 4 节和第 6 节),结果表明 1906~1956、1916~1966 和 1946~1996 年间(表 1)近地表温度的变化并非内部变率所致。不过由于未用到模拟信号的分布型,这里的检验标准相对较弱。此外,仅用我们所模拟出的自然强迫的组合并不能解释 1946 年以来的变暖现象,但在这一时间段内我们的观测资料覆盖度是最高的。而所有包含人为活动强迫的情况都通过了一致性检验。由于我们没有对模拟信号的幅值作任何限制,因此即便全球平均的模式敏感度或者强迫的幅值是错误的,上述结果依然成立。

表 1. 一致性检验

	F 检验 *				
	1906~1956	1916~1966	1926~1976	1936~1986	1946~1996
内部变率 †	**0.01**	**0.04**	0.69	0.76	**0.01**
G	0.18[G]	0.11	0.75	0.80	0.10[G]
GS	0.06[GS]	0.23	0.76	0.85	0.30[GS]
Sol†	0.11[Sol]	0.06	0.75	0.87	**0.03**
Vol†	**0.03**	0.08	0.74	0.80	**0.04**†
G&S	0.21[G]	0.17[G]	0.68	0.83	0.31[G,S]
G&Sol	0.23[G]	0.17[G,Sol]	0.68	0.81	0.07[G]
G&Vol	0.24[G]	0.09	0.67	0.73	0.08[G]
GS&Sol	0.13[Sol]	0.19[GS]	0.68	0.82	0.25[GS]
GS&Vol	0.05[GS]	0.22[GS]	0.68	0.80	0.25[GS]
Sol&Vol†	0.08[Sol]	0.09‡	0.67	0.82	**0.04**‡

* 这些列中给出的是最吻合的信号组合与利用自由度为 21 的 F 检验得到的观测结果一致的概率(P)。粗体标注的
   值与观测结果不一致。当检测到信号时,在 P 值旁边将其名称以上标的形式标出。表中给出了实际的 P 值,不过
   0.03 是控制试验可用时段下所能稳健地估算出的最小值。
† 表示信号组合至少有一次未能通过 F 检验,因此它并不足以解释二十世纪发生的温度变化。
‡ 在该信号组合中火山因素的幅值明显为负,因此没有物理意义。其他信号都没有明显的负振幅。

A risk in multivariate linear regression is signal degeneracy, which means that signals resemble linear combinations of one another. Applying tests for degeneracy (see pages 243–248 of ref. 25 and section 8 of Supplementary Information) to our signals and observations, we find that no more than two signal-amplitudes can be estimated simultaneously. While three-signal and four-signal combinations are possible explanations, results from them may be misleading. Thus we only discuss combinations of at most two signals.

A signal is formally detected in a signal-combination and period if that combination is consistent with observations and the uncertainty range for the signal is entirely positive (see sections 5 and 6 of Supplementary Information). This means that we have rejected the null hypothesis that the signal amplitude is zero (or that the simulated signal is of the wrong sign). In other words, it is highly likely that the signal is present in the observations.

All one- or two-signal combinations that include anthropogenic signals are, on the basis of the consistency test, possible explanations of twentieth-century temperature change, but we wish to focus on the "best" explanation. Ideally, this explanation would include all detectable signals and no others, provided that this combination is consistent with the observations. We find that three signals (G, S and Sol) are detected during the century (Table 1). The composite signal GS (in which the relative amplitudes of the greenhouse and sulphate signals are assumed to be as in the GS experiment, giving a single pattern of response to anthropogenic forcing) can also be detected. As signal degeneracy may make results from the three-signal combination, G&S&Sol, misleading, this leaves three two-signal combinations: G&S, GS&Sol and G&Sol. Before the most recent period, none of these three two-signal combinations are an obviously better fit to the observations than any other (Table 1). But in 1946–96, combinations including the influence of sulphate aerosols (S) fit the observations better than other explanations (Table 1), which all fail to pass the F-test at the 10% level, suggesting that some sulphate influence is required to account for recent changes. Thus we consider the two combinations containing S (G&S and GS&Sol), but we explore the sensitivity of the solar detection during 1906–56 to the size of the sulphate signal included in GS.

In G&S (unlike GS), the observations are allowed to determine the relative amplitude of the sulphate and greenhouse-gas signals. This allows for error in the prescribed sulphate forcing or modelled response which varies with each period. With these two signals we detect S in 1946–96 and G in 1906–56, 1916–66 and 1946–96 (Table 1). We find that during 1906–56 the amplitude of the sulphate aerosol signal relative to the greenhouse-gas signal is significantly less than that prescribed in GS.

Some other factor may be required to explain this discrepancy with GS rather than errors in the simulated anthropogenic signals. If the relative amplitude of the anthropogenic signals is as prescribed in the GS experiment, then Sol is detected: during 1906–56, some solar influence is required to explain the temperature changes (Table 1). If we reduce the amplitude of the sulphate signal in GS by 33% or more (making GS more like G), we detect

428

多元线性回归中存在一个风险，那就是信号的简并，即一个信号与其他信号的线性组合相似。通过对信号与实测值进行简并检验（见参考文献 25 中的第 243～248 页以及补充信息第 8 节），我们发现，可同时估算出的信号幅值不会多于两个。尽管三信号或四信号的组合也都是可能的解释，但依据它们得出的结论可能会存在误导。根据上述分析，我们只需讨论至多为两信号组合的情况即可。

倘若某信号组合与观测结果一致并且信号的不确定范围完全为正，那么该信号在此时间段内被以信号组合的形式检测出来（见补充信息第 5 和第 6 节）。也就是说，我们剔除了信号幅值为零（或者模拟信号符号错误）的无效假定。换句话说，信号极有可能出现于观测结果中。

根据一致性检验结果，所有包含人为因素信号的单信号或双信号组合均可能是引起二十世纪温度变化的原因，但我们希望关注的是"最佳"解释。理想的情况是，倘若该信号组合与实测值一致，则该种解释包含且只包含所有可检测信号。我们发现，二十世纪期间共检测到了三类信号（G、S 和 Sol）（表 1）。混合信号 GS 亦可检测出来（鉴于人为强迫的单一响应模态，其中假定温室气体和硫酸盐信号的相对幅值与 GS 实验中的值相同）。由于三信号的混合（G&S&Sol）导致的信号简并可能产生误导性的结论，因此这里讨论三类双信号组合：G&S、GS&Sol 以及 G&Sol。在最近一个时间段之前，上述三种双信号组合中的任一个与实测值的吻合程度并未明显优于其他两个（表 1）。而在 1946～1996 年间，包含硫酸盐气溶胶（S）因素的组合则相比其他组合与观测结果吻合得更好（表 1），其他组合均未通过 10% 显著性水平的 F 检验，说明需要考虑某些硫酸盐强迫对近期温度变化的影响。因此，我们考虑了两类含有 S 的信号组合（G&S 和 GS&Sol），但研究的是 1906～1956 年间 GS 中太阳辐射对硫酸盐信号大小的敏感度。

在 G&S（与 GS 不同）中，利用实测值可以确定出硫酸盐信号与温室气体信号的相对幅值。这就允许给定的硫酸盐强迫或模拟响应中的误差可随不同时段而变化。利用这两个信号我们检测了 1946～1996 年间的 S，以及 1906～1956、1916～1966 和 1946～1996 年间的 G（表 1）。结果发现，1906～1956 年间硫酸盐气溶胶信号与温室气体信号的相对幅值显著低于 GS 中的给定值。

上述关于 GS 的差异应该是由其他因素造成的，而非模拟人为因素信号的误差所致。倘若人为因素信号的相对幅值与 GS 实验给定的结果一致，那么应该能检测到 Sol：1906～1956 年间，温度的变化可能是由某些太阳辐射的影响引起的（表 1）。当把 GS 中硫酸盐信号的幅值降低 33% 或者更多（使 GS 更接近 G）时，我

the anthropogenic signal rather than the solar signal during 1906–56. GS is detected in 1946–96 with or without this change. The reduction of the relative amplitude of sulphate aerosols to greenhouse gases is within the possible uncertainty range[8]; thus our detection of a solar signal should be treated with some caution.

We reconstruct global-mean temperature changes by multiplying the simulated signals, x_i by the best-fit amplitudes (β_i) for 1906–56 and 1946–96. During 1906–56, in G&S, greenhouse gases warm, counterbalanced by a weaker aerosol effect than was imposed in GS (Fig. 1a). The warming peak around 1940 is accounted for by a combination of internal variability and a steadily increasing temperature due to anthropogenic forcings. In GS&Sol the observed early-century warming is explained by a combination of solar irradiance changes, internal variability and an increase in green-house gases (partly balanced by the radiative effect of sulphate aerosols), with solar activity largely responsible for the warming peak around 1940 (Fig. 1b). In both cases, from 1946 sulphate aerosols balance the effect of greenhouse gases giving little warming until the mid-1970s when the warming due to increasing greenhouse gases dominates (Fig. 1c, d).

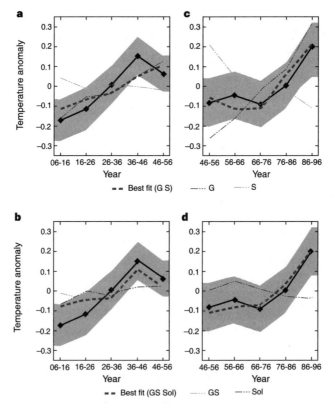

Fig. 1. Best-estimate contributions to global-mean temperature change in the twentieth century. Shown are reconstructions of temperature variations for 1906–56 (**a** and **b**) and 1946–96 (**c** and **d**) for G&S (**a** and **c**) and GS&Sol (**b** and **d**). Observed (thick black lines), best fit (dark grey dashed lines), and the uncertainty

们在 1906～1956 年间检测到的就是人为因素的信号而非太阳辐射信号了。而在
1946～1996 年间，无论是否做出上述改变都能检测到 GS。硫酸盐气溶胶与温室气
体相对辐值的降低在可能的不确定范围以内[8]，因此我们所检测到的太阳辐射信号
需谨慎处理。

我们将 1906～1956 年间和 1946～1996 年间的模拟信号 x_i 乘上其最佳拟合幅值
（β_i），重建了全球平均温度的变化。1906～1956 年间，在 G&S 中温室气体的变暖效
应被比 GS 中更弱的气溶胶效应所平衡（图 1a）。1940 年前后达到的变暖峰值是气候
内部变率与人类活动强迫引起的温度稳步上升共同作用的结果。在 GS&Sol 中所观
测到的二十世纪初的变暖现象可以解释为太阳辐照度变化、内部变率以及温室气体
增加（部分被硫酸盐气溶胶的辐射效应抵消）共同作用的结果，其中太阳活动是造成
1940 年前后的变暖峰值的主要原因（图 1b）。在两种信号组合（G&S，GS&Sol）中，
自 1946 年来，硫酸盐气溶胶抵消了温室气体的效应，使得升温幅度变得很小，直到
20 世纪 70 年代中期因温室气体增加导致的变暖效应逐步占据主导地位（图 1c 和 1d）。

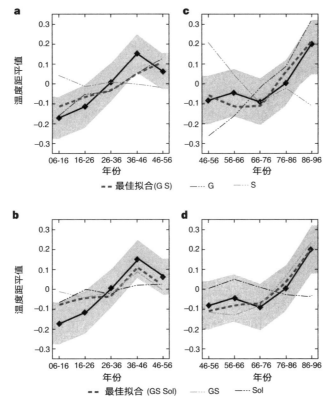

图 1. 二十世纪全球平均温度变化贡献率值的最优估算。图中所示为利用 1906～1956 年（a 和 b）和
1946～1996 年（c 和 d）间 G&S（a 和 c）和 GS&Sol（b 和 d）的重建温度变化序列。所有图均给出了实测
值（黑色粗线）、最佳拟合（深灰色虚线）以及内部变率的不确定范围（灰色阴影部分）。其中 a 和 c 给出

range due to internal variability (grey shading) are shown in all plots. Panels **a** and **c** show contributions from G (dashed red) and S (dashed green); panels **b** and **d** show the contributions from GS (dashed orange) and Sol (dashed blue). All time series were reconstructed from data in which the 50-year mean had first been removed.

In GS&Sol, solar influence is detectable during the 1906–56 period (Fig. 2), but contributes little to the linear trend in global-mean temperature (a best estimate of ~0.025 K per decade). Conversely, G&S shows that greenhouse gases could have contributed substantially to the warming during both the 1906–56 and 1916–66 periods (Fig. 2). In all other periods, the best-estimate solar contribution is negligible. For both cases the best estimate of the total anthropogenic warming is ~0.075 K per decade during 1946–96, within an approximate uncertainty range of 0.03–0.11 K per decade.

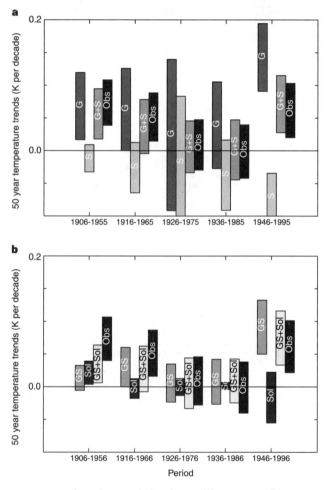

Fig. 2. 50-year temperature trends, and uncertainties, due to different causes. Black bars, observed 50-year trends (K per decade) in global-mean near-surface temperature with 5–95% range estimated from Control. **a**, Estimated contribution (and uncertainty range) to observed trends from G (red), S (green) and the sum (orange). **b**, Estimated contribution and uncertainty range from GS (orange), Sol (blue) and total (yellow). The signal name is centred at the estimated value and the bars cover the uncertainty range.

了 G(红色虚线)和 S(绿色虚线)的贡献率；b 和 d 给出了 GS(橙色虚线)和 Sol(蓝色虚线)的贡献率。所有时间序列都是根据减去 50 年平均值后的数据重建的。

在 GS&Sol 中，太阳辐射的影响可在 1906～1956 年时间段上检测到(图 2)，但对全球平均温度的线性趋势影响不大(最佳估计值约为每十年 0.025 K)。相反，G&S 则表明温室气体对 1906～1956 年间和 1916～1966 年间温度的升高有很大的贡献(图 2)。在所有其他的时间段内，最佳估计结果的太阳辐射的贡献率可以忽略不计。对于(G&S 和 GS&Sol)这两种情况，1946～1996 年间，由人类活动引起的总的温度升高值的最佳估计约为每十年 0.075 K，不确定范围大致为每十年 0.03～0.11 K。

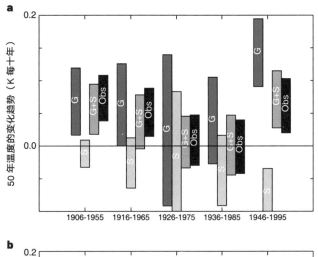

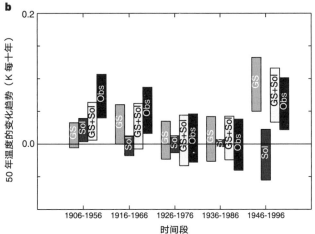

图 2. 不同因素影响下，50 年的温度变化趋势及其不确定性区间。黑色直条表示观测得到的 50 年全球平均近地表温度的变化趋势(单位为：K 每十年)，以及基于控制试验估算的 5%～95% 的不确定范围。a，G(红)、S(绿)和两者总和(橙)对观测趋势的贡献率(及不确定范围)。b，GS(橙)、Sol(蓝)及总(黄)的贡献率与不确定范围。直条中间标出了估算值的信号名称，而直条的长度则代表不确定范围。

To test the robustness of our results, we carried out six sensitivity studies in which we changed our analysis procedure (see section 10 of Supplementary Information). Detection of anthropogenic signals during 1906–56 (in G&S) and 1946–96 (in both cases) was unaltered by those changes. Detection of the 1946–96 sulphate signal is significant only at the 90% level if optimization is not used. Detection of the 1906–56 solar signal is significant at the 90% level only if greater weight is given to smaller spatial scales, but remains significant at the 95% level in all other studies. We also used a different solar irradiance reconstruction[26] from 1750, to produce an alternative Sol signal, and we reduced stratospheric ozone from 1974[4] to modify the GS signal. Our principal results are robust to both these changes. Furthermore, if we inflate the variance of Control by a factor of 4 we still detect an anthropogenic signal in the 1946–96 period in both of the two-signal cases considered.

Our simulations did not explicitly represent the effects of sulphate aerosols on cloud albedo[13] or lifetime[14] (though on the scales considered these are likely to be represented by the albedo changes imposed on the model), nor did we consider the climate effects of other aerosols, changes in the spectral distribution of the solar irradiance and possible associated changes in ozone[27,28]. Although our results are unaffected by error in the amplitude of the forcing, they could be sensitive to error in the patterns of response or other errors in the forcing, though our conclusions were unaltered by use of an alternative solar irradiance reconstruction. The test that we use to evaluate consistency between the observations and simulations could be misled by mutually compensating errors in model estimates of natural variability and the response of the model to external factors. We do not consider possible observational errors (likely to be small relative to internal climate variability) nor has our uncertainty analysis included any uncertainty due to the finite length of data used to derive the optimisation. For consistency with earlier work[21-24], we have used standard estimates of pattern amplitudes based on linear regression which are biased towards zero[25] when, as here, there is uncertainty in the signals.

Bearing in mind these caveats, we interpret our results as showing the following: first, the temperature changes over the twentieth century cannot be explained by any combination of natural internal variability and the response to natural forcings alone. Second, the recent warming, ~0.25 K, can be explained by the response of the climate to anthropogenic changes in greenhouse-gas concentrations partly offset by cooling due to anthropogenic sulphate aerosols, resulting in little net temperature change from 1946 to the mid-1970s. Last, the warming early this century can also be explained by anthropogenic causes and internal variability. However, solar irradiance changes could have made a significant contribution, ~0.125 K, if we assume little error in the relative amplitude of the forcing of sulphates and greenhouse gases prescribed in our model.

(**399**, 569-572; 1999)

为了检验结果的稳健性,我们通过调整分析步骤分别进行了六组敏感性实验(见补充信息第 10 节)。1906~1956(G&S 中)和 1946~1996(G&S 和 GS&Sol 两种情况下)人为因素信号的检测并没有因上述改变而发生变化。当不进行最优化处理时,1946~1996 年间硫酸盐气溶胶信号仅在 90% 的置信水平上为显著。仅当较小的空间尺度所占的权重较大时,1906~1956 年间检测到的太阳辐射信号在 90% 的置信水平上为显著,而所有其他研究显示该信号在 95% 的置信水平上仍显著。我们还通过采用不同的太阳辐照度(自 1750 年)重建序列[26]来替换 Sol 信号,并将 1974 年以来的平流层臭氧值减小[4]以改变 GS 信号。对于上述两种变化,我们的主要结果也都是稳健的。此外,即使将控制试验的方差乘以 4,在上述考虑的两种情况下,我们仍可在 1946~1996 年时段内检测到人为因素的信号。

我们的模拟研究未能显式地描述硫酸盐气溶胶对云层反照率[13]或者生命史[14]的影响(尽管在所研究的尺度范围内,它们可以由模式给定的反照率变化表示),同样,我们也不认为模拟结果能确切代表其他气溶胶的气候效应、太阳辐射光谱分布的变化以及可能伴随的臭氧层的变化[27,28]等等。尽管我们的结果并不受强迫幅值误差的影响,但该结果对于响应模态中的误差或强迫中的其他误差可能敏感。然而,当我们采用另一种太阳辐照度重建序列时,并不影响得到的结论。自然内部变率的模式估算值与模式对外部因素的响应之间的误差可相互补偿,因此我们对实测值与模拟值的一致性进行评估的检验有可能受其误导。此外,我们并未考虑可能存在的观测误差(相对于气候内部变率来说量级相对较小),在不确定分析中也未包含任何为了得到最优化组合而对数据长度进行限定所引起的不确定性。为了与早期的工作保持一致[21-24],我们采用了基于线性回归的模态幅值的标准估算法,当信号中存在不确定性时,幅值会偏向零[25]。

考虑上述因素,我们可将得到的结果总结如下:首先,二十世纪的温度变化并不能仅用自然内部变率与对自然强迫的响应的任何组合来解释。其次,近些年的增暖,温度升高了约 0.25 K,可解释为气候对温室气体浓度变化的响应,而温室气体浓度的变化是人类活动引起的。但由于人类活动还造成了硫酸盐气溶胶的变化,进而导致气候变冷,所以这两种人类活动引起的变化部分抵消,从而使得从 1946 年到 20 世纪 70 年代中期发生的温度净变化很小。最后,二十世纪初的变暖也可用人类活动与内部变率来解释。然而,如果假设我们的模式所给定的硫酸盐强迫与温室气体强迫的相对幅值的误差很小,太阳辐射的变化应该也具有重要作用(约为 0.125 K)。

(齐红艳 翻译;周天军 审稿)

Simon F. B. Tett[*], **Peter A. Stott**[*], **Myles R. Allen**[†], **William J. Ingram**[*] & **John F. B. Mitchell**[*]

[*] Hadley Centre for Climate Prediction and Research, Meteorological Office, London Road, Bracknell, Berkshire RG12 2SY, UK
[†] Space Science Department, Rutherford Appleton Laboratory, Chilton, OX11 0QX, UK and Department of Physics, University of Oxford

Received 5 October 1998; accepted 25 March 1999.

References:

1. Parker, D. E., Jones, P. D., Folland, C. K. & Bevan, A. Interdecadal changes of surface temperature since the late nineteenth century. *J. Geophys. Res.* **99**, 14373-14399 (1994).

2. Stouffer, R. J., Manabe, S. & Vinnikov, K. Y. Model assessment of the role of natural variability in recent global warming. *Nature* **367**, 634-636 (1994).

3. Santer, B. D. *et al.* A search for human influences on the thermal structure of the atmosphere. *Nature* **382**, 39-45 (1996).

4. Tett, S. F. B., Mitchell, J. F. B., Parker, D. E. & Allen, M. R. Human influence on the atmospheric vertical temperature structure: Detection and observations. *Science* **247**, 1170-1173 (1996).

5. Hegerl, G. C. *et al.* Multi-fingerprint detection and attribution analysis of greenhouse gas, greenhouse gas-plus-aerosol and solar forced climate change. *Clim. Dyn.* **13**, 613-634 (1997).

6. Johns, T. C. *et al.* The second Hadley Centre coupled ocean-atmosphere GCM: Model description, spinup and validation. *Clim. Dyn.* **13**, 103-134 (1997).

7. Tett, S. F. B., Johns, T. C. & Mitchell, J. F. B. Global and regional variability in a coupled AOGCM. *Clim. Dyn.* **13**, 303-323 (1997).

8. Shine, K. P., Fouquart, Y., Ramaswamy, V., Solomon, S. & Srinivasan, J. in *Climate Change 1995: the Science of Climate Change* (eds Houghton, J. T. *et al.*) 108-118 (Cambridge Univ. Press, 1996).

9. Mitchell, J. F. B., Johns, T. C., Gregory, J. M. & Tett, S. F. B. Climate response to increasing levels of greenhouse gases and sulphate aerosols. *Nature* **376**, 501-504 (1995).

10. Mitchell, J. F. B. & Johns, T. C. On modification of global warming by sulfate aerosols. *J. Clim.* **10**, 245-267 (1997).

11. Mitchell, J. F. B., Davis, R. A., Ingram, W. J. & Senior, C. A. On surface temperature, greenhouse gases, and aerosols: Models and observations. *J. Clim.* **10**, 2364-2386 (1995).

12. Langner, J. & Rodhe, H. A global three-dimensional model of the tropospheric sulfur cycle. *J. Atmos. Chem.* **13**, 225-263 (1991).

13. Twomey, S. A. Pollution and the planetary albedo. *Atmos. Environ.* **8**, 1251-1256 (1974).

14. Albrecht, B. A. Aerosols, cloud microphysics and fractional cloudiness. *Science* **245**, 1227-1230 (1989).

15. Sato, M., Hansen, J. E., McCormick, M. P. & Pollack, J. B. Stratospheric aerosol optical depths (1850-1990). *J. Geophys. Res.* **98**, 22987-22994 (1993).

16. Hoyt, D. V. & Schatten, K. H. A discussion of plausible solar irradiance variations, 1700-1992. *J. Geophys. Res.* **98**, 18895-18906 (1993).

17. Willson, R. C. Total solar irradiance trend during solar cycles 21 and 22. *Science* **277**, 1963-1965 (1997).

18. Stott, P. A. & Tett, S. F. B. Scale-dependent detection of climate change. *J. Clim.* **11**, 3282-3294 (1998).

19. Haywood, J., Stouffer, R., Wetherald, R., Manabe, S. & Ramaswamy, V. Transient response of a coupled model to estimated changes in greenhouse gas and sulfate concentrations. *Geophys. Res. Lett.* **24**, 1335-1338 (1997).

20. Ramaswamy, V. & Chen, C.-T. Linear additivity of climate response for combined albedo and greenhouse perturbations. *Geophys. Res. Lett.* **24**, 567-570 (1997).

21. Hasselmann, K. Optimal fingerprints for the detection of time-dependent climate change. *J. Clim.* **6**, 1957-1971 (1993).

22. Hasselmann, K. Multi-pattern fingerprint method for detection and attribution of climate change. *Clim. Dyn.* **13**, 601-611 (1997).

23. North, G. R., Kim, K.-K., Shen, S. S. P. & Hardin, J. W. Detection of forced climate signals. Part I: filter theory. *J. Clim.* **8**, 401-408 (1995).

24. Allen, M. R. & Tett, S. F. B. Checking for model consistency in optimal fingerprinting. *Clim. Dyn.* (in the press).

25. Mardia, K. V., Kent, J. T. & Bibby, J. M. *Multivariate Analysis* (Academic, London, 1979).

26. Lean, J., Beer, J. & Bradley, R. Reconstruction of solar irradiance since 1610: implications for climate change. *Geophys. Res. Lett.* **22**, 3195-3198 (1995).

27. Haigh, J. D. Impact of solar variability on climate. *Science* **272**, 981-984 (1996).

28. Haigh, J. D. The role of stratospheric ozone in modulating the solar radiative forcing of climate. *Nature* **370**, 544-546 (1994).

Supplementary information is available on *Nature*'s World-Wide Web site (http://www.nature.com) or as paper copy from the London editorial office of *Nature*.

Acknowledgements. S.F.B.T., P.A.S. and computer time were funded by the Department of the Environment, Transport and the Regions. W.J.I. and J.F.B.M. were supported by the UK Public Meteorological Service Research and Development programme. M.R.A. was supported by a research fellowship from the UK Natural Environment Research Council. Supplementary support was provided by the European Commission.

Correspondence and requests for materials should be addressed to S.B.F.T. (e-mail: sfbtett@meto.gov.uk).

Design and Synthesis of an Exceptionally Stable and Highly Porous Metal-organic Framework

Hailian Li *et al.*

Editor's Note

Materials with uniformly sized pores of widths comparable to the sizes of small molecules have proved immensely valuable in industrial chemistry, for example by serving as selective catalysts and filters for separating gases. Many such materials are aluminosilicates—either natural minerals called zeolites, or synthetic analogues. But porous frameworks built by rational assembly of molecular building blocks—organic molecules linked by metal ions—could offer more control over the structure and properties of these materials. Efforts to make them are generally hindered by the fact that the frameworks collapse when not filled with solvent, especially if heated. Here Omar Yaghi and coworkers report a metal-organic porous framework that remains robust up to 300 °C even when all solvent is removed.

Open metal-organic frameworks are widely regarded as promising materials for applications[1-15] in catalysis, separation, gas storage and molecular recognition. Compared to conventionally used microporous inorganic materials such as zeolites, these organic structures have the potential for more flexible rational design, through control of the architecture and functionalization of the pores. So far, the inability of these open frameworks to support permanent porosity and to avoid collapsing in the absence of guest molecules, such as solvents, has hindered further progress in the field[14,15]. Here we report the synthesis of a metal-organic framework which remains crystalline, as evidenced by X-ray single-crystal analyses, and stable when fully desolvated and when heated up to 300 °C. This synthesis is achieved by borrowing ideas from metal carboxylate cluster chemistry, where an organic dicarboxylate linker is used in a reaction that gives supertetrahedron clusters when capped with monocarboxylates. The rigid and divergent character of the added linker allows the articulation of the clusters into a three-dimensional framework resulting in a structure with higher apparent surface area and pore volume than most porous crystalline zeolites. This simple and potentially universal design strategy is currently being pursued in the synthesis of new phases and composites, and for gas-storage applications.

THE tetranuclear supertetrahedral cluster motif shown in Fig. 1 (top) is adopted by a number of metal carboxylates (acetate, benzoate and pivalate), where combination of Zn^{2+} and the appropriate carboxylic acid yields the oxide centred cluster as a distinct and well-defined unit[16]. Working towards an extended network based on this cluster, we viewed its Zn—O—C motif as a secondary building unit capable of assembly under similar

一种异常稳定且高度多孔的金属有机骨架的设计与合成

李海连等

编者按

孔隙大小均匀且孔径与小分子尺寸相当的材料，在工业化学中具有巨大的价值，例如可以作为选择性催化剂和分离气体的过滤器。这样的材料多是铝硅酸盐类，如称作沸石的天然矿石，或者人工合成的类似物。通过分子构筑单元合理组装所构建的多孔骨架——金属离子连接的有机分子——可以为此类物质的结构和性质提供更多的调控。但这方面的努力常常受阻，因为这类骨架在没有充满溶剂时，特别是加热时会坍塌。这里，奥马尔·亚吉和其同事报告了一种金属有机多孔骨架，它甚至能在所有溶剂被去除且加热到 300 ℃ 时还保持稳定。

　　普遍认为，开放式金属有机骨架是在催化、分离、储气和分子识别等应用[1-15]中均有良好前景的材料。与沸石等常规使用的无机微孔材料相比，通过对孔的结构与功能的调控，这些有机结构具有更便于进行灵活合理设计的潜在优势。但到目前为止，这些开放式结构在支持永久孔隙结构以及避免在没有诸如溶剂等客体分子存在时发生坍塌等方面还存在明显不足，这阻碍了该领域的进一步发展[14,15]。在本文中，我们报道一种金属有机骨架的合成，它在完全去溶剂化状态和加热到 300 ℃ 时仍保持晶体性质（由 X 射线单晶分析表明）和稳定性。本合成的成功在于借用了金属羧酸盐簇化学的想法，在制备超四面体簇的反应中，我们使用有机二羧酸盐作为连接体，并以单羧酸盐进行封端。连接体所具有的刚性和发散性特征使簇连接成一个三维骨架，可产生出一种与大多数多孔晶体型沸石相比具有更高的表观表面积和孔隙体积的结构。目前，我们正继续利用这种简单而又具有普适性的设计策略来合成具有新相和新组成的材料，并考虑其在储气方面的应用。

　　图 1（上部）中显示的四核超四面体簇基元是多种金属羧酸盐（醋酸盐、苯甲酸盐和三甲基乙酸盐）所具有的，其中 Zn^{2+} 与适当的羧酸结合产生出以氧化物为中心的簇，它是一个独特且结构明确的单元[16]。在利用该簇来获得扩展网络的工作中，我们把它的 Zn—O—C 基元看作是一个次级结构单元，在相似反应条件下，它能够在有

439

reaction conditions but in the presence of a dicarboxylate instead of a monocarboxylate. Our earlier work[18] involving the copolymerization of 1,4-benzenedicarboxylate (BDC) with metal ions pointed to its rigid, planar, and divergent attributes, which, when coupled with the rigidity of this secondary building unit, would be ideally suited in forming the target extended three-dimensional framework shown in Fig. 1 (bottom).

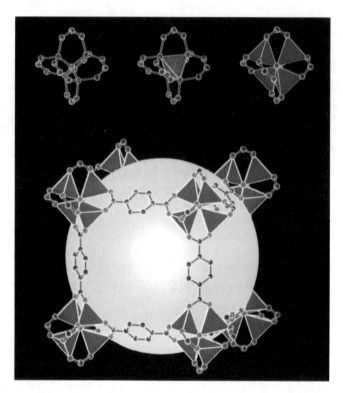

Fig. 1. Construction of the MOF-5 framework. Top, the $Zn_4(O)O_{12}C_6$ cluster. Left, as a ball and stick model (Zn, blue; O, green; C, grey). Middle, the same with the $Zn_4(O)$ tetrahedron indicated in green. Right, the same but now with the ZnO_4 tetrahedra indicated in blue. Bottom, one of the cavities in the $Zn_4(O)(BDC)_3$, MOF-5, framework. Eight clusters (only seven visible) constitute a unit cell and enclose a large cavity, indicated by a yellow sphere of diameter 18.5 Å in contact with 72 C atoms (grey).

We found that diffusion of triethylamine into a solution of zinc (II) nitrate and H_2BDC in N,N'-dimethylformamide (DMF)/chlorobenzene resulted in the deprotonation of H_2BDC and its reaction with Zn^{2+} ions. A small amount of hydrogen peroxide was added to the reaction mixture in order to facilitate the formation of O^{2-} expected at the center of the secondary building unit. Elemental microanalysis of the resulting colourless cubic crystals suggested that their formula was $Zn_4O(BDC)_3 \cdot (DMF)_8(C_6H_5Cl)$; here we call this material MOF-5, where MOF indicates metal organic framework. The presence of DMF and chlorobenzene guests was confirmed by solid-state ^{13}C NMR, which showed sharp characteristic peaks at the chemical shifts expected for the carbon atom resonances of these guests. Ion mass spectrometry, performed on the liberated guest vapours as the material was heated from 22 to 350 °C, showed their respective parent-ion masses.

二羧酸盐而非单羧酸盐存在时进行组装。我们在关于 1,4-苯二甲酸(BDC)与金属离子共聚的早期工作[18]中指出，BDC 的刚性、平面性和发散属性，再结合这种次级结构单元的刚性，很适用于构建我们所想要的扩展三维骨架，如图 1(下部)中所示。

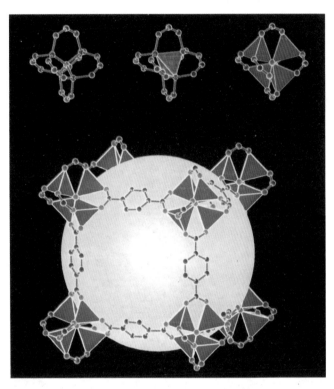

图 1. MOF-5 骨架的构造。上部，$Zn_4(O)O_{12}C_6$ 原子簇。左边是其球棍模型(Zn：蓝；O：绿；C：灰)。中间同样是该原子簇，其中 $Zn_4(O)$ 四面体用绿色标出。右边也是该原子簇，但这次是用蓝色标出 ZnO_4 四面体。下部，MOF-5 骨架，$Zn_4(O)(BDC)_3$ 中的一个空腔。八个原子簇(只能看到七个)构成一个晶胞并圈起一个大的空腔，空腔用黄色球体表示，其直径为 18.5 Å，与 72 个 C 原子(灰色)接触。

我们发现，三乙胺向硝酸锌(Ⅱ)和 H_2BDC 的 N,N'-二甲基甲酰胺(DMF)/氯苯溶液中扩散会导致 H_2BDC 去质子化并与 Zn^{2+} 反应。为促进次级结构单元的中心处 O^{2-} 的形成，我们向反应混合物中加入少量过氧化氢。对所得无色立方晶体进行元素微量分析，结果表明其化学式为 $Zn_4O(BDC)_3\cdot(DMF)_8(C_6H_5Cl)$；我们称该材料为 MOF-5，其中 MOF 表示金属有机骨架。通过固体 ^{13}C NMR 可确认 DMF 和氯苯客体分子的存在，核磁谱图显示出尖锐的特征峰，其化学位移与所预期的客体分子中碳原子的化学位移一致。将该物质从 22 ℃加热到 350 ℃使客体分子释放出来，由此得到的离子质谱显示出它们各自的母体离子的质量。

An X-ray single-crystal diffraction study on the as-synthesized materials reveals the expected topology (Fig. 2). The structure may be derived from a simple cubic six-connected net in two stages: first, the nodes (vertices) of the net are replaced by clusters of secondary building units; second, the links (edges) of the net are replaced by finite rods ("struts") of BDC molecules. The core of the cluster consists of a single O atom bonded to four Zn atoms, forming a regular Zn_4O tetrahedron. Each edge of each Zn tetrahedron is then capped by a $-CO_2$ group to form a $Zn_4(O)(CO_2)_6$ cluster. We note that a related zinc phosphate structure exists[17], in which the C atoms are replaced by the P atoms of PO_4 tetrahedra, which serves to link the clusters together. The important step taken in the present work is the replacement of the O_2-P-O_2 phosphate links by $O_2-C-C_6H_4-C-O_2$ (BDC) "struts", to give an extended network having a three-dimensional intersecting channel system with 12.94-Å spacing between the centres of adjacent clusters.

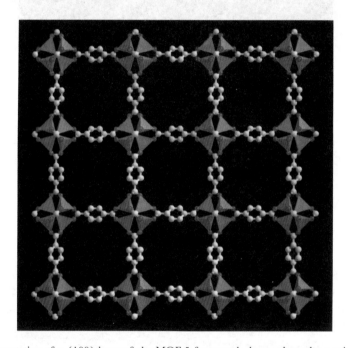

Fig. 2. Representation of a {100} layer of the MOF-5 framework shown along the a-axis (C, grey; O, green). The ZnO_4 tetrahedra are indicated in purple. Properties of MOF-5 are as follows. Single crystals of as-synthesized $Zn_4O(BDC)_3 \cdot (DMF)_8(C_6H_5Cl)$, MOF-5, are at 213 ± 2 K cubic, space group $Fm\text{-}3m$ with $a = 25.6690(3)$ Å, $V = 16,913.2(3)$ Å^3, and $Z = 8$. Analysis for MOF-5. (1) Elemental microanalysis. Calculated (%): C, 44.21; H, 5.02; N, 7.64; Zn, 17.82%. Found (%): C, 43.25; H, 5.29; N, 7.56; Zn, 17.04%. (2) Solid-state ^{13}C NMR. First, CP MAS. Guests $(CH_3)_2NC(O)H$: $(CH_3)_2-$, δ (p.p.m.) 37.81 and 32.71; $-C(O)-$, δ 164.93. C_6H_5Cl: δ 132.04, 130.41, 128.76. Framework, $-COO$, δ 175.93; $-C(COO)$, δ 139.11; $o\text{-}C_6H_4$, δ 130.41. Second, static. Guests $(CH_3)_2NC(O)H$: $(CH_3)_2-$, δ 37.63 and 32.59; $-C(O)-$, δ 164.67. C_6H_5Cl: δ 130.22 (broad). Framework, no peaks observed. (3) Mass spectrometry, 22–350 °C. $(CH_3)_2NC(O)H$: M^+, 73 a.m.u. $(C_3H_7NO^+)$. C_6H_5Cl: M^+, 77 $(C_6H_5^+)$; M^+, 112 a.m.u. $(C_6H_5{}^{35}Cl^+)$ and 114 a.m.u. $(C_6H_5{}^{37}Cl^+)$ in 3:1 intensity ratio. Properties of single crystals of fully desolvated MOF-5, $Zn_4O(BDC)_3$, are at 169 ± 2 K cubic, space group $Fm\text{-}3m$ with $a = 25.8849(3)$ Å, $V = 17,343.6(8)$ Å^3, and $Z = 8$ (for empirical formula, $C_{24}H_{12}O_{13}Zn_4$), R, $R_w = 0.023, 0.026$. Analysis for $Zn_4O(BDC)_3$. Elemental microanalysis. Calculated (%): C, 37.45; H, 1.57; N, 0.00%. Found (%): C, 36.94; H, 1.66; N, 0.26%. Solid-state ^{13}C NMR (CP MAS and static) showed no resonances due to any guests. Mass spectrometry (22–350 °C) showed no peaks, indicative of absence of any guests in the pores. Thermogravimetric analysis showed

所合成材料的 X 射线单晶衍射研究说明其拓扑结构与预期相符（图 2）。从简单的立方六联网络出发经过两步即可推导出其结构：首先，用次级结构单元簇取代网络的节点（顶点）；接着，用 BDC 分子的限定棒状部分（"支柱"）取代网络中的连接（棱）。一个单独的 O 原子与四个 Zn 原子键合在一起，处于簇的核心，形成一个规则的 Zn_4O 四面体结构。随后，每个 Zn 四面体的每一条棱被包上一个 $-CO_2$ 基团，形成一个 $Zn_4(O)(CO_2)_6$ 簇。我们注意到，存在着一种相关的磷酸锌结构[17]，其中 C 原子被将簇连接起来的 PO_4 四面体中的 P 原子所取代。当前研究采取的重要步骤是用 $O_2-C-C_6H_4-C-O_2$（BDC）"支柱"取代 O_2-P-O_2 磷酸盐连接基元，它给出了一种扩展网络，这种网络具有三维交叉孔道体系，其相邻簇中心之间的间距为 12.94 Å。

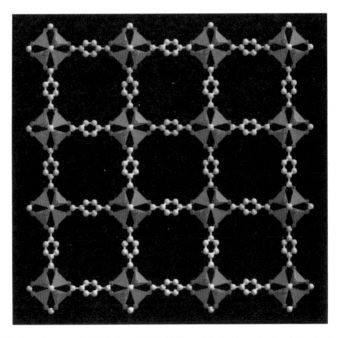

图 2. MOF-5 骨架中一个 {100} 层的图式，所示为沿着 a 轴方向（C，灰；O，绿）。ZnO_4 四面体用紫色显示。MOF-5 的性质如下。所合成的 $Zn_4O(BDC)_3 \cdot (DMF)_8(C_6H_5Cl)$ 单晶，MOF-5，213 ± 2 K 下测试为立方晶系，空间群 *Fm-3m*，$a = 25.6690(3)$ Å，$V = 16,913.2(3)$ Å^3，$Z = 8$。MOF-5 的分析结果如下。(1) 元素微量分析。计算值(%)：C，44.21；H，5.02；N，7.64；Zn，17.82%。测试值(%)：C，43.25；H，5.29；N，7.56；Zn，17.04%。(2) 固体 ^{13}C NMR。一、CP MAS（即交叉极化魔角旋转）方法。客体分子 $(CH_3)_2NC(O)H$：$(CH_3)_2-$，δ (ppm) 37.81 和 32.71；$-C(O)-$，δ 164.93。C_6H_5Cl：δ 132.04、130.41、128.76。骨架：$-COO$，δ 175.93；$-C(COO)$，δ 139.11；$o-C_6H_4$，δ 130.41。二、静态方法。客体分子 $(CH_3)_2NC(O)H$：$(CH_3)_2-$，δ 37.63 和 32.59；$-C(O)-$，δ 164.67。C_6H_5Cl：δ 130.22(宽包)。骨架：没有观测到峰。(3) 质谱，$22 \sim 350$ ℃。$(CH_3)_2NC(O)H$：M^+，73 amu $(C_3H_7NO^+)$。C_6H_5Cl：M^+，77$(C_6H_5^+)$；M^+，112 amu $(C_6H_5{}^{35}Cl^+)$ 和 114 amu $(C_6H_5{}^{37}Cl^+)$，强度比为 3:1。完全去溶剂的 MOF-5——$Zn_4O(BDC)_3$ 的单晶性质：169 ± 2 K 下测试为立方晶系，空间群 *Fm-3m*，$a = 25.8849(3)$ Å，$V = 17,343.6(8)$ Å^3，$Z = 8$(实验式 $C_{24}H_{12}O_{13}Zn_4$)，R，$R_w = 0.023$，0.026。$Zn_4O(BDC)_3$ 的分析结果如下。元素微量分析。计算值(%)：C，37.45；H，1.57；N，0.00%。测试值(%)：C，36.94；H，1.66；N，0.26%。固体 ^{13}C NMR(CP MAS 和静态方法)显示没有任何来自于客体分子的共振。质谱($22 \sim 350$ ℃)显示没有峰，表明孔隙中没有任何

no weight loss up to 410 °C. Single crystals of $Zn_4O(BDC)_3$, fully desolvated and heated at 300 °C, are at 149 ± 2 K cubic, space group *Fm-3m* with $a = 25.8496(3)$ Å.

The framework atoms in MOF-5 take up only a small fraction of the available space in the crystal. If overlapping spheres with van der Waals radii (1.5, 1.2, 1.7, 1.5 Å for Zn, H, C and O, respectively) are placed at the atomic positions, the space not so occupied is 80% of the crystal volume. The non-framework space is divided into two types of cavity lined with C and H atoms. The center of the smaller of these has 24 H atoms at 7.10 Å and 24 C atoms at 7.97 Å; the centre of the larger (Fig. 1) has 72 C atoms at 9.26 Å and 48 H atoms at 9.47 Å. The aperture joining the two cavities has 8 H at 5.10 Å and 8 C at 5.70 Å from its center. Allowing for the van der Waals radii, spheres of material with diameter 15.1 Å and 11.0 Å could fit in the large and small cavities, respectively, and the aperture would admit the passage of a sphere of diameter 8.0 Å. Assuming the same molar volume as in the respective liquids, we find that in the original material the guest molecules—8 × (8 DMF + 1 C_6H_5Cl)—have a volume of 9,512 $Å^3$, which is 55% of the unit cell volume. Indeed, we show below that 55–61% of the space is accessible to guest species.

Due to their high mobility, the guests can be fully exchanged with chloroform. More significant is that all the chloroform guests can be evacuated from the pores within 3 h at room temperature and 5×10^{-5} torr without loss of framework periodicity. In fact, the desolvated single crystals were checked by elemental microanalysis, ^{13}C NMR, mass spectrometry, and thermal gravimetry to confirm the absence of any guests, while optical microscopy was used to show that they maintained their transparency, morphology and crystallinity. This has allowed us to perform another X-ray single-crystal diffraction study on the fully desolvated form of this framework—a study that revealed mono-crystallinity with cell parameters and atomic positions coincident with those of the as-synthesized material. In fact, the highest peaks and lowest valleys in the ΔF map are 0.25 and −0.17 electrons per $Å^3$ scattered randomly both near the framework and the void space. These contrast with the relatively large values (1.56 and −0.45 electrons per $Å^3$) found in the as-synthesized crystals.

Further evidence for the stability of the framework was obtained by heating the fully desolvated crystals in air at 300 °C for 24 h, which had no effect on either their morphology or crystallinity as evidence by another X-ray single-crystal diffraction study. Here again, the cell parameters obtained were unaltered relative to those found for the unheated desolvated crystals, illustrating the rigidity and stability of the framework in the absence of guest molecules. These results are interesting as the density (0.59 g cm^{-3}) of the desolvated crystals is among the lowest recorded for any crystalline material.

客体分子。热重分析显示直到 410 ℃ 时没有重量损失。完全去溶剂并加热到 300 ℃ 处理过的 Zn₄O(BDC)₃ 单晶，149±2 K 下测试为立方晶系，空间群 *Fm-3m*，a = 25.8496(3) Å。

在晶体中，MOF-5 中的骨架原子只占可用空间中很小的一部分。若将相交在一起的具有范德华半径大小(对于 Zn、H、C 和 O 分别是 1.5、1.2、1.7 和 1.5 Å)的球体置于晶格原子所在位置上，则未被占据的空间为晶体体积的 80%。非骨架空间分为沿着 C 和 H 原子排列方向的两类空腔。其中较小空腔的中心含有位于 7.10 Å 的 24 个 H 原子和位于 7.97 Å 的 24 个 C 原子；较大空腔(图 1)的中心有位于 9.26 Å 的 72 个 C 原子和位于 9.47 Å 的 48 个 H 原子。连接这两种空腔的孔隙有距离其中心 5.10 Å 的 8 个 H 和 5.70 Å 的 8 个 C。考虑到范德华半径，较大和较小空腔中分别可以纳入直径为 15.1 Å 和 11.0 Å 的球形物体，而孔隙则允许直径为 8.0 Å 的球体通行。假定与各自相应的液体具有相同的摩尔体积，我们发现在原始材料中客体分子——8 × (8 DMF + 1 C₆H₅Cl)——具有 9,512 Å³ 的体积，占单位晶胞体积的 55%。实际上，如下所示，有 55% ~ 61% 的空间是可容纳客体分子的。

由于客体分子的高度运动性，它们能够被氯仿完全交换出。更重要的是，所有氯仿分子可以在室温和 5×10^{-5} 托的条件下于 3 h 内完全从孔隙中挥发出来，而骨架周期性没有任何损失。实际上，对去溶剂后的单晶，元素微量分析、¹³C NMR、质谱和热重分析可以确认其中没有任何客体分子。同时，光学显微镜观察表明，它们保持着原有的透明性、形貌和结晶度。这使得我们能够对该骨架的完全去溶剂形式进行新的 X 射线单晶衍射研究——结果表明其单晶性质具有与所合成物质一致的晶胞参数和原子位置。实际上，Δ*F* 图中最高的峰和最低的谷为每 Å³ 中 0.25 和 −0.17 个电子，随机散落于骨架和孔隙空间附近。这与在所合成晶体中发现的相对较大的数值(每 Å³ 中 1.56 和 −0.45 个电子)有所不同。

关于骨架稳定性的更多证据是通过将完全去溶剂的晶体在空气中于 300 ℃ 条件下加热 24 h 而获得，进一步的 X 射线单晶衍射研究证明无论是它们的形貌还是结晶度都没有受到影响。这里，所得晶胞参数相对于未加热的去溶剂晶体的晶胞参数并没有什么改变，这表明了该骨架在没有客体分子存在时的刚性和稳定性。这些结果是值得关注的，因为这种去溶剂晶体的密度(0.59 g·cm⁻³)在已报道的任何结晶物质密度中是属于最低的之列。

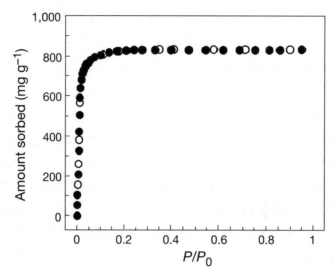

Fig. 3. Nitrogen gas sorption isotherm at 78 K for MOF-5 (filled circles, sorption; open circles, desorption). P/P_0 is the ratio of gas pressure (P) to saturation pressure (P_0), with $P_0 = 746$ torr.

Table 1. Gas and liquid vapour sorption in desolvated MOF-5

Sorbate	T (°C)	Amount sorbed (mg g^{-1})	Sorbate molecules per unit cell	Free volume (cm^3 g^{-1})	Free volume (cm^3 cm^{-3})
Ar	−194	1,492	230	1.03	0.61
N_2	−194	831	183	1.04	0.61
CH_2Cl_2	22	1,211	88	0.93	0.55
$CHCl_3$	22	1,367	71	0.94	0.55
C_6H_6	22	802	63	0.94	0.55
CCl_4	22	1,472	59	0.94	0.56
C_6H_{12}	22	703	51	0.92	0.54

To evaluate the pore volume and the apparent surface area of this framework, the gas and vapour sorption isotherms for the desolvated sample were measured using an electromicrogravimetric balance (Cahn 1000) set-up and an already published procedure[18]. The sorption of nitrogen revealed a reversible type I isotherm (Fig. 3), with concordant results obtained with large crystals (0.2 mm) and microcrystals (30 μm). The same sorption behaviour was observed for argon and organic vapours such as CH_2Cl_2, $CHCl_3$, CCl_4, C_6H_6 and C_6H_{12}. Similar to those of most zeolites, the isotherms are reversible and show no hysteresis upon desorption of gases from the pores. Using the Dubinin–Radushkevich equation, pore volumes of 0.61–0.54 cm^3 cm^{-3} were calculated and found to be nearly identical for all adsorbates shown (Table 1), which indicates the presence of uniform pores. Assuming a monolayer coverage of N_2 (which may not be strictly correct with such large cavities) the apparent Langmuir surface area was estimated at 2,900 m^2 g^{-1}. Zeolites, which generally have higher molar masses than $Zn_4O(BDC)_3$, have pore volumes that range from 0.18 cm^3 cm^{-3} for analcime to 0.47 cm^3 cm^{-3} for zeolite A[19].

(**402**, 276-279; 1999)

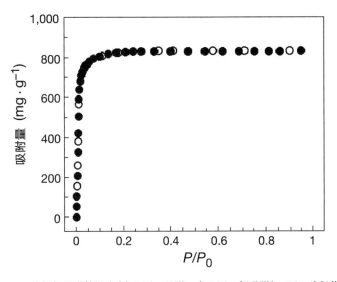

图 3. 78 K 时 MOF-5 的氮气吸附等温线 (实心圆，吸附；空心圆，解吸附)。P/P_0 为气体压强 (P) 与饱和压强 (P_0) 的比值，且 $P_0 = 746$ 托。

表 1. 去溶剂 MOF-5 中的气体和液体蒸气吸附

吸附物	T(℃)	吸附量 (mg·g^{-1})	每单位晶胞中的吸附分子数	自由体积 (cm^3·g^{-1})	自由体积 (cm^3·cm^{-3})
Ar	−194	1,492	230	1.03	0.61
N_2	−194	831	183	1.04	0.61
CH_2Cl_2	22	1,211	88	0.93	0.55
$CHCl_3$	22	1,367	71	0.94	0.55
C_6H_6	22	802	63	0.94	0.55
CCl_4	22	1,472	59	0.94	0.56
C_6H_{12}	22	703	51	0.92	0.54

为估计这种多孔骨架中的孔隙体积和表观表面积，利用电子微重分析天平 (Cahn 1000) 和一种已经发表的方法[18]，我们测定了去溶剂样品的气体和蒸气的吸附等温线。氮气的吸附数据揭示出这是一种可逆的 I 型等温线 (图 3)，无论对大晶体 (0.2 mm) 或微晶 (30 μm) 都得到了一致的结果。对于氩和诸如 CH_2Cl_2、$CHCl_3$、CCl_4、C_6H_6 和 C_6H_{12} 等有机物蒸气，也观测到了同样的吸附行为。与大多数沸石的等温线类似，该等温线是可逆的，而且没有显示出气体从孔隙中解吸附导致的滞后现象。利用 DR 方程计算出孔隙体积为 0.61 cm^3·cm^{-3} ～ 0.54 cm^3·cm^{-3}，我们发现，对于所示的 (表 1) 全部吸附物而言，孔隙体积几乎都是一样的，这表明均匀孔隙的存在。假设 N_2 为单层覆盖 (对于这样大的孔隙来说这可能不是严格准确的)，可估算出表观朗缪尔表面积为 2,900 m^2·g^{-1}。而沸石一般来说具有比 $Zn_4O(BDC)_3$ 更高的摩尔质量，其孔隙体积范围为从方沸石的 0.18 cm^3·cm^{-3} 到沸石 A 的 0.47 cm^3·cm^{-3}[19]。

（王耀杨 翻译；陈尔强 审稿）

Hailian Li[*], Mohamed Eddaoudi[†], M. O'Keeffe[*] and O. M. Yaghi[†]

[*] Materials Design and Discovery Group, Department of Chemistry and Biochemistry, Arizona State University, Tempe, Arizona 85287-1604, USA

[†] Department of Chemistry, University of Michigan, 930 North University, Ann Arbor, Michigan 48109-1055, USA

Received 20 May; accepted 17 September 1999.

References:

1. Kinoshita, Y., Matsubara, I., Higuchi, T. & Saito, Y. The crystal structure of bis(adiponitrilo) copper (I) nitrate. *Bull. Chem. Soc. Jpn* **32**, 1221-1226 (1959).

2. Kitgawa, S., Munakata, M. & Tanimura, T. Tetranuclear copper(I)-based infinite one-dimensional chain complex. *Chem. Lett.* 623-626 (1991).

3. Abrahams, B. F., Hoskins, B. F. & Robson, R. Infinite polymeric frameworks consisting of three dimensionally linked rod-like segments. *J. Am. Chem. Soc.* **113**, 3606-3607 (1991).

4. Zaworotko, M. J. Cooperative bonding affords a holesome story. *Nature* **386**, 220-221 (1997).

5. Fagan, P. J. & Ward, M. D. Building molecular crystals. *Sci, Am.* **267**, 48-54 (1992).

6. Carlucci, L., Ciani, G., Proserpio, D. M. & Sironi, A. Interpenetrating diamondoid frameworks of silver (I) cations linked by N,N'-bidentate molecular rods. *J. Chem. Soc. Chem. Commun.* 2755-2756 (1994).

7. Mallouk, T. E. Crowns get organized. *Nature* **387**, 350-351 (1997).

8. Gardner, G. B., Venkataraman, D., Moore, J. S. & Lee, S. Spontaneous assembly of a hinged coordination network. *Nature* **374**, 792-795 (1995).

9. Yaghi, O. M., Li, G. & Li, H. Selective binding and removal of guests in a microporous metal-organic framework. *Nature* **378**, 703-706 (1995).

10. Lu, J. *et al.* Coordination polymers of $Co(NCS)_2$ with pyrazine and 4,4-bipyridine: syntheses and structures. *Inorg. Chem.* **36**, 923-928 (1997).

11. Vaid, T. P., Lobkovsky, E. B. & Wolczanski, P. T. Covalent 3- and 2-dimensional titanium-quinone networks. *J. Am. Chem. Soc.* **119**, 8742-8743 (1997).

12. Liu, F. Q. & Tilley, T. D. A coordination network based on d^0 transition-metal centers: synthesis and structure of the [2,4]-connected layered compound $[(TiCl_4)_2Si(C_6H_4CN-p)4]\cdot1.5C_7H_8$. *J. Chem. Soc. Chem. Commun.* 103-104 (1998).

13. MacGillivray, L. R., Groeneman, R. H. & Atwood, J. L. Design and self-assembly of cavity-containing rectangular grids. *J. Am. Chem. Soc.* **120**, 2676-2677 (1998).

14. Yaghi, O. M., Li, H., Davis, C., Richardson, D. & Groy, T. L. Synthetic strategies, structure patterns, and emerging properties in the chemistry of modular porous solids. *Acc. Chem. Res.* **31**, 575-585 (1998).

15. Kepert, C. J. & Rosseinsky, M. J. Zeolite-like crystal structure of an empty microporous framework. *J. Chem. Soc. Chem. Commun.* 375-376 (1999).

16. Clegg, W. *et al.* Crystal structures of three basic zinc carboxylates together with infrared and FAB mass spectrometry studies in solution. *Inorg. Chim. Acta* **186**, 51-60 (1991).

17. Harrison, W. T. A. *et al.* Synthesis and characterization of a new family of thermally stable open framework zincophosphate/arsenate phases: $M_3Zn_4O(XO_4)_3\cdot nH_2O$ (M = Na,K,Rb,Li,..; X = P,As; n = ~3.5-6). Crystal structures of $Rb_3Zn_4O(PO_4)\cdot3.5H_2O$, $K_3Zn_4O(AsO_4)_3\cdot4H_2O$, and $Na_3Zn_4O(PO_4)_3\cdot6H_2O$. *Chem. Mater.* **8**, 691-700 (1996).

18. Li, H., Eddaouidi, M., Groy, T. L. & Yaghi, O. M. Establishing microporosity in open metal-organic frameworks: gas sorption isotherms for Zn(BDC) (BDC = 1,4-benzenedicarboxylate). *J. Am. Chem. Soc.* **120**, 8571-8672 (1998).

19. Breck, D. W. *Zeolite Molecular Sieves* (Wiley & Sons, New York, 1974).

Supplementary information is available on *Nature's* World-Wide Web site (http://www.nature.com) or as paper copy from the London editorial office of *Nature*.

Acknowledgements. We thank F. Hollander and R. Staples for X-ray structure determinations. This work was supported by the National Science Foundation (M.O.K. and O.M.Y.), the Department of Energy (O.M.Y.) and Nalco Chemical Company.

Correspondence and requests for materials should be addressed to O.M.Y. at the University of Michigan (e-mail: oyaghi@umich.edu).

The DNA Sequence of Human Chromosome 22

I. Dunham *et al.*

Editor's Note

This paper, a key landmark in the Human Genome Project, describes the first complete sequence of a human chromosome. This sequence of chromosome 22 offered insights into the way that genes are arranged along chromosomes and into how these genes might be controlled, paving the way for developments in medical diagnostics and therapeutics. The project, led by British-based geneticist Ian Dunham, involved an international consortium of sequencing centres, and the data were made freely available. The study shows that chromosome 22 is made up of 33.5 million "letters" and includes 679 genes, over half of which were previously unknown in humans. The complete human genome sequence was published four years later.

Knowledge of the complete genomic DNA sequence of an organism allows a systematic approach to defining its genetic components. The genomic sequence provides access to the complete structures of all genes, including those without known function, their control elements, and, by inference, the proteins they encode, as well as all other biologically important sequences. Furthermore, the sequence is a rich and permanent source of information for the design of further biological studies of the organism and for the study of evolution through cross-species sequence comparison. The power of this approach has been amply demonstrated by the determination of the sequences of a number of microbial and model organisms. The next step is to obtain the complete sequence of the entire human genome. Here we report the sequence of the euchromatic part of human chromosome 22. The sequence obtained consists of 12 contiguous segments spanning 33.4 megabases, contains at least 545 genes and 134 pseudogenes, and provides the first view of the complex chromosomal landscapes that will be found in the rest of the genome.

TWO alternative approaches have been proposed to determine the human genome sequence. In the clone by clone approach, a map of the genome is constructed using clones of a suitable size (for example, 100–200 kilobases (kb)), and then the sequence is determined for each of a representative set of clones that completely covers the map[1]. Alternatively, a whole genome shotgun[2] requires the sequencing of unmapped genomic clones, typically in a size range of 2–10 kb, followed by a monolithic assembly to produce the entire sequence. Although the merits of these two strategies continue to be debated[3], the public domain human genome sequencing project is following the clone by clone approach[4] because it is modular, allows efficient organization of distributed resources and sequencing capacities, avoids problems arising from distant repeats and results in early completion of significant units of the genome. Here we report the first sequencing

人类 22 号染色体的 DNA 序列

邓纳姆等

编者按

本文是人类基因组计划的一个关键里程碑,介绍了单条人类染色体的第一个完整序列。22 号染色体的序列帮助我们了解基因在染色体上的编排方式以及这些基因是如何被调控的,并为医学诊断和治疗的发展奠定了基础。本项目由英国遗传学家伊恩·邓纳姆领衔,多个测序中心组成的国际联盟共同完成,数据可免费获取。本研究表明,22 号染色体是由 3,350 万个"字母"组成,包含 679 个基因,其中半数以上之前在人类中是未知的。完整的人类基因组序列发表于 4 年后。

关于一种生物的完整基因组 DNA 序列的知识使我们可以用系统的方法确定其遗传组分。基因组序列可以提供所有基因的完整结构,包括功能未知的基因及其控制元件,通过推理还可得到它们编码的蛋白质,以及所有其他生物学上的重要序列。此外,对于针对生物体设计进一步的生物学研究和通过跨物种序列比对来进行演化研究而言,该物种的序列都是丰富且永久的信息来源。一系列微生物和模式生物序列的测定已经充分证明了这种方法的强大能力。下一步是获得整个人类基因组的完整序列。在本文中我们报道了人类 22 号染色体常染色质部分的序列。该序列由跨越 33.4 Mb 的 12 个连续片段组成,包含至少 545 个基因和 134 个假基因,并首次展示了在基因组其他部分将会发现的复杂的染色体概况。

目前有两种测定人类基因组序列的方法可供选择。连续克隆法用适当大小(如 100 ~ 200 千碱基(kb))的多个克隆构建基因组图谱,然后测定完全覆盖图谱的一套代表性克隆中每个克隆的序列[1]。另一种是全基因组鸟枪法[2],需要对未作图的基因组克隆进行测序,大小通常在 2 ~ 10 kb,然后通过整体组装得到整个序列。虽然这两种策略的优点仍然存在争议[3],但公开的人类基因组测序项目已经采取连续克隆法[4],因为这种方法是模块化的,可以有效地组织分散的资源及测序能力,避免远距离重复引起的问题,而且能够迅速完成基因组重要单元的测定。在本文中,我们报道了人类基因组计划的第一个测序里程碑:第一条人类染色体常染色质部分的

landmark of the human genome project, the operationally complete sequence of the euchromatic portion of a human chromosome.

Chromosome 22 is the second smallest of the human autosomes, comprising 1.6–1.8% of the genomic DNA[5]. It is one of five human acrocentric chromosomes, each of which shares substantial sequence similarity in the short arm, which encodes the tandemly repeated ribosomal RNA genes and a series of other tandem repeat sequence arrays. There is no evidence to indicate the presence of any protein coding genes on the short arm of chromosome 22 (22p). In contrast, direct[6] and indirect[7,8] mapping methods suggest that the long arm of the chromosome (22q) is rich in genes compared with other chromosomes. The relatively small size and the existence of a high-resolution framework map of the chromosome[9] suggested to us that sequencing human chromosome 22 would provide an excellent opportunity to show the feasibility of completing the sequence of a substantial unit of the human genome. In addition, alteration of gene dosage on part of 22q is responsible for the aetiology of a number of human congenital anomaly disorders including cat eye syndrome (CES, Mendelian Inheritance in Man (MIM) 115470, http://www.ncbi.nlm.nih.gov/omim/) and velocardiofacial/DiGeorge syndrome (VCFS, MIM 192430; DGS, MIM 188400). Other regions associated with human disease are the schizophrenia susceptibility locus[10,11], and the sequences involved in spinocerebellar ataxia 10 (SCA10)[12]. Making the sequence of human chromosome 22 freely available to the community early in the data collection phase has benefited studies of disease-related and other genes associated with this human chromosome[13-19].

Genomic Sequencing

To identify genomic clones as the substrate for sequencing chromosome 22, extensive clone maps of the chromosome were constructed using cosmids, fosmids, bacterial artificial chromosomes (BACs) and P1-derived artificial chromosomes (PACs). Clones representing parts of chromosome 22 were identified by screening BAC and PAC libraries representing more than 20 genome equivalents using sequence tagged site (STS) markers known to be derived from the chromosome, or by using cosmid and fosmid libraries derived from flow-sorted DNA from chromosome 22. Overlapping clone contigs were assembled on the basis of restriction enzyme fingerprints and STS-content data, and ordered relative to each other using the established framework map of the chromosome[9]. The resulting nascent contigs were extended and joined by iterative cycles of chromosome walking using sequences from the end of each contig. In two places, yeast artificial chromosome (YAC) clones were used to join or extend contigs (AL049708, AL049760). The sequence-ready map covers 22q in 11 clone contigs with 10 gaps and stretches from sequences containing known chromosome 22 centromeric tandem repeats to the 22q telomere[20].

In the final sequence, one additional gap that was intractable to sequencing is found 234 kb from the centromere (see below). The gaps between the clone contigs are located at the two ends of the map, in the 4.3 Mb adjacent to the centromere and in 7.3 Mb at the telomeric

可使用的完整序列。

22 号染色体是人类第二小的常染色体，占基因组 DNA 的 1.6% ~ 1.8%[5]。它是人类 5 条近端着丝粒染色体之一，这 5 条染色体具有序列高度相似的短臂，这些短臂编码串联重复的核糖体 RNA 基因和一系列其他串联重复序列阵列。目前没有证据表明 22 号染色体短臂(22p)上存在任何蛋白质编码基因。相反，直接[6]和间接[7,8]作图方法都表明该染色体的长臂(22q)与其他染色体相比拥有丰富的基因。人类 22 号染色体相对较小，并具有高分辨率框架图[9]，这表明对其进行测序是展示人类基因组大型单元的序列测定可行性的绝佳机会。此外，部分 22q 上基因数量的改变是引起很多人类先天性异常疾病的病因，这些疾病包括猫眼综合征(CES，人类孟德尔遗传数据库(MIM) 115470，http://www.ncbi.nlm.nih.gov/omim/)和腭心面/迪格奥尔格综合征(VCFS，MIM 192430；DGS，MIM 188400)。其他与人类疾病相关的区域包括精神分裂症易感性位点[10,11]和涉及脊髓小脑性共济失调 10(SCA10)的序列[12]。我们早在数据收集阶段就将人类 22 号染色体的序列免费对社会开放，这使得一些与该染色体相关的疾病和其他基因的研究因此受益[13-19]。

基因组测序

为了识别作为 22 号染色体测序底物的基因组克隆，科研人员用黏粒、F 黏粒、细菌人工染色体(BAC)和 P1 衍生人工染色体(PAC)构建了大量的该染色体的克隆图谱。使用已知的来自该染色体的序列标签位点(STS)标记的序列，对覆盖超过 20 倍基因组的 BAC 和 PAC 克隆文库进行筛选，或使用 22 号染色体流式分选的 DNA 的黏粒和 F 黏粒文库进行筛选，识别来自部分 22 号染色体的克隆。在限制性内切酶指纹和 STS 含量数据的基础上组装重叠的克隆重叠群，并用已建立的该染色体框架图对其进行相应排列[9]。通过用每个重叠群末端的序列进行染色体步移迭代循环，将上述生成的新重叠群延长并连接。其中两处重叠群(AL049708 和 AL049760)的连接或延长使用了酵母人工染色体(YAC)来进行。已知序列图以存在 10 个缺口的 11 个克隆重叠群覆盖 22q，从包含已知的 22 号染色体着丝粒串联重复的序列延伸至 22q 端粒[20]。

在最终序列中，距着丝粒 234 kb 处发现另一个缺口(见下文)，这个缺口很难进行测序。克隆重叠群之间的各个缺口分布于图的两端，在与着丝粒邻近的 4.3 Mb 处及端粒末端的 7.3 Mb 处。这些区域被一个 23 Mb 的中央重叠群分开。我们使用与重

end. These regions are separated by a central contig of 23 Mb. We have concluded that the gaps contain sequences that are unclonable with the available host-vector systems, as we were unable to detect clones containing the sequences in these gaps by screening more than 20 genome equivalents of bacterial clones using sequences adjacent to the contig ends.

The size of the seven gaps in the telomeric region has been estimated by DNA fibre fluorescence *in situ* hybridization (FISH). No gap in this region is judged to be larger than ~150 kb. For three of these gaps, a number of BAC and PAC clones that contain STSs on either side of the gap were shown to be deleted for at least a minimal core region by DNA fibre FISH. As these clones come from multiple donor DNA sources, these results are unlikely to be due to deletion in the DNA used to make the libraries. Furthermore, the same result was observed for the gap at 32,600 kb from the centromeric end of the sequence, when the DNA fibre FISH experiments were performed on DNA from two different lymphoblastoid cell lines. One possible explanation for this observation is that DNA fragments containing the gap sequences are initially cloned in the BAC library but clones that delete these sequences have a significant selective advantage as the library is propagated. As the observed size range of the cloned inserts in the BAC libraries ranges from 100 kb to more than 230 kb (http://bacpac.med.buffalo.edu/), such deletion events are not distinguishable on the basis of size from undeleted BACs. Additional analysis of the distribution of BAC end sequences from dbGSS (http://www.ncbi.nlm.nih.gov/dbGSS/index.html) suggests that the BAC coverage is sparser closer to the gaps and that this analysis did not identify any BACs spanning the gaps. The three remaining clone-map gaps in the proximal region of the long arm are in regions that may contain segments of previously characterized low-copy repeats[21]. These gaps could not be sized by DNA fibre FISH because of the extensive intra- and interchromosomal repeat sequences (see below) but were amenable to long-range restriction mapping. The gap between AP000529 and AP000530 was estimated to be shorter than 150 kb by comparison with a previously established long-range restriction map[22]. The gap closest to the centromere, which is less than 2 kb in size, could not be sequenced despite BAC clone coverage as it was unrepresented in plasmid or M13 libraries, and was intractable to all sequencing strategies applied. Detailed descriptions of several of the clone contigs have been published[21,23,24] or will be published elsewhere.

Each sequencing group took responsibility for completion of adjacent areas of the sequence as illustrated in Fig. 1. (Editorial note: this figure was original included as a fold-out insert. For details, see http://www.nature.com/articles/990031/figures/1) A set of minimally overlapping clones (the "tile path") was chosen from the physical map and sequenced using a combination of a random shotgun assembly, followed by directed sequencing to close gaps and resolve ambiguities ("finishing"). The major problems encountered during completion of the sequence in the directed sequencing phase were CpG islands, tandem repeats and apparent cloning biases. Directed sequencing using oligonucleotide primers, very short insert plasmid libraries, or identification of bridging clones by screening high complexity plasmid or M13 libraries solved these problems.

叠群末端相邻的序列对覆盖超过 20 倍基因组的细菌克隆进行筛选，仍无法检测到含有这些缺口序列的克隆，因此我们认为这些缺口包含用现有宿主−载体系统无法克隆的序列。

通过 DNA 纤维荧光原位杂交 (FISH) 估算端粒区域七个缺口的大小。经判断，这一区域的缺口均不超过 150 kb 左右。据 DNA 纤维荧光原位杂交显示，对其中三个缺口而言，一些在缺口的任意一侧包含 STS 的 BAC 和 PAC 克隆被删去了至少一处最小核心区。由于这些克隆有多个 DNA 供体来源，因此这些结果不可能是由于用于构建文库的 DNA 存在缺失。此外，当在两种不同成淋巴母细胞细胞系 DNA 上进行 DNA 纤维荧光原位杂交实验时，在序列中距着丝粒末端 32,600 kb 的缺口处也观察到同样的结果。这个现象的一种可能解释是，包含缺口序列的 DNA 片段最初在 BAC 文库中进行了克隆，但当文库扩增时，删除这些序列的克隆有显著的选择性优势。观察到的插入 BAC 文库的克隆片段的大小从 100 kb 到 230 kb 以上不等 (http://bacpac.med.buffalo.edu/)，因此这些缺失状况无法根据未缺失的 BAC 大小进行区分。对来自 dbGSS(http://www.ncbi.nlm.nih.gov/dbGSS/index.html) 的 BAC 末端序列的分布进一步分析表明，接近缺口处 BAC 覆盖稀疏，这一分析也没有发现任何跨过缺口的 BAC。剩下的三个在长臂近端区域的克隆图谱缺口位于可能包含早先表征过的低拷贝重复片段的区域[21]。由于大量染色体内和染色体间重复序列 (见下文) 的存在，这些缺口的大小无法通过 DNA 纤维荧光原位杂交测定，但可以通过长距离限制性作图得到。通过与先前建立的长距离限制图比较可知，AP000529 和 AP000530 之间的缺口估计小于 150 kb[22]。与着丝粒最近的缺口不到 2 kb，但由于在质粒和 M13 文库中没有表达，因此尽管被 BAC 克隆覆盖却仍无法测序，而且在所有测序策略中均无法处理。几个克隆重叠群的详细介绍已经发表[21,23,24]，或将另行发表。

序列相邻区域的测序由每个测序小组负责完成，如图 1 所示。(编者注：这张图在原文中是一个折叠插入图，细节请见 http://www.nature.com/articles/990031/figures/1) 从物理图中选出一套最小重叠克隆 (“覆瓦式”) 并采用随机鸟枪法组件的组合进行测序，继而为封闭缺口和消除模糊 (“精加工”) 进行直接测序。在直接测序阶段的序列完成期间遇到的主要问题是 CpG 岛、串联重复序列和明显的克隆偏好。这些问题通过以下方法进行解决：使用寡核苷酸引物进行直接测序、使用极短的插入质粒文库，或通过筛选高复杂性质粒或 M13 文库识别桥接克隆。

Fig. 1. The sequence of human chromosome 22. Coloured boxes depict the annotated features of the sequence of human chromosome 22, with the centromere to the left and the telomere to the right. Coordinates are in kilobases. Vertical yellow blocks indicate the positions of the gaps in the sequence and are proportional in size to the estimated size of each gap. From bottom to top the following features are displayed: positions of interspersed repetitive sequences including tandem repeats categorized by nucleotide repeat unit length (at this resolution *Alu* repeats are not visibly separated in some regions); the positions of the microsatellite markers in the genetic map of Dib *et al*[36]; the tiling path of genomic clones used to determine the sequence labelled by their GenBank/EMBL/DDBJ accession number and coloured according to the source of the sequence; and the annotated gene, pseudogene and CpG island content of the sequence. Transcripts and pseudogenes oriented 5' to 3' on the DNA strand from centromere to telomere are designated "+", those on the opposite strand "−". In the transcript rows, the annotated genes are subdivided by colour according to the criteria in the text. Annotated genes with approved gene symbols from the HUGO nomenclature committee are labelled. For details of all the genes with their positions in the reference sequence, see Supplementary Information, Table 1. In the case of the immunoglobulin variable region, the entire locus has been drawn as a single block; in reality, this is a complex of variable chain genes (see ref. 27). At the top is a graphical plot of the repeat density for the common interspersed repeats *Alu* and Line1, and the C+G base frequency across the sequence. Each is calculated as a percentage of the sequence using a sliding 100-kb window moved in 50-kb iterations. Since the production of Fig. 1, six accession codes have been updated. The new codes are AL050347 (for Z73987), AL096754 (for Z68686), AL049749 (for Z82197), Z75892 (for Z75891), AL078611 (for Z79997) and AL023733 (for AL023593).

The completed sequence covers 33.4 Mb of 22q with 11 gaps and has been estimated to be accurate to less than 1 error in 50,000 bases, by internal and external checking exercises[25]. The order and size of each of the contiguous pieces of sequence is detailed in Table 1. The largest contiguous segment stretches over 23 Mb. From our gap-size estimates, we calculate that we have completed 33,464 kb of a total region spanning 34,491 kb and that therefore the sequence is complete to 97% coverage of 22q. The complete sequence and analysis is available on the internet (http://www.sanger.ac.uk/HGP/Chr22 and http://www.genome.ou.edu/Chr22.html).

Table 1. Sequence contigs on chromosome 22

Contig*	Size (kb)	
AP000522–AP000529	234	
gap		1.9
AP000530–AP000542	406	
gap		~150
AP000543–AC006285	1,394	
gap		~150
AC008101–AC007663	1,790	
gap		~100
AC007731–AL049708	23,006	
gap		~50
AL118498–AL022339	767	
gap†		~50-100

图 1. 人类 22 号染色体的序列。彩色方块指示人类 22 号染色体序列的注释特征，其着丝粒在左侧，端粒在右侧。坐标单位为千碱基。垂直黄色区域指示序列中缺口的位置，其大小与每个缺口的估算大小成比例。自下而上显示的特征是：散布重复序列的位置，包括根据核苷酸重复单位长度分类的串联重复（在此分辨率下 *Alu* 重复在某些区域无法通过视觉区分）；迪卜等人的遗传图中微卫星标记物的位置[36]；基因组克隆覆瓦式路径用来决定 GenBank/EMBL/DDBJ 检索编号标记的序列，并根据序列来源用不同颜色显示；以及序列中注释的基因、假基因和 CpG 岛含量。在 DNA 链上，从着丝粒到端粒按 5′ 到 3′ 走向的转录本和假基因被命名为"+"，相应在反义链上的为"−"。在转录物各行中，注释基因根据文中的标准用颜色进行了细分。来自 HUGO 命名委员会核准的基因符号的注释基因进行了标记。所有基因及其在参考序列中的位置的详细情况见补充信息表 1。对于免疫球蛋白可变区，整个基因座已绘制为单一区域；实际上这是可变链基因的复合体（见参考文献 27）。顶部是常见散布重复 *Alu* 和 Line1 的重复密度，以及序列中 C+G 碱基频率的图表。二者都是使用 100 kb 的滑动窗口及 50 kb 的迭代进行计算得到的，并表示为序列的百分比。图 1 制成之后有 6 个检索号发生了更新。新编号为：AL050347（原 Z73987）、AL096754（原 Z68686）、AL049749（原 Z82197）、Z75892（原 Z75891）、AL078611（原 Z79997）和 AL023733（原 AL023593）。

全部的序列覆盖 22q 33.4 Mb，有 11 个缺口，通过内部和外部检查，序列准确度预计达到每 50,000 碱基中错误数小于 1[25]。每一个连续的序列片段的顺序和大小详见表 1。最大的连续片段延伸超过 23 Mb。根据对缺口大小的估算，我们已经完成了总长 34,491 kb 区域中的 33,464 kb，因此，序列对 22q 的覆盖率达到 97%。完整的序列及分析可以在网上获得（http://www.sanger.ac.uk/HGP/Chr22 和 http://www.genome.ou.edu/Chr22.html）。

表 1. 22 号染色体的序列重叠群

重叠群 *	大小（kb）
AP000522–AP000529	234
缺口	**1.9**
AP000530–AP000542	406
缺口	约 150
AP000543–AC006285	1,394
缺口	约 150
AC008101–AC007663	1,790
缺口	约 100
AC007731–AL049708	23,006
缺口	约 50
AL118498–AL022339	767
缺口 †	约 50～100

Continued

Contig*	Size (kb)	
Z85994–AL049811	1,528	
gap		**~150**
AL049853–AL096853	2,485	
gap‡		**~50**
AL096843–AL078607	190	
gap†		**~100**
AL078613–AL117328	993	
gap		**~100**
AL080240–AL022328	291	
gap†		**~100**
AL096767–AC002055	380	
Total sequence length	33,464	
Total length of 22q	34,491	

* Contigs are indicated by the first and last sequence in the orientation centromere to telomere, and are named by their GenBank/EMBL/DDBJ accession numbers.

† These gaps are spanned by BAC and/or PAC clones with deletions.

‡ This gap shows a complex duplication of AL096853 in DNA fibre FISH.

Sequence Analysis and Gene Content

Analysis of the genomic sequence of the model organisms has made extensive use of predictive computational analysis to identify genes[26-28]. In human DNA, identification of genes by these methods is more difficult because of extensive splicing, lower density of exons and the high proportion of interspersed repetitive sequences. The accuracy of *ab initio* gene prediction on vertebrate genomic sequence has been difficult to determine because of the lack of sequence that has been completely annotated by experiment. To determine the degree of overprediction made by such algorithms, all genes within a region need to be experimentally identified and annotated, however it is virtually impossible to know when this job is complete. A 1.4-Mb region of human genomic sequence around the BRCA2 locus has been subjected to extensive experimental investigation, and it is believed that the 170 exons identified is close to the total number expressed in the region.

The most recent calibration of *ab initio* methods against this region (R.B.S.K. and T.H., manuscript in preparation) shows that with the best methods[29,30] more than 30% of exon predictions do not overlap any experimental exons, in other words, they are overpredictions. Furthermore, having now applied this analysis to larger amounts of data (more than 15 Mb from the Sanger Annotated Genome Sequence Repository which can

重叠群 *	大小（kb）	
Z85994–AL049811	1,528	
缺口		约 150
AL049853–AL096853	2,485	
缺口 ‡		约 50
AL096843–AL078607	190	
缺口 †		约 100
AL078613–AL117328	993	
缺口		约 100
AL080240–AL022328	291	
缺口 †		约 100
AL096767–AC002055	380	
序列总长度	33,464	
22q 总长度	34,491	

* 重叠群按照从着丝粒到端粒方向的第一个和最后一个序列进行标记，并用其在 GenBank/EMBL/DDBJ 检索编号命名。
† 被删除的 BAC 和/或 PAC 克隆横跨这些缺口。
‡ 该缺口在 DNA 纤维荧光原位杂交中显示出复杂的 AL096853 重复。

序列分析和基因含量

模式生物的基因组序列分析已经广泛用于确定基因的预测计算分析中[26-28]。人类 DNA 中存在剪接多、外显子密度低和散在重复序列比例高的问题，因此通过这些方法鉴定基因更为困难。由于缺乏完全通过实验注释的序列，脊椎动物基因组中从头预测基因的准确性一直难以确定。为确定这些算法产生的过度预测的程度，一个区域内所有的基因都需要用实验方法鉴定并注释，但是这项任务何时能完成又是几乎不可知的。针对 BRCA2 基因座周围 1.4 Mb 的人类基因组序列区域的深入实验研究认为，鉴定出的 170 个外显子接近该区域中表达的总数。

针对这个区域的从头算法的最新校准（布鲁斯耶维奇和哈伯德，稿件准备中）表明，使用最佳方法[29,30]时，超过 30% 的外显子预测未与任何实验证实的外显子重合，换句话说，它们的预测是过度的。此外，将这一分析应用于更大规模的数据（超过 15 Mb，来自桑格注释基因组序列存储库，可作为 Genesafe 集合（http://www.hgmp.mrc.ac.uk/Genesafe/）的一部分而获得）之后证实，在序列不同区域基因模型预

be obtained as part of the Genesafe collection (http://www.hgmp.mrc.ac.uk/Genesafe/)), it is confirmed that prediction accuracy also varies considerably between different regions of sequence. It was hoped that these calibration efforts would lead to rules for reliable gene prediction based on *ab initio* methods alone, perhaps on the basis of combining several different methods, GC content and so on. However, so far this has not been possible. The same analysis also shows that although ~95% of genes are at least partially predicted by *ab initio* methods, few gene structures are completely correct (none in BRCA2) and more than 20% of experimental exons are not predicted at all. The comparison of *ab initio* predictions and the annotated gene structures (see below) in the chromosome 22 sequence is consistent with this, with 94% of annotated genes at least partially detected by a Genscan gene prediction, but only 20% of annotated genes having all exons predicted exactly. Sixteen per cent of all the exons in annotated genes were not predicted at all, although this is only 10% for internal exons (that is, not 5′ and 3′ ends). As a result, we do not consider that *ab initio* gene prediction software can currently be used directly to reliably annotate genes in human sequence, although it is useful when combined with other evidence (see below), for example, to define splice-site boundaries, and as a starting point for experimental studies.

Fortunately, a vast resource of experimental data on human genes in the form of complementary DNA and protein sequences and expressed sequence tags (ESTs) is available which can be used to identify genes within genomic DNA. Furthermore about 60% of human genes have distinctive CpG island sequences at their 5′ ends[31] which can also be used to identify potential genes. Thus, the approach we have taken to annotating genes in the chromosome 22 sequence relies on a combination of similarity searches against all available DNA and protein databases, as well as a series of *ab initio* predictions. Upon completion of the sequence of each clone in the tile path, the sequence was subjected to extensive computational analysis using a suite of similarity searches and prediction tools. Briefly, the sequences were analysed for repetitive sequence content, and the repeats were masked using RepeatMasker (http://ftp.genome.washington.edu/RM/RepeatMasker. html). Masked sequence was compared to public domain DNA and protein databases by similarity searches using the blast family of programs[32]. Unmasked sequence was analysed for C+G content and used to predict the presence of CpG islands, tandem repeat sequences, tRNA genes and exons. The completed analysis was assembled into contigs and visualized using implementations of ACEDB (http://www.sanger. ac.uk/Software/Acedb/). In addition, the contiguous masked sequence was analysed using gene prediction software[29,30].

Gene features were identified by a combination of human inspection and software procedures. Figure 1 shows the 679 gene sequences annotated across 22q. They were grouped according to the evidence that was used to identify them as follows: genes identical to known human gene or protein sequences, referred to as "known genes" (247); genes homologous, or containing a region of similarity, to gene or protein sequences from human or other species, referred to as "related genes" (150); sequences homologous to only ESTs, referred to as "predicted genes" (148); and sequences homologous to a known gene or protein, but with a disrupted open reading frame, referred to "pseudogenes" (134). (See

测的准确度有很大的差别。人们希望这些校准工作可以形成单纯基于从头算法或基于几种不同方法和 GC 含量等相结合的可靠的基因预测标准。然而迄今为止这还从未实现。同样的分析还表明，虽然大约 95% 的基因通过从头算法得到了至少部分的预测，但只有很少数基因的预测结构是完全正确的（在 BRCA2 中完全没有），而且 20% 以上实验验证的外显子根本无法被预测到。22 号染色体序列的从头算法预测和已注释的基因结构的对比（见下文）与此相符，通过 Genscan 基因预测，94% 的注释基因至少能部分检测到，但只有 20% 注释基因的所有外显子能准确预测。注释基因中所有外显子的 16% 完全未被预测，然而这只占内部外显子（即不是 5′ 和 3′ 端）的 10%。因此我们认为，目前基因从头预测软件还不能直接用于对人类序列基因进行可靠的注释，但它在结合其他证据时还是有用的（见下文），例如确定剪接位点边界和作为实验研究的起点。

幸运的是，大量关于人类基因的互补 DNA 和蛋白质序列及表达序列标签（EST）的实验数据资源可以用于确定基因组 DNA 中的基因。另外，约 60% 的人类基因在其 5′ 端有独特的 CpG 岛序列[31]，也可以用来确定可能存在的基因。因此，我们采取的用于注释 22 号染色体序列中基因的方法依赖于综合运用所有类似的可用的 DNA 和蛋白质数据库以及一系列从头预测。一旦完成覆瓦式分析中每个克隆的序列，该序列会用一套相似性的搜索和预测工具进行大量计算分析。简言之，分析序列的重复序列含量，并且用 RepeatMasker 对重复序列进行标记（http://ftp.genome.washington.edu/RM/RepeatMasker.html）。使用 blast 系列程序进行相似性搜索，将标记的序列与公共领域的 DNA 和蛋白质数据库进行比对[32]。分析未标记序列 C+G 的含量，并且用于预测 CpG 岛、串联重复序列、tRNA 基因和外显子的存在。将已完成的分析组装为重叠群，并通过 ACEDB 进行查询实现可视化（http://www.sanger.ac.uk/Software/Acedb/）。此外，用基因预测软件对连续标记的序列进行分析[29,30]。

基因特征通过人为识别检查和软件程序相结合确定。图 1 给出了 22q 中已注释的 679 个基因序列。根据识别它们的证据将其分组如下：与已知的人类基因或蛋白质序列一致的基因命名为"已知基因"（247 个）；与来自人类或其他物种的基因或蛋白质序列同源或含有相似区域的基因命名为"相关基因"（150 个）；只与 EST 同源的序列命名为"预测基因"（148 个）；与已知基因或蛋白质序列同源，但其可读框被中断的序列命名为"假基因"（134 个）。（关于这些基因的详细情况见补充信息表 1。）

Supplementary Information, Table 1, for details of these genes.) The *ab initio* gene prediction program, Genscan, predicted 817 genes (6,684 exons) in the contiguous sequence, of which 325 do not form part of the annotated genes categorized above. Given the calibration of *ab initio* prediction methods discussed above, we estimate that of the order of 100 of these will represent parts of "real" genes for which there is currently no supporting evidence in any sequence database, and that the remainder are likely to be false positives.

The total length of the sequence occupied by the annotated genes, including their introns, is 13.0 Mb (39% of the total sequence). Of this, only 204 kb contain pseudogenes. About 3% of the total sequence is occupied by the exons of these annotated genes. This contrasts sharply with the 41.9% of the sequence that represents tandem and interspersed repeat sequences. There is no significant bias towards genes encoded on one strand at the 5% level ($\chi^2 = 3.83$).

A striking feature of the genes detected is their variety in terms of both identity and structure. There are several gene families that appear to have arisen by tandem duplication. The immunoglobulin λ locus is a well-known example, but there also are other immunoglobulin-related genes on the chromosome outside the immunoglobulin λ region. These include the three genes of the immunoglobulin λ-like (IGLL) family plus a fourth possible member of the family (AC007050.7). There are five clustered immunoglobulin κ variable region pseudogenes in AC006548, and an immunoglobulin variable-related sequence (VpreB3) in AP000348. Much further away from the λ genes is a variable region pseudogene, ~123 kb telomeric of IGLL3 in sequence AL008721 (coordinates ~9,420–9,530 kb from the centromeric end of the sequence), and a cluster of two λ constant region pseudogenes and a variable region pseudogene in sequences AL008723/AL021937 (coordinates ~16,060–16,390 kb from the centromeric end).

Human chromosome 22 also contains other duplicated gene families that encode glutathione *S*-transferases, Ret-finger-like proteins[19], phorbolins or APOBECs, apolipoproteins and β-crystallins. In addition, there are families of genes that are interspersed among other genes and distributed over large chromosomal regions. The γ-glutamyl transferase genes represent a family that appears to have been duplicated in tandem along with other gene families, for instance the BCR-like genes, that span the 22q11 region and together form the well-known LCR22 (low-copy repeat 22) repeats (see below).

The size of individual genes encoded on this chromosome varies over a wide range. The analysis is incomplete as not all 5′ ends have been defined. However, the smallest complete genes are only of the order of 1 kb in length (for example, HMG1L10 is 1.13 kb), whereas the largest single gene (LARGE[15]) stretches over 583 kb. The mean genomic size of the genes is 19.2 kb (median 3.7 kb). Some complete gene structures appear to contain only single exons, whereas the largest number of exons in a gene (PIK4CA) is 54. The mean exon number is 5.4 (median 3). The mean exon size is 266 bp (median 135 bp). The smallest complete exon we have identified is 8 bp in the PITPNB gene. The largest single exon is 7.6 kb in the PKDREJ, which is an intron-less gene with a 6.7-kb open reading frame. In addition,

462

基因从头预测程序 Genscan 在连续序列中预测了 817 个基因(6,684 个外显子),其中 325 个不属于上面分类的注释基因。鉴于上文讨论的从头预测方法的校准,我们估计其中 100 个将代表部分目前在任何序列数据库中还没有支持性证据的"真正"基因,而其余则可能是假阳性。

这些注释基因包括其内含子所占的序列总长度为 13.0 Mb(占总序列的 39%),其中只有 204 kb 含有假基因。这些注释基因的外显子约占总序列的 3%,这与代表串联和散布重复序列的序列所占的比例(41.9%)形成鲜明对比。对于一条链上的编码基因,在 5% 的水平上不存在显著性偏差($\chi^2 = 3.83$)。

检测到的基因的一个显着特点是它们在一致性和结构上的多样性。有几个基因家族可能是通过串联重复产生的。免疫球蛋白 λ 位点是众所周知的例子,但在免疫球蛋白 λ 区域外的染色体上也有其他免疫球蛋白相关基因。其中包括免疫球蛋白 λ 样(IGLL)家族的 3 个基因以及该家族的第 4 个可能成员(AC007050.7)。AC006548 有五个成簇的免疫球蛋白 κ 可变区假基因,AP000348 有一个免疫球蛋白可变相关序列(VpreB3)。距离 λ 基因更远的地方,在序列 AL008721 中,IGLL3 端粒侧约 123 kb 处(距序列着丝粒端约 9,420 ~ 9,530 kb 处)有一个可变区假基因,在序列 AL008723/AL021937 中(距着丝粒端约 16,060 ~ 16,390 kb 处)有一个由两个 λ 恒定区假基因和一个可变区假基因形成的簇。

人类 22 号染色体还包含其他重复基因家族,这些基因家族编码谷胱甘肽 S-转移酶、Ret-指样蛋白[19]、phorbolin 或 APOBEC、载脂蛋白和 β-晶状体蛋白。此外,有些基因家族散布在其他基因中并分散在大的染色体区域内。γ-谷氨酰转移酶基因代表一个可能与其他基因家族串联重复的家族,如 BCR 样基因,该家族跨越 22q11 区域并共同组成了著名的 LCR22(低拷贝重复 22)重复(见下文)。

这条染色体编码的单个基因大小差别很大。并非所有编码基因的 5′ 端都已确定,所以该分析是不完整的。然而,最小的完整基因长度仅为 1 kb 量级(例如,HMG1L10 是 1.13 kb),而最大的单基因(LARGE[15])超过 583 kb。基因的平均大小是 19.2 kb(中位数为 3.7 kb)。一些完整的基因结构似乎只包含单一的外显子,而单个基因中外显子数量最多达 54 个(PIK4CA)。平均外显子数目是 5.4 个(中位数为 3 个)。外显子的平均大小为 266 bp(中位数为 135 bp)。我们已鉴定的最小的完整外显子为 8 bp,位于 PITPNB 基因中。最大的单一外显子为 7.6 kb,位于 PKDREJ 中,

two genes occur within the introns of other expressed genes. The 61-kb TIMP3 gene, which is involved in Sorsby fundus macular degeneration, lies within a 268-kb intron of the large SYN3 gene, and the 8.5-kb HCF2 gene lies within a 27.5-b intron of the PIK4CA gene. In each case, the genes within genes are oriented in the opposite transcriptional orientation to the outer gene. We also observe pseudogenes frequently lying within the introns of other functional genes.

Peptide sequences for the 482 annotated full-length and partial genes with an open reading frame of greater than or equal to 50 amino acids were analysed against the protein family (PFAM)[33], Prosite[34] and SWISS-PROT[35] databases. These data were processed and displayed in an implementation of ACEDB. Overall, 240 (50%) predicted proteins had matching domains in the PFAM database encompassing a total of 164 different PFAM domains. Of the residues making up these 482 proteins, 25% were part of a PFAM domain. This compares with PFAM's residue coverage of SWISS-PROT/TrEMBL, which is more than 45% and indicates that the human genome is enriched in new protein sequences. Sixty-two PFAM domains were found to match more than one protein, including ten predicted proteins containing the eukaryotic protein kinase domain (PF00069), nine matching the Src homology domain 3 (PF00018) and eight matching the RhoGAP domain (PF00620). Fourteen predicted proteins contain zinc-finger domains (See Supplementary Information, Table 2, for details of the PFAM domains identified in the predicted proteins).

Nineteen per cent of the coding sequences identified were designated as pseudogenes because they had significant similarity to known genes or proteins but had disrupted protein coding reading frames. Because 82% of the pseudogenes contained single blocks of homology and lacked the characteristic intron-exon structure of the putative parent gene, they probably are processed pseudogenes. Of the remaining spliced pseudogenes, most represent segments of duplicated gene families such as the immunoglobulin κ variable genes, the β-crystallins, CYP2D7 and CYP2D8, and the GGT and BCR genes. The pseudogenes are distributed over the entire sequence, interspersed with and sometimes occurring within the introns of annotated expressed genes. However, there also is a dense cluster of 26 pseudogenes in the 1.5-Mb region immediately adjacent to the centromere; the significance of this cluster is currently unclear.

Given that the sequence of 33.4 Mb of chromosome 22q represents 1.1% of the genome and encodes 679 genes, then, if the distribution of genes on the other chromosomes is similar, the minimum number of genes in the entire human genome would be at least 61,000. Previous work has suggested that chromosome 22 is gene rich[6] by a factor of 1.38 (http://www.ncbi.nlm.nih.gov/genemap/page.cgi?F = GeneDistrib.html), which would reduce this estimate to ~45,000 genes. It is important, however, to recognize that the analysis described here only provides a minimum estimate for the gene content of chromosome 22q, and that further studies will probably reveal additional coding sequences that could not be identified with the current approaches.

464

该基因是一个有 6.7 kb 可读框的无内含子的基因。此外，有两个基因出现在其他表达基因的内含子中。与索斯比眼底黄斑变性有关的 61 kb 的 TIMP3 基因位于较大的 SYN3 基因的一个 268 kb 的内含子中，而 8.5 kb 的 HCF2 基因位于 PIK4CA 基因的 27.5 b 的内含子中。在这两种情况中，内部基因的转录方向与外部基因相反。我们还观察到，假基因经常位于其他功能性基因的内含子中。

使用蛋白质家族（PFAM）[33]、Prosite[34] 和 SWISS-PROT[35] 数据库，对可读框大于等于 50 个氨基酸的 482 个已注释的全长和部分基因的多肽序列进行了分析。这些数据在执行 ACEDB 任务查询中被处理和显示。总体而言，240 个（50%）预测蛋白质在总数为 164 个不同 PFAM 域的 PFAM 数据库中有匹配结构域。在组成这 482 个蛋白质的残基中，有 25% 属于一个 PFAM 结构域。与此相比，PFAM 的残基覆盖 SWISS-PROT/TrEMBL 超过 45%，表明人类基因组在新的蛋白质序列中被富集。有 62 个 PFAM 结构域可匹配一个以上的蛋白质，包括 10 个含有真核蛋白质激酶结构域（PF00069）的测预蛋白质，9 个匹配 Src 同源结构域 3（PF00018）的测预蛋白质和 8 个匹配 RhoGAP 域（PF00620）的测预蛋白质。14 个预测蛋白质含有锌指结构域（在预测蛋白中鉴定 PFAM 域的详细信息见补充信息表 2）。

已鉴定的编码序列中有 19% 被认为是假基因，其原因是它们与已知基因或蛋白有显著相似性，但蛋白质编码可读框被破坏。由于 82% 的假基因含单一区域同源性，而且缺乏假定的亲本基因特征性的内含子–外显子结构，因此它们可能是被加工过的假基因。在其余的剪接假基因中，大多数代表重复的基因家族片段，如免疫球蛋白 κ 可变基因、β–晶状体蛋白、CYP2D7 和 CYP2D8，以及 GGT 和 BCR 基因。假基因分布在整个序列中，分散存在于已注释的表达基因的内含子之间，有时存在于其内部。然而，还有一簇密集的 26 个假基因位于毗邻着丝粒的 1.5 Mb 区域内；这个基因簇的意义目前还不清楚。

考虑到染色体 22q 上 33.4 Mb 的序列代表人类基因组的 1.1%，并编码 679 个基因，因此，如果基因在其他染色体上的分布与此相似，那么整个人类基因组的最小基因总数可能至少为 61,000 个。之前的工作已经显示，22 号染色体基因丰度较高[6]，系数为 1.38（http://www.ncbi.nlm.nih.gov/genemap/page.cgi?F = GeneDistrib.html），据此，该估值将降为约 45,000 个基因。尽管认识到此处描述的分析只提供 22q 染色体最低基因含量的估值很重要，进一步研究可能会揭示更多不能用现有方法鉴定的编码序列。

Two lines of evidence point to the existence of additional genes that are not detected in this analysis. First, the 553 predicted CpG islands, which typically lie at the true 5' ends of about 60% of human genes[31], are in excess of 60% of the number of genes identified (60% = 327, excluding pseudogenes); 282 of the genes identified have CpG islands at or close to the 5' end (within 5-kb upstream of the first exon, or 1-kb downstream). Thus, there could be up to 271 additional genes associated with CpG islands undetected in the sequence. Second, there are 325 putative genes predicted by the *ab initio* gene prediction program, Genscan, that are not in regions already containing annotated transcripts. We estimate (see above) that roughly 100 of these will represent parts of real genes. Identifying additional genes will require further computational and experimental studies. These studies are continuing and entail testing candidate sequences for possible messenger RNA expression, implementing new gene prediction software able to detect the regions around or near CpG islands that currently have no identified transcript, and further analysis of sequences that are conserved between human and mouse. Furthermore, full-length cDNA sequences that accumulate in the sequence databases of human and other species will be used to refine the gene structures.

The Long-range Chromosome Landscape

Critical to the utility of the genomic sequence to genetic studies is the integration of established genetic maps. The positions of the commonly used microsatellite markers from the Genethon genetic map[36] are given in Fig. 1. The correlation of the order of markers between the genetic map and the sequence is good, within the limitations of genetic mapping. Only a single marker (D22S1175) is discrepant between the two data sets, and this lies in a sequence that is repeated twice on the chromosome (AL021937, see below). In the telomeric region, four of the Genethon markers must lie in our sequence gaps, and we were unable to identify clones from all libraries tested for these. Comparison of genetic distance against physical distance for all the microsatellites whose order is maintained between the datasets shows a mean value of 1.87 cM Mb^{-1}. However, the relationship between genetic and physical distance across the chromosome partitions into two types of region, areas of high and low recombination (Fig. 2). The areas of high recombination may represent recombinational hot spots, although we have not yet been able to identify any specific sequence characteristics common to these areas.

The mean G+C content of the sequence is 47.8%. This is significantly higher than the G+C content calculated for the sum of all human genomic sequence determined so far (42%). Although this result was expected from previous indirect measurements of the G+C content of chromosome 22[7,8,37], the distribution is not uniform, but regionally segmented as illustrated in Fig. 1. There are clear fluctuations in the base content, resulting in areas that are relatively G+C rich and others that are relatively G+C poor. On chromosome 22 these regions stretch over several megabases. For example, the 2 Mb of sequence closest to the centromeric end of the sequence is relatively G+C poor, with the G+C content dropping below 40%. Similarly, the area between 16,000 and 18,800 kb from the

466

有两个系列的证据指出存在着更多在这一分析中没有检测到的基因。首先，CpG 岛通常位于约 60% 的人类基因的真正 5′ 端[31]，而 553 个预测 CpG 岛超过已鉴定基因数量的 60%（不包括假基因，则 60% = 327 个）；有 282 个基因被发现在 5′ 端或接近 5′ 端处（第一外显子上游 5 kb 或下游 1 kb 内）有 CpG 岛。因此，在序列中可能还有与 CpG 岛有关的多达 271 个额外基因未检测到。其次，有 325 个由基因从头预测程序 Genscan 预测的假定基因未包含在已注释转录物的区域内。我们估计（见上文），其中大约 100 个代表部分真正的基因。要鉴定更多的基因需要进一步的计算与实验研究。这些研究仍在继续，并需要检测候选序列，寻找可能存在的信使 RNA 表达，应用新的基因预测软件，以检测目前没有鉴定到转录物的 CpG 岛的周围或附近区域，并进一步分析人类与小鼠之间的保守序列。此外，在人类和其他物种序列数据库中积累的全长 cDNA 序列将用于完善基因结构。

长距离染色体概况

用基因组序列进行遗传研究的关键是整合已有的遗传图谱。Genethon 遗传图谱常用的微卫星标记的位置[36] 如图 1。遗传图谱和序列之间标记顺序的相关性良好，并在遗传作图的要求限制之内。只有一个标记（D22S1175）在两个数据集之间有差异，它位于一个在染色体中重复两次的序列内（AL021937，见下文）。在端粒区域，一定有四个 Genethon 标志位于我们的序列缺口中，我们无法从为这些缺口而测试的所有文库中识别出克隆。将在这些数据集之间保持顺序的所有微卫星的遗传距离与物理距离进行对比显示其平均值为 $1.87 \text{ cM} \cdot \text{Mb}^{-1}$。然而，整个染色体中遗传距离和物理距离之间的关系分成两个区域类型：高重组区和低重组区（见图 2）。高重组区可能代表重组热点，不过我们还未能确定任何这些区域常见的特定序列特征。

序列的平均 G+C 含量为 47.8%。显著高于目前确定的所有人类基因组序列计算得到的 G+C 含量（42%）。虽然从先前间接测量的 22 号染色体的 G+C 含量[7,8,37] 可以预测这一结果，但分布并不一致，其区域分割如图 1 所示。碱基含量存在明显波动，导致某些区域的 G+C 含量相对较高，而其他区域的 G+C 含量相对较低。在 22 号染色体上，这些区域可延伸达几 Mb。例如，最接近序列着丝粒末端的 2 Mb 序列的 G+C 含量相对较低，降至 40% 以下。同样，距序列着丝粒末端 16,000 到 18,800 kb 的区域 G+C 含量也始终低于 45%。G+C 丰富的区域（如距序列着丝粒

centromeric end of the sequence is consistently below 45% G+C. The G+C rich regions often reach more than 55% G+C (for example, at 20,100–23,400 kb from the centromeric end of the sequence). This fluctuation appears to be consistent with previous observations that vertebrate genomes are segmented into "isochores" of distinct G+C content[38] and is similar to the structure seen in the human major histocompatibility complex (MHC) sequence[39]. Isochores correlate with both genes and chromosome structure. The G+C rich isochores are rich in genes and *Alu* repeats, and are located in the G+C rich chromosomal R-bands, whereas the G+C poor isochores are relatively depleted in genes and *Alu* repeats, and are located in the G-bands[8,37,40]. The G+C poor regions of chromosome 22 are depleted in genes and relatively poor in *Alu* sequences. For example, the region between 16,000 and 18,800 kb from the centromeric end contains just three genes, two of which are greater than 400 kb in length. The G+C poor regions also are depleted in CpG islands, which are clustered in the gene-rich, G+C rich regions. Although it is tempting to correlate the sequence features that we see with the chromosome banding patterns, we believe that high-resolution mapping of the chromosome band boundaries will be required to assign definitively these to genomic sequence.

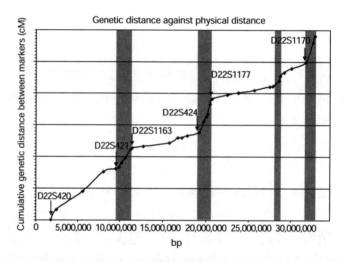

Fig. 2. The relationship between physical and genetic distance. The sex-averaged genetic distances of Dib *et al.*[36] were obtained from ftp://ftp.genethon.fr/pub/Gmap/Nature-1995/ and the cumulative intermarker distances for unambiguously ordered markers (in cM) were plotted against the positions of the microsatellite markers in the genomic sequence. It should be stressed that the y axis does not represent the true genetic distance between distant markers but the sum of the local intermarker distances. The positions of selected genetic markers are labelled. Grey regions are indicative of areas of relatively increased recombination per unit physical distance.

Over 41.9% of the chromosome 22 sequence comprises interspersed and tandem repeat family sequences (Table 2). The density of repeats across the sequence is plotted in Fig. 1. There is variation in the density of *Alu* repeats and some of the regions with low *Alu* density correlate with the G+C poor regions, for example, in the region 16,000–18,800 kb from the centromeric end, and these data support the relationship of isochores with *Alu* distribution. However, in other areas the relationship is less clear. We provide a World-Wide Web interface to the long-range analyses presented here and to further analysis of the

末端 20,100 ~ 23,400 kb 的区域)中,其含量往往超过 55%。这种波动似乎与以前的结果一致,即脊椎动物基因组分割为不同 G+C 含量的"等容线"[38],与人类主要组织相容性复合物(MHC)序列中的结构相似[39]。等容线与基因和染色体结构有关。G+C 丰富的等容线富含基因和 *Alu* 重复,位于 G+C 丰富的染色体 R 带,而 G+C 稀少的等容线所含的基因和 *Alu* 重复相对较少,且位于 G 带[8,37,40]。22 号染色体 G+C 稀少的区域不含基因且 *Alu* 序列也相对较少。例如,距着丝粒端 16,000 和 18,800 之间的区域只包含三个基因,其中两个基因长度大于 400 kb。G+C 稀少的区域也缺少 CpG 岛,后者集中在基因和 G+C 丰富的区域。虽然将我们看到的序列特征与染色体带型相结合是很吸引人的,但我们认为,要将这些序列特征明确地分配到基因组序列中,需要对染色体带的界限进行高分辨率作图。

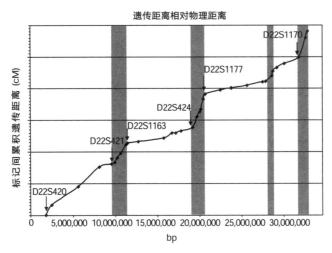

图 2. 物理和遗传距离之间的关系。迪卜等人研究的性别平均遗传距离[36] 来自 ftp://ftp.genethon.fr/pub/Gmap/Nature-1995/,将顺序明确的累积的标记间距离(以 cM 为单位)相对基因组序列中微卫星标记的位置进行作图。应当指出,y 轴并不代表远距离标记物间的真正遗传距离,但代表局部标记间的距离总和。选定的遗传标记物的位置已标出。灰色区域指示每单位物理距离相对增加的重组区域。

22 号染色体上超过 41.9% 的序列包含散在和串联重复家族序列(见表 2)。序列中的重复密度绘制在图 1 中。*Alu* 重复的密度存在差异,一些低 *Alu* 密度的区域与 G+C 稀少区域(如距着丝粒末端 16,000 ~ 18,800 kb 的区域)有关,这些数据支持等容线与 *Alu* 分布的关系。但是,这种关系在其他区域不太明显。我们为本文进行的长距离分析以及对序列的许多其他重复类型和特征的进一步分析提供一个万维网界

many other repeat types and features of the sequence at http://www.sanger.ac.uk/cgi-bin/cwa/22cwa.pl. The 1-Mb region closest to the centromere contains several interesting repeat sequence features that may be typical of other pericentromeric regions. In addition to the density of pseudogenes described above, there is a large 120-kb block of tandemly repeated satellite sequence (D22Z3) centred 500 kb from the centromeric sequence start (not shown in Fig. 1, but evident from the absence of *Alu* and LINE1 sequences at this point). There is also a cluster of satellite II repeats 80-kb telomeric of the D22Z3 sequences. Isolated alphoid satellite repeats are found closer to the centromeric end of the sequence. Furthermore, this pericentromeric 1 Mb closest to the centromere contains many sequences that are shared with a number of different chromosomes, particularly chromosomes 2 and 14. During map construction, 33 out of 37 STSs designed from sequence that was free of high-copy repeats amplified from more than one chromosome in somatic cell hybrid panel analysis.

Table 2. The interspersed repeat content of human chromosome 22

Repeat type	Total number	Coverage (bp)	Coverage (%)
Alu	20,188	5,621,998	16.80
HERV	255	160,697	0.48
Line1	8,043	3,256,913	9.73
Line2	6,381	1,273,571	3.81
LTR	848	256,412	0.77
MER	3,757	763,390	2.28
MIR	8,426	1,063,419	3.18
MLT	2,483	605,813	1.81
THE	304	93,159	0.28
Other	2,313	625,562	1.87
Dinucleotide	1,775	133,765	0.40
Trinucleotide	166	18,410	0.06
Quadranucleotide	404	47,691	0.14
Pentanucleotide	16	1,612	0.0048
Other tandem	305	102,245	0.31
Total	55,664	14,024,657	41.91

Low-copy Repeats on Chromosome 22

To detect intra- and interchromosomal repeats, we compared the entire sequence of chromosome 22 to itself, and also to all other existing human genomic DNA sequence using Blastn[32] after masking high and medium frequency repeats. The results of the

面：http://www.sanger.ac.uk/cgi-bin/cwa/22cwa.pl。最接近着丝粒的 1 Mb 区域包含几个有趣的可能代表其他着丝粒周边区域的重复序列特征。除了上述假基因密度，有一个大的 120 kb 的串联重复卫星序列（D22Z3）集中在距着丝粒序列起始位置的500 kb 处（图 1 中未显示，但根据这一位置缺乏 *Alu* 及 LINE1 序列可明显看出）。还有 ·组卫星 II 重复位于 D22Z3 序列的端粒 80 kb 处。孤立的 alphoid 卫星重复接近序列的着丝粒末端。此外，最接近着丝粒周边的 1 Mb 区域含有许多不同染色体（特别是染色体 2 和 14）共有的序列。在染色体图谱构建过程中，通过无高拷贝重复的序列设计的 37 个 STS 中，有 33 个在体细胞杂交分析中从不止一处染色体发生了扩增。

表 2. 人类 22 号染色体散布的重复情况

重复类型	总数目	覆盖范围（bp）	覆盖范围（%）
Alu	20,188	5,621,998	16.80
HERV	255	160,697	0.48
Line1	8,043	3,256,913	9.73
Line2	6,381	1,273,571	3.81
LTR	848	256,412	0.77
MER	3,757	763,390	2.28
MIR	8,426	1,063,419	3.18
MLT	2,483	605,813	1.81
THE	304	93,159	0.28
其他	2,313	625,562	1.87
二核苷酸	1,775	133,765	0.40
三核苷酸	166	18,410	0.06
四核苷酸	404	47,691	0.14
五核苷酸	16	1,612	0.0048
其他串联	305	102,245	0.31
总数	55,664	14,024,657	41.91

22 号染色体上的低拷贝重复

为检测染色体内和染色体间重复，在标记高、中频重复之后，我们用 Blastn 将22 号染色体的整个序列与其本身进行比对，也与所有其他现有的人类基因组 DNA

intrachromosomal sequence analysis were plotted as a dot matrix (Fig. 3) and reveal a series of interesting features. Locally duplicated gene families lie close to the diagonal axis of the plot. The most striking is the immunoglobulin λ locus that comprises a cluster of 36 potentially functional V-λ gene segments, 56 V-λ pseudogenes, and 27 partial V-λ pseudogenes ("relics"), together with 7 each of the J and C λ segments[24]. Other duplicated gene families that are visible from the dot matrix plot include the clustered genes for glutathione *S*-transferases, β-crystallins, apolipoproteins, phorbolins or APOBECs, the lectins LGALS1 and LGALS2 and the CYP2Ds. A partial inverted duplication of CSF2RB is also observed.

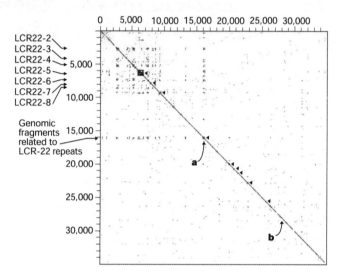

Fig. 3. Intrachromosomal repeats on human chromosome 22. High- and medium-copy repeats and low complexity sequence were masked using RepeatMasker and Dust, and masked sequences were compared using Blastn. The results were filtered to identify regions of more than 50% identity to the query sequence, and were plotted in a 2D matrix with a line proportional to the size of the region of identity. Localized gene family repeats are indicated by arrowheads along the diagonal. From the top, these are the immunoglobulin λ locus, the glutathione *S*-transferase genes, the β-crystallin genes, the Ret-finger-protein-like genes, the apolipoprotein genes, the colony-stimulating factor receptor (CSF2RB) inverted partial duplication, the lectins LGALS1 and LGALS2, the APOBEC genes and the CYP2D genes. Two 60-kb regions of more than 90% homology are labelled "a" (AL008723/AL021937) and "b" (AL031595/AL022339). Seven low-copy repeat regions (LCR22) and a region containing related genomic fragments are indicated at the left margin.

Much more striking are the long-range duplications, which are visible away from the diagonal axis. For example, a 60-kb segment of more than 90% similarity is seen between sequences AL008723/AL021937(at ~16,060–16,390 kb from the centromeric end) and AL031595/AL022339 (at ~27,970–28,110 kb from the centromeric end) separated by almost 12 Mb. The 22q11 region is particularly rich in repeated clusters[41]. Previous work described a low-copy repeat family in 22q11 that might mediate recombination events leading to the chromosomal rearrangements seen in cat eye, velocardiofacial and DiGeorge syndromes[21,42]. The availability of the entire DNA sequence allows detailed dissection of the molecular structure of these low-copy repeats (LCR22s). Edelmann *et al.* described eight

序列进行比对[32]。染色体内序列分析的结果绘制为点阵图（图3），并显示出一系列有趣的特征。局部重复基因家族位于接近图对角线轴处。最引人注目的是免疫球蛋白 λ 位点，它包含由 36 个潜在功能 V-λ 基因片段、56 个 V-λ 假基因和 27 个部分 V-λ 假基因（"残迹"）形成的一个簇，以及 J 和 C λ 片段形成的 7 个簇[24]。其他重复基因家族在点阵图上都可以看到，包括以下成簇基因：谷胱甘肽 S-转移酶、β-晶状体蛋白、载脂蛋白、phorbolin 或 APOBEC、凝集素 LGALS1 和 LGALS2，以及 CYP2D。此外还观察到 CSF2RB 的部分反向重复。

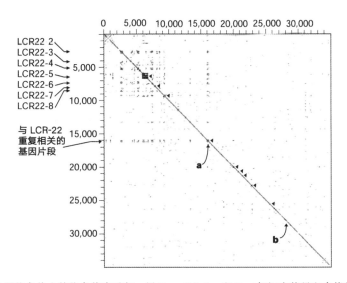

图 3. 人类 22 号染色体上的染色体内重复。用 RepeatMasker 和 Dust 标记高拷贝和中拷贝重复以及低复杂性序列，用 Blastn 比较被标记的序列。结果进行过滤以确定与所查询序列同一性多于 50% 的区域，并用与同一性区域大小成比例的直线绘制在二维矩阵中。沿对角线的箭头指示局部基因家族重复。从上方开始依次是免疫球蛋白 λ 位点、谷胱甘肽 S-转移酶基因、β-晶状体蛋白基因、Ret 指蛋白样基因、载脂蛋白基因、集落刺激因子受体（CSF2RB）倒置部分重复、凝集素 LGALS1 和 LGALS2、APOBEC 基因和 CYP2D 基因。同源性大于 90% 的两个 60 kb 区域标记为 "a"（AL008723/AL021937）和 "b"（AL031595/AL022339）。7 个低拷贝重复区域（LCR22）和一个包含相关基因组片段的区域显示在左边空白处。

更引人注目的是远离对角线轴的长距离重复。例如，在相距接近 12 Mb 的 AL008723/AL021937 序列（距着丝粒末端约 16,060 ~ 16,390 kb）和 AL031595/AL022339 序列（距着丝粒末端约 27,970 ~ 28,110 kb）之间可见一个相似性 90% 以上的 60 kb 的片段。22q11 区域尤其富含重复簇[41]。以往的工作介绍了 22q11 中有可能介导重组事件的一个低拷贝重复家族，这些重组事件导致猫眼、腭心面和迪格奥尔格综合征中观察到的染色体重排[21,42]。整个 DNA 序列的获得使我们可以详细分析这些低拷贝重复序列（LCR22）的分子结构。埃德尔曼等描述了 8 个 LCR22 区域[21,42]。我们无法

LCR22 regions[21,42]. We were unable to find the LCR22 repeat closest to the centromere, but it may lie in the gap at 700 kb from the centromeric end of the sequence. The other LCR22 regions are distributed over 6.5 Mb of 22q11. Analysis of the sequence shows that each LCR22 contains a set of genes or pseudogenes (Fig. 4). For example, five of the LCR22s contain copies of the γ-glutamyl transferase genes and γ-glutamy-transferase-related genes. There is also evidence that a more distant sequence at ~16,000 kb from the centromeric start of the genomic sequence shares certain sequences with the LCR22 repeats. This similarity involves related genomic fragments including parts of the Ret-finger-protein-like genes, and the IGLC and IGLV genes.

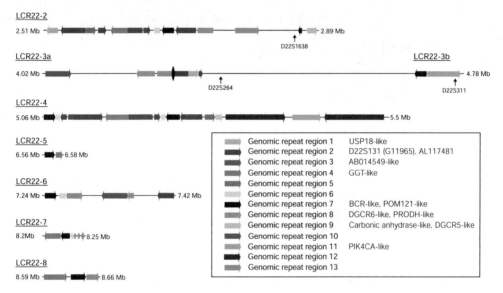

Fig. 4. Sequence composition of the LCR22 repeats. Illustration of the sequence composition of seven LCR22 repeats. The span of each LCR22 region is shown in megabases from the centromere. Coloured arrows indicate the extent of one of the thirteen genomic repeat regions and the orientation of the repeat. The known gene and marker content of these genomic repeat regions is indicated in the key. The black oval indicates the position of the gap in the sequence in LCR22-3.

Regions of Conserved Synteny with the Mouse

The genomic organization of different mammalian species is well known to be conserved[43]. Comparison of genetic and physical maps across species can aid in predicting gene locations in other species, identifying candidate disease genes[13], and revealing various other features relevant to the study of genome organization and evolution. For all the cross-species relationships, that between man and mouse has been most studied. We have examined the relationship of the human chromosome 22 genes to their mouse orthologues.

Of the 160 genes we identified in the human chromosome 22 sequence that have orthologues in mouse, 113 of the murine orthologues have known mouse chromosomal

发现最接近着丝粒的 LCR22 重复，但它可能位于距序列着丝粒末端 700 kb 的缺口中。其他 LCR22 区域分布在 22q11 上超过 6.5 Mb 的范围内。序列分析显示，每个 LCR22 包含一组基因或假基因（图 4）。例如，5 个 LCR22 包含 γ–谷氨酰转移酶基因和 γ–谷氨酰转移酶相关基因的拷贝。还有证据表明，一处更远的，距基因组序列着丝粒起点约 16,000 kb 的序列与 LCR22 重复存在某些共有序列。这种相似性涉及相关的基因组片段，包括部分 Ret 指蛋白样基因以及 IGLC 和 IGLV 基因。

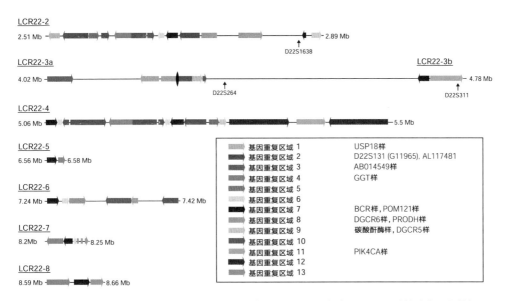

图 4. LCR22 重复的序列组成。7 个 LCR22 重复的序列组成图示。每个 LCR22 区域的跨度从着丝粒开始计算，以 Mb 为单位。彩色箭头表示 13 个基因重复区域中一个的长度和重复方向。这些基因组重复区域的已知基因和标记物含量在图例中注明。黑色椭圆表示序列 LCR22-3 中缺口的位置。

与小鼠具有保守共线性的区域

目前已经清楚地知道，不同哺乳动物物种的基因组的组成是保守的[43]。比较物种间的遗传图谱和物理图谱有助于预测其他物种的基因位置，确定可能的疾病基因[13]，并揭示与基因组的组成和演化等研究相关的各种其他特征。在所有跨物种关系中，人与小鼠之间的关系研究得最多。我们之前已研究过人类 22 号染色体基因与小鼠直系同源基因之间的关系。

我们在人类 22 号染色体序列中鉴定出 160 个基因在小鼠中有直系同源基因，其中 113 个小鼠直系同源基因在小鼠染色体上的位置是已知的（数据可在以下网址获

locations (data available at http://www.sanger.ac.uk/HGP/Chr22/Mouse/). Examination of these mouse chromosomal locations mapped onto the human chromosome 22 sequence confirms the conserved linkage groups corresponding to human chromosome 22 on mouse chromosomes 6, 16, 10, 5, 11, 8 and 15[18,44-46] (Fig. 5). Furthermore, these studies allow placement of the sites of evolutionary rearrangements that have disrupted the conservation of synteny more accurately at the DNA sequence scale. For example, the breakdown of synteny between the mouse 8C1 block and the mouse 15E block occurs between the equivalents of the human HMOX1 and MB genes, which are separated by less than 160 kb that also contains a conspicuous 41-copy 18-nucleotide tandem repeat. A clear prediction from these data is that, for the most part, the unmapped murine orthologues of the human genes lie within these established linkage groups, along with the orthologues of the human genes that currently lack mouse counterparts. Exploitation of the chromosome 22 sequence may hasten the determination of the mouse genomic sequence in these regions.

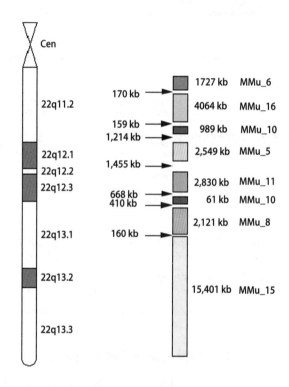

Fig. 5. Regions of conserved synteny between human chromosome 22 and the mouse genome. Regions of mouse chromosomes with conserved synteny to human chromosome 22 are shown as adjacent coloured blocks, determined by the mouse map position of mouse orthologues to human chromosome 22 genes. The size of human chromosome 22 corresponding to each mouse chromosomal region is indicated in kb, as well as the size of the gap between the last orthologue in each conserved block. These data are available at http://www.sanger.ac.uk/Chr22/Mouse.

得：http://www.sanger.ac.uk/HGP/Chr22/Mouse/）。将小鼠染色体中的位置绘制到人类 22 号染色体序列上的检查，确认了小鼠 6、16、10、5、11、8 和 15 号染色体上的与人类 22 号染色体相对应的保守连锁群[18,44-46]（图 5）。此外，这些研究使我们能够在 DNA 序列的尺度上更准确地定位这些已经破坏了共线性的保守序列的演化重排。例如，小鼠 8C1 模块和 15E 模块之间共线性的破坏发生在相应的人 HMOX1 和 MB 基因之间的位置，这两个基因之间的距离不到 160 kb，其中包含一个明显的 41 拷贝的 18 个核苷酸串联重复。根据这些数据可以做出明确的预测：在大多数情况下，未作图的人类基因的小鼠直系同源基因与目前尚无小鼠对应物的人类基因的直系同源基因共同位于这些既定连锁群内。对 22 号染色体序列的开发利用有望加快确定这些区域的小鼠基因组序列。

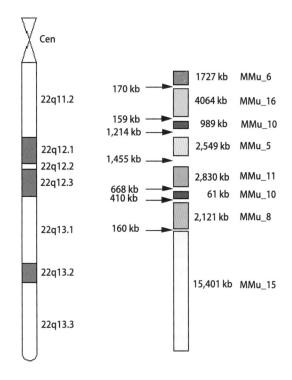

图 5. 人类 22 号染色体与小鼠基因组间具有保守共线性的区域。小鼠染色体上与人类 22 号染色体具有保守共线性的区域用相邻的彩色方块标出，通过小鼠与人类 22 号染色体基因的直系同源基因在小鼠图谱中的位置测得。对应于每个小鼠染色体区域的人类 22 号染色体的大小，以及每个保守模块中最后一个直系同源基因之间的缺口的大小，均以 kb 为单位。这些数据可在以下网站获得：http://www.sanger.ac.uk/Chr22/Mouse。

Conclusions

We have shown that the clone by clone strategy is capable of generating long-range continuity sufficient to establish the operationally complete genomic sequence of a chromosome. In doing so, we have generated the largest contiguous segment of DNA sequence to our knowledge to date. The analysis of the sequence gives a foretaste of the information that will be revealed from the remaining chromosomes.

We were unable to obtain sequence over 11 small gaps using the available cloning systems. It may be possible that additional approaches such as using combinations of cloning systems with small insert sizes and low-copy number could reduce the size of these gaps. Direct cloning of restriction fragments that cross these gaps into small insert plasmid or M13 libraries, or direct sequencing approaches might eventually provide access to all the sequence in the gaps. However, closing these gaps is certain to require considerable time and effort, and might be considered as a specialist activity outside the core genome-sequencing efforts. It also is probable that the sequence features responsible for several of these gaps are unlikely to be specific to chromosome 22. In the best case, similar unclonable sequences might be restricted to the centromeric and telomeric regions of the other chromosomes and areas with large tandem repeats, and it will be possible to obtain large contiguous segments for the bulk of the euchromatic genome.

Over the course of the project, the emerging sequence of chromosome 22 has been made available in advance of its final completion through the internet sites of the consortium groups and the public sequence databases[47]. The benefits of this policy can be seen in both the regular requests received from investigators for materials and information that arise as the result of sequence homology searches, and the publications that have used the data[14-19]. The genome project will continue to pursue this data release policy as we move closer to the anticipated completed sequence of humans, mice and other complex genomes[47,48].

Methods

The methods for construction of clone maps have been previously described[24,49,50] and can also be found at http://www.sanger.ac.uk/HGP/methods/. Details of sequencing methods and software are available at http://www.sanger.ac.uk/HGP/methods/, http://www.genome.ou.edu/proto.html, http://www-alis.tokyo.jst.go.jp/HGS/team_KU/team.html and in the literature[1,24].

(**402**, 489-495; 1999)

I. Dunham, N. Shimizu, B. A. Roe, S. Chissoe *et al.*[†]
[†] A full list of authors and affiliations is given in the original paper.

Received 5 November; accepted 11 November 1999.

结　　论

我们已经证实，连续克隆这一方法产生的长距离连续性足以建立一条可操作的完整的染色体基因组序列。在此过程中，我们已经生成了迄今所知最大的连续 DNA 序列片段。对序列的分析预示了其他染色体将要揭示的信息。

使用现有的克隆系统，我们无法获取横跨 11 个小缺口的序列。其他方法，例如将具有小插入尺寸和低拷贝数目的克隆系统结合使用，有望缩减这些缺口的大小。将跨越这些缺口的限制性片段直接克隆到小插入片段质粒或 M13 文库中或直接进行测序，可能最终能够得到缺口中所有的序列。然而，弥合这些缺口必然需要大量的时间和精力，并且可能会被视为核心的基因组测序工作以外的专项活动。也有可能其中一些缺口的序列特征不是 22 号染色体特有的。最理想的情况是，类似的不能克隆的序列会局限于其他染色体的着丝粒和端粒区域，以及存在大串联重复的区域，这样我们就有可能获得大量基因组常染色质大的连续片段。

在项目执行过程中，22 号染色体的序列在最终完成前已通过合作单位的网站和公共序列数据库对外公布[47]。研究者在检索序列同源性后通常要求获取相应的材料和信息，也有一些出版物使用了我们的数据，由此可见该政策确实具有一定的益处[14-19]。在我们获取期望的完整的人类、小鼠和其他复杂基因组的序列的过程中，基因组计划将继续奉行这一数据发布政策[47,48]。

方　　法

用于构建克隆图谱的方法此前已描述过[24,49,50]，也可以在 http://www.sanger.ac.uk/HGP/methods/ 找到。测序方法和软件的详细信息可在 http://www.sanger.ac.uk/HGP/methods/，http://www.genome.ou.edu/proto.html 和 http://www-alis.tokyo.jst.go.jp/HGS/team_KU/team.html 以及文献中找到[1,24]。

（李梅 翻译；肖景发 审稿）

References:

1. The Sanger Centre & The Genome Sequencing Centre. Toward a complete human genome sequence. *Genome Res.* **8**, 1097-1108 (1998).

2. Weber, J. L. & Myers, E. W. Human whole-genome shotgun sequencing. *Genome Res.* **7**, 401-409 (1997).

3. Green, P. Against a whole-genome shotgun. *Genome Res.* **7**, 410-417 (1997).

4. Collins, F. S. *et al.* New goals for the U.S. human genome project: 1998-2003. *Science* **282**, 682-689 (1998).

5. Morton, N. E. Parameters of the human genome. *Proc. Natl Acad. Sci. USA* **88**, 7474-7476 (1991).

6. Deloukas, P. *et al.* A physical map of 30,000 human genes. *Science* **282**, 744-746 (1998).

7. Craig, J. M. & Bickmore, W. A. The distribution of CpG islands in mammalian chromosomes. *Nature Genet.* **7**, 376-382 (1994).

8. Saccone, S., Caccio, S., Kusuda, J., Andreozzi, L. & Bernardi, G. Identification of the gene-richest bands in human chromosomes. *Gene* **174**, 85-94 (1996).

9. Collins, J. E. *et al.* A high-density YAC contig map of human chromosome 22. *Nature* **377**, 367-379 (1995).

10. Pulver, A. E. *et al.* Psychotic illness in patients diagnosed with velo-cardio-facial syndrome and their relatives. *J. Nerv. Ment. Dis.* **182**, 476-478 (1994).

11. Gill, M. *et al.* A combined analysis of D22S278 marker alleles in affected sib-pairs: support for a susceptibility locus for schizophrenia at chromosome 22q12. Schizophrenia Collaborative Linkage Group (Chromosome 22). *Am. J. Med. Genet.* **67**, 40-45 (1996).

12. Zu, L., Figueroa, K. P., Grewal, R. & Pulst, S. M. Mapping of a new autosomal dominant spinocerebellar ataxia to chromosome 22. *Am. J. Hum. Genet.* **64**, 594-599 (1999).

13. Southard-Smith, E. M. *et al.* Comparative analyses of the dominant megacolon-SOX10 genomic interval in mouse and human. *Mamm. Genome* **10**, 744-749 (1999).

14. Nishino, I., Spinazzola, A. & Hirano, M. Thymidine phosphorylase gene mutations in MNGIE, a human mitochondrial disorder. *Science* **283**, 689-692 (1999).

15. Peyrard, M. *et al.* The human LARGE gene from 22q12.3-q13.1 is a new, distinct member of the glycosyltransferase gene family. *Proc. natl Acad. Sci. USA* **96**, 598-603 (1999).

16. Kao, H. T. *et al.* A third member of the synapsin gene family. *Proc. Natl Acad. Sci. USA* **95**, 4667-4672 (1998).

17. Mittman, S., Guo, J., Emerick, M. C. & Agnew, W. S. Structure and alternative splicing of the gene encoding alpha1I, a human brain T calcium channel alpha1 subunit. *Neurosci. Lett.* **269**, 121-124 (1999).

18. Seroussi, E. *et al.* TOM1 genes map to human chromosome 22q13.1 and mouse chromosome 8C1 and encode proteins similar to the endosomal proteins HGS and STAM. *Genomics* **57**, 380-388 (1999).

19. Seroussi, E. *et al.* Duplications on human chromosome 22 reveal a novel ret finger protein-like gene family with sense and endogenous antisense transcripts. *Genome Res.* **9**, 803-814 (1999).

20. Ning, Y., Rosenberg, M., Biesecker, L. G. & Ledbetter, D. H. Isolation of the human chromosome 22q telomere and its application to detection of cryptic chromosomal abnormalities. *Hum. Genet.* **97**, 765-769 (1996).

21. Edelmann, L., Pandita, R. K. & Morrow, B. E. Low-copy repeats mediate the common 3-Mb deletion in patients with velo-cardio-facial syndrome. *Am. J. Hum. Genet.* **64**, 1076-1086 (1999).

22. McDermid, H. E. *et al.* Long-range mapping and construction of a YAC contig within the cat eye syndrome critical region. *Genome Res.* **6**, 1149-1159 (1996).

23. Johnson, A. *et al.* A 1.5-Mb contig within the cat eye syndrome critical region at human chromosome 22q11.2. *Genomics* **57**, 306-309 (1999).

24. Kawasaki, K. *et al.* One-megabase sequence analysis of the human immunoglobulin lambda gene locus. *Genome Res.* **7**, 250-261 (1997).

25. Felsenfeld, A., Peterson, J., Schloss, J. & Guyer, M. Assessing the quality of the DNA sequence from the Human Genome Project. *Genome Res.* **9**, 1-4 (1999).

26. Mewes, H. W. *et al.* Overview of the yeast genome. *Nature* **387**, 7-65 (1997).

27. Blattner, F. R. *et al.* The complete genome sequence of Escherichia coli K-12. *Science* **277**, 1453-1474 (1997).

28. The C. elegans Sequencing Consortium. Genome sequence of the nematode C. elegans: a platform for investigating biology. *Science* **282**, 2012-2018 (1998).

29. Solovyev, V. & Salamov, A. The Gene-Finder computer tools for analysis of human and model organisms genome sequences. *Ismb* **5**, 294-302 (1997).

30. Burge, C. & Karlin, S. Prediction of complete gene structures in human genomic DNA. *J. Mol. Biol.* **268**, 78-94 (1997).

31. Cross, S. H. & Bird, A. P. CpG islands and genes. *Curr. Opin. Genet. Dev.* **5**, 309-314 (1995).

32. Altschul, S. F., Gish, W., Miller, W., Myers, E. W. & Lipman, D. J. Basic local alignment search tool. *J. Mol. Biol.* **215**, 403-410 (1990).

33. Bateman, A. *et al.* Pfam 3.1: 1313 multiple alignments and profile HMMs match the majority of proteins. *Nucleic Acids Res.* **27**, 260-262 (1999).

34. Hofmann, K., Bucher, P., Falquet, L. & Bairoch, A. The PROSITE database, its status in 1999. *Nucleic Acis Res.* **27**, 215-219 (1999).

35. Bairoch, A. & Apweiler, R. The SWISS-PROT protein sequence data bank and its supplement TrEMBL in 1999. *Nucleic Acids Res.* **27**, 49-54 (1999).

36. Dib, C. *et al.* A comprehensive genetic map of the human genome based on 5,264 microsatellites. *Nature* **380**, 152-154 (1996).

37. Holmquist, G. P. Chromosome bands, their chromatin flavors, and their functional features. *Am. J. Hum. Genet.* **51**, 17-37 (1992).

38. Bernardi, G. *et al.* The mosaic genome of warm-blooded vertebrates. *Science* **228**, 953-958 (1985).

39. The MHC sequencing consortium. Complete sequence and gene map of a human major histocompatibility complex. *Nature* **401**, 921-923 (1999).

40. Bernardi, G. The isochore organization of the human genome. *Annu. Rev. Genet.* **23**, 637-661 (1989).

41. Collins, J. E., Mungall, A. J., Badcock, K. L., Fay, J. M. & Dunham, I. The organization of the gamma-glutamyl transferase genes and other low copy repeats in human chromosome 22q11. *Genome Res.* **7**, 522-531 (1997).

42. Edelmann, L. *et al.* A common molecular basis for rearrangement disorders on chromosome 22q11. *Hum. Mol. Genet.* **8**, 1157-1167 (1999).

43. Eppig, J. T. & Nadeau, J. H. Comparative maps: the mammalian jigsaw puzzle. *Curr. Opin. Genet. Dev.* **5**, 709-716 (1995).

44. Bucan, M. *et al.* Comparative mapping of 9 human chromosome 22q loci in the laboratory mouse. *Hum. Mol. Genet.* **2**, 1245-1252 (1993).

45. Carver, E. A. & Stubbs, L. Zooming in on the human-mouse comparative map: genome conservation re-examined on a high-resolution scale. *Genome Res.* **7**, 1123-1137 (1997).

46. Puech, A. *et al.* Comparative mapping of the human 22q11 chromosomal region and the orthologous region in mice reveals complex changes in gene organization. *Proc. Natl Acad. Sci.* USA **94**, 14608-14613 (1997).

47. Bentley, D. R. Genomic sequence information should be released immediately and freely in the public domain. *Science* **274**, 533-534 (1996).

48. Guyer, M. Statement on the rapid release of genomic DNA sequence. *Genome Res.* **8**, 413 (1998).

49. Dunham, I., Dewar, K., Kim, U.-J. & Ross, M. T. in *Genome Analysis: A Laboratory Manual Series, Volume 3: Cloning Systems* (eds Birren, B. *et al.*) 1-86 (Cold Spring Harbor Laboratory Press, Cold Spring Harbor, New York, 1999).

50. Asakawa, S. *et al.* Human BAC library: construction and rapid screening. *Gene* **191**, 69-79 (1997).

Supplementary information is available on *Nature*'s World-Wide Web site (http://www.nature.com) or as paper copy from the London editorial office of *Nature*.

Correspondence and requests for materials should be addressed to I.D. (e-mail: id1@sanger.ac.uk).

Molecular Analysis of Neanderthal DNA from the Northern Caucasus

I.V. Ovchinnikov *et al.*

Editor's Note

The only thing better than having a sequence of DNA from an extinct species is having a second one, for comparative purposes. Mitochondrial DNA extracted from the original bones of type specimen Neanderthal 1 in 1997 showed that Neanderthals were not closely related to modern humans. But how secure was this single set of data? Here Igor Ovchinnikov and colleagues describe mitochondrial DNA extracted from 29,000-year-old Neanderthal remains from the northern Caucasus. Although far removed in space and time from the other Neanderthal sequence, the results still showed that the Caucasus DNA grouped more closely with the original Neanderthal finds, further ruling out admixture with modern humans.

The expansion of premodern humans into western and eastern Europe ~40,000 years before the present led to the eventual replacement of the Neanderthals by modern humans ~28,000 years ago[1]. Here we report the second mitochondrial DNA (mtDNA) analysis of a Neanderthal, and the first such analysis on clearly dated Neanderthal remains. The specimen is from one of the eastern-most Neanderthal populations, recovered from Mezmaiskaya Cave in the northern Caucasus[2]. Radiocarbon dating estimated the specimen to be ~29,000 years old and therefore from one of the latest living Neanderthals[3]. The sequence shows 3.48% divergence from the Feldhofer Neanderthal[4]. Phylogenetic analysis places the two Neanderthals from the Caucasus and western Germany together in a clade that is distinct from modern humans, suggesting that their mtDNA types have not contributed to the modern human mtDNA pool. Comparison with modern populations provides no evidence for the multiregional hypothesis of modern human evolution.

THE first successful extraction and sequencing of the mtDNA hypervariable regions (I and II (HVRI & HVRII)) was performed on the Neanderthal type specimen from Feldhofer Cave, the Neander valley, Germany[4,5]. Phylogenetic analysis of the sequence placed the Neanderthal mtDNA outside the mtDNA pool of modern humans. This was regarded as a breakthrough in the study of modern human evolution, providing molecular evidence that Neanderthals did not contribute mtDNA to modern humans. From this sequence the divergence of Neanderthals and modern humans was estimated to have occurred between 317,000 and 741,000 years ago[4,5]. However, these estimates were based on the molecular analysis of a single specimen. The shortage of potentially well preserved Neanderthal material[6] and limited access to Neanderthal remains for destructive analysis have hindered the analysis of additional specimens, but genetic

高加索北部尼安德特人 DNA 的分子分析

奥夫钦尼科夫等

编者按

比从已灭绝的物种得到一个DNA序列更好的事情只有一件,那就是获得第二个序列,以便进行对比。1997 年从尼安德特人 1 的原始骨骼标本中提取的线粒体 DNA 显示,尼安德特人与现代人的关系并不密切。但这仅有的一组数据的可信度怎样呢? 在本文中,伊戈尔·奥夫钦尼科夫及其同事描述了从高加索北部绝对年龄为 29,000 年的尼安德特人遗骸中提取的线粒体 DNA。尽管在时空上与另一个尼安德特人序列相距甚远,但是结果仍表明在高加索发现的该 DNA 与最初发现的尼安德特人关系更近,并进一步排除了与现代人有基因交流的可能性。

距今约 40,000 年前,古老型智人扩散至西欧和东欧,并最终导致尼安德特人在约 28,000 年前被现代人取代[1]。在本文中我们报道了第二例对尼安德特人线粒体 DNA(mtDNA)的分子分析,这也是首例对有清晰年代尼安德特人遗骸进行的此类分析。标本来自尼安德特人最东部的种群之一,发现于高加索北部的梅兹迈斯卡亚洞[2]。根据放射性碳同位素年代测定,该标本绝对年龄约为 29,000 年,因此是生存最晚的尼安德特人之一[3]。该序列显示其与费尔德霍费尔原始的尼安德特人之间存在 3.48% 的差异[4]。系统发育分析将高加索和德国西部的这两个尼安德特人置于有别于现代人的同一分支,表明他们的 mtDNA 类型对现代人的 mtDNA 库没有贡献。与现代人种群进行的比较没有为现代人演化的多地起源说提供证据。

第一次对 mtDNA 高变区(I 和 II,即 HVRI 和 HVRII)的成功提取和测序是用德国尼安德谷费尔德霍费尔洞出土的尼安德特人模式标本完成的[4,5]。该序列的系统发育分析将此尼安德特人的 mtDNA 置于现代人的 mtDNA 库之外。这曾被认为是现代人类演化研究的一项突破,为证明尼安德特人对现代人的 mtDNA 没有贡献提供了分子证据。根据这一序列推测尼安德特人和现代人的分化发生在 741,000 年至 317,000 年前[4,5]。然而,这些推测是以单一标本的分子分析为基础的。保存完好的尼安德特人原材料的缺乏[6]以及对尼安德特人遗骸进行破坏性分析的局限性阻碍了对更多样本的分析。然而,为了理解尼安德特人的分子多样性及不同尼安德特人种

characterization of additional Neanderthals is essential to understand their molecular diversity and the relationship between different Neanderthal populations, and to assess their relationship to modern humans further.

The Caucasus, which is located on the southeastern boundary between Europe and Asia, is one of the areas through which pre-modern humans and anatomically modern *Homo sapiens* may have entered Europe from the Near East and Africa. Neanderthals invaded the region at an unknown point in time[7,8] and may have occupied the region alongside modern humans from ~40,000 years before the present (B.P.). During the excavation of the Mezmaiskaya Cave[2], which is located in the northern Caucasus (Fig. 1), a fragmentary skeleton of an infant was found that contained a set of morphological characteristics which indicated clear affinities to the Neanderthals of western and central Europe[2]. Mitochondrial DNA analysis was undertaken using one of this Neanderthal's ribs.

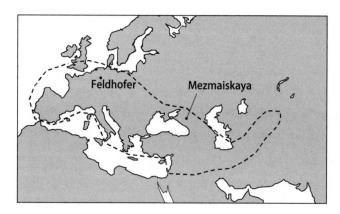

Fig. 1. The area within which Neanderthal remains have been found (dotted line). The Mezmaiskaya Cave is situated 1,310 m above sea level within the northern Caucasus at 44° 10′ N 40° 00′ E on the bank of the Sukhoy Kurdzhips river. The location of the Feldhofer Cave in Germany is also shown.

The preservation of collagen-type debris was used as an indicator of macromolecule preservation in the bone. The amount of collagen-type debris extracted[9] from 130 mg of the Mezmaiskaya Neanderthal rib fragment was 22% of the average level extracted from modern bones, and the extracted collagen contained 41.6% carbon and 14.7% nitrogen. This is within the values recovered from prehistoric samples displaying good preservation[10]. These data suggested that there were low levels of diagenetic modification. The high collagen yield made it possible to date the Neanderthal infant to $29,195 \pm 965$ (Ua-14512) years B.P. by using a radio-carbon accelerator. This date does not agree with the previously published dates of more than 45,000 (Le-3841) and $40,660 \pm 1,600$ (Le-3599) years for the Mousterian layers in the Mezmaiskaya Cave[2] with which the skeleton was associated. The most likely reason for this discrepancy is the incorrect identification of the poorly defined layers in this area of the cave. The value obtained from the bone itself rather than from associated material gives the most reliable date for this individual.

Two sections of one rib (90 mg and 123 mg) were used for DNA extraction in two

群之间的关系，并进一步评估他们与现代人的关系，了解更多尼安德特人的遗传特征是至关重要的。

高加索位于欧洲和亚洲的东南交界处，可能是古老型智人和早期现代人从近东和非洲进入欧洲途经的地区之一。尼安德特人在某个未知的时间侵入了这一地区[7,8]，并且在距今（BP）约 40,000 年与现代人共同占据这一地区。梅兹迈斯卡亚洞[2]位于高加索北部（图 1），在挖掘该洞期间发现了一具破碎的幼儿骨架，其蕴含的一组形态学特征展现了与西欧和中欧尼安德特人明确的亲缘关系[2]。线粒体 DNA 分析正是使用这个尼安德特人的一根肋骨进行的。

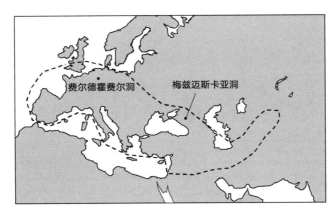

图 1. 发现尼安德特人遗骸的区域（虚线所示）。梅兹迈斯卡亚洞海拔 1,310 米，位于高加索北部苏霍伊库尔吉普斯河畔，北纬 44°10′，东经 40°00′。德国费尔德霍费尔洞的位置也在图中标明。

骨骼中胶原蛋白型片断的保存状况被用作大分子保存状况的指标。从 130 mg 梅兹迈斯卡亚尼安德特人肋骨碎片提取的胶原蛋白型片断[9]为现代骨骼平均提取水平的 22%，提取出的胶原蛋白含碳量为 41.6%，含氮量为 14.7%。这一数值处于保存良好的史前样本的数值范围内[10]。这些数据表明成岩改造作用水平较低。该肋骨碎片的高胶原含量使得我们可以利用放射性碳加速器测定该尼安德特幼儿的年龄为距今 29,195±965(Ua-14512) 年。这一年代与之前发表的梅兹迈斯卡亚洞[2]此骨架出土的莫斯特文化层年代为距今 45,000(Le-3841) 年和 40,660±1,600(Le-3599) 年是不一致的。导致这一差异最可能的原因是对该洞穴中这一区域不明确的骨架出土层位的错误鉴定。从骨骼本身而非间接关联材料得到的数值才能给出该个体最可信的年代信息。

两个独立的实验室分别选取了一段肋骨的两个切片（90 mg 和 123 mg）来提取

independent laboratories. In the Glasgow laboratory, a total of 345 base pairs (bp) of HVRI was determined from two overlapping polymerase chain reaction (PCR) fragments with lengths of 232 and 256 bp. Forty PCR amplification cycles produced sufficient product to enable direct sequencing. Products from independent PCR amplifications were also cloned into a TA vector and sequenced (Fig. 2).

```
              1 1 1 1 1 1 1 1 1 1 1 1 1 1 1 1 1 1 1 1 1 1 1 1 1 1 1 1 1 1 1 1   1 1 1 1 1 1 1 1
              6 6 6 6 6 6 6 6 6 6 6 6 6 6 6 6 6 6 6 6 6 6 6 6 6 6 6 6 6 6 6 6   6 6 6 6 6 6 6 6
              0 0 0 0 1 1 1 1 1 1 1 1 1 1 1 1 1 1 1 1 2 2 2 2 2 2 2 2 2 2 2 2   2 2 3 3 3 3 3 3
              3 7 8 9 0 0 1 1 1 2 3 4 5 5 6 8 8 8 0 2 3 3 4 4 5 5 5 6 6 6 7 9   1 2 4 6 6 9
              7 8 6 3 7 8 1 2 8 9 9 8 4 6 9 2 3 9 9 3 0 4 3 4 0 6 8 1 2 3.1 8 9  1 0 4 2 5 3
-------------------------------------------------------------------------------------------------
Reference     A A T T C C C C G G A C T G C A A T T C A C T G C C A C C - C A T C C T C C

Direct 1      . C . . . . . . A T T . A T C C C T G T . A . A .

P1            . C . . . . . . A T T . A T C C C T G T . A . A . . T A T

P2            . C . . . . . . A T T . A T C C C T G T . A . A . . T A T

P3            . C . . . . A A T T . A T C C C T G T . A . A . . T A T

Direct 2                                        T G T . A . A . . T A T G C T T C . .

577.1                                           T G T . A . A . T T A T G C T T C . T

557.2                                           T G T . A T A . . T A T G C T T C . .

581.2                                           T G T . A . A . . T A T G C T T C . .

581.3                                           T G T C A . A . . T A T G C T T C T .

Mezmaiskaya   . C . . . . . A T T . A T C C C T G T . A . A . . T A T G C T T C . .

Feldhofer     G G . C T T T T . A T T C . T . C C C T G T . A . A G . T A T G C T . C . .
```

Fig. 2. Variable sites in the DNA sequences of the PCR fragments obtained by direct sequencing (Direct 1 and 2) and cloned PCR products derived from the Neanderthal from Mezmaiskaya Cave. The human reference sequence[11] and the sequence of the Neanderthal from Feldhofer Cave[4] are included for comparison. A full stop indicates that the sequence is the same as the reference. The sequence that could be duplicated in the Stockholm laboratory is shown in bold within the compiled Mezmaiskaya sequence.

The authenticity of the DNA sequence obtained in the Glasgow laboratory is supported by a number of factors. First, a section of the mtDNA was isolated and sequenced with congruent results in the Stockholm laboratory. Second, the PCR products were generated using Neanderthal-specific primer pairs that, under the amplification conditions, failed to amplify any fragments using modern DNA controls from individuals of different ethnic origins. Third, the retrieval of the sequence was not dependent on the primers used. Fourth, the low level of diagenetic modification indicated that the sample could theoretically contain amplifiable DNA. Last, and most convincingly, the sequence is similar to, and after phylogenetic analysis clusters strongly with, the previously analysed Neanderthal sequence[4].

Comparison of the 345-bp fragment of HVRI with the Anderson reference sequence[11] and the Neanderthal from Feldhofer Cave[4] revealed 22 differences (17 transitions, 4 transversions and 1 insertion) and 12 differences (11 transitions and 1 transversion), respectively. The Feldhofer Neanderthal HVRI contained 27 differences to Anderson reference sequence[11] (over the equivalent 345 bp[4]). The two Neanderthals share 19 substitutions relative to the reference sequence. The cloned PCR products contained all the substitutions that were detected by direct sequencing; six other non-reproducible substitutions occurred in seven different clones. No

DNA。格拉斯哥实验室通过长度分别为 232 bp 和 256 bp 的两个重叠的聚合酶链式反应(PCR)测定了片段总长度为 345 个碱基对(bp)的 HVRI。通过 40 个 PCR 扩增循环产生了足够的产物，可以进行直接测序。同时也将独立 PCR 扩增的产物克隆到 TA 载体中并进行了测序(图 2)。

```
           1 1 1 1 1 1 1 1 1 1 1 1 1 1 1 1 1 1 1 1 1 1 1 1 1 1 1 1 1 1 1 1 1 1 1 1 1 1 1
           6 6 6 6 6 6 6 6 6 6 6 6 6 6 6 6 6 6 6 6 6 6 6 6 6 6 6 6 6 6 6 6 6 6 6 6 6 6 6
           0 0 0 0 1 1 1 1 1 1 1 1 1 1 1 1 1 1 2 2 2 2 2 2 2 2 2 2 2 2 2 2 2 3 3 3 3 3
           3 7 8 9 0 0 1 1 1 2 3 4 5 6 8 8 8 0 2 3 3 4 4 5 5 5 6 6 7 9 1 2 4 6 6 9
           7 8 6 3 7 8 1 2 8 9 9 8 4 6 9 2 3 9 9 3 0 4 3 4 0 6 8 1 2 3.1 8 9 1 0 4 2 5 3

参考序列      A A T T C C C C G G A C T G C A A T T C A C T G C C A C C - C A T C C T C C
Direct 1     . C . . . . . A T T . A T C C C C T G T . A . A .
P1           . C . . . . . A T T . A T C C C C T G T . A . A . . T A T
P2           . C . . . . . A T T . A T C C C C T G T . A . A . . T A T
P3           . C . . . . A A T T . A T C C C C T G T . A . A . . T A T
Direct 2                               T G T . A . A . . T A T G C T T C . .
577.1                                 T G T . A . A . T T A T G C T T C . T
557.2                                 T G T . A T A . T G C T T C . T
581.2                                 T G T . A . A . . T A T G C T T C . T
581.3                                 T G T C A . . T A T G C T T C T
梅兹迈斯卡亚   . C . . . . . A T T . A T C C C C **T G T** . **A** . **A** . . **T A T G C T** T C .
费尔德霍费尔  G G . C T T T T . A T T C . T . C C C T G T . A . A G . T A T G C T . C . .
```

图 2. 通过直接测序(Direct 1 和 2) PCR 片段及梅兹迈斯卡亚洞尼安德特人的克隆 PCR 产物，得到 DNA 序列上的可变位点。现代人参考序列[11]及费尔德霍费尔洞的尼安德特人序列[4]都包含在内，进行比对。圆点表示序列与参考序列相同。在斯德哥尔摩实验室中可被复制的序列在梅兹迈斯卡亚合成序列中用黑体表示。

有多个因素支持格拉斯哥实验室得到的 DNA 序列的可靠性。首先，一段线粒体 DNA 被分离并测序，取得了与斯德哥尔摩实验室一致的结果。第二，PCR 产物是使用针对尼安德特人的特异性引物对生成的，在扩增条件下，这些引物无法扩增来自不同种族个体的现代 DNA 对照片段。第三，序列的获取不取决于所使用的引物。第四，低水平的成岩改造作用表明这一样本理论上含有可扩增的 DNA。最后且最有说服力的是，该序列与以前分析过的尼安德特人序列[4]相似，且系统发育分析显示它们强烈类聚。

该 345 bp 的 HVRI 片段与安德森参考序列[11]和费尔德霍费尔洞的尼安德特人[4]序列的分别对比显示出 22 处碱基差异(17 处转换，4 处颠换及 1 处插入)和 12 处碱基差异(11 处转换及 1 处颠换)。费尔德霍费尔尼安德特人的 HVRI 与安德森参考序列[11]相比有 27 处碱基差异(在相应的 345 bp 中[4])。与参考序列相比，这两个尼安德特人共同拥有 19 处碱基替换。克隆的 PCR 产物包含了直接测序检测到的所有碱基替换；六处其他不可重复的替换出现于七个不同的克隆中。在格拉斯哥实验室中，

modern sequences were found in the Glasgow laboratory either by direct sequencing or by sequencing cloned PCR products. The Stockholm laboratory experienced problems with contamination: most of the cloned PCR products that they analysed contained sequences that are found in the modern human mtDNA pool, with two haplotypes predominant. However, three clones contained DNA that was the same as the sequence determined in Glasgow (two of these contained non-reproducible substitutions).

The preservation of 256-bp DNA fragments in bone that is ~29,000 years old, that has not been preserved in permafrost and that contained sufficient DNA to enable direct DNA sequencing after amplification is unprecedented and may be attributed to specific features of the microenvironment of the limestone cavern[2]. The retrieval of mtDNA showed a positive correlation to the preservation of collagen content and the skeletal morphology.

Phylogenetic analysis using both distance and parsimony optimizations places the two Neanderthal sequences together, in a distinct clade, basal to modern humans. Neighbour-joining analysis supports this separation (Fig. 3a). Parsimony analysis, which makes minimal assumptions about the model of evolution and optimizes the fit between the tree and data, produced similar results (Fig. 3b).

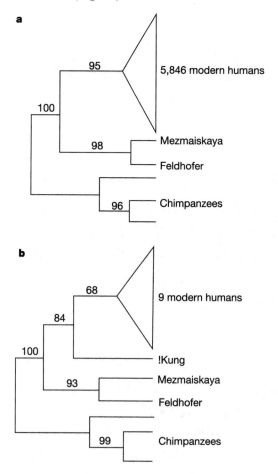

Fig. 3. Phylogenetic relationship of the two Neanderthals and modern humans. **a**, A neighbour-joining

直接测序和克隆 PCR 产物测序都没有发现现代序列。斯德哥尔摩实验室则遇到了污染问题：他们分析的大部分克隆 PCR 产物都含有现代人线粒体 DNA 库中存在的序列，主要为两个单倍型。不过，有三个克隆含有的 DNA 与格拉斯哥实验室确定的序列相同（其中两个含有不可重复的替换）。

年龄约 29,000 年的骨骼中保存了 256 bp 的 DNA 片段，没有保存在永久冻土中却含有足够的 DNA 可供扩增后直接 DNA 测序，这种史无前例的情况或许可以归因于这一石灰岩洞穴中特殊的微环境[2]。线粒体 DNA 的获得表明胶原蛋白成分的保存状况与骨骼形态是正相关的。

使用距离和简约性优化法则进行的系统发育分析将这两个尼安德特人的序列归并到一起，构成一个独立的分支，并位于现代人基底位置。邻接分析支持这一分化（图 3a）。简约性分析得到了类似的结果，该分析对演化模式的假设最少，并会优化系统树与数据之间的拟合（图 3b）。

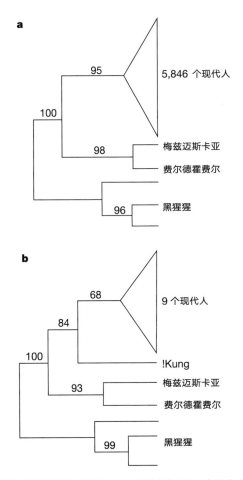

图 3. 两个尼安德特人与现代人的系统发育关系。**a**，使用来自 5,846 个现代人的共 1,897 个单倍型计算

tree computed using a total of 1,897 haplotypes derived from 5,846 modern humans[19]. **b**, A maximum parsimony branch and bound search with the two Neanderthal sequences along with the sequences of one !Kung, three other Africans, three Asians and three Europeans, all randomly selected[19]. This result is congruent with four additional data sets that were analysed. In both analyses, three chimpanzee sequences[13] were used as an outgroup. The numbers in both diagrams refer to the bootstrap frequencies (%) obtained from 1,000 replicates.

The level of pairwise difference found between the two Neanderthals was higher than the average values found in random samples of 300 Caucasoids (5.28 ± 2.24) and Mongoloids (6.27 ± 2.29)—less than 1% of Caucasoid and Mongoloid pairs differ at 12 or more positions—but comparable to a random sample of 300 Africans (8.36 ± 3.2), where 37% of pairs differed at 12 or more positions. When analysing ancient DNA there is the possibility of misincorporating nucleotides in the early stages of PCR, especially when the target DNA is possibly damaged and present in low copy number[12]. As both Neanderthals were analysed in replicate and the results were consistent, however, errors of this type can be discounted.

The Feldhofer and Mezmaiskaya Neanderthals were separated geographically by over 2,500 km. Given that these two individuals contained closely related mtDNA, which is phylogenetically distinct from modern humans, and displays only a moderate level of sequence diversity compared with some primates[13], these data provide further support for the hypothesis of a very low gene flow between the Neanderthals and modern humans. In particular, these data reduce the likelihood that Neanderthals contained enough mtDNA sequence diversity to encompass modern human diversity.

The "out-of-Africa" hypothesis for the origin of modern humans predicts equal distances between the Neanderthal sequences and all modern sequences. We observed this in our analysis—the average pairwise differences between the Neanderthals and 300 randomly selected Africans, Mongoloids and Caucasoids were calculated to be 23.09 ± 2.86, 23.27 ± 4.06 and 25.45 ± 3.27, respectively.

We estimated the age of the most recent common ancestor (MRCA) of the mtDNA of the eastern and western Neanderthals to be 151,000–352,000 years. This coincides with the time of emergence of the Neanderthal lineage in the palaeontological records[14]. The divergence of modern human and Neanderthal mtDNA was estimated to be between 365,000 and 853,000 years. Using the same model, we estimated the age of the earliest modern human divergences in mtDNA to be between 106,000 and 246,000 B.P.

The results obtained from this specimen suggest that some other Neanderthal samples may be amenable to molecular analysis. To obtain a more complete picture of the relationship of Neanderthals to modern humans, additional Neanderthals and early modern humans must be analysed, especially from the regions where they may have co-existed. The excellent preservation of this specimen leads to the potential of analysing the entire Neanderthal mitochondrial genome.

出的邻接树[19]。**b**，两个尼安德特人序列与随机选择的一个 !Kung 人、三个其他非洲人、三个亚洲人和三个欧洲人的序列采用分支限界法搜索到的最简约系统树[19]。这一结果与另外四组数据的分析结果一致。以上两个分析使用三只黑猩猩的序列[13]作为外群。两个图中的数字代表 1,000 次重复得到的自举频率(%)。

两个尼安德特人的配对差异水平高于随机选择的 300 个高加索人种(5.28 ±2.24)和蒙古人种(6.27±2.29)样本的平均值——不到 1% 的高加索人种和蒙古人种配对组合有 12 个或更多位点存在差异——但是与一个 300 个非洲人(8.36±3.2)的随机样本相当，该非洲样本在 12 个及更多位点有 37% 的配对组合存在差异。在分析古 DNA 时，PCR 初始阶段可能会发生核苷酸被错误插入的情况，特别是当标靶 DNA 可能损坏且拷贝数较低时[12]。因为这两个尼安德特人都进行了重复分析而且结果是一致的，所以该类失误可以被忽略。

费尔德霍费尔尼安德特人和梅兹迈斯卡亚尼安德特人在地理上相距 2,500 多千米。考虑到这两个个体拥有密切相关的线粒体 DNA，且在系统发育上不同于现代人，与某些灵长类相比[13]仅显示出中等水平的序列多样性，这些证据为尼安德特人和现代人之间基因交流很少这一假说提供了进一步的支持。需要特别指出的是，这些证据同时反驳了尼安德特人拥有足够的线粒体 DNA 序列多样性且足以覆盖现代人类多样性这一说法。

关于现代人类起源的"走出非洲"的假说认为尼安德特人序列与所有现代序列间的差异距离相等。在我们的分析中也观察到了这一点——经计算，尼安德特人与 300 个随机选择的非洲人、蒙古人和高加索人间的平均配对差异分别为 23.09 ±2.86、23.27±4.06 和 25.45±3.27。

我们推测东西方尼安德特人线粒体 DNA 的最近共同祖先(MRCA)的年龄约为距今 151,000 至 352,000 年。这与古生物学记录中尼安德特人谱系出现的时间是一致的[14]。据推测现代人和尼安德特人线粒体 DNA 的分化是在距今 365,000 至 853,000 年。使用同一模型，我们推测最早的现代人类线粒体 DNA 分化发生在距今 106,000 至 246,000 年。

从该标本得到的结果表明，一些其他尼安德特人标本也可以用来进行分子分析。为了得到尼安德特人与现代人类关系更完整的图谱，必须分析其他尼安德特人和早期现代人类，特别是来自他们共同生存地区的标本。该标本极佳的保存状况使得对尼安德特人整个线粒体基因组进行分析成为可能。

Methods

DNA extraction, PCR, cloning and sequencing

The DNA extraction methods used in Glasgow[4] and Stockholm[15] have been described. We took precautions to prevent contamination from modern DNA[4]. The Neanderthal-specific primer pairs, NL16,055 and NH16,262, and NL16,209 and NH16,400 (5'-TGATTTCACGGAGGATGGTGA-3') were used in Glasgow and the primers L16,212 (5'-ATGCTTACAAGCAAGCACA-3') and H16,332 (5'-TTGACTGTAATGTGCTATG-3') were used in Stockholm. The annealing temperatures for the primer pairs NL16,055–NH16,262, NL16,209–NH16,400 and L16,212–H16,332 were 50°C, 60°C and 50°C respectively; 40 cycles were used for the first two pairs, 55 cycles for the third. AmpliTaq Gold (Perkin Elmer Cetus) was used in all PCRs. PCR products were purified using the QIAquick Gel Extraction kit (Qiagen) before direct sequencing using the Dye Terminator sequencing kit (Perkin Elmer) or cloning into the TA vector (Invitrogen) before sequencing with the same kit using the M13 and T7 primers.

Sequence analyses

The neighbour-joining and the maximum parsimony branch and bound trees were both constructed using PAUP* 4.0 (ref. 16). For the neighbour-joining analysis, the Tamura-Nei DNA substitution model[17] was used with a gamma distribution of 0.4 (ref. 18), for all other parameters the defaults provided by PAUP* 4.0 were used. The MRCA was calculated using the described methods and assumptions[5]. PAUP* 4.0 was used to calculate pairwise differences between sequences: the data sets used for this were constructed by randomly selecting appropriate samples from a published data set[19].

(**404**, 490-493; 2000)

Igor V. Ovchinnikov[*†‡], **Anders Götherström**[§], **Galina P. Romanova**[∥], **Vitaliy M. Kharitonov**[¶], **Kerstin Lidén**[§] & **William Goodwin**[*]

[*] Human Identification Centre, University of Glasgow, Glasgow G12 8QQ, Scotland, UK

[†] Institute of Gerontology, Moscow 129226, Russia

[§] Archaeological Research Laboratory, Stockholm University, 106 91 Stockholm, Sweden

[∥] Institute of Archaeology, Moscow 117036, Russia

[¶] Institute and Museum of Anthropology, Moscow State University, Moscow 103009, Russia

[‡] Present address: Department of Medicine, Columbia University, New York, New York 10032 USA

Received 15 November 1999; accepted 31 January 2000.

References:

1. Stringer, C. B. & Mackie, R. *African Exodus: the Origin of Modern Humanity* (Cape, London, 1996).

2. Golovanova, L. V., Hoffecker, J. F., Kharitonov, V. M. & Romanova, G. P. Mezmaiskaya Cave: A Neanderthal occupation in the Northern Caucasus. *Curr. Anthropol.* **40**, 77-86 (1999).

3. Smith, F. H., Trinkaus, E., Pettitt, P. B., Karavanic, I. & Paunovic, M. Direct radiocarbon dates for Vindija G1 and Velika Pecina Late Pleistocene hominid remains.

方　法

DNA 提取、PCR、克隆及测序

　　格拉斯哥[4]和斯德哥尔摩[15]实验室使用的 DNA 提取方法已经介绍过。我们采取了一些预防措施以防止来自现代 DNA 的污染[4]。格拉斯哥实验室使用的尼安德特人特异性引物对包括 NL16,055 和 NH16,262，以及 NL16,209 和 NH16,400(5′-TGATTTCACGGAGGATGGTGA-3′)；斯德哥尔摩实验室使用的引物对为 L16,212(5′-ATGCTTACAAGCAAGCACA-3′)和 H16,332(5′-TTGACTGTAATGTGCTATG-3′)。引物对 NL16,055–NH16,262、NL16,209 NH16,400 和 L16,212–H16,332 的退火温度分别为 50℃、60℃和 50℃；前两对引物扩增了 40 个循环，第三对扩增了 55 个循环。所有 PCR 均使用 AmpliTaq Gold(Perkin Elmer Cetus)试剂。用 QIAquick 凝胶回收试剂盒(Qiagen)对 PCR 产物进行纯化，然后使用 Dye Terminator 测序试剂盒(Perkin Elmer)进行直接测序，或者使用 M13 和 T7 引物，用相同的试剂盒在测序之前克隆到 TA 载体(Invitrogen)上。

序列分析

　　使用 PAUP* 4.0 构建邻接分析和最简约分支界限树(参考文献 16)。邻接分析中使用 Tamura-Nei DNA 替换模型[17]，γ 分布取 0.4(参考文献 18)，其他所有参数使用 PAUP* 4.0 提供的默认值。使用文献中描述的方法和假设计算 MRCA[5]。使用 PAUP* 4.0 计算序列之间的成对差异：进行该计算的数据集是通过从已发表的数据集中随机选择适当样本而构建的[19]。

　　　　　　　　　　　　　　　　　　　　　(刘皓芳 翻译；张颖奇 审稿)

Proc. Natl Acad. Sci. USA **96**, 12281-12286 (1999).

4. Krings, M. *et al.* Neandertal DNA sequence and the origin of modern humans. *Cell* **90**, 19-30 (1997).

5. Krings, M., Geisert, H., Schmitz, R. W., Krainitzki, H. & Pääbo, S. DNA sequence of the mitochondrial hypervariable region II from the Neandertal type specimen. *Proc. Natl Acad. Sci.* USA **96**, 5581-5585 (1999).

6. Cooper, A. *et al.* Neandertal genetics. *Science* **277**, 1021-1023 (1997).

7. Gabunia, L. & Vekua, A. A Plio-Pleistocene hominid from Dmanisi, East Georgia, Caucasus. *Nature* **373**, 509-512 (1995).

8. Kozlowski, J. K. in *Neandertals and Modern Humans in Western Asia* 461-482 (Plenum, New York-London, 1998).

9. Brown, T. A., Nelson, D. E., Vogel, J. S. & Southon, J. R. Improved collagen extraction by modified Longin method. *Radiocarbon* **30**, 171-177 (1988).

10. DeNiro, M. J. Postmortem preservation and alteration of *in vivo* bone collagen isotope ratios in relation to palaeodietary reconstruction. *Nature* **317**, 806-809 (1985).

11. Anderson, S. *et al.* Sequence and organisation of the human mitochondrial genome. *Nature* **290**, 457-474 (1981).

12. Höss, M. *et al.* DNA damage and DNA sequence retrieval from ancient tissue. *Nucleic Acids Res.* **24**, 1304-1307 (1996).

13. Gagneux, P. *et al.* Mitochondrial sequences show diverse evolutionary histories of African hominoids. *Proc. Natl Acad. Sci.* USA **96**, 5077-5082 (1999).

14. Gamble, C. in *Prehistoric Europe* 5-41 (Oxford Univ. Press, Oxford, 1998).

15. Lidén, K., Götherström, A. & Eriksson, E. Diet, gender and rank. *ISKOS* **11**, 158-164 (1997).

16. Swofford, D. L. *PAUP*: Phylogenetic Analysis Using Parsimony (* and Other Methods)* Version 4. (Sinauer Associates, Sunderland, Massachusetts, 1998).

17. Tamura, K. & Nei, M. Estimation of the number of nucleotide substitutions in the control region of mitochondrial DNA in humans and chimpanzees. *J. Mol. Evol.* **10**, 512-526 (1993).

18. Excoffier, L. & Yang, Z. Substitution rate variation among sites in mitochondrial hypervariable region I of humans and chimpanzees. *Mol. Biol. Evol.* **16**, 1357-1368 (1999).

19. Burckhardt, F., von Haeseler, A. & Meyer, S. HvrBase: compilation of mtDNA control region sequences from primates. *Nucleic Acids Res.* **27**, 138-142 (1999).

Acknowledgements. We are indebted to L. V. Golovanova for the excavations in Mezmaiskaya Cave that provided materials for analysis. We thank V. P. Ljubin and P. Vanezis for encouragement and support; B. L. Cohen for numerous discussions; J. L. Harley, O. I. Ovtchinnikova, E. B. Druzina and J. Wakefield for technical help and assistance; R. Page for help with the phylogenetic analysis; and P. Beerli, A. Cooper, M. Cusack, M. Nordborg and M. Ruvolo for useful comments. I.V.O. thanks his host G. Curry. I.V.O. was supported by a Royal Society/NATO Fellowship. We thank the Swedish Royal Academy of Sciences and the Swedish Research Council for Natural Sciences for partial financial support.

Correspondence and requests for material should be addressed to W.G. (e-mail: w.goodwin@formed.gla.ac.uk).

A Refugium for Relicts?

M. Manabe *et al.*

Editor's Note

The Yixian Formation of Liaoning Province in north-east China has yielded one of the finest assemblages of fossils, including feathered dinosaurs and early flowering plants. Estimates of the age of the Yixian varied over a 25-million-year timespan, from the Late Jurassic to Early Cretaceous. Evidence placing the age at 124 million years—the younger, Cretaceous end of the spectrum—made the Yixian look like a refuge for ancient relics: a real-life Jurassic Park. This view was put forward in a News and Views article in *Nature* by Zhexi Luo of the Carnegie Museum of Natural History. But it is countered in this paper by palaeontologist Makoto Manabe and colleagues, who show that Cretaceous forms from Japan also have a Jurassic cast, making the picture more complex than Luo had indicated.

LUO[1] suggests that the vertebrate fauna from the Yixian Formation (Liaoning Province, China) shows that this region of eastern Asia was a refugium, in which several typically Late Jurassic lineages (compsognathid theropod dinosaurs, "rhamphorhynchoid" pterosaurs, primitive mammals) survived into the Early Cretaceous[1] (Fig. 1). Data from slightly older sediments in the Japanese Early Cretaceous, however, suggest that the faunal composition of this region can only be partly explained by the concept of a refugium.

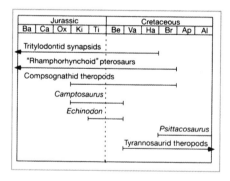

Fig. 1. Stratigraphic ranges of clades that include taxa recovered from the Yixian Formation, China, and the Kuwajima and Itsuki Formations, Japan[1,4,9]. Data on *Camptosaurus* and *Echinodon* are from ref. 13. Arrows, lineage extends beyond the time range shown here; solid bars, first and last occurrences. Al, Albian; Ap, Aptian; Ba, Bathonian; Be, Berriasian; Br, Barremian; Ca, Callovian; Ha, Hauterivian; Ki, Kimmeridgian; Ox, Oxfordian; Ti, Tithonian; Va, Valanginian.

The Kuwajima Formation of Ishikawa Prefecture, central Japan, is yielding an important Early Cretaceous vertebrate fauna. This unit is a lateral equivalent of the Okurodani

残遗种庇护所？

真锅真等

编者按

中国东北部的辽宁省义县组已经产出了最完美的化石群之一，这些化石包括带羽毛的恐龙和早期的被子植物。对义县组的年代估测有着超过 2,500 万年的差异，从晚侏罗世到早白垩世。有证据表明该地层年龄为 1 亿 2,400 万年，对应于更年轻的白垩纪，这使得义县看起来似乎是古代孑遗的避难所：一个真实的侏罗纪公园。这个观点由美国卡内基自然历史博物馆的罗哲西博士发表于《自然》的"新闻与观点"栏目中。但是在本文中，它被古生物学家真锅真和他的同事们反驳了，他们指出日本的白垩纪形成时也有一个"侏罗纪公园"，使得情况远比罗哲西提出的要复杂得多。

罗哲西[1]认为义县组（中国辽宁省）的脊椎动物群说明该亚洲东部地区是一个残遗种庇护所。在这个区域内，晚侏罗世的几个典型谱系（兽脚亚目的美颌龙类、翼龙目的"喙嘴翼龙类"、原始的哺乳类）存活到早白垩世[1]（图 1）。但是，日本早白垩世的一些更古老沉积物的数据显示，这个区域的动物群组成不能仅通过残遗种庇护所的概念进行解释。

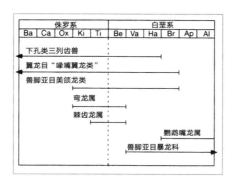

图 1. 包括中国义县组和日本桑岛组、五木村组恢复的分类单元在内的分支的地层分布图[1,4,9]。弯龙属和棘齿龙属的数据来自参考文献 13。箭头：谱系延伸的时间范围超过图示；实心棒：最先和最后出现。Al：阿尔必阶；Ap：阿普特阶；Ba：巴通阶；Be：贝里阿斯阶；Br：巴列姆阶；Ca：卡洛维阶；Ha：欧特里沃阶；Ki：基默里奇阶；Ox：牛津阶；Ti：提塘阶；Va 代表凡兰吟阶。

日本中部石川县桑岛组正在出产一个重要的早白垩世脊椎动物群。桑岛组与邻近的岐阜县大黑谷组层位相同[2]。地层学、生物地层学和放射性测量的数据表明大

Formation that outcrops in neighbouring Gifu Prefecture[2]. Stratigraphic, biostratigraphic and radiometric data show that the Okurodani Formation is basal Cretaceous (Valanginian or Hauterivian) in age[3]. The Kuwajima Formation has yielded more than one hundred isolated teeth of a new genus of tritylodontid synapsid[4]. Before these discoveries, tritylodontids were thought to have become extinct sometime in the Middle or early Late Jurassic, as the youngest-known tritylodontid (*Bienotheroides*) was recovered from late Middle Jurassic deposits. This discovery supports the concept of an East Asian refugium, but other evidence suggests that different factors may have had an equally strong influence on faunal composition.

A theropod dinosaur referable to the unnamed clade Oviraptorosauria + Therizinosauroidea[5] has also been found in the tritylodontid locality. This clade is best known from the Late Cretaceous of mainland Asia, although several taxa referable to this clade are known from the late Early Cretaceous of Liaoning (*Beipiaosaurus*[6] and *Caudipteryx*[7]), and possibly from the Early Jurassic of Yunnan Province, China[8]. The Japanese material, consisting of a single manual ungual (Fig. 2) with a pronounced posterodorsal lip (a feature synapomorphic of this group of theropods[5]), is one of the earliest representatives of this group. The Itsuki Formation of Fukui Prefecture, a lateral equivalent of the Okurodani and Kuwajima Formations[2], has produced an isolated tyrannosaurid tooth, identifiable by its D-shaped cross-section—a synapomorphy of tyrannosaurids[9].

Fig. 2. Manual ungual of a theropod dinosaur from the Kuwajima Formation (Valanginian or Hauterivian) of Shiramine, Ishikawa Prefecture, Japan. Note the prominent lip posterodorsal to the articular surface of the ungual, a synapomorphy of the clade Oviraptorosauria + Therizinosauroidea[5]. Scale bar, 5 mm.

These Japanese discoveries, combined with the presence of late Early Cretaceous taxa in the Yixian Formation (such as the ornithischian dinosaur *Psittacosaurus*[10]), suggest that several dinosaur clades (such as tyrannosaurids and psittacosaurids) may have originated and diversified in eastern Asia while a number of other lineages (tritylodontid synapsids, compsognathid dinosaurs and "rhamphorhynchoid" pterosaurs) persisted in this region. Moreover, the presence of hypsilophodontid and iguanodontid ornithopod dinosaurs in the Japanese Early Cretaceous[11] suggests faunal connections with western Asia and Europe. The historical biogeography of this region appears to be much more complex than was thought previously.

Alternatively, the so-called relict taxa in eastern Asia may indicate that faunal turnover at the Jurassic–Cretaceous boundary was not as marked as has been suggested[12]. The presence of camptosaurid (*Camptosaurus*) and heterodontosaurid (*Echinodon*) ornithopods

黑谷组在白垩系基部（凡兰吟阶或者欧特里沃阶）[3]。桑岛组发现了一个下孔类的三列齿兽新属[4]的一百多颗独立的牙齿。在这些发现之前，由于最晚的三列齿兽（似卞氏兽）在中侏罗世晚期沉积物中发现，因此人们认为三列齿兽在中侏罗世或晚侏罗世早期的某个时间早已灭绝。尽管桑岛组这个发现支持亚洲东部残遗种庇护所的观点，但是其他证据表明不同的因素可能同样强烈地影响了动物群的组成。

一种未命名的兽脚亚目恐龙也在三列齿兽的产地发现，可能属于窃蛋龙类和镰刀龙类共同组成的一个未命名分支[5]。这个分支大部分属种主要发现自晚白垩世的亚洲大陆，只有少数几个分类单元在中国辽宁省的早白垩世晚期地层（北票龙[6]和尾羽龙[7]）以及可能在中国云南省的早侏罗世地层[8]被发现。日本发现的材料包括一个带有明显后背向的唇突（该兽脚类类群的共同衍征[5]）的单一手爪尖（图2），这是该类群最早的代表之一。在福井县五木村组（与大黑谷组和桑岛组层位相同[2]）发现了一颗单独的恐龙牙齿。通过 D 形横断面——暴龙的共同衍征[9]可以将其鉴定为暴龙牙齿。

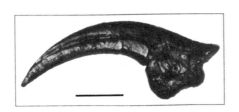

图 2. 日本石川县白峰村桑岛组（凡兰吟阶或欧特里沃阶）发现的兽脚亚目恐龙的手爪尖。注意爪尖关节面后背向的明显的唇突，此为窃蛋龙类 + 镰刀龙类分支[5]的共同衍征。比例尺为 5 毫米。

日本的这些发现，结合义县组早白垩世晚期存在的分类单元（如鸟臀目的鹦鹉嘴龙[10]），表明一些恐龙分支（例如暴龙科和鹦鹉嘴龙科）可能从亚洲东部起源并且发生分化，而许多其他谱系（下孔类的三列齿兽、美颌龙类恐龙和翼龙目"喙嘴翼龙类"）在该地区一直存在。此外，日本早白垩世的棱齿龙类和鸟脚亚目的禽龙类的发现[11]揭示在亚洲西部和欧洲之间存在动物群的联系。该地区的历史生物地理学似乎比之前预想的要复杂得多。

或者说，亚洲东部所谓的残遗分类单元可能表明侏罗纪–白垩纪边界的动物群转化不像人们猜测的那么明显[12]。在欧洲早白垩世动物群中存在的鸟脚亚目恐龙——弯龙类（弯龙属）和异齿龙类（棘齿龙属）[13]表现出与北美洲晚侏罗世莫里森组

in European Early Cretaceous faunas[13] indicates faunal similarities to the Late Jurassic Morrison Formation of North America. The presence of "Late Jurassic" taxa in eastern Asia may simply represent another example of this more gradual Jurassic–Cretaceous faunal transition (Fig. 1), although more evidence is needed to distinguish between these alternatives.

(**404**, 953; 2000)

Makoto Manabe[*], **Paul M. Barrett**[†], **Shinji Isaji**[‡]

[*] Department of Geology, National Science Museum, 3-23-1 Hyakunin-cho, Shinjuku-ku, Tokyo 169-0073, Japan

[†] Department of Zoology, University of Oxford, South Parks Road, Oxford OX1 3PS, UK e-mail: paul.barrett@zoo.ox.ac.uk

[‡] Chiba Prefectural Museum of Natural History, Chiba 260-0682, Japan

References:

1. Luo, Z. *Nature* **400**, 23-25 (1999).

2. Maeda, S. *J. Collect. Arts Sci. Chiba Univ.* **3**, 369-426 (1961).

3. Evans, S. E. *et al. New Mexico Mus. Nat. Hist. Sci. Bull.* **14**, 183-186 (1998).

4. Setoguchi, T., Matsuoka, H. & Matsuda, M. in *Proc. 7th Annu. Meet. Chinese Soc. Vert. Paleontol.* (eds Wang, Y.-Q. & Deng, T.) 117-124 (China Ocean, Beijing, 1999).

5. Makovicky, P. J. & Sues, H.-D. *Am. Mus. Novitates* **3240**, 1-27 (1998).

6. Xu, X., Tang, Z.-L. & Wang, X.-L. *Nature* **399**, 350-354 (1999).

7. Sereno, P. C. *Science* **284**, 2137-2147 (1999).

8. Zhao, X.-J. & Xu, X. *Nature* **394**, 234-235 (1998).

9. Manabe, M. *J. Paleontol.* **73**, 1176-1178 (1999).

10. Xu, X. & Wang, X.-L. *Vertebrata PalAsiatica* **36**, 147-158 (1998).

11. Manabe, M. & Hasegawa, Y. in *6th Symp. Mesozoic Terrest. Ecosyst. Biotas* (eds Sun, A.-L. & Wang, Y.-Q.) 179 (China Ocean, 1995).

12. Bakker, R. T. *Nature* **274**, 661-663 (1978).

13. Norman, D. B. & Barrett, P. M. *Spec. Pap. Palaeontol.* (in the press).

的动物群的相似性。尽管我们需要更多的证据去区分这两种可能性，但是亚洲东部"晚侏罗世"分类单元的存在可能仅仅代表侏罗纪向白垩纪的动物群逐渐演变的另外一个例证（图1）。

（吕静 张茜楠 翻译；汪筱林 审稿）

A Flat Universe from High-resolution Maps of the Cosmic Microwave Background Radiation

P. de Bernardis *et al.*

Editor's Note

In the early Universe, before subatomic particles united into atoms, photons, electrons and atomic nuclei were sloshing around in a kind of primordial soup. As with sloshing water in a bathtub, there were certain naturally resonant frequencies. Radiation from this time carries echoes of those resonances, preserved as characteristic ripples in the cosmic microwave background radiation. Paolo de Bernardis and his colleagues launched a high-altitude balloon in the Antarctic to measure these ripples. Here they report their findings: the ripples are best explained by a "flat" Universe, in which the sum of standard matter, dark matter and dark energy is equal to a "critical density", poised between positive (closed) and negative (open) curvature. Subsequently the WMAP satellite confirmed this conclusion.

The blackbody radiation left over from the Big Bang has been transformed by the expansion of the Universe into the nearly isotropic 2.73 K cosmic microwave background. Tiny inhomogeneities in the early Universe left their imprint on the microwave background in the form of small anisotropies in its temperature. These anisotropies contain information about basic cosmological parameters, particularly the total energy density and curvature of the Universe. Here we report the first images of resolved structure in the microwave background anisotropies over a significant part of the sky. Maps at four frequencies clearly distinguish the microwave background from foreground emission. We compute the angular power spectrum of the microwave background, and find a peak at Legendre multipole $l_{peak} = (197 \pm 6)$, with an amplitude $\Delta T_{200} = (69 \pm 8)$ μK. This is consistent with that expected for cold dark matter models in a flat (euclidean) Universe, as favoured by standard inflationary models.

PHOTONS in the early Universe were tightly coupled to ionized matter through Thomson scattering. This coupling ceased about 300,000 years after the Big Bang, when the Universe cooled sufficiently to form neutral hydrogen. Since then, the primordial photons have travelled freely through the Universe, redshifting to microwave frequencies as the Universe expanded. We observe those photons today as the cosmic microwave background (CMB). An image of the early Universe remains imprinted in the temperature anisotropy of the CMB. Anisotropies on angular scales larger than ~2° are dominated by the gravitational redshift the photons undergo as they leave the density fluctuations present at decoupling[1,2]. Anisotropies on smaller angular scales are enhanced

高分辨率的宇宙微波背景辐射观测揭示我们的宇宙是平直的

贝尔纳迪斯等

编者按

在宇宙早期，亚原子粒子结合成原子之前，光子、电子以及原子核在某种原始汤中游荡。类似于浴缸中受到扰动的水，其中自然而然存在一些共振的频率。这个时期的辐射携带了这些共振的回音，以典型涟漪的形式保留在宇宙微波背景辐射上。保罗·德·贝尔纳迪斯和他的合作者们在南极洲发射了一个高海拔的气球来探测这些涟漪。这里报道了他们的结果：涟漪能够通过"平直"宇宙模型得到最好的解释。在这个模型中，标准物质、暗物质以及暗能量的总密度等于"临界密度"，介于正的（封闭的）以及负的（开的）曲率之间。随后，WMAP 卫星证实了这个结论。

随着宇宙的膨胀，大爆炸遗留下来的黑体辐射逐步演化成为几乎各向同性、温度为 2.73 K 的宇宙微波背景。早期宇宙的不均匀性在微波背景上留下了微小的各向异性的痕迹。这些各向异性蕴含着宇宙学基本参数的信息，尤其是宇宙整体能量密度与曲率。在本工作中，我们首次展示了一块足够大的天区上微波背景各向异性结构的高分辨率图像。四个波段下的图像可以清楚地将微波背景辐射与前景辐射区分开。我们计算了微波背景的角功率谱，并在勒让德阶数 $l_{peak} = (197 \pm 6)$ 处发现了一个峰，其幅度为 $\Delta T_{200} = (69 \pm 8)$ μK。该结果同平直（欧几里得）宇宙下冷暗物质模型的理论预期相吻合，也被标准的暴胀模型所支持。

在宇宙早期，光子通过汤姆孙散射与电离物质紧密耦合在一起。这种耦合状态一直持续到大爆炸之后 300,000 年，这个时候宇宙冷却到足以形成中性氢。自那时候开始，原初光子就在宇宙中自由地传播，随着宇宙膨胀而红移到微波波段。这就是我们今天所观测到的宇宙微波背景（CMB）。而 CMB 中的温度各向异性则保留了早期宇宙的基本图像。角尺度大于约 2° 的各向异性信息由退耦时离开密度扰动的光子经历的引力红移所主导[1,2]。而更小角尺度上各向异性的程度因退耦前光子–重子

503

by oscillations of the photon–baryon fluid before decoupling[3]. These oscillations are driven by the primordial density fluctuations, and their nature depends on the matter content of the Universe.

In a spherical harmonic expansion of the CMB temperature field, the angular power spectrum specifies the contributions to the fluctuations on the sky coming from different multipoles l, each corresponding to the angular scale $\theta = \pi/l$. Density fluctuations over spatial scales comparable to the acoustic horizon at decoupling produce a peak in the angular power spectrum of the CMB, occurring at multipole l_{peak}. The exact value of l_{peak} depends on both the linear size of the acoustic horizon and on the angular diameter distance from the observer to decoupling. Both these quantities are sensitive to a number of cosmological parameters (see, for example, ref. 4), but l_{peak} primarily depends on the total density of the Universe, Ω_0. In models with a density Ω_0 near 1, $l_{peak} \approx 200/\Omega_0^{1/2}$. A precise measurement of l_{peak} can efficiently constrain the density and thus the curvature of the Universe.

Observations of CMB anisotropies require extremely sensitive and stable instruments. The DMR[5] instrument on the COBE satellite mapped the sky with an angular resolution of $\sim 7°$, yielding measurements of the angular power spectrum at multipoles $l < 20$. Since then, experiments with finer angular resolution[6-16] have detected CMB fluctuations on smaller scales and have produced evidence for the presence of a peak in the angular power spectrum at $l_{peak} \approx 200$.

Here we present high-resolution, high signal-to-noise maps of the CMB over a significant fraction of the sky, and derive the angular power spectrum of the CMB from $l = 50$ to 600. This power spectrum is dominated by a peak at multipole $l_{peak} = (197 \pm 6)$ (1σ error). The existence of this peak strongly supports inflationary models for the early Universe, and is consistent with a flat, euclidean Universe.

The Instrument

The Boomerang (balloon observations of millimetric extragalactic radiation and geomagnetics) experiment is a microwave telescope that is carried to an altitude of ~ 38 km by a balloon. Boomerang combines the high sensitivity and broad frequency coverage pioneered by an earlier generation of balloon-borne experiments with the long (~ 10 days) integration time available in a long-duration balloon flight over Antarctica. The data described here were obtained with a focal plane array of 16 bolometric detectors cooled to 0.3 K. Single-mode feedhorns provide two 18′ full-width at half-maximum (FWHM) beams at 90 GHz and two 10′ FWHM beams at 150 GHz. Four multi-band photometers each provide a 10.5′, 14′ and 13′ FWHM beam at 150, 240 and 400 GHz respectively. The average in-flight sensitivity to CMB anisotropies was 140, 170, 210 and 2,700 $\mu K\ s^{1/2}$ at 90, 150, 240 and 400 GHz, respectively. The entire optical system is heavily baffled against terrestrial radiation. Large sunshields improve rejection of radiation from $> 60°$ in azimuth from the telescope boresight. The rejection has been measured to be greater than

504

流体振荡而加剧[3]。这些振荡由原初密度扰动驱动，而它们的本质依赖于宇宙中的物质成分。

对 CMB 温度场进行球谐展开，其角功率谱表征了全天不同阶数 l（对应角尺度 $\theta = \pi/l$）处扰动的强度。与退耦时声学视界相比拟的空间尺度上的密度扰动在 CMB 的角功率谱上产生一个峰，对应阶数为 l_{peak}。l_{peak} 的精确取值依赖于声学视界的线性尺度以及观测者（现在）到退耦时期的角直径距离。虽然这些量与一系列宇宙学参数密切相关（比如参考文献 4），但是 l_{peak} 主要取决于宇宙整体密度 Ω_0。对于密度 Ω_0 接近 1 的宇宙学模型，$l_{peak} \approx 200/\Omega_0^{1/2}$。因此，对 l_{peak} 进行精确测量可有效地限制宇宙整体的密度，从而可以限制宇宙的曲率。

对 CMB 各向异性的观测需要极高分辨率和稳定性的观测仪器。COBE 卫星上的 DMR[5] 设备的巡天角分辨率约为 7°，这使得对 $l < 20$ 的角功率谱的测量成为可能。自此之后，具有更高角分辨率的实验[6-16]探测到更小尺度上的 CMB 扰动，并且得到的证据表明在角功率谱上 l_{peak} 约 200 处存在一个峰。

这里我们展示了一块足够大天区的 CMB 高分辨率高信噪比的图像，得到了 $l = 50$ 到 600 的 CMB 角功率谱。在这个范围内，功率谱在 $l_{peak} = 197 \pm 6$（1σ 误差范围）处存在一个峰。这个峰的存在极大地支持了早期宇宙的暴胀模型，与一个平直的欧几里得宇宙的理论预言相吻合。

观 测 设 备

Boomerang（毫米波段河外辐射和地磁场球载观测）实验利用球载微波望远镜在海拔约 38 千米的高空进行观测。Boomerang 在上一代球载实验开创的高分辨率和宽波段覆盖特性的基础上，更具在南极上空长时间的飞行中可用积分时间长（约 10 天）的能力。本工作所采用的数据来源于由 16 个冷却到 0.3 K 的热辐射探测器所组成的焦平面阵列。通过单模喇叭馈源，我们在 90 GHz 处得到了两个半峰全宽为 18 角分的波束，并在 150 GHz 处得到了两个半峰全宽为 10 角分的波束。而四个多波段光度计均在 150、240 以及 400 GHz 处得到了半峰全宽分别为 10.5、14 以及 13 角分的波束。CMB 各向异性在 90、150、240 以及 400 GHz 处运行中的平均灵敏度分别为 140、170、210 以及 2,700 $\mu K \cdot s^{1/2}$。整个光学系统可以很好地抵御地球辐射的影响。大型遮阳板有助于遮挡来自望远镜视轴方向且方位角大于 60° 的辐射。我们测

80 dB at all angles occupied by the Sun during the CMB observations. Further details on the instrument can be found in refs 17–21.

Observations

Boomerang was launched from McMurdo Station (Antarctica) on 29 December 1998, at 3:30 GMT. Observations began 3 hours later, and continued uninterrupted during the 259-hour flight. The payload approximately followed the 79° S parallel at an altitude that varied daily between 37 and 38.5 km, returning within 50 km of the launch site.

We concentrated our observations on a target region, centred at roughly right ascension (RA) 5 h, declination (dec.) −45°, that is uniquely free of contamination by thermal emission from interstellar dust[22] and that is approximately opposite the Sun during the austral summer. We mapped this region by repeatedly scanning the telescope through 60° at fixed elevation and at constant speed. Two scan speeds (1° s^{-1} and 2° s^{-1} in azimuth) were used to facilitate tests for systematic effects. As the telescope scanned, degree-scale variations in the CMB generated sub-audio frequency signals in the output of the detector[23]. The stability of the detector system was sufficient to allow sensitive measurements on angular scales up to tens of degrees on the sky. The scan speed was sufficiently rapid with respect to sky rotation that identical structures were observed by detectors in the same row in each scan. Detectors in different rows observed the same structures delayed in time by a few minutes.

At intervals of several hours, the telescope elevation was interchanged between 40°, 45° and 50° in order to increase the sky coverage and to provide further systematic tests. Sky rotation caused the scan centre to move and the scan direction to rotate on the celestial sphere. A map from a single day at a single elevation covered roughly 22° in declination and contained scans rotated by ±11° on the sky, providing a cross-linked scan pattern. Over most of the region mapped, each sky pixel was observed many times on different days, both at 1° s^{-1} and 2° s^{-1} scan speed, with different topography, solar elongation and atmospheric conditions, allowing strong tests for any contaminating signal not fixed on the celestial sphere.

The pointing of the telescope has been reconstructed with an accuracy of 2′ r.m.s. using data from a Sun sensor and rate gyros. This precision has been confirmed by analysing the observed positions of bright compact H II regions in the Galactic plane (RCW38[24], RCW57, IRAS08576 and IRAS1022) and of radio-bright point sources visible in the target region (the QSO 0483–436, the BL Lac object 0521–365 and the blazar 0537–441).

量发现，在整个 CMB 观测中各角度的太阳光均被遮挡了 80 dB 以上。更多关于设备的细节请参看参考文献 17～21。

观　　测

格林尼治标准时间 1998 年 12 月 29 日 3 时 30 分，Boomerang 从南极洲麦克默多站发射。观测始于 3 小时后，并在 259 小时的飞行过程中保持不间断。探测器每天大致在离发射场 50 千米范围内沿着南纬 79°飞行，海拔高度在 37～38.5 千米范围内。

我们将观测集中在以赤经 5 h、赤纬 −45°为中心的目标区域内。这块区域非常独特，不仅可以免受星际尘埃热辐射的污染[22]，且在整个南半球夏季都会背对着太阳。我们利用望远镜以 60°倾角和恒定的速度进行重复扫描来对该区域成图。为帮助测试其中的系统误差，我们使用了两种扫描速度（每秒方位角 1°和 2°）。随着望远镜对天区进行扫描，CMB 的度尺度上的变化在探测器输出中产生了亚音频信号[23]。整个探测器系统的稳定性很好，使得对天区进行大到数十度角尺度的高灵敏度的测量成为可能。所采用的扫描速度相对于天空旋转而言是足够快的，这样在每次扫描过程中同一行的探测器观测到的是相同的结构。而不同行的探测器则会在延迟几分钟之后也观测到这些相同的结构。

为了增加天区覆盖并且进行更多系统误差的测试，望远镜的倾角每隔几小时会在 40°、45°以及 50°之间交替变化。而天空的旋转不仅会使扫描中心发生移动，同时还会使扫描方向在天球上发生旋转。同一天同一倾角的图像大致覆盖赤纬 22°的范围，包含天空中旋转了 ±11°的扫描模式，这就形成了交联的扫描模式。对于大部分成图的区域而言，每个天空格点都会在不同时间里以两种扫描速度（每秒 1°和 2°）在不同地形、太阳距角以及大气条件下被观测多次，这就使得我们可以对天球上各种可能的污染信号进行强有力的测试。

利用太阳传感器和速率陀螺的数据，望远镜的指向已进行重构，其均方根精度为 2 角分。通过对银道面上明亮的致密 H ⅱ 区（RCW38[24]、RCW57、IRAS08576 以及 IRAS1022）以及目标天区可见的射电明亮点源（类星体 QSO 0483−436、BL Lac 天体 0521−365 以及耀变体 0537−441）观测位置的分析，我们进一步确认了望远镜的指向精度。

Calibrations

The beam pattern for each detector was mapped before flight using a thermal source. The main lobe at 90, 150 and 400 GHz is accurately modelled by a gaussian function. The 240 GHz beams are well modelled by a combination of two gaussians. The beams have small shoulders (less than 1% of the total solid angle), due to aberrations in the optical system. The beam-widths were confirmed in flight via observations of compact sources. By fitting radial profiles to these sources we determine the effective angular resolution, which includes the physical beamwidth and the effects of the 2′ r.m.s. pointing jitter. The effective FWHM angular resolution of the 150 GHz data that we use here to calculate the CMB power spectrum is $(10 \pm 1)'$, where the error is dominated by uncertainty in the pointing jitter.

We calibrated the 90, 150 and 240 GHz channels from their measured response to the CMB dipole. The dipole anisotropy has been accurately (0.7%) measured by COBE-DMR[25], fills the beam and has the same spectrum as the CMB anisotropies at smaller angular scales, making it the ideal calibrator for CMB experiments. The dipole signal is typically ~3 mK peak-to-peak in each 60° scan, much larger than the detector noise, and appears in the output of the detectors at $f = 0.008$ Hz and $f = 0.016$ Hz in the 1° s^{-1} and 2° s^{-1} scan speeds, respectively. The accuracy of the calibration is dominated by two systematic effects: uncertainties in the low-frequency transfer function of the electronics, and low-frequency, scan-synchronous signals. Each of these is significantly different at the two scan speeds. We found that the dipole-fitted amplitudes derived from separate analysis of the 1° s^{-1} and 2° s^{-1} data agree to within $\pm 10\%$ for every channel, and thus we assign a 10% uncertainty in the absolute calibration.

From Detector Signals to CMB Maps

The time-ordered data comprises 5.4×10^{7} 16-bit samples for each channel. These data are flagged for cosmic-ray events, elevation changes, focal-plane temperature instabilities, and electromagnetic interference events. In general, about 5% of the data for each channel are flagged and not used in the subsequent analysis. The gaps resulting from this editing are filled with a constrained realization of noise in order to minimize their effect in the subsequent filtering of the data. The data are deconvolved by the bolometer and electronics transfer functions to recover uniform gain at all frequencies.

The noise power spectrum of the data and the maximum-likelihood maps[26-28] were calculated using an iterative technique[29] that separates the sky signal from the noise in the time-ordered data. In this process, the statistical weights of frequencies corresponding to angular scales larger than 10° on the sky are set to zero to filter out the largest-scale modes of the map. The maps were pixelized according to the HEALPix pixelization scheme[30].

校　　准

在进行飞行试验前我们利用一个热源对每一个探测器的波束模式进行了成像。在 90、150 以及 400 GHz 的主瓣可以精确地用高斯函数进行拟合，而在 240 GHz 的波束则可以用两个高斯函数进行叠加。因为光学系统存在光行差，波束存在小的肩状分布（小于整个立体角的 1%）。在整个飞行过程中通过对致密源进行观测进一步确认了波束宽度。通过对这些源的径向轮廓进行拟合，我们得到了有效的角分辨率，这其中不仅包含其本身波束宽度的贡献，还包含 2 角分均方根指向抖动带来的影响。本工作中我们计算 CMB 功率谱所采用的 150 GHz 数据的有效半峰全宽角分辨率为 (10 ± 1) 角分，其误差主要来源于指向抖动的不确定性。

我们通过测量 90、150 以及 240 GHz 波段对 CMB 偶极子的响应对相应波段的观测进行校准。偶极子各向异性充满整个波束，并且在更小的角尺度上与 CMB 各向异性拥有相同的谱结构，而之前 COBE-DMR 已经对偶极子的各向异性进行了准确的测量（0.7%）[25]，因此偶极子各向异性是校准 CMB 实验观测的理想工具。每次 60° 扫描的典型偶极子信号总幅度大约为 3 mK，比探测器噪声大很多，且在每秒 1° 以及 2° 的扫描速度下分别在探测器输出的 $f = 0.008$ Hz 以及 $f = 0.016$ Hz 处出现。校准的准确性主要受到两个系统误差的影响：电子设备低频转移函数的不确定性以及低频扫描同步信号的不确定性。我们发现，对不同波段以及不同扫描速度（每秒 1° 以及每秒 2°）下的数据单独分析得到的偶极子拟合幅度在 ±10% 范围内相互吻合。因此我们绝对校准的不确定度定为 10%。

从探测器信号到 CMB 图像

每个通道的时间序列数据包含 5.4×10^7 个 16 位样本。我们将受到宇宙线事件、倾角变化、焦平面温度不稳定性和电磁干扰事件影响的数据进行了标记。整体来说，每个通道大约有 5% 的数据被标记，这些数据不用于后续分析。为了降低后续数据平滑过滤过程中上述操作所带来的影响，我们利用一个约束噪声来填补这个过程所造成的数据缺失。随后，通过测辐射热计和电子设备转移函数对数据进行解卷积，我们重新得到了在所有频段具有均匀增益的数据样本。

通过一个可以在时间序列数据中将天空信号和噪声分离的迭代技术[29]，我们计算得到了数据的噪声功率谱以及最大似然图像[26-28]。在这个过程中，我们将对应天空角尺度大于 10° 的频率的统计权重设为 0，以此来滤除图像中的最大尺度模式。整个图像根据 HEALPix 像素化方法进行像素化[30]。

Figure 1 shows the maps obtained in this way at each of the four frequencies. The 400 GHz map is dominated by emission from interstellar dust that is well correlated with that observed by the IRAS and COBE/DIRBE satellites. The 90, 150 and 240 GHz maps are dominated by degree-scale structures that are resolved with high signal-to-noise ratio. A qualitative but powerful test of the hypothesis that these structures are CMB anisotropy is provided by subtracting one map from another. The structures evident in all three maps disappear in both the $90-150$ GHz difference and in the $240-150$ GHz difference, as expected for emission that has the same spectrum as the CMB dipole anisotropy used to calibrate the maps.

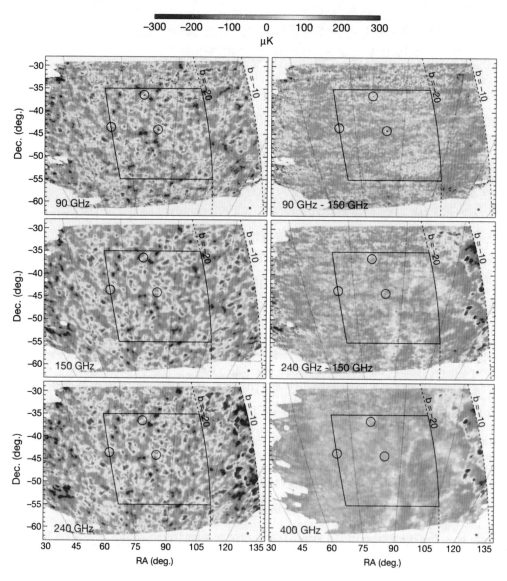

Fig. 1. Boomerang sky maps (equatorial coordinates). The sky maps at 90, 150 and 240 GHz (left panels) are shown with a common colour scale, using a thermodynamic temperature scale chosen such that CMB anisotropies will have the same amplitude in the three maps. Only the colour scale of the 400 GHz map

图 1 中显示了利用上述方法得到的四个频段的图像。400 GHz 的图像由星际尘埃的辐射主导，与 IRAS 和 COBE/DIRBE 卫星的观测结果存在很好的相关性。而 90、150 以及 240 GHz 的图像则由高信噪比的度尺度结构所主导。假设这些结构是 CMB 各向异性的，一个定性但有效检验这一假设的方法是将这些图像相互去减。那些在三个图像中均明显存在的结构在 90 – 150 GHz 的残差图以及 240 – 150 GHz 的残差图中都消失了，这从辐射中可以预见出，这些辐射与用来校准图像的 CMB 偶极子各向异性具有相同谱结构。

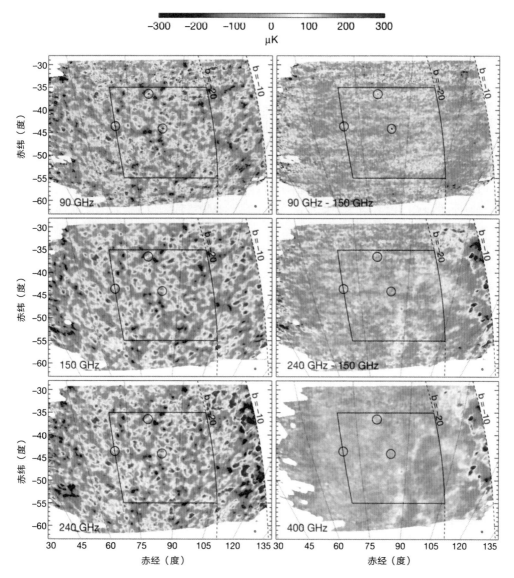

图 1. Boomerang 天图（赤道坐标）。左图中 90、150 以及 240 GHz 的天图具有相同的颜色显示标度，均采用相同的热力学温标，这样在三幅天图中 CMB 各向异性具有相同的幅度。只有右下角的 400 GHz 天

(bottom right) is 14 times larger than the others: this has been done to facilitate comparison of emission from interstellar dust (ISD), which dominates this map, with ISD emission present in the lower-frequency maps. The maps at 90 and 400 GHz are each from a single detector, while maps at 150 and 240 GHz have each been obtained by co-adding data from three detectors. For purposes of presentation, the maps have been smoothed with gaussian filters to obtain FWHM effective resolution of 22.5' (small circle in the bottom right side of each panel). Structures along the scan direction larger than 10° are not present in the maps. Several features are immediately evident. Most strikingly, the maps at 90, 150 and 240 GHz are dominated by degree-scale structures that fill the map, have well-correlated morphology and are identical in amplitude in all three maps. These structures are not visible at 400 GHz. The 400 GHz map is dominated by diffuse emission which is correlated with the ISD emission mapped by IRAS/DIRBE[22]. This emission is strongly concentrated towards the right-hand edge of the maps, near the plane of the Galaxy. The same structures are evident in the 90, 150 and 240 GHz maps at Galactic latitude $b > -15°$, albeit with an amplitude that decreases steeply with decreasing frequency. The large-scale gradient evident especially near the right edge of the 240 GHz map is a result of high-pass-filtering the very large signals near the Galactic plane (not shown). This effect is negligible in the rest of the map. The two top right panels show maps constructed by differencing the 150 and 90 GHz maps and the 240 and 150 GHz maps. The difference maps contain none of the structures that dominate the maps at 90, 150 and 240 GHz, indicating that these structures do indeed have the ratios of brightness that are unique to the CMB. The morphology of the residual structures in the 240–150 GHz map is well-correlated with the 400 GHz map, as is expected if the residuals are due to the ISD emission. Three compact sources of emission are visible in the lower-frequency maps, as indicated by the circles. These are known radio-bright quasars from the SEST pointing catalogue at 230 GHz. The boxed area has been used for computing the angular power spectrum shown in Fig. 2.

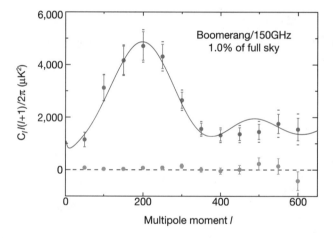

Fig. 2. Angular power spectrum measured by Boomerang at 150 GHz. Each point is the power averaged over $\Delta l = 50$ and has negligible correlations with the adjacent points. The error bars indicate the uncertainty due to noise and cosmic/sampling variance. The errors are dominated by cosmic/sampling variance at $l < 350$; they grow at large l due to the signal attenuation caused by the combined effects[39] of the 10' beam and the 14' pixelization (0.87 at $l = 200$ and 0.33 at $l = 600$). The current $\pm 10\%$ uncertainty in the calibration corresponds to an overall re-scaling of the y-axis by $\pm 20\%$, and is not shown. The current 1' uncertainty in the angular resolution of the measurement creates an additional uncertainty—indicated by the distance between the ends of the red error bars and the blue horizontal lines—that is completely correlated and is largest (11%) at $l = 600$. The green points show the power spectrum of a difference map obtained as follows. We divided the data into two parts corresponding to the first and second halves of the timestream. We made two maps (A and B) from these halves, and the green points show the power spectrum computed from the difference map, $(A-B)/2$. Signals originating from the sky should disappear in this map, so this is a test for contamination in the data (see text). The solid curve has parameters $(\Omega_b, \Omega_m, \Omega_\Lambda, n_s, h) = (0.05, 0.31, 0.75, 0.95, 0.70)$. It is the best-fit model for the Boomerang test flight data[15,16], and is shown for comparison only. The model that best fits the new data reported here will be presented elsewhere.

图的颜色标度是其他图像的 14 倍，这么做的原因是为了方便将此天图与其他更低频段天图的星际尘埃辐射进行对比。对 400 GHz 天图而言，星际尘埃的辐射占主导。90 和 400 GHz 的天图的数据各由单独探测器获得，而 150 和 240 GHz 的天图则是将三个探测器的数据进行叠加之后所得。为了便于展示，所有天图均经过高斯滤波平滑处理，使得有效半峰全宽分辨率为 22.5 角分（每幅图右下角的小圆圈的大小）。沿扫描方向大于 10° 的结构没有显示在天图中。图中有几个特征非常明显。其中最为显著的是，90、150 以及 240 GHz 的天图由充满天图的度尺度结构主导，在三幅天图中它们的形态存在很好的相关性，且幅度完全相同。然而在 400 GHz 的天图中却看不到这些结构。400 GHz 天图由弥散的辐射所主导，与 IRAS/DIRBE[22] 观测得到的星际尘埃辐射存在很好的相关性。该辐射主要集中在天图的右边缘，靠近银道面的区域。在 90、150 以及 240 GHz 的天图对应银纬 b > −15° 的区域中，这些结构同样显著，尽管随着频率的下降其幅度急剧减小。在 240 GHz 的天图中，特别是右边缘附近存在较为明显的大尺度梯度，这是由对银道面（未显示）附近非常高的信号进行高通滤波所造成的结果。这个效应对天图其他的区域所造成的影响可以忽略不计。右上的两幅图分别显示了 150 GHz 天图同 90 GHz 天图相减以及 240 GHz 同 150 GHz 天图相减所得的残差图。这些残差图中并不存在那些在 90、150 以及 240 GHz 天图中占主导的结构，这意味着这些结构确实具有 CMB 所特有的亮度比例。而 240−150 GHz 残差图中剩余结构的形态分布同 400 GHz 天图存在很好的相关性，这与我们假设残差主要源自星际尘埃辐射所得到的理论预期相符。在低频段的天图中，三个致密的辐射源清晰可见，具体位置已用圆圈标记。它们均为 230 GHz 的 SEST 星表中已知的射电明亮类星体。方框内的区域被用来计算图 2 中显示的角功率谱。

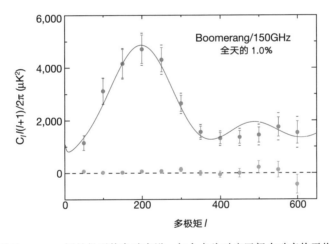

图 2. 150 GHz 处 Boomerang 测量得到的角功率谱。每个点为对应区间内功率的平均值，区间宽度 Δl = 50，相邻两点间的相关性可以忽略不计。误差棒反映了因噪声以及宇宙/样本方差所造成的不确定性。对于 l < 350 的数据点而言，其误差主要来自宇宙/样本方差的不确定性；由于将 10 角分的波束同 14 角分的格点相结合导致的信号衰减[39]（l = 200 时为 0.87，而 l = 600 时为 0.33），误差会随着 l 的增加而变大。目前校准过程带来的 ±10% 的不确定性对应 y 轴整体改变 ±20%，这在本图中并未显示。而目前测量中角分辨率中存在的 1 角分的不确定性还会导致一个额外的不确定因素（表现在红色误差棒端与蓝色水平线之间的距离），这些不确定因素完全相关，且在 l = 600 处最大（可达 11%）。绿色的点显示了用下面方法得到的残差图的功率谱。我们将数据按时间流分为两半，分别独立成图（A 和 B），而绿色的点显示了残差图 (A−B)/2 的功率谱。在这幅残差图中所有来自天空的信号都将消失，因此这是检验数据中污染源的一个测试。实线对应的宇宙学参数 (Ω_b, Ω_m, Ω_Λ, n_s, h) = (0.05, 0.31, 0.75, 0.95, 0.70)。这是与 Boomerang 测试飞行数据[15,16]最相符的拟合参数，在此显示只作比较之用。与本工作新数据最相符的宇宙学模型将在别处讨论。

To quantify this conclusion, we performed a "colour index" analysis of our data. We selected the ~18,000 14′ pixels at Galactic latitude $b < -15°$, and made scatter plots of 90 GHz versus 150 GHz and 240 GHz versus 150 GHz. A linear fit to these scatter plots gives slopes of 1.00 ± 0.15 and 1.10 ± 0.16, respectively (including our present 10% calibration error), consistent with a CMB spectrum. For comparison, free–free emission with spectral index -2.35 would produce slopes of 2.3 and 0.85, and was therefore rejected with $> 99\%$ confidence; emission from interstellar dust with temperature $T_d = 15$ K and spectral index of emissivity $\alpha = 1$ would produce slopes of 0.40 and 2.9. For any combination of $T_d > 7$ K and $1 < \alpha < 2$, the dust hypothesis is rejected with $> 99\%$ confidence. We conclude that the dominant source of structure that we detect at 90, 150 and 240 GHz is CMB anisotropy.

We further argue that the 150 GHz map at $b < -15°$ is free of significant contamination by any known astrophysical foreground. Galactic synchrotron and free–free emission is negligible at this frequency[31]. Contamination from extragalactic point sources is also small[32]; extrapolation of fluxes from the PMN survey[33] limits the contribution by point sources (including the three above-mentioned radio-bright sources) to the angular power spectrum derived below to $< 0.7\%$ at $l = 200$ and $< 20\%$ at $l = 600$. The astrophysical foreground that is expected to dominate at 150 GHz is thermal emission from interstellar dust. We placed a quantitative limit on this source of contamination as follows. We assumed that dust properties are similar at high ($b < -20°$) and moderate ($-20° < b < -5°$) Galactic latitudes. We selected the pixels at moderate Galactic latitudes and correlated the structure observed in each of our four bands with the IRAS/DIRBE map, which is dominated by dust in cirrus clouds. The best-fit slope of each of the scatter plots measures the ratios of the dust signal in the Boomerang channels to the dust signal in the IRAS/DIRBE map. We found that the 400 GHz map is very well correlated to the IRAS/DIRBE map, and that dust at $b < -20°$ can account for at most 10% of the signal variance at 240 GHz, 3% at 150 GHz and 0.5% at 90 GHz.

Angular Power Spectra

We compared the angular power spectrum of the structures evident in Fig. 1 with theoretical predictions. In doing so, we separated and removed the power due to statistical noise and systematic artefacts from the power due to the CMB anisotropies in the maps. The maximum-likelihood angular power spectrum of the maps was computed using the MADCAP[34] software package, whose algorithms fully take into account receiver noise and filtering.

Full analysis of our entire data set is under way. Because of the computational intensity of this process, we report here the results of a complete analysis of a limited portion of the data chosen as follows. We analysed the most sensitive of the 150 GHz detectors. We restricted the sky coverage to an area with RA $> 70°$, $b < -20°$ and $-55° <$ dec. $< -35°$, and we used only the ~50% of the data from this detector that was obtained at a scan speed of $1° \text{ s}^{-1}$. We used a relatively coarse pixelization of 8,000 14-arcmin pixels as a compromise

514

为了将这个结论定量化，我们对数据进行了"色指数"分析。我们挑选了银纬 $b < -15°$ 的大约 18,000 个 14 角分的格点，并对其作 90 GHz 相对于 150 GHz 以及 240 GHz 相对于 150 GHz 的散点图。对这些散点图进行线性拟合，其斜率分别为 1.00 ± 0.15 以及 1.10 ± 0.16（其中包含我们之前提及的 10% 的校准误差），与 CMB 谱结果相吻合。作为比较，谱指数为 -2.35 的自由–自由辐射产生的斜率分别为 2.3 和 0.85，因此以 $>99\%$ 的置信度被排除；温度为 $T_d = 15$ K、发射率谱指数为 $\alpha = 1$ 的星际尘埃辐射产生的斜率分别为 0.40 和 2.9。满足 $T_d > 7$ K 以及 $1 < \alpha < 2$ 范围中的尘埃辐射都会以 $>99\%$ 的置信度被排除。因此我们得出结论：我们在 90、150 以及 240 GHz 图像上探测到的结构，其主要来源为 CMB 各向异性。

进一步，我们认为银纬 $b < -15°$ 的 150 GHz 图像几乎免受任何已知前景天体的严重污染。银河系同步辐射以及自由–自由辐射在这个频段可以忽略不计[31]。河外点源的污染也很小[32]；对 PMN 巡天[33] 的光通量进行外推，可以限制点源（包括三个上面提及的射电明亮天体）对角功率谱的贡献：$l = 200$ 处低至 $<0.7\%$，而 $l = 600$ 处低至 $<20\%$。在 150 GHz 频段，我们预测最主要的前景天体应该是星际尘埃的热辐射。我们采用如下方法对这一污染源进行定量化限制。首先我们假设中（$-20° < b < -5°$）、高（$b < -20°$）银纬的尘埃性质类似。我们选取中等银纬的天区图像格点，将我们四个波段所观测到的由卷云尘埃主导的结构同 IRAS/DIRBE 图像进行相关分析。这几个散点图的最佳拟合斜率表征了 Boomerang 各通道测量得到的尘埃信号与 IRAS/DIRBE 图像中的尘埃信号的比值。我们发现 400 GHz 图像同 IRAS/DIRBE 图像存在很好的相关性，同时银纬 $b < -20°$ 的尘埃分别至多可以贡献 240 GHz、150 GHz 以及 90 GHz 图像中 10%、3% 以及 0.5% 的信号变化。

角 功 率 谱

我们将图 1 中明显结构的角功率谱同理论预言进行比较。在这个过程中，我们从图像中将统计噪声以及系统误差的贡献和 CMB 各向异性的信号分离并去除。利用 MADCAP[34] 软件包计算了这些图像的最大似然角功率谱，该软件的算法充分考虑了接收器噪声以及滤波过程所带来的影响。

我们正在对整个数据进行详尽的分析。由于整个过程的计算量很大，这里我们只展示如下所选的有限的一部分数据的完整分析结果。我们选择分析最为敏感的 150 GHz 探测器所得到的数据。我们将天区范围限制在赤经 $>70°$，$b < -20°$ 以及 $-55° <$ 赤纬 $< -35°$ 的区域内，并只分析该探测器扫描速度为每秒 $1°$ 所得到的近一半的数据。作为计算速度与高的多级覆盖的折中选择，我们使用一套相对粗糙且包含 8,000 个 14 角分的格点。最后，我们将分析限制在 $l \leqslant 600$ 的范围内，因为在该

between computation speed and coverage of high multipoles. Finally, we limited our analysis to $l \lesssim 600$ for which the effects of pixel shape and size and our present uncertainty in the beam size $(1')$ are small and can be accurately modelled.

The angular power spectrum determined in this way is shown in Fig. 2 and reported in Table 1. The power spectrum is dominated by a peak at $l_{peak} \approx 200$, as predicted by inflationary cold dark matter models. These models additionally predict the presence of secondary peaks. The data at high l limit the amplitude, but do not exclude the presence of a secondary peak. The errors in the angular power spectrum are dominated at low multipoles ($l \lesssim 350$) by the cosmic/sampling variance, and at higher multipoles by detector noise.

Table 1. Angular power spectrum of CMB anisotropy

l range	150 GHz ([1st half] + [2nd half])/2	150 GHz ([1st half] − [2nd half])/2
26–75	$1,140 \pm 280$	63 ± 32
76–125	$3,110 \pm 490$	16 ± 20
126–175	$4,160 \pm 540$	17 ± 28
176–225	$4,700 \pm 540$	59 ± 44
226–275	$4,300 \pm 460$	68 ± 59
276–325	$2,640 \pm 310$	130 ± 82
326–375	$1,550 \pm 220$	-7 ± 92
376–425	$1,310 \pm 220$	-60 ± 120
426–475	$1,360 \pm 250$	0 ± 160
476–525	$1,440 \pm 290$	220 ± 230
526–575	$1,750 \pm 370$	130 ± 300
576–625	$1,540 \pm 430$	-430 ± 360

Shown are measurements of the angular power spectrum of the cosmic microwave background at 150 GHz, and tests for systematic effects. The values listed are for $\Delta T_l^2 = l(l+1)c_l/2\pi$, in μK^2. Here $c_l = \langle a_{lm}^2 \rangle$, and a_{lm} are the coefficients of the spherical harmonic decomposition of the CMB temperature field: $\Delta T(\theta, \phi) = \Sigma a_{lm} Y_{lm} (\theta, \phi)$. The stated 1σ errors include statistical and cosmic/sample variance, and do not include a 10% calibration uncertainty.

The CMB angular power spectrum shown in Fig. 2 was derive from 4.1 days of observation. As a test of the stability of the result, we made independent maps from the first and second halves of these data. The payload travels several hundred kilometres, and the Sun moves $2°$ on the sky, between these maps. Comparing them provides a stringent test for contamination from sidelobe pickup and thermal effects. The angular power spectrum calculated for the difference map is shown in Fig. 2. The reduced χ^2 of this power spectrum with respect to zero signal is 1.11 (12 degrees of freedom), indicating that the difference map is consistent with zero contamination.

范围内格点形状、大小以及目前波束尺寸（1 角分）的不确定性所带来的影响很小而且能够利用模型准确进行描述。

图 2 及表 1 中展示了用这种方法得到的角功率谱结果。该功率谱在 $l_{peak} \approx 200$ 处存在一个主峰，这与冷暗物质暴胀模型的预言相符。这些模型还预言了次峰的存在。高 l 的数据的幅度受到了限制，但并不排除次峰存在的可能性。角功率谱中低的多级区域（$l \leqslant 350$）的误差主要来自宇宙/样本方差，而高的多级区域的误差则主要来自探测器噪声。

表 1. CMB 各向异性的角功率谱

l 范围	150 GHz（[前一半]+[后一半]）/2	150 GHz（[前一半]−[后一半]）/2
26~75	1,140±280	63±32
76~125	3,110±490	16±20
126~175	4,160±540	17±28
176~225	4,700±540	59±44
226~275	4,300±460	68±59
276~325	2,640±310	130±82
326~375	1,550±220	−7±92
376~425	1,310±220	−60±120
426~475	1,360±250	0±160
476~525	1,440±290	220±230
526~575	1,750±370	130±300
576~625	1,540±430	−430±360

这里显示的是 150 GHz 处测量得到的宇宙微波背景的角功率谱以及系统误差的测试结果。这里显示的值对应 $\Delta T_l^2 = l(l+1) c_l/2\pi$，单位为 μK^2。这里 $c_l = \langle a_{l_m}^2 \rangle$，$a_{l_m}$ 为 CMB 温度场的球谐分解系数：$\Delta T(\theta, \phi) = \Sigma a_{l_m} Y_{l_m}(\theta, \phi)$。所展示的 1σ 误差包含统计误差以及宇宙/样本方差，但没有包含 10% 的校准不确定性。

图 2 中显示的 CMB 角功率谱是基于 4.1 天的观测数据分析所得。为了测试结果的稳定性，我们将所有数据按时间分为两半，分别独立成图。在这两套图像之间，观测设备移动了数百千米，而太阳在天空中移动了 2°。通过比较这些图像可以对旁瓣污染以及热效应等污染因素进行严格的测试。图 2 还展示了两套图像残差结果的功率谱。该功率谱相对于零信号的约化 χ^2 为 1.11（12 个自由度），这个结果表明残差结果与零污染结果相吻合。

A Peak at $l \approx 200$ Implies a Flat Universe

The location of the first peak in the angular power spectrum of the CMB is well measured by the Boomerang data set. From a parabolic fit to the data at $l = 50$ to 300 in the angular power spectrum, we find $l_{peak} = (197 \pm 6)$ (1σ error). The parabolic fit does not bias the determination of the peak multipole: applying this method to Monte Carlo realizations of theoretical power spectra we recover the correct peak location for a variety of cosmological models. Finally, the peak location is independent of the details of the data calibration, which obviously affect only the height of the peak and not its location. The height of the peak is $\Delta T_{200} = (69 \pm 4) \pm 7$ μK (1σ statistical and calibration errors, respectively).

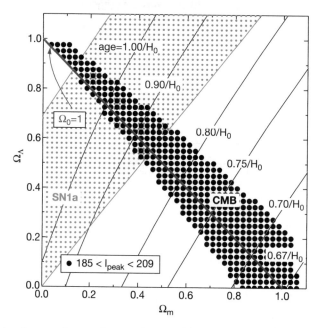

Fig. 3. Observational constraints on Ω_m and Ω_Λ. All the cosmological models (from our data base) consistent with the position of the peak in the angular power spectrum measured by Boomerang (95% confidence intervals) define an "allowed" region in the Ω_m–Ω_Λ plane (marked by large black dots). Such a region is elongated around the $\Omega_0 = 1$ line identifying a flat geometry, euclidean Universe. The blue lines define the age of the Universe for the considered models. The green-dotted region is consistent (95% confidence contour) with the recent results of the high-redshift supernovae surveys[40,41].

The data are inconsistent with current models based on topological defects (see, for example, ref. 35) but are consistent with a subset of cold dark matter models. We generated a database of cold dark matter models[36,37], varying six cosmological parameters (the range of variation is given in parentheses): the non-relativistic matter density, Ω_m (0.05–2); the cosmological constant, Ω_Λ (0–1); the Hubble constant, h (0.5–0.8); the baryon density, $h^2\Omega_b$ (0.013–0.025); the primordial scalar spectral index, n_s (0.8–1.3); and the overall normalization A (free parameter) of the primordial density fluctuation power spectrum. We compared these models with the power spectrum we report here to place constraints

$l \approx 200$ 的峰意味着一个平直宇宙

通过 Boomerang 数据，我们很好地测量到 CMB 角功率谱第一个峰的位置。通过对角功率谱 $l = 50 \sim 300$ 范围内的数据进行抛物线拟合，我们发现 $l_{peak} = (197 \pm 6)$（1σ 误差范围）。这种抛物线拟合不会对多级峰值的测量带来偏差：将这种方法应用于理论功率谱的蒙特卡洛实现，我们恢复出一系列宇宙学模型正确的峰值位置。最后可知，峰值位置不受数据校准细节的影响，因为数据校准细节仅仅影响峰的高度而并不影响它的位置。峰的高度为 $\Delta T_{200} = (69 \pm 4) \pm 7~\mu K$（分别对应 1σ 的统计误差和校准误差）。

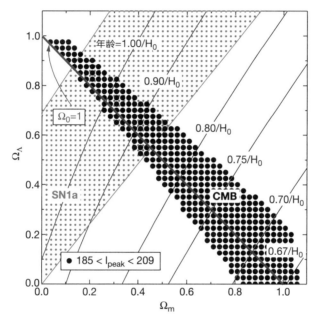

图 3. 对 Ω_m 和 Ω_Λ 的观测限制。所有与 Boomerang 测量得到的角功率谱的峰位置一致（95% 置信区间）的宇宙学模型（来自我们模型库）在 Ω_m-Ω_Λ 平面定义了一个"允许"区域（用大的黑色点进行标记）。这样的一个区域沿 $\Omega_0 = 1$ 的线进行延伸，这意味着宇宙是平直的，符合欧几里得几何。蓝色的线表征了所考虑的模型的宇宙年龄。绿色点的区域在 95% 置信度上符合最近高红移超新星巡天观测结果[40,41]。

这些数据与目前基于拓扑缺陷所建立的模型（比如参考文献 35）预言不吻合，但同冷暗物质模型的一个子集的理论预言相吻合。我们通过改变六个宇宙学参数构建了冷暗物质模型的模型库[36,37]（括号中显示参数的变化范围）：非相对论性的物质密度 Ω_m（$0.05 \sim 2$）、宇宙学常数 Ω_Λ（$0 \sim 1$）、哈勃常数 h（$0.5 \sim 0.8$）、重子密度 $h^2\Omega_b$（$0.013 \sim 0.025$）、原初标量谱指数 n_s（$0.8 \sim 1.3$）以及原初密度扰动功率谱的归一化因子 A（自由参数）。通过将这些模型同我们得到的功率谱进行比较，我们可以在这个 6 参数空间中对所允许的区域进行限制。图 3 中我们用大的黑色圆点标记了 Ω_m-Ω_Λ

on allowed regions in this 6-parameter space. In Figure 3 we mark with large black dots the region of the Ω_m–Ω_Λ plane where some combination of the remaining four parameters within the ranges defined by our model space gives a power spectrum consistent with our 95% confidence interval for l_{peak}. This region is quite narrow, and elongated along the "flat Universe" line $\Omega_m + \Omega_\Lambda = 1$. The width of this region is determined by degeneracy in the models, which produce closely similar spectra for different values of the parameters[38]. We further evaluated the likelihood of the models given the Boomerang measurement and the same priors (constraints on the values of the cosmological parameters) as in ref. 16. Marginalizing over all the other parameters, we found the following 95% confidence interval for $\Omega_0 = \Omega_m + \Omega_\Lambda$: $0.88 < \Omega_0 < 1.12$. This provides evidence for a euclidean geometry of the Universe. Our data clearly show the presence of power beyond the peak at $l = 197$, corresponding to smaller-scale structures. The consequences of this fact will be fully analysed elsewhere.

(**404**, 955-959; 2000)

P. de Bernardis[1], P. A. R. Ade[2], J. J. Bock[3], J. R. Bond[4], J. Borrill[5,12], A. Boscaleri[6], K. Coble[7], B. P. Crill[8], G. De Gasperis[9], P. C. Farese[7], P. G. Ferreira[10], K. Ganga[8,11], M. Giacometti[1], E. Hivon[8], V. V. Hristov[8], A. Iacoangeli[1], A. H. Jaffe[12], A. E. Lange[8], L. Martinis[13], S. Masi[1], P. V. Mason[8], P. D. Mauskopf[14,15], A. Melchiorri[1], L. Miglio[16], T. Montroy[7], C. B. Netterfield[16], E. Pascale[6], F. Piacentini[1], D. Pogosyan[4], S. Prunet[4], S. Rao[17], G. Romeo[17], J. E. Ruhl[7], F. Scaramuzzi[13], D. Sforna[1] & N. Vittorio[9]

[1] Dipartimento di Fisica, Universita' di Roma "La Sapienza", P.le A. Moro 2, 00185 Roma, Italy

[2] Department of Physics, Queen Mary and Westfield College, Mile End Road, London E1 4NS, UK

[3] Jet Propulsion Laboratory, Pasadena, California 91109, USA

[4] CITA University of Toronto, Toronto M5S 3H8, Canada

[5] NERSC-LBNL, Berkeley, California 94720, USA

[6] IROE–CNR, Via Panciatichi 64, 50127 Firenze, Italy

[7] Department of Physics, University of California at Santa Barbara, Santa Barbara, California 93106, USA

[8] California Institute of Technology, Mail Code 59-33, Pasadena, California 91125, USA

[9] Dipartimento di Fisica, Universita' di Roma Tor Vergata, Via della Ricerca Scientifica 1, 00133 Roma, Italy

[10] Astrophysics, University of Oxford, Keble Road, OX1 3RH, UK

[11] PCC, College de France, 11 pl. Marcelin Berthelot, 75231 Paris Cedex 05, France

[12] Center for Particle Astrophysics, University of California at Berkeley, 301 Le Conte Hall, Berkeley, California 94720, USA

[13] ENEA Centro Ricerche di Frascati, Via E. Fermi 45, 00044 Frascati, Italy

[14] Physics and Astronomy Department, Cardiff University, Cardiff CF2 3YB, UK

[15] Department of Physics and Astronomy, University of Massachusetts, Amherst, Massachusetts 01003, USA

[16] Department of Physics and Astronomy, University of Toronto, Toronto M5S 3H8, Canada

[17] Istituto Nazionale di Geofisica, Via di Vigna Murata 605, 00143, Roma, Italy

Received 24 March; accepted 3 April 2000.

References:

1. Sachs, R. K. & Wolfe, A. M. Perturbations of a cosmological model and angular variations of the microwave background. *Astrophys. J.* **147**, 73-90 (1967).

2. Weinberg S., *Gravitation and Cosmology* (Wiley & Sons, New York, 1972).

3. Hu, W., Sugiyama, N. & Silk, J. The physics of cosmic microwave background anisotropies. *Nature* **386**, 37-43 (1997).

4. Bond, J. R., Efstathiou, G. & Tegmark, M. Forecasting cosmic parameter errors from microwave background anisotropy experiments. *Mon. Not. R. Astron. Soc.* **291**, L33-L41 (1997).

5. Hinshaw, G. *et al.* Band power spectra in the COBE-DMR four-year anisotropy map. *Astrophys. J.* **464**, L17-L20 (1996).

6. Scott, P. F. *et al.* Measurement of structure in the cosmic background radiation with the Cambridge cosmic anisotropy telescope. *Astrophys. J.* **461**, L1-L4 (1996).

平面上"允许"的区域；该区域中，在我们参数空间定义的范围内，通过对剩余四个参数进行组合，所得的功率谱在 95% 的置信区间内与得到的 l_{peak} 结果相吻合。这个区域相当窄，而且沿着"平直宇宙"线 $\Omega_m + \Omega_\Lambda = 1$ 进行延伸。这个区域的宽度由模型的简并性决定，当这些参数取不同的数值时可以得到非常相似的光谱[38]。我们进一步基于 Boomerang 的测量结果以及同参考文献 16 中相同的先验信息（对宇宙学参数值的限制）评估了不同模型的似然性。通过对所有其他参数进行边缘化分析，我们得到了对 $\Omega_0 = \Omega_m + \Omega_\Lambda$ 的欧几里得几何学在 95% 置信区间内的限制结果：$0.88 < \Omega_0 < 1.12$。这为宇宙模型的欧几里得几何学提供了依据。我们的结果也清楚地显示出在峰值 $l = 197$ 位置之外存在功率，这对应着更小尺度的结构。我们将在别处对这一现象展开详细的分析。

（刘项琨 翻译；李然 审稿）

7. Netterfield, C. B. *et al.* A measurement of the angular power spectrum of the anisotropy in the cosmic microwave background. *Astrophys. J.* **474**, 47-66 (1997).

8. Leitch, E. M. *et al.* A measurement of anisotropy in the cosmic microwave background on 7-22 arcminute scales. *Astrophys. J.* (submitted); also as preprint astro-ph/9807312 at ⟨http://xxx.lanl.gov⟩ (1998).

9. Wilson, G. W. *et al.* New CMB power spectrum constraints from MSAMI. *Astrophys. J.* (submitted); also as preprint astro-ph/9902047 at ⟨http://xxx.lanl.gov⟩ (1999).

10. Baker, J. C. *et al.* Detection of cosmic microwave background structure in a second field with the cosmic anisotropy telescope. *Mon. Not. R. Astron. Soc.* (submitted); also as preprint astro-ph/9904415 at ⟨http://xxx.lanl.gov⟩ (1999).

11. Peterson, J. B. *et al.* First results from Viper: detection of small-scale anisotropy at 40 GHZ. Preprint astro-ph/9910503 at ⟨http://xxx.lanl.gov⟩ (1999).

12. Coble, K. *et al.* Anisotropy in the cosmic microwave background at degree angular scales: Python V results. *Astrophys. J.* **519**, L5-L8 (1999).

13. Torbet, E. *et al.* A measurement of the angular power spectrum of the microwave background made from the high Chilean Andes. *Astrophys. J.* **521**, 79-82 (1999).

14. Miller, A. D. *et al.* A measurement of the angular power spectrum of the CMB from $l = 100$ to 400. *Astrophys. J.* (submitted); also as preprint astro-ph/9906421 at ⟨http://xxx.lanl.gov⟩ (1999).

15. Mauskopf, P. *et al.* Measurement of a peak in the CMB power spectrum from the test flight of BOOMERanG. *Astrophys. J.* (submitted); also as preprint astro-ph/9911444 at ⟨http://xxx.lanl. gov⟩ (1999).

16. Melchiorri, A. *et al.* A measurement of Ω from the North American test flight of BOOMERanG. *Astrophys. J.* (submitted); also as preprint astro-ph/9911445 at ⟨http://xxx.lanl.gov⟩ (1999).

17. de Bernardis, P. *et al.* Mapping the CMB sky: the BOOMERanG experiment. *New Astron. Rev.* **43**, 289-296 (1999).

18. Mauskopf, P. *et al.* Composite infrared bolometers with Si_3N_4 micromesh absorbers. *Appl. Opt.* **36**, 765-771 (1997).

19. Bock, J. *et al.* Silicon nitride micromesh bolometer arrays for SPIRE. *Proc. SPIE* **3357**; 297-304 (1998).

20. Masi, S. *et al.* A self contained ^{3}He refrigerator suitable for long duration balloon experiments. *Cryogenics* **38**, 319-324 (1998).

21. Masi, S. *et al.* A long duration cryostat suitable for balloon borne photometry. *Cryogenics* **39**, 217-224 (1999).

22. Schlegel, D. J., Finkbeiner, D. P. & Davis, M. Maps of dust IR emission for use in estimation of reddening and CMBR foregrounds. *Astrophys. J.* **500**, 525-553 (1998).

23. Delabrouille, J., Gorski, K. M. & Hivon, E. Circular scans for CMB anisotropy observation and analysis. *Mon. Not. R. Astron. Soc.* **298**, 445-450 (1998).

24. Cheung, L. H. *et al.* 1.0 millimeter maps and radial density distributions of southern HII/molecular cloud complexes. *Astrophys. J.* **240**, 74 83 (1980).

25. Kogut, A. *et al.* Dipole anisotropy in the COBE DMR first-year sky maps. *Astrophys. J.* **419**, 1-6 (1993).

26. Tegmark, M. CMB mapping experiments: a designer's guide. *Phys. Rev. D* **56**, 4514-4529 (1997).

27. Bond, J. R., Crittenden, R., Jaffe, A. H. & Knox, L. E. Computing challenges of the cosmic microwave background. *Comput. Sci. Eng.* **21**, 1-21 (1999).

28. Borrill, J. in *Proc. 3K Cosmology EC-TMR Conf.* (eds Langlois, D., Ansari, R. & Vittorio, N.) 277 (American Institute of Physics Conf. Proc. Vol. 476, Woodbury, New York, 1999).

29. Prunet, S. *et al.* in *Proc. Conf. Energy Density in the Universe* (eds Langlois, D., Ansari, R. & Bartlett, J.) (Editiones Frontieres, Paris, 2000).

30. Gorski, K. M., Hivon, E. & Wandelt, B. D. in *Proc. MPA/ESO Conf.* (eds Banday, A. J., Sheth, R. K. & Da Costa, L.) (European Southern Observatory, Garching); see also ⟨http://www.tac.dk/~healpix/⟩.

31. Kogut, A. in *Microwave Foregrounds* (eds de Oliveira Costa, A. & Tegmark, M.) 91-99 (Astron. Soc. Pacif. Conf. Series. Vol 181, San Francisco, 1999).

32. Toffolatti, L. *et al.* Extragalactic source counts and contributions to the anisotropies of the CMB. *Mon. Not. R. Astron. Soc.* **297**, 117-127 (1998).

33. Wright, A. E. *et al.* The Parkes-MIT-NRAO (PMN) surveys II. Source catalog for the southern survey. *Astrophys. J. Supp. Ser.* **91**, 111-308 (1994); see also ⟨http://astron.berkeley.edu/wombat/foregrounds/radio.html⟩.

34. Borrill, J. in *Proc. 5th European SGI/Cray MPP Workshop* (CINECA, Bologna, 1999); Preprint astro-ph/9911389 at ⟨xxx.lanl.gov⟩ (1999); see also ⟨http://cfpa. berkeley.edu/~borrill/cmb/madcap.html⟩.

35. Durrer, R., Kunz, M. & Melchiorri, A. *Phys. Rev. D* **59**, 1-26 (1999).

36. Seljak, U. & Zaldarriaga, M. A line of sight approach to cosmic microwave background anisotropies. *Astrophys. J.* **437**, 469-477 (1996).

37. Lewis, A., Challinor, A. & Lasenby, A. Efficient computation of CMB anisotropies in closed FRW models. Preprint astro-ph/9911177 at ⟨http://xxx.lanl.gov⟩ (1999).

38. Efstathiou, G. & Bond, R. Cosmic confusion: degeneracies among cosmological parameters derived from measurements of microwave background anisotropies. *Mon. Not. R. Astron. Soc.* **304**, 75-97 (1998).

39. Wright, E. *et al.* Comments on the statistical analysis of excess variance in the COBE-DMR maps. *Astrophys. J.* **420**, 1-8 (1994).

40. Perlmutter, S. *et al.* Measurements of and Λ from 42 high-redshift supernovae. *Astrophys. J.* **517**, 565-586 (1999).

41. Schmidt, B. P. *et al.* The high-Z supernova search: measuring cosmic deceleration and global curvature of the Universe using type Ia supernovae. *Astrophys. J.* **507**, 46-63 (1998).

Acknowledgements. The Boomerang experiment was supported by Programma Nazionale di Ricerche in Antartide, Universita' di Roma "La Sapienza", and Agenzia Spaziale Italiana in Italy, by the NSF and NASA in the USA, and by PPARC in the UK. We thank the staff of the National Scientific Ballooning Facility, and the United States Antarctic Program personnel in McMurdo for their preflight support and an effective LDB flight. DOE/NERSC provided the supercomputing facilities.

Correspondence and requests for materials should be addressed to P. d. B. (e-mail: debernardis@roma1.infn.it). Details of the experiment and numerical data sets are available at the web sites ⟨http://oberon.roma1.infn.it/boomerang⟩ and ⟨http://www.physics.ucsb.edu/~boomerang⟩.

The Accelerations of Stars Orbiting the Milky Way's Central Black Hole

A. M. Ghez *et al.*

Editor's Note

It had been widely believed for several years that black holes with masses of a million or more solar masses reside at the centres of most galaxies. But proving that notion was challenging. Here Andrea Ghez and colleagues track the motions of stars very close to the centre of the Milky Way, seeing them move in their orbits. Subsequent work has tracked these stars through their entire orbits; one travels around the Galactic Centre at the tremendous speed of over 5,000 km/s. These studies strongly suggest that the stars are gravitationally influenced by an object such as a supermassive black hole; the only (increasingly unlikely) alternative is a massive star made of some kind of exotic matter.

Recent measurements[1-4] of the velocities of stars near the centre of the Milky Way have provided the strongest evidence for the presence of a supermassive black hole in a galaxy[5], but the observational uncertainties poorly constrain many of the black hole's properties. Determining the accelerations of stars in their orbits around the centre provides much more precise information about the position and mass of the black hole. Here we report measurements of the accelerations of three stars located ~0.005 pc (projected on the sky) from the central radio source Sagittarius A* (Sgr A*); these accelerations are comparable to those experienced by the Earth as it orbits the Sun. These data increase the inferred minimum mass density in the central region of the Galaxy by an order of magnitude relative to previous results, and localize the dark mass to within 0.05 ± 0.04 arcsec of the nominal position of Sgr A*. In addition, the orbital period of one of the observed stars could be as short as 15 years, allowing us the opportunity in the near future to observe an entire period.

IN 1995, we initiated a programme of high-resolution 2.2-μm (K band) imaging of the inner 5 arcsec × 5 arcsec of the Galaxy's central stellar cluster with the W. M. Keck 10-m telescope on Mauna Kea, Hawaii (1 arcsec = 0.04 pc at the distance to the Galactic Centre, 8 kpc, ref. 6). From each observation, several thousand short exposure ($t_{\exp} = 0.137$ s) frames were collected, using the facility near-infrared camera, NIRC[7,8], and combined to produce a final diffraction-limited image having an angular resolution of 0.05 arcsec. Between 1995 and 1997, images were obtained once a year with the aim of detecting the stars' velocities in the plane of the sky. The results from these measurements are detailed in ref. 4; in summary, two-dimensional velocities were measured for 90 stars with simple

环绕银河系中心黑洞运动的恒星的加速度

盖兹等

编者按

近年来，人们普遍相信在大部分的星系中心都存在质量超过百万太阳质量的黑洞。然而要想证明这一观点却是极具挑战性的。在本文中，安德烈娅·盖兹和她的同事们追踪了那些非常接近银河系中心的恒星的运动，以观测这些恒星的运动轨迹。其中有一颗围绕银河系中心运动的恒星，其速度竟可高达 5,000 km/s。这些研究结果表明银河系中心很有可能存在一个超大质量的黑洞，而这些恒星正是在这个超大质量黑洞的引力影响下围绕银河系中心运动。除此之外，唯一的可能性（越来越不可能）就是银河系中心存在一个由某种奇异物质组成的大质量恒星。

近年来对银河系中心附近恒星速度的测量[1-4]为星系中心存在超大质量黑洞提供了最有力的证据[5]，但是观测误差使得我们很难对黑洞的许多性质给出强有力的限制。确定恒星环绕中心轨道的加速度，将向我们提供更精确的黑洞位置和质量信息。在本文中，我们报道了距离中心射电源人马座 A*（简称 Sgr A*）约 0.005 秒差距（pc）（投影到天球上）处三颗恒星加速度的测量结果；这些加速度和地球环绕太阳经历的加速度相似。相对以前的结果，从这些数据推导出的银河系中心的最小质量密度增加了一个量级，并将暗质量区域限定在 Sgr A* 标定位置的 0.05 ± 0.04 角秒以内。此外，其中一颗恒星的轨道周期可能只有短短的 15 年，这让我们在未来有机会观测它的整个运动轨道周期。

我们在 1995 年启动了一个项目，利用夏威夷冒纳凯阿山的 10 米凯克望远镜在 2.2 μm 波段（K 波段）对银河系中心星团内部 5 角秒 × 5 角秒的区域做高分辨率的成图观测（距离银河系中心 8 kpc 处，1 角秒 = 0.04 pc，参考文献 6）。对于每次观测，我们使用近红外照相机设备 NIRC[7,8]收集了几千幅短曝光（$t_{\exp} = 0.137\,\mathrm{s}$）图像，并将这些图像合在一起最终得到一幅角分辨率为 0.05 角秒的衍射极限图像。为了测量恒星在天球切面上的速度，我们从 1995 年到 1997 年每年成图一次。这些测量的具体结果罗列在参考文献 4 中；总的来说，我们对 90 颗恒星的位置与时间关系进行简单的线性拟合得到它们的二维速度。这些速度，有的可以高达 1,400 km·s⁻¹，

525

linear fits to the positions as a function of time. These velocities, which reach up to 1,400 km s^{-1}, implied the existence of a 2.6×10^6 $M_\odot$ black hole coincident (± 0.1 arcsec) with the nominal location of Sgr A* (ref. 9), the unusual radio source[10-12] long believed to be the counterpart of the putative black hole.

The new observations presented here were obtained several times a year from 1997 to 1999 to improve the sensitivity to accelerations. With nine independent measurements, we now fit the positions of stars as a function of time with second order polynomials (see Fig. 1). The resulting velocity uncertainties are reduced by a factor of 3 compared to our earlier work, primarily as a result of the increased time baseline, and, in the central square arcsecond, by a factor of 6 compared to that presented in ref. 3, owing to our higher angular resolution. Among the 90 stars in our original proper motion sample[4], we have now detected significant accelerations for three stars, S0-1, S0-2 and S0-4 (see Table 1); specifically, these are the sources for which the reduced chi-squared value for a quadratic fit is smaller than that for the linear model of their motions by more than 1. These three stars are independently distinguished in our sample, being among the fastest moving stars ($v = 570$ to $1,350$ km s^{-1}) and among the closest to the nominal position of Sgr A* ($r_{1995} = 0.004$–0.013 pc). With accelerations of 2–5 milliarcsec yr^{-2}, or equivalently $(3$–$6) \times 10^{-6}$ km s^{-2}, they are experiencing accelerations similar to the Earth in its orbit about the Sun.

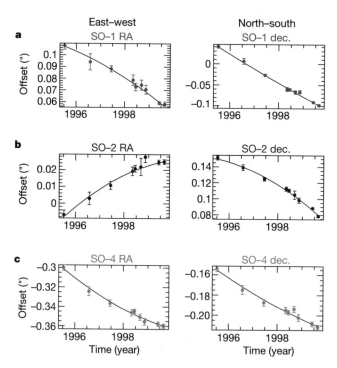

Fig. 1. East–west and north–south positional offsets from the nominal location of Sgr A* versus time for (**a**) S0-1, (**b**) S0-2, and (**c**) S0-4. The offset range shown is scaled to the points in each plot and therefore varies from $\sim$−0.4″ to $\sim$0.15″. Each of these stars is located with a precision of about 1–5 mas in the individual maps and the alignment of these positions, which is carried out by minimizing

意味着在 Sgr A* 的标定位置（参考文献 9）处（±0.1 角秒）很可能存在一个质量为 $2.6 \times 10^6 \, M_\odot$ 的黑洞。而 Sgr A* 这个不寻常的射电源[10-12]一直以来被认为是假定黑洞的对应体。

这里所采用的新数据是从 1997 年到 1999 年每年观测好几次得到的，为的是提高测量加速度的灵敏度。利用 9 次独立的测量结果，我们现在用二阶多项式来拟合恒星的位置–时间关系（见图 1）。主要因为时间基线变长，得到的速度误差减少为我们早期工作的 1/3；又因为角分辨率的提高，在图像中央的平方角秒区域内，速度误差减少为参考文献 3 的 1/6。在原初样本中有自行信息的 90 颗恒星中[4]，我们现在已经探测到 S0-1、S0-2 和 S0-4 这三颗恒星的显著加速度（见表 1）；特别是，用二次多项式拟合这些源的运动所得的约化 χ^2 值比线性拟合的约化 χ^2 值要少 1 以上。这三颗星作为运动最快（$v = 570 \sim 1{,}350 \, \mathrm{km \cdot s^{-1}}$）且最接近 Sgr A* 标定位置的恒星（$r_{1995} = 0.004 \sim 0.013 \, \mathrm{pc}$），在我们的样本中被单独区分出来。它们的加速度为 $2 \sim 5$ $\mathrm{milliarcsec \cdot yr^{-2}}$，或等效于 $(3 \sim 6) \times 10^{-6} \, \mathrm{km \cdot s^{-2}}$，正在经历与地球环绕太阳类似的加速运动过程。

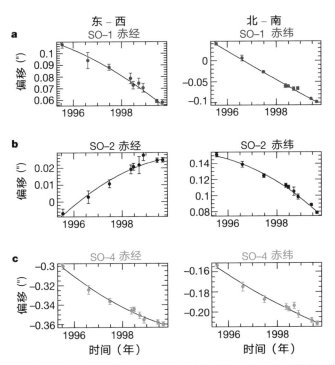

图 1. 恒星（**a**）S0-1、（**b**）S0-2 和（**c**）S0-4 相对于 Sgr A* 标定位置在东–西、北–南方向上的偏移随时间的变化。每幅图中偏移范围都拟合在数据点上，其变化范围约在 −0.4″ 到 0.15″ 之间。这三颗恒星在各自图中的位置精度为 1 ~ 5 毫角秒，按参考文献 4 描述的方法，我们通过将所有恒星的净位移最小化来将位置对准，半抽样自举重采样方法[22]带来大约 3 毫角秒的对准误差（对准误差在图中间最小，随着半径

the net displacement of all stars as described in ref. 4, has an uncertainty of about 3 mas based on the half sample bootstrap resampling method[22] (the alignment uncertainty is at a minimum at the centre of the map and grows linearly with radius). These two uncertainty terms are added in quadrature; the results are used in the fitting process and depicted as errorbars here. In each plot, the solid line shows the second order polynomial used to derive the acceleration term. These plots demonstrate that we are able to measure, for the first time, accelerations of 2 to 5 mas yr^{-2} (0.3–0.6 cm s^{-2}) for stars orbiting a supermassive black hole.

Table 1. Measurements for stars with significant accelerations

	S0-1 (S1)	S0-2 (S2)	S0-4 (S8)
Radius from Sgr A*-magnitude (milliparsecs)	4.42 ± 0.05	5.83 ± 0.04	13.15 ± 0.04
Radius from Sgr A*-position angle (degrees)	290.1 ± 0.7	3.1 ± 0.4	117.3 ± 0.2
Velocity-magnitude (km s^{-1})	$1,350 \pm 40$	570 ± 20	990 ± 30
Velocity-position angle (degrees)	168 ± 2	241 ± 2	129 ± 2
Acceleration-magnitude (milliarcsec yr^{-2})	2.4 ± 0.7	5.4 ± 0.3	3.2 ± 0.5
Acceleration-position angle (degrees)	80 ± 15	154 ± 4	294 ± 7

The primary nomenclature adopted here and in the text is from ref. 4; in parentheses star names from ref. 1 are also given. All quantities listed are derived from the second-order polynomial fit to the data and are given for epoch 1995.53. All position angles are measured east of north. All uncertainties are 1σ and are determined by the jackknife resampling method[22]. The radius uncertainties listed include only the relative positional uncertainties and not the uncertainty in the origin used (the nominal position of Sgr A*; see text for offset to measured dynamical centre).

Acceleration vectors, in principle, are more precise tools than the velocity vectors for studying the central mass distribution. Even projected onto the plane of the sky, each acceleration vector should be oriented in the direction of the central mass, assuming a spherically symmetric potential. Thus, the intersection of multiple acceleration vectors is the location of the dark mass. Figure 2 shows the acceleration vector's direction for the three stars. With in 1σ, these vectors do indeed overlap, and furthermore, the intersection point lies a mere 0.05 ± 0.03 arcsec east and 0.02 ± 0.03 arcsec south of the nominal position of Sgr A*, consistent with the identification of Sgr A* as the carrier of the mass. Previously, with statistical treatments of velocities only, the dynamical centre was located to within ± 0.1 arcsec (1σ) of Sgr A*'s position[4]. This velocity-based measurement is unaffected by the increased time baseline, as its uncertainty is dominated by the limited number of stars at a given radius. The accelerations thus improve the localization of our Galaxy's dynamical centre by a factor of 3, which is essential for reliably associating any near-infrared source with the black hole given the complexity of the region.

的增加而线性增大）。对这两个误差项取平方和开根号，得到拟合时所用的总误差，并在图中用误差棒描述。每幅图中，实线表示用二阶多项式拟合的最佳结果，用来推导恒星的加速度。这些结果显示，对环绕超大质量黑洞运动的恒星，我们首次能够测得它们量级在 $2 \sim 5$ mas·yr^{-2}($0.3 \sim 0.6$ cm·s^{-2})的运动加速度。

表 1. 有明显加速度的恒星的测量结果

	S0-1(S1)	S0-2(S2)	S0-4(S8)
相对 Sgr A* 的位置－大小 (milliparsec)	4.42±0.05	5.83±0.04	13.15±0.04
相对 Sgr A* 的位置－位置角 (度)	290.1±0.7	3.1±0.4	117.3±0.2
速度－大小 (km·s^{-1})	1,350±40	570±20	990±30
速度－位置角 (度)	168±2	241±2	129±2
加速度－大小 (milliarcsec·yr^{-2})	2.4±0.7	5.4±0.3	3.2±0.5
加速度－位置角 (度)	80±15	154±4	294±7

这里和正文中采用的主要命名方式来自参考文献 4；参考文献 1 所给出的恒星名字也在圆括号里给出。所有列出量均来源于二阶多项式拟合所得数据，对应历元为 1995.53。所有方位角由北向东起量。误差都是 1σ，由刀切法确定[22]。列出的半径误差仅包括相对位置误差，没有包括中心的位置误差(Sgr A* 的标定位置；关于测量到的动力学中心与标定位置的偏移的具体细节请见正文)。

原则上来说，比起速度矢量，利用加速度矢量可以更精确地研究中心质量分布。假设一个球对称的引力势，就算投影到天球切面上，每个加速度矢量还将指向中心质量的方向。因此，多个加速度矢量的交叉点就是暗质量所在的位置。图 2 显示了这三颗恒星的加速度矢量方向。这些矢量在 1σ 误差范围内确实发生重叠，而且交叉点正好位于 Sgr A* 标定位置往东 0.05 ± 0.03 角秒、往南 0.02 ± 0.03 角秒处，与暗质量位于 Sgr A* 的结论一致。之前，在仅有对速度的相关统计分析的情况下，人们大致将动力学中心确定在位于 Sgr A* 位置 ±0.1 角秒(1σ)的范围内[4]。这种基于速度的测量并不依赖时间基线的长短，因为它的误差主要来自给定半径内恒星数目的限制。因此，对恒星进行加速度矢量分析将银河系动力学中心位置的定位精度提高为原来的 3 倍。考虑到这个区域的复杂度，这点对于研究近红外源和黑洞的可靠成协关系来说很重要。

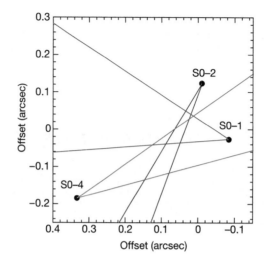

Fig. 2. The acceleration uncertainty cones and their intersections. The cones' edges represent the directions for which the accelerations deviate by 1σ from their best-fit values and their vertices are the time-averaged positions, measured relative to the nominal position of Sgr A*, rather than the positions listed in Table 1. If the accelerations are caused by a single supermassive black hole, or even a spherically symmetric mass distribution, these vectors should intersect at a common location, the centre of the mass. The 1σ intersection point lies 0.05 ± 0.03 arcsec east and 0.02 ± 0.03 arcsec south of the nominal position of Sgr A*. The existence of an intersection point suggests that there is indeed a common origin for the acceleration and pinpoints the location of the black hole to within 0.03 arcsec.

Like the directions, the magnitudes of the acceleration vectors also constrain the central black hole's properties. In three dimensions, the acceleration and radius vectors (a_{3D} and r_{3D}, respectively) provide a direct measure of the enclosed mass simply by $a_{3D} = G \times M / r_{3D}^2$. For a central potential, the acceleration and radius vectors are co-aligned and, with a projection angle to the plane of the sky θ, the two-dimensional projections place the following lower limit on the central mass: $M \cos^3(\theta) = a_{2D} \times r_{2D}^2 / G$. This analysis is independent of the star's orbital parameters, although θ is in fact a lower limit for the orbital inclination angle. Figure 3 shows the minimum mass implied by each star's two-dimensional acceleration as a function of projected radius, with dashed curves displaying how the implied mass and radius grow with projection angle. For each point, the uncertainty in the position of the dynamical centre dominates the minimum mass uncertainties. Also plotted are the results from the statistical analysis of the velocity vectors measured in the plane of the sky, which imply a dark mass of $(2.3$–$3.3) \times 10^6 \, M_\odot$ inside a radius of 0.015 pc[3,4]. If, as has been assumed, this dark mass is in the form of a single supermassive black hole, the enclosed mass should remain level at smaller radii. Projection angles of 51–56° and 25–37° for S0-1 and S0-2, respectively, would yield the mass inferred from velocities and place these stars at a mere ~0.008 pc (solid line portions of the limiting mass curves in Fig. 3), thus increasing the dark mass density implied by velocities by an order of magnitude to $8 \times 10^{12} \, M_\odot \, \mathrm{pc}^{-3}$. With smaller projection angles, these two stars also allow for the enclosed mass to decrease at smaller radii, as would occur in the presence of an extended distribution of dark matter surrounding a less massive black hole[13-17]. In contrast to the agreement between the mass distribution inferred from

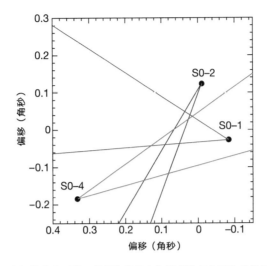

图 2. 加速度误差锥以及它们的交叠区域。锥的边缘代表加速度偏离它们的最佳拟合值 1σ 的方向，顶点则是它们相对 Sgr A* 标定位置的时间平均位置，而不是表 1 中所列举的位置。假如加速运动是由单个超大质量黑洞或者甚至是一个球形对称质量分布所引起的，这些矢量应该会在质量中心处相交。1σ 交叉点位于 Sgr A* 标定位置东边 0.05±0.03 角秒和南边 0.02±0.03 角秒处。交叉点的存在表明加速度确实来自一个共同的来源，并将黑洞位置精确定位在 0.03 角秒内。

　　和方向一样，加速度矢量的大小也能够用来限制中心黑洞的性质。在三维空间中，利用加速度和径向矢量（分别是 a_{3D} 和 r_{3D}），仅通过 $a_{3D} = G \times M/r_{3D}^2$，我们就可以直接测量所考虑半径范围内的质量。在中心引力势作用下，加速度和径向矢量的方向一致，若它们在天球切面上的投影角为 θ，那么两维投影将给出中心质量的下限为：$M\cos^3(\theta) = a_{2D} \times r_{2D}^2/G$。这个分析与恒星的轨道参数无关，尽管 θ 实际上就是轨道倾角的下限。图 3 显示由每颗恒星的二维加速度所推导的中心黑洞的最小质量随恒星与中心的投影距离的变化，其中虚线显示了导出的质量和距离随投影角的变化规律。对图中每个点，这个质量下限的误差主要来自动力学中心的位置误差。同时图中还画上了对投影在天球切面上的速度矢量的统计分析结果，表明在半径 0.015 pc 内存在 $(2.3 \sim 3.3) \times 10^6 \, M_{\odot}$ 的暗质量[3,4]。如果和假设的一样，暗质量是一个超大质量黑洞，那么这些暗质量应该被限制在一个更小半径之内。S0-1 和 S0-2 的投影角分别为 51°~56° 和 25°~37°，这些投影角使得我们可以从速度推导得质量并且将这些恒星的位置限定在距离中心仅仅约 0.008 pc 以内的区域（图 3 中质量下限曲线的实线部分），因此由速度推导得出的暗质量密度将增加一个量级，达到 $8 \times 10^{12} \, M_{\odot} \cdot pc^{-3}$。假如这两颗恒星具有更小的投影角，那么这两颗恒星的观测结果也同样允许出现所考虑半径范围内暗质量随着半径减小而减少的可能性，就和在一个质量较小的黑洞周围存在一个延展暗物质分布所造成的情形一样[13-17]。虽然从 S0-1 和 S0-2 的速度和加速度推断得到的质量分布相互吻合，可是从 S0-4 加速度

the velocities and the accelerations for S0-1 and S0-2, the minimum mass implied by S0-4's acceleration is inconsistent at least at the 1σ level. We note, in support of the validity of S0-4's acceleration vector, that its orientation is consistent with the intersection of the S0-1 and S0-2 acceleration vectors (see Fig. 2). Nonetheless, continued monitoring of S0-4 will be important for assessing this possible discrepancy. Overall, the individual magnitudes of the three acceleration vectors support the existence of a central black hole with a mass of approximately $3 \times 10^6\, M_\odot$.

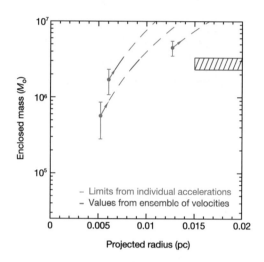

Fig. 3. The minimum enclosed mass implied by each star's acceleration measurement versus projected distance from the newly determined dynamical centre position. If the projection between the plane of the sky and both the radius and the acceleration vectors is θ, then the true mass increases as $1/\cos^3(\theta)$ and the true acceleration and radius vectors increase as $1/\cos(\theta)$. Also plotted is the mass range implied from a statistical analysis of the approximately 100 velocity vectors that have been measured in the plane of the sky[3,4].

Using acceleration measurements, it is now possible to constrain the individual orbits. We checked the conclusions of the previous studies[3,4] by assuming that the stars are bound to a central mass of $(2.2–3.3) \times 10^6\, M_\odot$ located within 0.03 arcsec of the nominal position of Sgr A*. Excellent orbital fits were found for S0-1 and S0-2 for the entire range of masses and for true foci within 0.01 arcsec of the nominal position of Sgr A*, suggesting that we now have comparable accuracy in determining the infrared location of Sgr A* (ref. 9) and the dynamical centre of our Galaxy (see Fig. 4). The orbital solutions for these stars have eccentricities ranging from 0 (circular) to 0.9 for S0-1 and 0.5 to 0.9 for S0-2 and periods in the range 35–1,200 and 15–550 years, respectively, raising the possibility of seeing a star make a complete journey around the centre of the Galaxy within the foreseeable future. Although the fits are not yet unique, they impose a maximum orbital distance from the black hole, or apoapse, of 0.1 pc. This suggests that S0-1 and S0-2 might have formed locally. If these stars are indeed young[2], then their small apoapse distance presents a challenge to classical star formation theories in light of the strong tidal forces created by the central black hole and might require a collisional or compressional star formation scenario[18,19]. However, dynamical friction may be able to act on a short enough timescale to bring these stars in from a much larger distance[20]. More accurate orbital

推出的暗质量下限在 1σ 水平内，和其他两颗星得出的结果不一致。值得注意的是，S0-4 的加速度矢量方向在 S0-1 和 S0-2 的加速度矢量的交叠范围内（见图 2），这表明 S0-4 的加速度矢量是正确的。虽然如此，继续监测 S0-4 对判断这个差异是否存在非常重要。总体来看，三个加速度矢量的大小都支持银河系中心存在一个质量约为 $3 \times 10^6\, M_{\odot}$ 的超大质量黑洞。

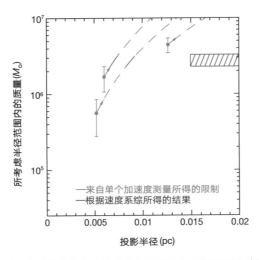

图 3. 根据每颗恒星的加速度测量值推导出的所考虑半径范围内的质量下限同新确定的动力学中心位置的投影距离的关系图。假如径向矢量和加速度矢量在天球切面的投影角均为 θ，那么真实的质量随着 $1/\cos^3(\theta)$ 增加，真实的加速度和径向矢量随着 $1/\cos(\theta)$ 增加。对在天球切面上测量到的接近 100 个速度矢量进行统计分析所得到的质量范围也画在图中[3,4]。

通过测量加速度，我们现在能够分别限制这三颗恒星各自的轨道。通过假设恒星束缚在位于 Sgr A* 标定位置 0.03 角秒内且质量为 $(2.3 \sim 3.3) \times 10^6\, M_{\odot}$ 的中心质量周围，我们检查了前人的研究结论[3,4]。假定轨道实焦点在 Sgr A* 标定位置 0.01 角秒以内，对于整个可能的中心质量范围，我们都可以对 S0-1 和 S0-2 的轨道给出很好的拟合结果，这表明我们现在确定的 Sgr A* 红外位置（参考文献 9）和银河系动力学中心的精确度都相当好（见图 4）。在得到的轨道解当中，S0-1 解出的轨道偏心率分布在 0（圆）到 0.9 之间，S0-2 的轨道偏心率分布在 0.5 到 0.9 之间，周期分别分布在 35 ~ 1,200 年和 15 ~ 550 年的范围之内。这意味着在可预见的将来，也许我们能够完整地看到某颗恒星环绕银河系中心的运动轨迹。虽然所得到的拟合结果并不是唯一的，但是它们给出了离黑洞的最大轨道距离（或远质心点）为 0.1 pc。这表明 S0-1 和 S0-2 可能是在中心附近形成。如果这些恒星实际上很年轻[2]，那么这些恒星拥有这么小的远质心点距离将对经典恒星形成理论提出挑战。考虑到中心黑洞产生的强潮汐力，我们可能需要一个在碰撞或压缩条件下的恒星形成理论[18,19]。不过，

parameters are needed to fully address these problems and others such as the distance to the Galactic Centre[21]. The determinations of these parameters will be considerably improved when radial velocities are measured for these stars using adaptive optics techniques, as the current solutions based on the proper motion data alone predict radial velocities ranging from 200 and 2,000 km s^{-1}.

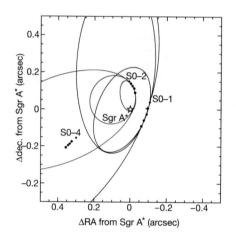

Fig. 4. The measured motion of S0-1, S0-2 and S0-4 and several allowed orbital solutions. Only the measurements obtained in June of each of the 5 years are shown. S0-1 and S0-2 are moving clockwise about Sgr A* and S0-4 is travelling radially outward. In the orbital analysis, two constraints are applied, a central mass of $2.6 \times 10^6 M_\odot$ and a true focus located at the position of Sgr A*. Displayed are orbital solutions with periods of 17, 80 and 505 years for S0-2 and 63, 200 and 966 years for S0-1.

(**407**, 349-351; 2000)

A. M. Ghez, M. Morris, E. E. Becklin, A. Tanner & T. Kremenek
Department of Physics and Astronomy, UCLA, Los Angeles, California 90095-1562, USA

Received 24 May; accepted 19 July 2000.

References:

1. Eckart, A. & Genzel, R. Stellar proper motions in the central 0.1 pc of the Galaxy. *Mon. Not. R. Astron. Soc.* **284**, 576-598 (1997).

2. Genzel, R., Eckart, A., Ott, T. & Eisenhauer, F. On the nature of the dark mass in the centre of the Milky Way. *Mon. Not. R. Astron. Soc.* **291-234**, 219 (1997).

3. Genzel, R., Pichon, C., Eckart, A., Gerhard, O. E. & Ott, T. Stellar dynamics in the galactic centre: Proper motions and anisotropy. *Mon. Not. R. Astron. Soc.* (in the press).

4. Ghez, A. M., Klein, B. L., Morris, M. & Becklin, E. E. High proper-motion stars in the vicinity of Sagittarius A*: Evidence for a supermassive black hole at the center of our galaxy. *Astrophys. J.* **509**, 678-686 (1998).

5. Maoz, E. Dynamical constraints on alternatives to supermassive black holes in galactic nuclei. *Astrophys. J.* **494**, L181-L184 (1998).

6. Reid, M. J. The distance to the center of the Galaxy. *Annu. Rev. Astron. Astrophys.* **31**, 345-372 (1993).

7. Matthews, K. & Soifer, B. T. in *Astronomy with Infrared Arrays: The Next Generation* (ed. McLean, I.) Vol. 190, 239-246 (Astrophysics and Space Sciences Library, Kluwer Academic, Dordrecht, 1994).

8. Matthews, K., Ghez, A. M., Weinberger, A. J. & Neugebauer, G. The diffraction-limited images from the W. M. Keck Telescope. *Proc. Astron. Soc. Pacif.* **108**, 615-619 (1996).

9. Menten, K. M., Reid, M. J., Eckart, A. & Genzel, R. The position of Sgr A*: Accurate alignment of the radio and infrared reference frames at the galactic center. *Astrophys. J.* **475**, L111-L114 (1997).

动力学摩擦也可能在足够短的时标内将这些恒星从更远距离处拉进来[20]。要解决这些问题，以及对其他诸如距银河系中心的距离[21]等问题有更加准确的理解，我们还需要更精确的轨道参数的测量。应用自适应光学技术测量这些恒星的视向速度，将极大地提高这些恒星轨道参数的测量精度，因为目前仅依赖自行数据得到的恒星视向速度只能限制在 200 到 2,000 km·s⁻¹ 的广阔范围内。

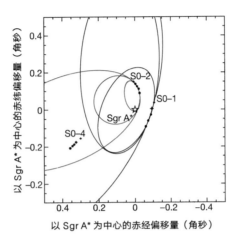

图 4. 测量到的 S0-1、S0-2 和 S0-4 的运动轨迹和几个可能的轨道解。图中只显示 5 年中每年 6 月份的观测。S0-1 和 S0-2 围着 Sgr A* 顺时针运动，S0-4 沿径向向外运动。在轨道分析中我们应用了中心质量为 $2.6 \times 10^6 \, M_\odot$ 和实焦点位于 Sgr A* 处这两个限制条件。图中显示了 S0-2 的周期分别为 17、80 和 505 年的轨道解，以及 S0-1 的周期分别为 63、200 和 966 年的轨道解。

（肖莉 翻译；周礼勇 审稿）

10. Lo, K. Y., Shen, Z.-Q., Zhao, J.-H. & Ho, P. T. Intrinsic size of SGR A*: 72 Schwarzschild radii. *Astrophys. J.* **508**, L61-L64 (1998).

11. Reid, M. J., Readhead, A. C. S., Vermeulen, R. C. & Treuhaft, R. N. The proper motion of Sagittarius A*. I. First VLBA results. *Astrophys. J.* **524**, 816-823 (1999).

12. Backer, D. C. & Sramek, R. A. Proper motion of the compact, nonthermal radio source in the galactic center, Sagittarius A*. *Astrophys. J.* **524**, 805-815 (1999).

13. Salati, P. & Silk, J. A stellar probe of dark matter annihilation in galactic nuclei. *Astrophys. J.* **338**, 24-31 (1989).

14. Tsiklauri, D. & Viollier, R. D. Dark matter concentration in the galactic center. *Astrophys. J.* **500**, 591-595 (1998).

15. Gondolo, P. & Silk, J. Dark matter annihilation at the galactic center. *Phys. Rev. Lett.* **83**, 1719-1722 (1999).

16. Torres, D. F., Capozziello, S. & Lambiase, G. A supermassive boson star at the galactic center? *Astrophys. J.* (in the press).

17. Romanowsky, A. & Kochanek, C. in *Proceedings of Dynamics of Star Clusters and the Milky Way* (ASP Conference Series, Astronomical Society of the Pacific, San Francisco, in the press).

18. Sanders, R. H. The case against a massive black hole at the Galactic Centre. *Nature* **359**, 131-132 (1992).

19. Morris, M., Ghez, A. M. & Becklin, E. E. The galactic center black hole: clues for the evolution of black holes in galactic nuclei. *Adv. Space Res.* **23**, 959-968 (1999).

20. Gerhard, O. The galactic center He I stars: remains of a dissolved young cluster? *Astrophys. J.* (submitted).

21. Salim, S. & Gould, A. Sagittarius A* "Visual Binaries": A direct measurement of the galactocentric distance. *Astrophys. J.* **523**, 633-641 (1999).

22. Babu, G. J. & Feigelson, E. D. *Astrostatistics* (Chapman & Hall, London, 1996).

Acknowledgements. This work was supported by the National Science Foundation and the Packard Foundation. We are grateful to J. Larkin for exchanging telescope time; O. Gerhard, M. Jura, and A. Weinberger for useful input; telescope observing assistants J. Aycock, T. Chelminiak, G. Puniwai, C. Sorenson, W. Wack, M. Whittle and software/instrument specialists A. Conrad and B. Goodrich for their help during the observations. The data presented here were obtained at the W. M. Keck Observatory, which is operated as a scientific partnership among the California Institute of Technology, the University of California and the National Aeronautics and Space Administration. The Observatory was made possible by the financial support of the W. M. Keck Foundation.

Correspondence and requests for materials should be addressed to A.G. (e-mail: ghez@astro.ucla.edu).

Evidence from Detrital Zircons for the Existence of Continental Crust and Oceans on the Earth 4.4 Gyr Ago

S. A. Wilde *et al.*

Editor's Note

The Earth is thought to have formed about 4.56 Gyr ago. Initially it was entirely molten, but as it cooled, the chemical components of the planet differentiated and its surface solidified. Although tectonic movements have recycled the earliest crust, some remnants do remain, offering evidence for when it was first formed. Here geologist Simon Wilde and coworkers report the oldest crustal material so far known, showing that it was formed at about 4.4 Gyr ago. It comprises a mineral called zircon, which can be dated from radioactive decay in the uranium-thorium-lead series. The zircon seems to have formed from material in the crust interacting with surface water, suggesting that oceans may have existed by that time.

No crustal rocks are known to have survived since the time of the intense meteor bombardment that affected Earth[1] between its formation about 4,550 Myr ago and 4,030 Myr, the age of the oldest known components in the Acasta Gneiss of northwestern Canada[2]. But evidence of an even older crust is provided by detrital zircons in metamorphosed sediments at Mt Narryer[3] and Jack Hills[4-8] in the Narryer Gneiss Terrane[9], Yilgarn Craton, Western Australia, where grains as old as ~4,276 Myr have been found[4]. Here we report, based on a detailed micro-analytical study of Jack Hills zircons[10], the discovery of a detrital zircon with an age as old as $4,404 \pm 8$ Myr—about 130 million years older than any previously identified on Earth. We found that the zircon is zoned with respect to rare earth elements and oxygen isotope ratios ($\delta^{18}O$ values from 7.4 to 5.0‰), indicating that it formed from an evolving magmatic source. The evolved chemistry, high $\delta^{18}O$ value and micro-inclusions of SiO_2 are consistent with growth from a granitic melt[11,2] with a $\delta^{18}O$ value from 8.5 to 9.5‰. Magmatic oxygen isotope ratios in this range point toward the involvement of supracrustal material that has undergone low-temperature interaction with a liquid hydrosphere. This zircon thus represents the earliest evidence for continental crust and oceans on the Earth.

WE extracted a new aliquot of zircon from the original heavy mineral concentrate of the Jack Hills conglomerate from which the oldest documented crystal (~4,276 Myr; ref. 4) was obtained, and identified several grains with ages in excess of 4 Gyr. One of these is a deep purple zircon (W74/2-36), measuring 220 by 160 μm, that is a broken fragment of a larger crystal (Fig. 1). It shows few internal complexities or inclusions, a

44 亿年前地球上存在大陆壳与海洋：
来自碎屑锆石的证据

怀尔德等

编者按

地球被认为形成于 45.6 亿年前。在地球形成之初，熔浆遍布，当它冷却之后，这个星球发生了化学组成的分离和表层的固化。尽管之后的多期构造运动已使最早期陆壳遭受再循环作用，但仍有部分残留物质保存了下来，为其何时形成提供了证据。本文中，地质学家西蒙·怀尔德及其合作者报道了迄今所知的最古老陆壳物质，并揭示其形成于 44 亿年前。它含有一种被称为锆石的矿物，这种矿物可以通过铀钍铅体系的放射性衰变来定年。锆石似乎由地壳与水面交汇处的物质形成，这表明当时可能已经出现了海洋。

地球自约 45.5 亿年前诞生直到 40.3 亿年前这一期间遭受了强烈的陨石撞击，这期间的地壳岩石没有保留下来[1]。目前已知的最古老岩石年龄为 40.3 亿年，仅见于加拿大西北部的阿卡斯塔片麻岩[2]。但是，在西澳大利亚伊尔岗克拉通纳里耶片麻岩地体[9]的纳里耶山[3]和杰克山[4-8]变质沉积物中，42.76 亿年碎屑锆石的发现提供了更古老地壳物质存在的证据[4]。基于对杰克山锆石[10]详细微区分析研究，本文报道了年龄为 (4,404 ± 8) Myr (编者注：Myr 为百万年) 的碎屑锆石——比之前在地球上识别出的最古老年龄纪录还要早约 1.3 亿年。锆石具有与稀土元素和氧同位素比值 ($\delta^{18}O$ 值：7.4‰ ~ 5.0‰) 有关的环带结构，表明其生长过程中，岩浆源处于演化过程中。演化的化学特征、高的 $\delta^{18}O$ 值以及 SiO_2 包体的存在都表明锆石形成于 $\delta^{18}O$ 值为 8.5‰ ~ 9.5‰ 的花岗质熔体[11,2]。这一范围的岩浆氧同位素比值表明，花岗质岩浆形成过程中有与液态水圈存在过低温相互作用的地表物质的参与。因此，该锆石提供了地球上存在最古老陆壳和海洋的证据。

杰克山砾岩中曾发现的最古老锆石的年龄约为 42.76 亿年[4]。我们从原重矿物样中重新取了部分碎屑锆石进行研究，识别出数粒年龄超过 40 亿年的锆石颗粒。其中一粒 (W74/2-36) 呈深紫色，尺寸为 220 μm × 160 μm，是一粒原本更大锆石晶体的碎片 (图 1)。锆石内部结构简单，含少量包体，这些特征同样存在于杰克山砾岩

feature previously noted in zircons older than 4 Gyr from the Jack Hills conglomerate[4,13].

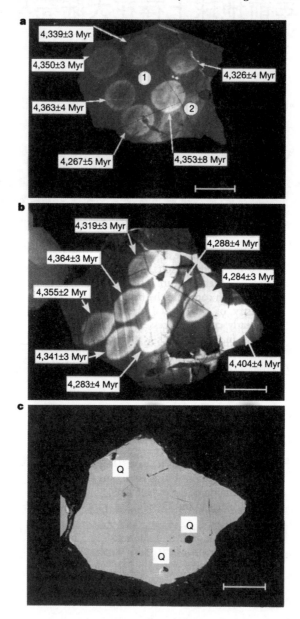

Fig.1. Cathodoluminescence and back-scattered electron images of zircon crystal W74/2-36. Scale bars are 50 μm. **a**, Image taken subsequent to first SHRIMP analysis. We note that lighter circular areas are the SHRIMP analytical sites and the values record the $^{207}Pb/^{206}Pb$ age (1σ) of each site. The two white areas represent the approximate location of the oxygen analytical spots, with δ^{18}O of 5‰ at point 1 and 7.4‰ at point 2. **b**, Cathodoluminescence image taken after the second SHRIMP analytical session showing the sites of analysis (as in **a**) and the $^{207}Pb/^{206}Pb$ ages (1σ) for each spot. We note the larger area showing light luminescence, the well defined cracks and the oscillatory zoning in both the light and dull portions of the crystal. We note the circular outline of the SHRIMP analytical sites and the lighter rectangular area, near the pointed termination, which corresponds to the dark area in the cathodoluminescence image. **c**, Back-scattered electron image obtained after the second SHRIMP analytical session. The black areas (marked Q) are SiO_2 inclusions.

老于 40 亿年的锆石中 [4,13]。

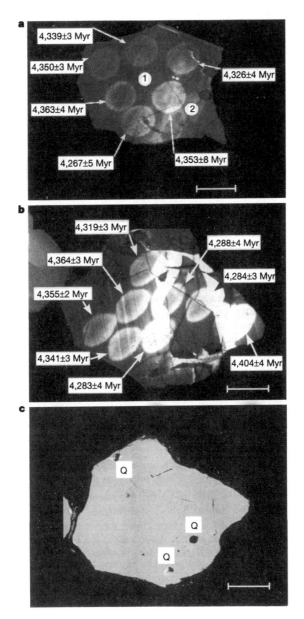

图 1. 锆石晶体 W74/2-36 的阴极发光图像和背散射电子图像（比例尺为 50 μm）。**a**，第一次 SHRIMP 分析后的阴极发光图像。较亮的圆形区域为分析点，旁边数字为 $^{207}Pb/^{206}Pb$ 年龄（1σ）。两个小圆圈为氧同位素分析的大致位置，分析点 1 的 $\delta^{18}O$ 值为 5‰，分析点 2 的 $\delta^{18}O$ 值为 7.4‰。**b**，第二次 SHRIMP 分析后的阴极发光图像，同图 **a** 一样，显示了分析点位置，并标出了 $^{207}Pb/^{206}Pb$ 年龄（1σ）。与第一次分析相比，阴极发光高亮度区域的范围更大，在阴极发光的亮部和暗部都存在清晰的裂隙和振荡环带。注意靠近锆石右侧突出部位的 SHRIMP 分析点圆形轮廓和更亮的矩形区域，后者与阴极发光图像中的暗色区域相对应。**c**，第二次 SHRIMP 分析后的背散射电子图像。黑色区域（以 Q 标识）为 SiO_2 包裹体。

We performed U-Th-Pb zircon analyses on the Perth Consortium SHRIMP II ion microprobe at Curtin University of Technology in Western Australia, following standard operating techniques[14-16]. Six analyses of grain W74/2-36 were initially made and five of these gave $^{207}Pb/^{206}Pb$ ages in excess of 4.3 Gyr (Table 1); the oldest and most concordant (97% concordance, Table 1) had an age of 4,363 ± 8 Myr (2σ), approximately 90 Myr older than the oldest known terrestrial material[4]. Other sites yielded more discordant ages and fell on a discordia line that passed through zero, indicative of recent lead loss (Fig. 2). A detailed cathodoluminescence (CL) image made after analysis indicated that all but the oldest site overlapped cracks within the crystal (Fig. 1a). However, crack-free domains greater than 50 μm in diameter were present and, after oxygen isotope and rare earth element (REE) analysis, approximately 20 μm was ground off the surface of the zircon to remove all previous analytical pits; the grain was then re-analysed on SHRIMP II in an attempt to obtain more concordant data. Care was taken to locate new analytical sites away from identifiable cracks and this proved partially successful, with four analyses obtained from such areas (Fig. 1b). Three of these sites yielded data that are 100, 98 and 95% concordant (Table 1), and confirm the initial estimate of the age, giving $^{207}Pb/^{206}Pb$ ages of 4,355 ± 4 Myr, 4,341 ± 6 Myr and 4,364 ± 6 Myr (2σ), respectively (Table 1 and Fig. 2). Importantly, the fourth site, located near the pointed, broken termination of the crystal (right corner of Fig. 1b), was also nearly concordant (99%) and this gave a $^{207}Pb/^{206}Pb$ age of 4,404 ± 8 Myr (2σ) (Fig. 2), about 40 Myr older than the other results and only about 150 Myr younger than the oldest high-temperature inclusions known in meteorites[17] (~4,560 Myr), which constrain the maximum age for the Earth.

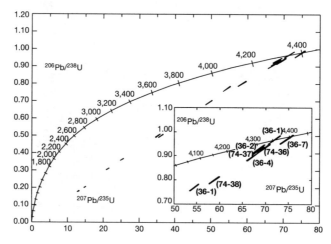

Fig. 2. Combined concordia plot for grain W74/2-36, showing the U-Pb results obtained during the two analytical sessions. The inset shows the most concordant data points together with their analysis number (as in Table 1). Error boxes are shown at 1σ.

The chemistry of this crystal therefore provides important insights into early crustal processes. Because the internal cracks that are evident in CL and back-scattered electron (BSE) images (Fig. 1) may affect the chemistry, they were investigated using a high-resolution scanning electron microscope (SEM) and electron microprobe. Unlike the cracks in the

利用西澳大利亚科廷技术大学珀斯联盟的 SHRIMP II 离子探针，我们遵循标准操作技术 [14-16] 对锆石进行了 U-Th-Pb 同位素分析。第一次分析中，在锆石 W74/2-36 上选取了 6 个点，其中 5 个分析点的 $^{207}Pb/^{206}Pb$ 年龄超过 43 亿年（表 1）；最古老且最谐和（谐和度 97%，表 1）的年龄为 $(4,363 \pm 8)$ Myr(2σ)，比已知的最古老陆壳物质大约早 9,000 万年 [4]。其他分析点的年龄更不谐和，落在一条过原点的不一致线上，反映了近代铅丢失（图 2）。根据分析后重照的高清阴极发光（CL）图像，除最古老年龄分析点外，其他所有分析点均位于晶体内部的裂隙之上（图 1a）。不过，锆石中也存在直径大于 50 μm 的无裂隙区域。在完成氧同位素和稀土元素（REE）分析后，磨除锆石表面大约 20 μm 的厚度以去掉之前测试造成的凹坑，然后再次进行 SHRIMP 分析以获得更谐和的年龄数据。谨慎选择的新分析点远离可见裂隙，这一尝试是部分成功的，有 4 个分析点落在了无裂隙的区域（图 1b）。其中 3 个 $^{207}Pb/^{206}Pb$ 年龄分别为 $(4,355 \pm 4)$ Myr、$(4,341 \pm 6)$ Myr 和 $(4,364 \pm 6)$ Myr(2σ)（表 1 和图 2），谐和度分别为 100%、98% 和 95%（表 1），与第一次分析获得的年龄结果一致。重要的是，在右侧突出的晶体碎片末端（图 1b），即第 4 个分析点的年龄几乎完全谐和（谐和度为 99%），并给出了 $(4,404 \pm 8)$ Myr(2σ) 的 $^{207}Pb/^{206}Pb$ 年龄（图 2）。比其他年龄大约老 4,000 万年，后者比已知的陨石 [17] 中最古老高温包裹体（约 45.60 亿年）仅年轻约 1.5 亿年，这限定了地球年龄的最大值。

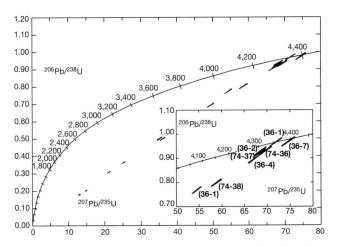

图 2. 锆石 W74/2-36 的 U-Pb 谐和图，给出了两次分析的 U-Pb 结果。插图给出了最谐和的分析点及编号（与表 1 的一致）。误差框为 1σ。

这粒锆石的化学特征可为研究早期地壳过程提供重要信息。在阴极发光图像和背散射电子（BSE）图像中，内部裂隙清楚可见（图 1），由于它们可能会影响元素分布特征，因此用高分辨率扫描电镜（SEM）和电子探针进行了研究。锆石 W74/2-36

Table 1. Shrimp U-Pb-Th isotopic analytical data

Spot	U (p.p.m.)	Th (p.p.m.)	Th/U ratio	Pb (p.p.m.)	f206%	$^{204}Pb/^{206}Pb$	$^{207}Pb^*/^{206}Pb^*$	$^{208}Pb^*/^{232}Th$	$^{208}Pb^*/^{206}Pb^*$	$^{206}Pb^*/^{238}U$	$^{207}Pb^*/^{235}U$	Conc. (%)	$^{206}Pb^*/^{238}U$ age (Myr)	$^{207}Pb^*/^{235}U$ age (Myr)	$^{208}Pb^*/^{232}Th$ age (Myr)	$^{207}Pb^*/^{206}Pb$ age (Myr)
							Session 1									
74-36	258	178	0.69	355	0.22	0.00014	0.5432 ± 1	0.222 ± 5	0.1645 ± 12	0.928 ± 18	69.5 ± 1.4	97	4,233 ± 61	4,321 ± 20	4,058 ± 80	4,363 ± 4
74-37	449	347	0.77	615	0.048	0.00003	0.5386 ± 1	0.225 ± 5	0.1895 ± 8	0.919 ± 18	68.2 ± 1.3	97	4,201 ± 59	4,303 ± 20	4,109 ± 75	4,350 ± 3
74-38	450	1,224	2.72	544	0.098	0.00006	0.5345 ± 1	0.065 ± 1	0.2227 ± 9	0.797 ± 15	58.8 ± 1.2	87	3,780 ± 55	4,153 ± 20	1,277 ± 25	4,339 ± 3
74-58	487	1,731	3.56	217	1.142	0.00071	0.5088 ± 2	0.012 ± 0	0.1419 ± 2	0.304 ± 6	21.4 ± 0.4	40	1,713 ± 29	3,155 ± 19	244 ± 6	4,267 ± 5
74-59	454	765	1.68	335	0.781	0.00049	0.5297 ± 1	0.048 ± 1	0.1637 ± 17	0.496 ± 10	36.2 ± 0.7	60	2,595 ± 41	3,672 ± 20	952 ± 21	4,326 ± 4
74-60	390	1,336	3.43	120	1.9	0.00119	0.5394 ± 3	0.010 ± 0	0.1641 ± 41	0.200 ± 4	14.9 ± 0.3	27	1,177 ± 21	2,808 ± 20	193 ± 6	4,353 ± 8
							Session 2									
36-1	361	262	0.73	515	0.002	0.00001	0.5401 ± 8	0.232 ± 5	0.1743 ± 6	0.965 ± 18	71.9 ± 1.4	100	4,356 ± 60	4,355 ± 19	4,211 ± 74	4,355 ± 2
36-2	258	174	0.67	351	0.01	0.00006	0.5353 ± 10	0.226 ± 5	0.1641 ± 7	0.929 ± 18	68.6 ± 1.3	98	4,235 ± 59	4,307 ± 19	4,125 ± 75	4,341 ± 3
36-3	184	117	0.64	204	0	0	0.5143 ± 12	0.194 ± 4	0.1612 ± 10	0.768 ± 15	54.5 ± 1.1	86	3,674 ± 53	4,078 ± 20	3,588 ± 68	4,283 ± 4
36-4	250	188	0.75	333	0	0	0.5436 ± 10	0.218 ± 4	0.1831 ± 7	0.897 ± 17	67.2 ± 1.3	95	4,126 ± 58	4,287 ± 19	3,986 ± 72	4,364 ± 3
36-5	208	304	1.46	153	0	0	0.5163 ± 13	0.067 ± 1	0.1959 ± 10	0.498 ± 9	35.4 ± 0.7	61	2,604 ± 40	3,651 ± 19	1,306 ± 25	4,288 ± 4
36-6	346	940	2.71	182	0.001	0	0.5147 ± 11	0.026 ± 1	0.2005 ± 9	0.356 ± 7	25.2 ± 0.5	46	1,961 ± 32	3,317 ± 19	524 ± 10	4,284 ± 3
36-7	127	75	0.59	180	0	0	0.5587 ± 14	0.237 ± 5	0.1437 ± 9	0.968 ± 19	74.6 ± 1.5	99	4,364 ± 61	4,392 ± 20	4,291 ± 83	4,404 ± 4
36-8	329	339	1.03	337	0.013	0.00001	0.5272 ± 9	0.141 ± 3	0.2134 ± 8	0.663 ± 13	49.6 ± 0.9	78	3,355 ± 49	3,985 ± 19	2,672 ± 49	4,319 ± 3

* Common lead corrected using ^{204}Pb.

Data for zircon grain W74/2-36. f206% is (common ^{206}Pb/total ^{206}Pb) × 100. Conc. (%) is percentage concordance defined as $[(^{206}Pb/^{238}U$ age)/$(^{207}Pb/^{206}Pb$ age)] × 100. All errors are quoted at 1σ level. Errors given for Pb/Pb and Pb/Th ratios are based on counting statistics. Errors in Pb/U ratios also include an estimate of the Pb/U reproducibility error based on multiple analyses of the standard zircon CZ3 during the analytical sessions. During the first analytical session, all spots were given a unique number. In the second session, the grain was identified as W74-36 and each site was numbered consecutively, 36-1 to 36-8.

表 1. Shrimp U-Pb-Th 同位素分析数据

数据点	U (ppm)	Th (ppm)	Th/U ratio	Pb (ppm)	f206%	$^{204}Pb/^{206}Pb$	$^{207}Pb^*/^{206}Pb^*$	$^{208}Pb^*/^{232}Th$	$^{206}Pb^*/^{238}U$	$^{207}Pb^*/^{235}U$	Conc. (%)	$^{206}Pb^*/^{238}U$ age (Myr)	$^{207}Pb^*/^{235}U$ age (Myr)	$^{208}Pb^*/^{232}Th$ age (Myr)	$^{207}Pb^*/^{206}Pb$ age (Myr)
							第一次分析								
74-36	258	178	0.69	355	0.22	0.00014	0.5432±1	0.222±5	0.928±18	69.5±1.4	97	4,233±61	4,321±20	4,058±80	4,363±4
74-37	449	347	0.77	615	0.048	0.00003	0.5386±1	0.225±5	0.919±18	68.2±1.3	97	4,201±59	4,303±20	4,109±75	4,350±3
74-38	450	1,224	2.72	544	0.098	0.00006	0.5345±1	0.065±1	0.797±15	58.8±1.2	87	3,780±55	4,153±20	1,277±25	4,339±3
74-58	487	1,731	3.56	217	1.142	0.00071	0.5088±2	0.012±0	0.304±6	21.4±0.4	40	1,713±29	3,155±19	244±6	4,267±5
74-59	454	765	1.68	335	0.781	0.00049	0.5297±1	0.048±1	0.496±10	36.2±0.7	60	2,595±41	3,672±20	952±21	4,326±4
74-60	390	1,336	3.43	120	1.9	0.00119	0.5394±3	0.010±0	0.200±4	14.9±0.3	27	1,177±21	2,808±20	193±6	4,353±8
							第二次分析								
36-1	361	262	0.73	515	0.002	0.00001	0.5401±8	0.232±5	0.965±18	71.9±1.4	100	4,356±60	4,355±19	4,211±74	4,355±2
36-2	258	174	0.67	351	0.01	0.00006	0.5353±10	0.226±5	0.929±18	68.6±1.3	98	4,235±59	4,307±19	4,125±75	4,341±3
36-3	184	117	0.64	204	0	0	0.5143±12	0.194±4	0.768±15	54.5±1.1	86	3,674±53	4,078±20	3,588±68	4,283±4
36-4	250	188	0.75	333	0	0	0.5436±10	0.218±4	0.897±17	67.2±1.3	95	4,126±58	4,287±19	3,986±72	4,364±3
36-5	208	304	1.46	153	0	0	0.5163±13	0.067±1	0.498±9	35.4±0.7	61	2,604±40	3,651±19	1,306±25	4,288±4
36-6	346	940	2.71	182	0.001	0	0.5147±11	0.026±1	0.356±7	25.2±0.5	46	1,961±32	3,317±19	524±10	4,284±3
36-7	127	75	0.59	180	0	0	0.5587±14	0.237±5	0.968±19	74.6±1.5	99	4,364±61	4,392±20	4,291±83	4,404±4
36-8	329	339	1.03	337	0.013	0.00001	0.5272±9	0.141±3	0.683±13	49.6±0.9	78	3,355±49	3,985±19	2,672±49	4,319±3

* 由 ^{204}Pb 作普通铅校正。

锆石颗粒 W74/2-36 的数据。f206% =（普通 ^{206}Pb/ 总 ^{206}Pb）×100。Conc.(%) 是百分比谐和度，定义为 [（$^{206}Pb/^{238}U$ 年龄）/（$^{207}Pb/^{206}Pb$ 年龄）] ×100。所有误差为 1σ。Pb/Pb 和 Pb/Th 比值的误差基于计数统计。Pb/U 比值的误差也包括了分析过程中基于标准锆石 CZ3 的多次分析对 Pb/U 再现性误差的估计。第一次分析中，每个分析点对应一个特定的编号。第二次分析中，锆石颗粒被标记为 W74/2-36，每个分析点从 36-1 至 36-8 连续编号。

xenocrysts older than 4 Gyr that were recently discovered in granitoids from the Narryer and Murchison Terranes of Western Australia[18], which are infilled with recrystallized zircon, the cracks in W74/2-36 are open. They contain thin (generally less than 300 nm thick), discontinuous films of phosphates, resulting in slightly elevated values of Fe, Al, P and Y. These are, however, too small to have a significant effect on the oxygen budget of 25-μm analytical sites; thus they do not compromise our interpretation of the oxygen data.

Grain W74/2-36 has U and Th values ranging from 127–487 and 75–1,731 p.p.m., respectively (Table 1), reaching values considerably higher than those recorded from the ~4,276-Myr-old crystal[4]. There is no correlation between the ^{207}Pb/^{206}Pb age of the various analytical sites and the U, Th and Pb contents, or the Th/U ratio (Table 1), although the oldest site has the lowest values for all of these except Pb. These results suggest complex micrometre to sub-micrometre variations in composition, as indicated by several detailed studies[18-20], and also preclude any estimate of the composition of the host rock based on isotopic characteristics. The Th/U ratio varies from 0.59 to 3.56, with those sites in the range 0.59 to 0.77 having ^{208}Pb/^{232}Th ages greater than 3,588 Myr and being the least discordant (Table 1). The large differences in Pb/U for similar ^{207}Pb/^{206}Pb ages (Table 1) indicate recent lead loss. This is especially evident within the combined light and dark rectangular area visible in the CL image, which may have formed part of the central region of the original zircon (Fig. 1b). The three spots from the second analytical run (36-3, 36-5 and 36-6), which partially overlap this zone, define a weighted ^{207}Pb/^{206}Pb age of $4,289 \pm 7$ Myr (2σ) that is younger than the ^{207}Pb/^{206}Pb age of all other sites (Table 1). A similar feature was noted in the core of a xenocrystic zircon older than 4.0 Gyr in granitic gneiss from Churla Well in the Narryer Terrane[18] and was interpreted as reflecting an ancient episode of lead loss resulting from an event at high temperature. Unlike Churla Well, our data show no evidence of subsequent lead loss during Precambrian time, suggesting that the zircon was not affected by later igneous or metamorphic processes before its deposition in the conglomerate.

We interpret the concordant ^{207}Pb/^{206}Pb age of $4,404 \pm 8$ Myr (2σ) as recording the time of crystallization of zircon W74/2-36. Although this is represented by only one analysis, it is concordant (99%), is not affected by cracks, and there is no evidence from the analytical data that this site, rather than any other measured, may be anomalous. The crystal was subsequently affected by one or more episodes of ancient lead loss, but the timing of these is unresolved. The concordant and near-concordant ages of $4,364 \pm 6$ Myr, $4,355 \pm 4$ Myr and $4,341 \pm 6$ Myr (2σ) may represent actual geological events, possibly triggered by meteor bombardment[18], or they may indicate slight changes in ^{207}Pb/^{206}Pb without generating much discordance[2], because of their position on the concordia curve (Fig. 2). The ^{207}Pb/^{206}Pb age of $4,289 \pm 7$ Myr (2σ) for the three more discordant points from the light and dark area in the CL image (Fig. 1b) were most probably reset at this time by a high-temperature event, and subsequently underwent recent lead loss.

的裂隙是开放的，这不同于最近发现于西澳大利亚纳里耶和默奇森地体的花岗质岩石中老于 40 亿年的捕虏锆石，后者的裂隙都被重结晶锆石充填[18]。锆石 W74/2-36 的裂隙中存在不连续的磷酸盐薄膜（通常小于 300 nm），导致 Fe、Al、P 和 Y 含量轻微升高。然而这些裂隙太小，完全不足以对大小为 25 μm 的分析点的氧同位素分析产生影响，因此也不会影响我们对氧同位素数据的解释。

锆石 W74/2-36 的 U 和 Th 含量分别为 127 ~ 487 ppm 和 75 ~ 1,731 ppm（表 1），明显高于早期发现的年龄约 42.76 亿年的锆石[4]。不同分析点的 $^{207}Pb/^{206}Pb$ 年龄与 U、Th、Pb 含量或 Th/U 比值之间不具有相关性（表 1），尽管除了 Pb 以外，最古老年龄分析点的其他元素含量都最低。与其他一些详细研究[18-20]一样，这些结果暗示了锆石内部的化学组成在微米到亚微米尺度上存在复杂的变化，也排除了基于锆石同位素特征对母岩做出组成判断的可能。Th/U 比值范围为 0.59 ~ 3.56，其中 Th/U 比值为 0.59 ~ 0.77 的分析点 $^{208}Pb/^{232}Th$ 年龄大于 35.88 亿年，并具有最高的谐和度（表 1）。锆石 $^{207}Pb/^{206}Pb$ 年龄相似而 Pb/U 比值存在巨大差异（表 1），表明近代铅丢失的存在。这一现象在阴极发光图像中亮暗结合的矩形区域内尤为明显，这些矩形区域可能构成了原始锆石中央区域的一部分（图 1b）。第二次分析的 3 个点（36-3，36-5 和 36-6）与该区域部分重合，给出了（4,289±7）Myr（2σ）的 $^{207}Pb/^{206}Pb$ 加权平均年龄，比其他所有分析点的 $^{207}Pb/^{206}Pb$ 年龄都要年轻（表 1）。类似特征也存在于纳里耶地体 Churla Well 花岗质片麻岩的老于 40 亿年锆石捕虏晶的核部[18]，年龄被解释为一次高温事件所导致的古铅丢失的时代。与 Churla Well 花岗质片麻岩不同的是，我们的数据显示，在前寒武纪时期不存在后继的铅丢失，表明该锆石在沉积进入砾石之前并未再次受到岩浆或变质过程的影响。

我们将（4,404±8）Myr 这一谐和的 $^{207}Pb/^{206}Pb$ 年龄解释为锆石 W74/2-36 的结晶时间。尽管只有一个分析点，但没受裂隙的影响，谐和度很高（99%），而且从分析数据看，没有证据表明该分析点存在异常情况，而其他分析点没有异常。锆石在后来受到了一次或多次古铅丢失事件的影响，但其时间仍未确定。谐和或近于谐和的 3 个年龄值（4,364±6）Myr、（4,355±4）Myr 和（4,341±6）Myr（2σ）也许代表了可能由陨石轰击所触发的真实地质事件，或者可能指示了 $^{207}Pb/^{206}Pb$ 比值的轻微变化，但没有导致明显的不谐和[2]，因为分析点仍位于谐和线上（图 2）。位于阴极发光图像（图 1b）中明暗结合区域而谐和度更差的 3 个分析点，其（4,289±7）Myr（2σ）的 $^{207}Pb/^{206}Pb$ 年龄很可能代表了一次高温事件的重置时间，并随后经历了近代铅丢失。

Most of the crystal exhibits weak banding in the CL images (Fig. 1a and b), indicating elemental variation marked by differences in luminescence which usually extend fully across the crystal, suggesting that the zircon was originally much larger. The development of banding parallel to the two sharp crystal boundaries in the top left-hand corner of the image and also within the bright area (Fig. 1b) suggests that it reflects oscillatory zoning of magmatic origin. The bright area almost entirely encloses the dark central region, which is devoid of any CL pattern.

Variation in the CL image is confirmed by the REE data (Fig. 3), which were obtained from the floors of five of the initial SHRIMP analytical pits. The analyses yielded prominent positive Ce anomalies, negative Eu anomalies, and heavy (H)REE-enrichment relative to light (L)REE contents; these are features typical of zircons analysed by ion microprobe[11] and from other Jack Hills detrital zircons[13]. The crystal is zoned with respect to LREE (La by about 10-fold), with greater LREE enrichment within and around the bright portion of the crystal. Calculated partition coefficients[11] yield magma compositions with marked LREE-enrichment for this part of the crystal (Lu/La_n (magma) is 238 to 616) and negative Eu anomalies, features indicative of evolved granitic melts. The crystal's growth in a silica-saturated environment is further supported by the identification of three areas, up to 20-μm in diameter, with SiO_2 inclusions (Fig. 1c).

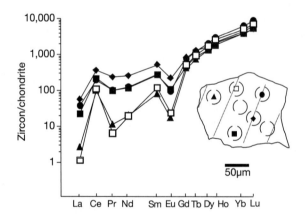

Fig. 3. Rare earth element data for Jack Hills zircon W74-2/36 measured by ion microprobe. We note that spacing of elements along the x-axis is based on radii of 3^+ cations (in Å).

Four analyses of oxygen isotope ratio were made on the Edinburgh Cameca ion microprobe on two spots close to two of the initial SHRIMP sites. One spot was located near the $4,339 \pm 3$ Myr site in the low LREE region that is dull and banded in CL (Fig. 1a). This has a $\delta^{18}O$ of 5.0 ± 0.7‰ (1σ), which is indistinguishable from the $\delta^{18}O$ values of mantle-derived zircon[21] and zircon from juvenile, 3.0 to 2.7 Gyr tonalite-trondhjemite-granodiorite plutonic rocks of the Superior Province[22]. The second spot was located in the high LREE region of the grain, which is brighter in the CL image (Fig. 1a), close to the site yielding a $^{207}Pb/^{206}Pb$ age of $4,353 \pm 8$ Myr (2σ). This has a $\delta^{18}O$ value of 7.4 ± 0.7‰ (1σ), consistent with equilibrium with a magma of about 8.5 to 9.5‰. Magmatic oxygen isotope ratios in

在阴极发光图中，锆石晶体的大部分显示出微弱的环带（图 1a 和 1b），指示了由阴极发光亮度变化所标记的元素组成变化。这种变化通常贯穿整个晶体，表明锆石的原始粒度要大得多。在图像的左上角及明亮区域，带状结构发育，并与两条锆石平直边界相平行（图 1b），它们为岩浆成因的振荡环带。明亮区域几乎完全包围了没有任何阴极发光结构的暗色中央区域。

在 5 个原 SHRIMP 分析点位置进行了稀土元素元素分析，结果印证了阴极发光图像所显示的锆石结构变化。锆石具有明显的正 Ce 异常和负 Eu 异常，重稀土元素（HREE）相对于轻稀土元素（LREE）更为富集。这些是以往离子探针分析已揭示出的锆石的典型特征[11]，也是其他来自杰克山的碎屑锆石的典型特征[13]。锆石的轻稀土元素组成显示环带结构（La 的丰度相差 10 倍），阴极发光高亮度区域及其附近轻稀土元素更为富集。计算出的分配系数[11]表明，锆石晶体的这些部位（Lu/La_n（岩浆）的范围为 238 ~ 616）显示与之平衡的岩浆具有轻稀土元素明显富集的组成特征，同时存在负 Eu 异常，表明锆石结晶于已具有一定演化的花岗质熔体。在锆石晶体中识别出 3 个直径达 20 μm 的 SiO_2 包体（图 1c），也支持了其生长于 Si 饱和环境的认识。

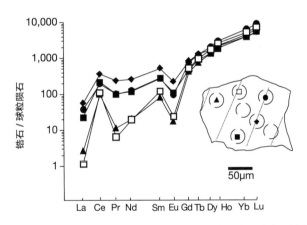

图 3. 杰克山锆石 W74-2/36 的稀土元素数据，由离子探针方法测定。各元素在 x 轴上的间距基于 3^+ 价阳离子的半径，以 Å 为单位。

在两个 SHRIMP 分析点的附近选取了两个点，在爱丁堡 Cameca 离子探针上进行了 4 次氧同位素分析。其中一个位于年龄为 (4,339 ± 3) Myr 的分析点附近轻稀土元素含量低的位置，在阴极发光图像中颜色灰暗，呈环带结构（图 1a）。这一点的 $\delta^{18}O$ 值为 5.0‰ ± 0.7‰（1σ），氧同位素组成与地幔来源的岩浆锆石[21]难以区别，也与苏必利尔省 30 亿到 27 亿年新生英云闪长岩–奥长花岗岩–花岗闪长岩侵入体的岩浆锆石[22]难以区别。第二个点位于阴极发光图像中更明亮（图 1a）轻稀土元素含量高的区域，紧挨着 $^{207}Pb/^{206}Pb$ 年龄为 (4,353 ± 8) Myr（2σ）的分析点位置。该点的 $\delta^{18}O$ 值为

this range are indicative of involvement of supracrustal material which has undergone low-temperature interaction with a liquid hydrosphere; there is no known primitive or mantle reservoir of this composition.

REE and oxygen isotope compositions in grain W74/2-36 provide unique insights into the Earth's crust at 4.4 Gyr ago. They indicate formation in a granitic melt derived by partial melting of pre-existing crust. Calculated magma compositions show characteristics distinctive of evolved melts (such as LREE enrichment, prominent negative Eu anomalies, silica saturation), and are not consistent with the REEs of primitive, mafic rocks[10]. The variation from relatively high to low LREEs, coupled with a 2.4‰ decrease in the oxygen isotope ratio, is consistent with igneous zoning associated with melting and assimilation and subsequent equilibration with a mantle-derived melt. The initially high-$\delta^{18}O$ magma (~9‰) could result from melting of ocean crust that was hydrothermally altered at low temperatures[23], altered or evolved continental crust, or sediments. The high $\delta^{18}O$ value cannot result from closed-system fractional crystallization by a normal mantle melt. In the case of simple melting without assimilation, zircons in the melt would have a similar $\delta^{18}O$ value to zircons in its protolith. An alternative hypothesis, involving assimilation coupled with fractional crystallization (AFC) by a normal basalt of high $\delta^{18}O$ wallrock, is unlikely, as it would require either energetically unrealistic amounts of assimilation ($>50\%$) or wallrocks with average $\delta^{18}O$ values above 12‰ which are uncommon even in the late Archaean[22]. The correlation of high $\delta^{18}O$ values and enriched LREEs is most consistent with the melting of continental crust or sediments, but whichever of these processes dominated, a large reservoir of liquid water is required on the surface of the Earth.

The zoning of zircon W74/2-36 provides further insights into magmatic processes at 4.4 Gyr ago. Such zones could form during AFC, but we suggest that the zircon first crystallized in a magma derived from the melting of existing continental crust at $4,404 \pm 8$ Myr ago, which subsequently formed a magmatic over-growth when mixed with, or remelted by, a more primitive magma between $4,364 \pm 6$ Myr ago and $4,289 \pm 7$ Myr ago. This indicates that at 4.4 Gyr ago there were already intermediate to granitic, high-$\delta^{18}O$ continental rocks to contaminate the magma in which grain W74/2-36 grew.

The existence of liquid water at 4.4 Gyr ago could have important implications for the evolution of life. Microfossils as old as 3.5 Gyr are known[24]. Metasediments and carbonaceous materials with biogenic carbon isotope ratios as low as $\delta^{13}C = -28$‰ are known at 3.8 Gyr ago[25]. Zircon crystal W74/2-36 is over 500 Myr older than this organic matter and if liquid water was available to cause the evolved geochemistry that we have measured, then such water was also available for possible biological processes. High-energy asteroid bombardment before 3.9 Gyr ago is consistent with periodic formation and destruction of early oceans and the possibility that primitive life, if it evolved in the oceans, was globally extinguished more than once.

550

$7.4‰±0.7‰(1\sigma)$，大致与 $\delta^{18}O$ 值约为 8.5‰～9.5‰ 的岩浆平衡。这一氧同位素比值表明，岩浆过程中有与液态水圈发生过低温反应的地表物质参与。已知的原始储库或地幔储库都不具有这种氧同位素组成特征。

锆石 W74/2-36 的稀土元素和氧同位素组成使人们对 44 亿年前的地壳有了独特的认识，刻画了先存地壳发生部分熔融进而形成花岗岩熔体的过程。计算所得的岩浆成分显示了一定程度的演化特征，例如轻稀土元素富集、明显的负 Eu 异常以及 Si 饱和等，与原始基性岩的稀土元素特征 [10] 存在明显区别。轻稀土元素相对含量从高到低的变化，以及氧同位素比值 2.4‰ 的下跌，表明火成岩浆作用具有分带性，这与熔融作用、同化作用以及其后与幔源熔体的平衡作用有关。高 $\delta^{18}O$ 值的初始岩浆（～9‰）可由遭受过低温热液蚀变的洋壳发生熔融而形成 [23]，也可由蚀变或演化的陆壳或沉积物发生熔融而形成，但不可能由正常地幔熔体在封闭系统内通过结晶分异而形成。在缺乏同化作用的简单熔融情况下，熔体中的锆石与其母岩中的锆石应有相似的 $\delta^{18}O$ 值。另一种假设，即存在高 $\delta^{18}O$ 围岩的正常玄武质岩浆同时经历同化作用和结晶分异作用（AFC 过程），也是不可能的，因为这需要极高的同化量（>50%，从能量角度讲这并不现实），或者围岩的平均 $\delta^{18}O$ 值要高于 12‰，而这即使在晚太古代也是不常见的 [22]。高 $\delta^{18}O$ 值和轻稀土元素富集之间的相关性与大陆地壳或沉积物的熔融最吻合，无论哪个过程占主导地位，都需要地球表面存在大规模的液态水。

锆石 W74/2-36 的环带结构为 44 亿年前的岩浆活动提供了新的相关信息。这些环带可能形成于结晶分异过程，但我们认为，锆石首先是在 (4,404±8) Myr 前从已存在的陆壳熔融而来的岩浆中结晶而出。随后在 (4,364±6) Myr 前和 (4,289±7) Myr 前这一期间，由于与更为原始的岩浆进行混合或由其引发再次熔融，形成了岩浆锆石增生边。这表明，在 44 亿年前就已存在中性至花岗质的、具有高 $\delta^{18}O$ 值的大陆岩石，正是这些大陆岩石与原始岩浆发生混染，生长出了锆石 W74/2-36 的岩浆锆石增生边。

44 亿年前液态水的存在可能对生命演化有重要启示。目前已经发现 35 亿年前的微体化石 [24]。38 亿年前便存在具有生物成因碳同位素比值（$\delta^{13}C$ 低至 −28‰）的变质沉积物和碳质物质 [25]。锆石 W74/2-36 比这些有机物还要早 5 亿年以上。如果液态水能影响地球演化过程的化学特征（如本文确定的那样），那么也同样有助于可能的生物演化进程。39 亿年以前的高能小行星撞击与早期海洋的周期性消长相吻合，如果原始生命的进化发生在海洋中，那它们曾不止一次地发生过全球性灭绝。

Table 2. Rare earth element and oxygen isotope data

Trace element data (p.p.m.)												
Spot number	La	Ce	Pt	Nd	Sm	Eu	Gd	Tb	Dy	Ho	Yb	Lu
m2-13	5.5	134	9.4	55	41.4	6.3	100	34	338	106	666	135
m2-14	9.2	122	9.5	59	42.6	5.6	155	47	518	173	1065	225
m2-17	0.6	61	1.1	9	12.3	1.0	85	28	347	124	768	159
m2-30	13.6	226	21.8	119	78.4	12.7	170	49	424	119	672	130
m2-31	0.3	69	0.6	9	18.4	1.4	108	36	438	149	877	182

Oxygen isotope data			
Spot number	Analysis number	Measured $^{18}O/^{16}O$ ($\times 10^{-3}$)	$\delta^{18}O$ (VSMOW)
m2-1	2(15)	1.9378	4.8
m2-2d	2(16)	1.9388	5.3
			Avg. 5.0
m2-3	2(21)	1.9414	6.8
m2-4d	2(22)	1.9436	8
			Avg. 7.4

Data were measured by ion microprobe for crystal W74-2/36.

Methods

Zircon data collection

Approximately 100 zircon grains were hand picked from a previously prepared +135-μm concentrate and mounted onto double-sided adhesive tape, along with pieces of the Curtin University Sri Lankan gem zircon standard (CZ3) with a conventionally-measured U-Pb age of 564 Myr (ref. 26). They were enclosed in epoxy resin disks, ground and polished so as to effectively cut all zircon grains in half, and then gold coated. Samples were imaged by cathodoluminescence (CL), resulting in a map that allowed us to identify grains, as well as providing information on internal structure that could be tested during analysis. During the first analytical session, a mass resolution of 5,400 was obtained and the error associated with the measurement of Pb/U isotopic ratios for the standard, at 1 standard deviation, was 1.96% for seven standards. After discovery of a grain with an age in excess of 4.3 Gyr, and following the collection of oxygen isotope and trace element data, the sample was reground, repolished and gold coated and, on the basis of a new CL image, the grain was reanalysed on SHRIMP II, with a total of eight new sites selected so that most were away from cracks in the crystal. During this session, the mass resolution was 4,885 and the error on the standard, at 1 standard deviation, was 1.86% for six standards. The relationship between measured Pb/U and UO/U ratios on SHRIMP follows a power-law equation with the exponent equal to two (ref. 27). The Pb/U ratios on the unknowns were normalized to those measured on the standard zircon (CZ3 $- (^{206}Pb/^{238}U = 0.0914)$). Both data sets were reduced following the methods of Nelson[16], using the single-stage model Broken Hill common Pb correction for mass ^{204}Pb, since this

表 2. 稀土元素和氧同位素数据

稀土元素数据（ppm）												
数据点编号	La	Ce	Pt	Nd	Sm	Eu	Gd	Tb	Dy	Ho	Yb	Lu
m2-13	5.5	134	9.4	55	41.4	6.3	100	34	338	106	666	135
m2-14	9.2	122	9.5	59	42.6	5.6	155	47	518	173	1065	225
m2-17	0.6	61	1.1	9	12.3	1.0	85	28	347	124	768	159
m2-30	13.6	226	21.8	119	78.4	12.7	170	49	424	119	672	130
m2-31	0.3	69	0.6	9	18.4	1.4	108	36	438	149	877	182

氧同位素数据			
数据点编号	分析编号	测量过的 $^{18}O/^{16}O(\times 10^{-3})$	$\delta^{18}O(VSMOW)$
m2-1	2(15)	1.9378	4.8
m2-2d	2(16)	1.9388	5.3
			平均 5.0
m2-3	2(21)	1.9414	6.8
m2-4d	2(22)	1.9436	8
			平均 7.4

数据通过离子探针测定锆石 W74-2/36 得到。

方　法

锆石数据收集

从已有的 +135-μm 精矿中手工挑选出大约 100 粒锆石，连同科廷大学的斯里兰卡宝石级标准锆石（CZ3，用传统方法测得的 U-Pb 年龄为 5.64 亿年）碎片[26]一起放在双面胶带上，封装在环氧树脂盘内，经研磨和抛光，露出所有锆石颗粒的中部，然后镀金。再照阴极发光图像，以便识别颗粒并提供有助于分析测试的颗粒内部结构信息。在第一次分析中，质量分辨率为 5,400，七个标准样品 Pb/U 同位素比值的分析误差为 1.96%（标准差为 1）。在发现年龄超过 43 亿年的一粒锆石并测定其氧同位素和微量元素数据后，锆石样品靶被重新研磨、抛光和镀金，在新的阴极发光图像基础上，避开晶体中的裂隙，选择八个新的位置，在 SHRIMP II 上对该粒锆石进行了重新分析。在这一分析过程中，质量分辨率为 4,885，六个标准样品的分析误差为 1.86%（标准差为 1）。SHRIMP 测得的 Pb/U 和 UO/U 比值之间的关系遵循指数为 2 的幂律方程[27]。未知分析点的 Pb/U 比值通过标准锆石的测量值（CZ3 － ($^{206}Pb/^{238}U = 0.0914$)）进行标准化。两组数据均按照纳尔逊的方法[16]进行处理，使用单阶段模型布罗肯山普通铅作 ^{204}Pb 校正，因为普通铅主要是在镀金过程中带入的[16]。每一次分析中，分析点的平均直径为 30 μm，每一个点在分析前，对分析位置作扫描清洗，清洗范围为

is considered to be introduced chiefly through the gold coating[16]. The analytical spot size averaged 30 μm during each analytical run and each spot was rastered over 100 μm for five minutes before analysis to remove common Pb on the surface or contamination from the gold coating. All stated uncertainties and data listed in Table 1, and the error bars shown in Fig. 2, are at 1σ; ages discussed in the text are all 2σ.

Oxygen isotope analysis

Four analyses were made by Cameca ims 4f ion microprobe on two spots on grain W74/2- 36 using an energy offset of 350 eV. These were analysed for a total of 2×10^6 counts of ^{18}O for each analysis, yielding precision close to 0.7‰ (1σ), based on gaussian counting statistics[28]. Internal precision for each analysis was ± 0.3 and $\pm 0.6‰$ (1σ), comparing 80-cycle analysis halves on each full analysis, in agreement with theoretical counting statistics. The two point-analyses of the zircon crystal were interspersed with 11 analyses of Kim-5, a homogeneous zircon standard $(\delta^{18}O = 5.04 \pm 0.07‰$ VSMOW by laser fluorination) mounted in a separate standard block. Five standard zircons with 1.06 to 1.52 wt% HfO_2 were also analysed by ion microprobe and laser fluorination[10], and no statistically significant dependence of instrumental mass fractionation on HfO_2 content was found[29] for the narrow range of Hf in these samples.

Rare earth element analysis

In situ determination of REEs was performed by Cameca ims 4f ion microprobe[11]. Five ion microprobe REE analyses of the zircon crystal were made using a 14.5-keV primary beam of O⁻ defocused to an approximately 20 to 30 μm spot. Positive secondary ions were collected using an energy offset of 125 eV. Analyses were standardized to the SRM-610 glass standard. Energy filtering and strategies to avoid and correct for isobaric interferences were used[11].

(**409**, 175-178; 2001)

Simon A. Wilde*, **John W. Valley**†, **William H. Peck**†‡ & **Colin M. Graham**§

* School of Applied Geology, Curtin University of Technology, GPO Box U1987, Perth, Australia
† Department of Geology & Geophysics, University of Wisconsin, Madison, Wisconsin 53706, USA
§ Department of Geology & Geophysics, University of Edinburgh, Edinburgh, EH9 3JW, UK
‡ Present address: Department of Geology, Colgate University, Hamilton, New York 13346, USA

Received 1 August; accepted 24 November 2000.

References:

1. Ryder, G. Chronology of early bombardment in the inner solar system. *Geol. Soc. Am. Abstr. Progm* **21** A299 (1992).

2. Bowring, S. A. & Williams, I. S. Priscoan (4.00-4.03) orthogneisses from northwestern Canada. *Contrib. Mineral. Petrol.* **134**, 3-16 (1999).

3. Froude, D. O. *et al.* Ion microprobe identification of 4,100-4,200 Myr-old terrestrial zircons. *Nature* **304**, 616-618 (1983).

4. Compston, W. & Pidgeon, R. T. Jack Hills, evidence of more very old detrital zircons in Western Australia. *Nature* **321**, 766-769 (1986).

5. Wilde, S. A. & Pidgeon, R. T. in *3rd International Archaean Symposium (Perth), Excursion Guidebook* (eds Ho, S. E., Glover, J. E., Myers, J. S. & Muhling, J. R.) 82-95 (University of Western Australia Extension Publication, Vol. 21, Perth, 1990).

6. Kober, B., Pidgeon, R. T. & Lippolt, H. J. Single-zircon dating by step-wise Pb-evaporation constrains the Archean history of detrital zircons from the Jack Hills, Western Australia. *Earth Planet. Sci. Lett.* **91**, 286-296 (1989).

100 μm，持续时间为 5 分钟，以除去样品靶表面的普通铅或镀金带来的污染。表 1 中提到的不确定性和列出的数据以及图 2 中的误差棒都为 1 个标准差，而文中讨论的所有年龄的误差都为 2 个标准差。

氧同位素分析

利用 Cameca ims 4f 离子探针，对锆石 W74/2-36 上的两个点进行了 4 次氧同位素分析，所用的能量补偿为 350 eV。每次分析 ^{18}O 的总计数为 2×10^6，基于高斯计数统计 [28]，分析精确度接近 0.7‰（1σ）。每次分析的内部精度为 ±0.3‰ 和 ±0.6‰（1σ），分别对应于全分析和数量减半的 80 组分析，其精度与理论计数统计一致。在对锆石 W74/2-36 两个点进行分析的过程中，插入了 11 次标准锆石 Kim-5 的分析，该标准锆石的氧同位素组成均一（$\delta^{18}O$ = 5.04‰ ±0.07‰ VSMOW，激光氟化分析），被固定于一个独立放置标样的位置。同时也利用离子探针和激光氟化法对五个 HfO_2 含量为 1.06 wt% ~ 1.52 wt% 的标准锆石进行了分析 [10]。在这样 Hf 含量变化不大的情况下，未发现仪器质量分馏效应与 HfO_2 含量之间存在具有统计意义上的相关性 [29]。

稀土元素分析

稀土元素的原位测定由 Cameca ims 4f 离子探针 [11] 完成。使用 14.5 keV 的 O^- 一次粒子束，散焦至大约 20 ~ 30 μm 的束斑范围。对锆石 W74/2-36 进行了 5 次离子探针稀土元素分析。能量补偿为 125 eV，对正二次离子进行收集。利用 SRM-610 玻璃标准样品对分析结果进行标准化。此外还使用了能量过滤及其他方法来避免并校正同质异位素的干扰 [11]。

（卢皓 翻译；万渝生 刘守偈 审稿）

7. Amelin, Y. V. Geochronology of the Jack Hills detrital zircons by precise U-Pb isotope dilution analysis of crystal fragments. *Chem. Geol.* **146**, 25-38 (1998).

8. Amelin, Y., Lee, D.-C., Halliday, A. N. & Pidgeon, R. T. Nature of the Earth's earliest crust from hafnium isotopes in single detrital zircons. *Nature* **399**, 252-255 (1999).

9. Myers, J. S. in *Early Precambrian Processes* (eds Coward, M. P. & Ries, A. C.) 143-154 (Geological Society of London Special Publication No. 95, 1995).

10. Peck, W. H., Valley, J. W., Wilde, S. A. & Graham, C. M. Oxygen isotope ratios and rare earth elements in 3.3 to > 4.0 Ga zircons: ion microprobe evidence for Early Archaean high $\delta^{18}O$ continental crust. *Geochim. Cosmochim. Acta* (submitted).

11. Hinton, R. W. & Upton, B. G. J. The chemistry of zircon; variations within and between large crystals from syenite and alkali basalt xenoliths. *Geochim. Cosmochim. Acta* **55**, 3287-3302 (1991).

12. Valley, J. W., Chiarenzelli, J. R. & McLelland, J. M. Oxygen isotope geochemistry of zircon. *Earth Planet. Sci. Lett.* **126**, 187-206 (1994).

13. Maas, R., Kinny, P. D., Williams, I. S., Froude, D. O. & Compston, W. The earth's oldest known crust: a geochronological and geochemical study of 3900-4200 Ma old detrital zircons from Mt. Narryer and Jack Hills, Western Australia. *Geochim. Cosmochim. Acta* **56**, 1281-1300 (1992).

14. Compston, W., Williams, I. S., Kirschvink, J. L., Zhang, Z. & Ma, G. Zircon U-Pb ages for the Early Cambrian time-scale. *J. Geol. Soc. Lond.* **149**, 171-184 (1992).

15. Williams, I. S. in *Applications of Microanalytical Techniques to Understanding Mineralizing Processes* (eds McKibben, M. A., Shanks III, W. C. & Ridley, W. I.) 1-95 (Reviews in Economic Geology, Vol. 7, Society of Economic Geologists, Littleton, Colorado, 1998).

16. Nelson, D. R. Compilation of SHRIMP U-Pb geochronology data, 1996. *Geol. Surv. Western Australia Rec.* **1997/2**, 1-11 (1997).

17. Taylor, S. R. *Solar System Evolution: A New Perspective* 289 (Cambridge Univ. Press, Cambridge, 1992).

18. Nelson, D. R., Robinson, B. W. & Myers, J. S. Complex geological histories extending from > 4.0 Ga deciphered from xenocryst zircon microstructures. *Earth Planet. Sci. Lett.* **181**, 89-102 (2000).

19. Williams, I. S., Compston, W., Black, L. P., Ireland, T. R. & Foster, J. J. Unsupported radiogenic Pb in zircon: a cause of anomalously high Pb-Pb, U-Pb and Th-Pb ages. *Contrib. Mineral. Petrol.* **88**, 322-327 (1984).

20. Mattinson, J. M. A study of complex discordance in zircons using step-wise dissolution techniques. *Contrib. Mineral. Petrol.* **16**, 117-129 (1994).

21. Valley, J. W., Kinny, P. D., Schulze, D. J. & Spicuzza, M. J. Zircon megacrysts from kimberlite: oxygen isotope variability among mantle melts. *Contrib. Mineral. Petrol.* **133**, 1-11 (1998).

22. King, E. M., Valley, J. W., Davis, D. W. & Edwards, G. R. Oxygen isotope ratios of Archean plutonic zircons from granite-greenstone belts of the Superior province: indicator of magmatic source. *Precamb. Res.* **92**, 365-387 (1998).

23. Muehlenbachs, K. in *Stable Isotopes* (eds Valley, J. W. *et al.*) *MSA Rev. Min.* **16**, 425-444 (1986).

24. Schopf, J. W. Microfossils in the early Archean Apex Chert: New evidence for the antiquity of life. *Science* **260**, 640-646 (1993).

25. Hayes, J. M., Kaplan, I. R. & Wedeking, K. W. in *Earth's Earliest Biosphere; its Origin and evolution* 93-134 (Princeton Univ. Press, Princeton, NJ, 1983).

26. Pidgeon, R. T. *et al.* in *Eighth Int. Conf. Geochron., Cosmochron. Isotope Geol.* (eds Lanphere, M. A., Dalrymple, G. B. & Turrin, B. D.) 251 (US Geological Survey Circular 1107, Denver, Colorado, 1994).

27. Claoué-Long, J. C., Compston, W., Roberts, J. & Fanning, C. M. in *Geochronology, Time Scales and Global Stratigraphic Correlation* (eds Berggren, W. A., Kent, D. V., Aubry, M.-P. & Hardenbol, J.) 3-21 (Soc. of Sedimentary Geology, SEPM Sp. Publ. 4, 1995).

28. Valley, J. W., Graham, C. M., Harte, B., Eiler, J. M. & Kinny, P. D. in *Applications of Microanalysis to Understanding Mineralizing Processes* (eds McKibben, M. A., Shanks III, W. C. & Ridley, W. I.) 73-98 (xReviews in Economic Geology, Vol. 7, Society of Economic Geologists, Littleton, Colorado, 1998).

29. Eiler, J. M., Graham, C. M. & Valley, J. W. SIMS analysis of oxygen isotopes: matrix effects in complex minerals and glasses. *Chem. Geol.* **138**, 221-244 (1997).

Acknowledgements. We thank A. Nemchin for assistance with the cathodoluminescence imaging, J. Craven for assistance in stable isotope analysis by ion microprobe and J. Fournelle for assistance with electron microprobe analysis. Initial fieldwork was supported by the Australian Research Council and analytical work by NERC, NSF and the US Department of Energy. D. Nelson and K. McNamara kindly commented on the manuscript.

Correspondence and requests for materials should be addressed to S.A.W. (e-mail: wildes@lithos.curtin.edu.au).

Observation of Coherent Optical Information Storage in an Atomic Medium Using Halted Light Pulses

C. Liu *et al.*

Editor's Note

Laser light shining on an opaque medium can sometimes make it transparent to other light, a phenomenon known as electromagnetically induced transparency. The optical properties of the medium change so dramatically that light pulses can be slowed and compressed by many orders of magnitude. Here Chien Liu and colleagues report on experiments in which they used the effect to bring laser pulses to a complete standstill in a cold gas of sodium atoms held in a magnetic trap. They also show that information may be frozen into the atomic medium for as long as 1 microsecond, and then recovered. Liu and colleagues suggest that this technique could be useful in the development of quantum computers and other quantum information systems.

Electromagnetically induced transparency[1-3] is a quantum interference effect that permits the propagation of light through an otherwise opaque atomic medium; a "coupling" laser is used to create the interference necessary to allow the transmission of resonant pulses from a "probe" laser. This technique has been used[4-6] to slow and spatially compress light pulses by seven orders of magnitude, resulting in their complete localization and containment within an atomic cloud[4]. Here we use electromagnetically induced transparency to bring laser pulses to a complete stop in a magnetically trapped, cold cloud of sodium atoms. Within the spatially localized pulse region, the atoms are in a superposition state determined by the amplitudes and phases of the coupling and probe laser fields. Upon sudden turn-off of the coupling laser, the compressed probe pulse is effectively stopped; coherent information initially contained in the laser fields is "frozen" in the atomic medium for up to 1 ms. The coupling laser is turned back on at a later time and the probe pulse is regenerated: the stored coherence is read out and transferred back into the radiation field. We present a theoretical model that reveals that the system is self-adjusting to minimize dissipative loss during the "read" and "write" operations. We anticipate applications of this phenomenon for quantum information processing.

WITH the coupling and probe lasers used in the experiment, the atoms are accurately modelled as three-level atoms interacting with the two laser fields (Fig. 1a). Under perfect electromagnetically-induced transparency (EIT) conditions (two-photon resonance), a stationary eigenstate exists for the system of a three-level atom and resonant laser fields, where the atom is in a "dark", coherent superposition of states $|1\rangle$ and $|2\rangle$:

运用停止光脉冲来观测原子介质中的相干光学信息存储

刘谦（音译）等

编者按

照射在不透明介质上的激光有时会使介质对其他光线透明，这种现象称为电磁感应透明。介质的光学性质变化非常显著，以至于光脉冲可以被放慢和压缩许多个数量级。本文中，刘谦（音译）与其同事们报道了使用这种效应的实验，在这些实验中，他们使激光脉冲完全停滞在磁阱中的冷钠原子气团中。他们还展示了信息可冻结在原子介质中长达1微秒，然后恢复。刘谦与其同事们认为这种技术可以用于量子计算机和其他量子信息系统的开发。

电磁感应透明[1-3]是一种量子干涉效应，它允许光在其他状态下非透明的原子介质中传播；"耦合"激光被用来产生必要的干涉，以实现从"探测"激光发出的共振脉冲的传输。这种技术[4-6]已经被用于减缓及空间压缩光脉冲达七个数量级之多，从而导致其在原子云内的完全局域化和容纳[4]。本文运用电磁感应透明来使激光脉冲在磁阱捕获的冷钠原子云中完全停止。在空间局域化脉冲区域内，原子处于由耦合激光场和探测激光场的振幅和相位所决定的叠加态。当突然关闭耦合激光时，压缩的探测光束会有效地停止；最初存储在激光场中的相干信息就被"冷冻"在原子介质中，最长可达1 ms。随后耦合激光被重新开启，探测脉冲便再次产生：所存储的相干信息被读取出来并转移回辐射场。本文提出了一个理论模型，揭示了该系统的自调节特征使得"读"和"写"操作过程中耗散损失最小。我们预期这种现象可以被应用于量子信息处理。

通过实验中所使用的耦合和探测激光，原子被精确地建模为与两激光场相互作用的三能级原子（图1a）。在完全电磁感应透明（EIT）条件下（双光子共振），对于三能级原子和共振激光场组成的系统存在一稳定的本征态，其中原子处于态 |1) 和 |2) 的相干叠加"暗"状态：

$$|D\rangle = \frac{\Omega_c|1\rangle - \Omega_p|2\rangle \exp\left[i(\mathbf{k}_p - \mathbf{k}_c)\cdot\mathbf{r} - i(\omega_p - \omega_c)t\right]}{\sqrt{\Omega_c^2 + \Omega_p^2}} \tag{1}$$

Here Ω_p and Ω_c are the Rabi frequencies, $\mathbf{k}_p$ and $\mathbf{k}_c$ the wavevectors, and ω_p and ω_c the optical angular frequencies of the probe and coupling lasers, respectively. The Rabi frequencies are defined as $\Omega_{p,c} \equiv e\,\mathbf{E}_{p,c}\cdot\mathbf{r}_{13,23}/\hbar$, where e is the electron charge, $\mathbf{E}_{p,c}$ are the slowly varying envelopes of probe and coupling field amplitudes, and $e\,\mathbf{r}_{13,23}$ are the electric dipole moments of the atomic transitions. The dark state does not couple to the radiatively decaying state $|3\rangle$, which eliminates absorption of the laser fields[1-3].

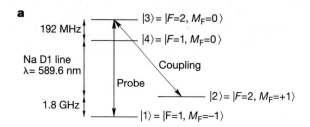

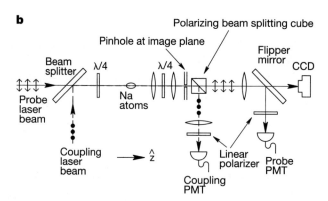

Fig. 1. Experimental set-up and procedure. **a**, States $|1\rangle$, $|2\rangle$ and $|3\rangle$ form the three-level EIT system. The cooled atoms are initially magnetically trapped in state $|1\rangle = |3S, F=1, M_F=-1\rangle$. Stimulated photon exchanges between the probe and coupling laser fields create a "dark" superposition of states $|1\rangle$ and $|2\rangle$, which renders the medium transparent for the resonant probe pulses. **b**, We apply a 2.2-mm diameter, σ^--polarized coupling laser, resonant with the $|3S, F=2, M_F=+1\rangle \rightarrow |3P, F=2, M_F=0\rangle$ transition, and a co-propagating, 1.2-mm diameter σ^+-polarized probe pulse tuned to the $|3S, F=1, M_F=-1\rangle \rightarrow |3P, F=2, M_F=0\rangle$ transition. The two laser beams start out with orthogonal linear polarizations (two-headed arrows and filled circles show the directions of linear polarization of the probe and coupling lasers, respectively). They are combined with a beam splitter, circularly polarized with a quarter-wave plate ($\lambda/4$), and then injected into the atom cloud. After leaving the cloud, the laser beams pass a second quarter-wave plate and regain their original linear polarizations before being separated with a polarizing beam-splitting cube. The atom cloud is imaged first onto an external image plane and then onto a CCD (charge-coupled device) camera. A pinhole is placed in the external image plane and positioned at the centre of the cloud image. With the pinhole and flipper mirror in place, only those portions of the probe and coupling laser beams that have passed through the central region of the cloud are selected and monitored simultaneously by two photomultiplier tubes (PMTs). States $|1\rangle$ and $|2\rangle$ have identical first-order Zeeman shifts so the two-photon resonance is maintained across the trapped atom clouds. Cold atoms and co-propagating lasers eliminate Doppler effects. However, off-resonance transitions to state $|4\rangle$ prevent perfect transmission of the light pulses in this case.

$$|D\rangle = \frac{\Omega_c|1\rangle - \Omega_p|2\rangle \exp[i(\mathbf{k}_p - \mathbf{k}_c) \cdot \mathbf{r} - i(\omega_p - \omega_c)t]}{\sqrt{\Omega_c^2 + \Omega_p^2}} \tag{1}$$

这里 Ω_p 和 Ω_c 分别为探测和耦合激光的拉比频率，$\mathbf{k}_p$ 和 $\mathbf{k}_c$ 为波矢，ω_p 和 ω_c 为光学角频率。拉比频率定义为 $\Omega_{p,c} \equiv e\mathbf{E}_{p,c} \cdot \mathbf{r}_{13,23}/\hbar,$，其中 e 为电子电荷，$\mathbf{E}_{p,c}$ 为探测和耦合场振幅的缓慢变化包络，$e\mathbf{r}_{13,23}$ 为原子跃迁的电偶极矩。暗态不会与辐射衰减态 $|3\rangle$ 耦合，从而抵消了激光场的吸收[1-3]。

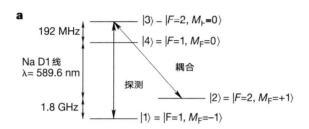

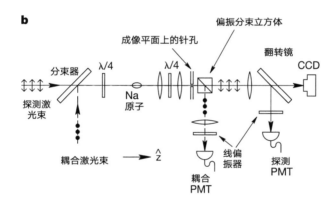

图 1. 实验装置和实验过程。**a**，量子态 $|1\rangle$，$|2\rangle$，$|3\rangle$ 构成一个三能级 EIT 系统。被降温的原子最初被磁阱捕获于态 $|1\rangle = |3S, F = 1, M_F = -1\rangle$。探测和耦合激光场之间的受激光子交换产生了态 $|1\rangle$ 和 $|2\rangle$ 的"暗"叠加态，这使得介质对共振探测脉冲是透明的。**b**，我们施加一个直径 2.2 mm 的 σ^- 偏振耦合激光，频率与 $|3S, F = 2, M_F = +1\rangle \to |3P, F = 2, M_F = 0\rangle$ 态的跃迁共振，还施加一个同向传播的，1.2 mm 直径，频率调谐至 $|3S, F = 1, M_F = -1\rangle \to |3P, F = 2, M_F = 0\rangle$ 跃迁的 σ^+ 偏振探测激光。两激光束以正交的线性偏振方向发出（双箭头和实心圆分别代表探测和耦合光束的偏振方向）。它们通过分束器合成，经四分之一（λ/4）波片形成圆偏振，然后射入原子云。从原子云中射出后，激光束又经过四分之一（λ/4）波片，在被偏振分束立方体分离之前重获原来各自的线偏振态。原子云首先成像在一外置的成像平面，然后成像在 CCD（电荷耦合器件）摄像机上。一针孔被置于外置的成像平面上，位于云图像的中心。通过针孔和翻转镜的适当摆放，只有通过了云中心区域的探测和耦合激光束被采集并且同时由两个光电倍增管（PMT）监测。量子态 $|1\rangle$ 和 $|2\rangle$ 有相同的一阶塞曼位移，所以双光子共振在被捕获的原子云中保持下来。冷原子和同向传播的激光消除了多普勒效应。但是，在这种情况下，由于存在到量子态 $|4\rangle$ 的非共振跃迁，光脉冲无法获得理想的透射率。

Atoms are prepared (magnetically trapped) in a particular internal quantum state $|1\rangle$ (Fig. 1a). The atom cloud is first illuminated by a coupling laser, resonant with the $|2\rangle$–$|3\rangle$ transition. With only the coupling laser on and all atoms in $|1\rangle$, the system is in a dark state (equation (1) with $\Omega_p = 0$). A probe laser pulse, tuned to the $|1\rangle$–$|3\rangle$ transition and co-propagating with the coupling laser, is subsequently sent through the atomic medium. Atoms within the pulse region are driven into the dark-state superposition of states $|1\rangle$ and $|2\rangle$, determined by the ratio of the instantaneous Rabi frequencies of the laser fields (equation (1)).

The presence of the coupling laser field creates transparency, a very steep refractive index profile, and low group velocity, V_g, for the probe pulse[1-10]. As the pulse enters the atomic medium, it is spatially compressed by a factor c/V_g whereas its peak electric amplitude remains constant during the slow-down[4,7].

The experiment is performed with the apparatus described in refs 4 and 11. Figure 1 shows the new optical set-up and atomic energy levels involved. A typical cloud of 11 million sodium atoms is cooled to 0.9 µK, which is just above the critical temperature for Bose–Einstein condensation. The cloud has a length of 339 µm in the z direction, a width of 55 µm in the transverse directions, and a peak density of 11 µm^{-3}. Those portions of the co-propagating probe and coupling laser beams that have passed through the 15-µm-diameter centre region of the cloud are selected and monitored simultaneously by two photomultiplier tubes (PMTs).

Figure 2a shows typical signals detected by the PMTs. The dashed curve is the measured intensity of the coupling laser, which is turned on a few microseconds before the probe pulse. The open circles indicate a gaussian-shaped reference probe pulse recorded in the absence of atoms (1/e full width is 5.70 µs). The filled circles show a probe pulse measured after it has passed through a cold atom cloud, and the solid curve is a gaussian fit to the data. The delay of this probe pulse, relative to the reference pulse, is 11.8 µs corresponding to a group velocity of 28 m s^{-1}, a reduction by a factor of 10^7 from its vacuum value. The measured delay agrees with the theoretical prediction of 12.2 µs based on a measured coupling Rabi frequency Ω_c of 2.57 MHz $\times 2\pi$ and an observed atomic column density of 3,670 µm^{-2}.

原子制备（磁阱捕获）时处于一特定的内量子态 $|1\rangle$（图 1a）。原子云最初受一耦合激光辐射，与 $|2\rangle$–$|3\rangle$ 跃迁共振。只有耦合激光开启，并且所有的原子都处于 $|1\rangle$ 态时，系统处于暗态中（等式（1）中 $\Omega_p = 0$）。一探测激光脉冲，被调谐至 $|1\rangle$–$|3\rangle$ 跃迁，与耦合激光同向传播，随后被发送穿过原子介质。在脉冲区域的原子被驱动至 $|1\rangle$ 和 $|2\rangle$ 的暗叠加态，该叠加态由激光场的瞬时拉比频率的比率决定（等式（1））。

耦合激光场的存在产生了透明现象，并导致探测脉冲的折射率图谱变得非常陡峭，群速度 V_g 变得很低[1-10]。当脉冲进入原子介质时，其空间尺寸被压缩了 c/V_g 倍，而在减速期间其峰值电子振幅保持恒定[4,7]。

实验装置如参考文献 4 和 11 所述。图 1 给出了新的光学装置以及所涉及的原子能级。一典型的原子云包含 1,100 万钠原子，它被冷却至 0.9 μK，刚好在玻色–爱因斯坦凝聚临界温度以上。原子云在 z 方向长 339 μm，横向宽度 55 μm，峰值密度为 11 μm^{-3}。选择同向传播的探测和耦合激光光束中通过原子云中心直径 15 μm 区域的部分，并用两个光电倍增管（PMT）同时监测。

图 2a 给出了 PMT 探测到的典型信号。虚线表示耦合激光的测量强度，比探测脉冲早几微秒打开。空心圆为没有原子存在时（1/e 全宽为 5.70 μs）所记录的高斯型参考探测脉冲。实心圆为探测脉冲通过一冷原子云后所测得的信号，实线为数据的高斯拟合。这个探测脉冲与参考脉冲相比的延迟为 11.8 μs，对应群速度为 28 m·s^{-1}，相当于真空中速度的 $1/10^7$。实验测得的延迟与理论预测值 12.2 μs 相符合。该理论预测值是基于测量所得的耦合拉比频率 Ω_c（2.57 MHz×2π）以及观测所得的原子柱密度（3,670 μm^{-2}）获得的。

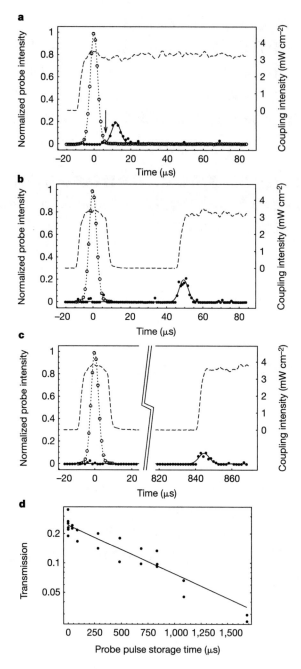

Fig. 2. Measurements of delayed and revived probe pulses. Open circles (fitted to the dotted gaussian curves) show reference pulses obtained as the average of 100 probe pulses recorded in the absence of atoms. Dashed curves and filled circles (fitted to the solid gaussian curves) show simultaneously measured intensities of coupling and probe pulses that have propagated under EIT conditions through a 339-μm-long atom cloud cooled to 0.9 μK. The measured probe intensities are normalized to the peak intensity of the reference pulses (typically, $\Omega_p/\Omega_c = 0.3$ at the peak). **a**, Probe pulse delayed by 11.8 μs. The arrow at 6.3 μs indicates the time when the probe pulse is spatially compressed and contained completely within the atomic cloud. (The intersection of the back edge of the reference pulse and the front edge of the delayed pulse defines a moment when the tail of the probe pulse has just entered the cloud and the leading

564

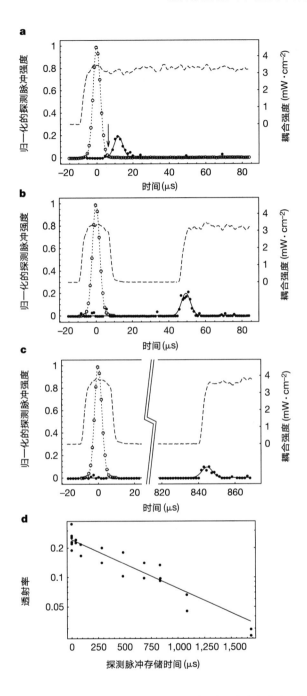

图 2. 探测脉冲延迟和再生的测量。空心圆（高斯拟合用点线表示）对应参考脉冲，为没有原子时所记录的 100 个探测脉冲的平均值。虚线和实心圆（高斯拟合用实线表示）对应在 EIT 条件下同时测量耦合和探测脉冲经由 339 μm 长，降温至 0.9 μK 的原子云传播所得到的强度。测量所得的探测强度相对参考脉冲的峰值强度进行了归一化（一般地，峰值处的 $\Omega_p/\Omega_c = 0.3$）。a，探测脉冲延迟 11.8 μs。6.3 μs 处的箭头表示探测脉冲空间上被压缩并被完全包含在原子云中的时刻（参考脉冲的后沿与延迟脉冲的前沿之间的交叉点决定了脉冲尾部刚刚进入云而前沿又恰好要离开的时刻）。b, c，耦合场在 $t = 6.3$ μs 关闭，

edge is just about to exit.) **b**, **c**, Revival of a probe pulse after the coupling field is turned off at $t = 6.3$ µs and turned back on at $t = 44.3$ µs and $t = 839.3$ µs, respectively. During the time interval when the coupling laser is off, coherent information imprinted by the probe pulse, is stored in the atomic medium. Upon subsequent turn-on of the coupling field, the probe pulse is regenerated through coherent stimulation. The time constants for the probe and coupling PMT amplifiers are 0.3 µs and 3 µs, respectively. The actual turn on/off time for the coupling field is 1 µs, as measured with a fast photodiode. **d**, Measured transmission of the probe pulse energy versus storage time. The solid line is a fit to the data, which gives a $1/e$ decay time of 0.9 ms for the atomic coherence.

At time $t = 6.3$ µs, indicated by the arrow in Fig. 2a, the probe pulse is spatially compressed and contained completely within the atomic cloud. The probe pulse in free space is 3.4 km long and contains 27,000 photons within a 15-µm diameter at its centre. It is compressed in the atomic medium to match the size of the cloud (339 µm), and the remaining optical energy in the probe field is only 1/400 of a free-space photon. Essentially all of the probe energy has been transferred through stimulated emission into the coupling laser field and the atomic medium, and coherent optical information has been imprinted on the atoms (equation (1)).

To store this coherent information, we turn off the coupling field abruptly when the probe pulse is contained within the cloud. The stored information is read out at a later time by turning the coupling laser back on. A result is shown in Fig. 2b. The dashed curve shows the coupling laser's turn-off at $t = 6.3$ µs and its subsequent turn-on at 44.3 µs. The filled circles represent the measured probe intensity. As seen from the data, when the coupling laser is turned back on the probe pulse is regenerated: we can stop and controllably regenerate the probe pulse. Similar effects have been predicted in a recent theoretical paper[12].

When the probe pulse is contained within the medium, the coherence of the laser fields is already imprinted on the atoms. As the coupling laser is turned off, the probe field is depleted to maintain the dark state (equation (1)) and (negligible) atomic amplitude is transferred from state $|1\rangle$ to state $|2\rangle$ through stimulated photon exchange between the two light fields. Because of the extremely low energy remaining in the compressed probe pulse, as noted above, it is completely depleted before the atomic population amplitudes have changed by an appreciable amount. When the coupling laser is turned back on, the process reverses and the probe pulse is regenerated through stimulated emission into the probe field. It propagates subsequently under EIT conditions as if the coupling beam had never been turned off.

During the storage time, information about the amplitude of the probe field is contained in the population amplitudes defining the atomic dark states. Information about the mode vector of the probe field is contained in the relative phase between different atoms in the macroscopic sample. The use of cold atoms minimizes thermal motion and the associated smearing of the relative phase during the storage time. (We obtain storage times that are up to 50 times larger than the time it takes an atom to travel one laser wavelength. As seen from equation (1), the difference between the wavevectors of the two laser fields determines the wavelength of the periodic phase pattern imprinted on the medium, which is 10^5 times larger than the individual laser wavelengths.)

566

又分别在 $t = 44.3~\mu s$ 和 $t = 839.3~\mu s$ 重新开启之后探测脉冲的再生。在耦合激光关闭的时间间隔期间内，探测脉冲所携带的相干信息就存储在原子介质中。随着耦合场再次开启，探测脉冲通过相干激励再次产生。探测和耦合 PMT 放大器的时间常数分别为 0.3 μs 和 3 μs。根据快速光电二极管的测量，耦合场的实际开启/关闭时间为 1 μs。**d**，测量所得的探测脉冲能量透射率与存储时间的关系。实线是对数据进行的拟合结果，给出了对于原子相干的 $1/e$ 衰减时间 0.9 ms。

当时间 $t = 6.3~\mu s$ 时，如图 2a 中的箭头所示，探测脉冲空间上被压缩并被完全包含在原子云中。这个探测脉冲在自由空间中长 3.4 km，在其中心 15 μm 直径范围内含有 27,000 个光子。为了与云尺寸（339 μm）相匹配，它在原子介质中被压缩，且探测场所剩余的光能仅为自由空间光子的 $1/400$。实质上来讲几乎所有的探测能量都通过受激辐射被转移至耦合激光场和原子介质，而相干光学信息也被印记于原子之上（等式(1)）。

为了存储这些相干信息，当探测脉冲刚好被原子云包含时我们突然关闭耦合场。所存储的信息在随后耦合激光开启时被读取出来。图 2b 给出了一结果图。虚线显示耦合激光在 $t = 6.3~\mu s$ 时关闭，随后在 44.3 μs 时开启。实心圆表示测得的探测强度。从这些数据中可以看出，当耦合激光再一次被开启时探测脉冲也再次产生，也就是说我们可以停止并控制探测脉冲的再生。类似的现象已经被近期的理论文章所预测[12]。

当探测脉冲被包含在介质中，激光场的相干信息已经被印记在原子上。当耦合激光关闭时，探测场被耗尽以保持暗态（等式(1)），而（可忽略的）原子振幅通过两光场间的受激光子交换从 $|1\rangle$ 态转移到 $|2\rangle$ 态。如上所述，由于被压缩的探测脉冲剩余的能量相当低，在原子布居振幅发生可观的改变之前，能量已经完全被耗尽。当耦合激光被重新开启时，逆过程发生，探测脉冲通过受激辐射到探测场而再生。随后探测脉冲在 EIT 条件下传播，就像耦合光束从未被关闭一样。

在存储时间内，探测场的振幅信息包含在定义原子暗态的布居振幅之中。探测场的模式矢量信息包含在宏观样本中不同原子间的相对相位之中。冷原子的运用使热运动降至最低，也就使存储过程中相应的相对相位拖尾效应达到最小。（我们获得存储时间最长达到一个原子通过一个激光波长所需时间的 50 倍。通过等式(1)可以得出，两激光场波矢的差决定了印记在介质中的周期性相位图样的波长，该波长是单个激光波长的 10^5 倍。）

The regenerated probe pulse in Fig. 2b has the same shape as the "normal" EIT pulse shown in Fig. 2a. Figure 2c shows a case where the optical coherence is stored in the atomic medium for more than 800 μs before it is read out by the coupling laser. Here the amplitude of the revived probe pulse is reduced compared to that of the pulse in Fig. 2b. Figure 2d shows the measured transmission for a series of pulses as a function of their storage time in the atom cloud. The data are consistent with an exponential decay with a $1/e$ decay time of 0.9 ms, comparable to the calculated mean free time of 0.5 ms between elastic collisions in the atom cloud with a density of 11 μm^{-3}. Further studies of the decoherence mechanisms are planned but are beyond the scope of this Letter.

We have verified experimentally that the probe pulse is regenerated through stimulated rather than spontaneous emission. To do this, we prepared all atoms in state $|2\rangle$ and subsequently turned on the coupling laser alone. The coupling laser was completely absorbed for tens of microseconds without generating any signal in the probe PMT.

In Fig. 3a–c, we show three PMT signal traces recorded under similar conditions except that we vary the intensity, I_{c2}, of the coupling laser when it is turned back on. When I_{c2} is larger than the original coupling intensity, I_{c1}, the amplitude of the revived probe pulse increases and its temporal width decreases (Fig. 3a). For $I_{c2} < I_{c1}$, the opposite occurs (Fig. 3c). These results support our physical picture of the process. The stored atomic coherence dictates the ratio of the Rabi frequencies of the coupling and revived probe fields, as well as the spatial width of the regenerated pulse. In Fig. 3d we show that with a large I_{c2}, the peak intensity of the revived probe pulse exceeds that of the original input pulse by 40%.

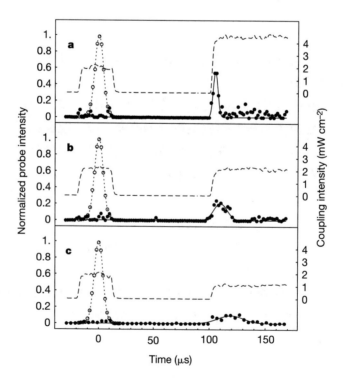

图 2b 中的再生探测脉冲与图 2a 中所示的"正常"EIT 脉冲波形相同。图 2c 显示了光学相干信息在被耦合激光读取前，存储在原子介质中超过 800 μs 的情况。这里的再生探测脉冲振幅与图 2b 相比有所降低。图 2d 为测量所得的一系列脉冲在原子云中的透射率与存储时间的函数关系图。这些数据给出了 0.9 ms 的 1/e 衰减时间，与计算所得的在密度为 11 μm⁻³ 的原子云中弹性碰撞的平均自由时间 0.5 ms 相当。关于退相干机制的进一步研究正在计划中，但不在本文讨论范围之内。

我们已经在实验上验证了探测脉冲的再生是由于受激辐射而非自发辐射。为了实现这个目的，我们准备了所有原子处于 $|2\rangle$ 的态，随后单独开启耦合激光。耦合激光在几十个毫秒内被完全吸收，但没有在 PMT 上产生任何信号。

图 3a ~ 3c 显示了在相似条件下记录的三种 PMT 信号轨迹，不同之处在于我们改变了当耦合激光再次开启时的强度 I_{c2}。当 I_{c2} 比最初的耦合强度 I_{c1} 大时，再生的探测脉冲振幅增大，其时间宽度减小（图 3a）。对于 $I_{c2} < I_{c1}$，情况相反（图 3c）。这些结果支持我们为整个过程建立的物理图像。存储的原子相干信息限定了耦合与再生探测场的拉比频率比率，以及再生脉冲的空间宽度。在图 3d 中，当 I_{c2} 很大时，再生的探测脉冲峰值强度超过初始输入脉冲的 40%。

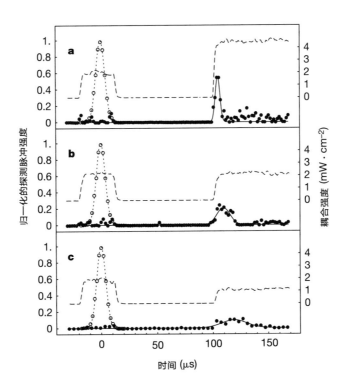

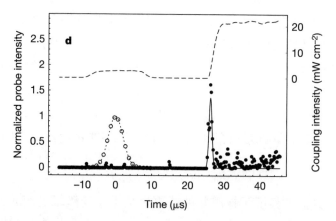

Fig. 3. Measurements of revived probe pulses for varying intensities (I_{c2}) of the second coupling pulse. The intensity (I_{c1}) of the first coupling pulse is held constant. **a–c**, The figures are recorded for I_{c2}/I_{c1} ratios of 2, 1, and 0.5, respectively. A series of data show that the height and the inverse temporal width of the revived pulses are each proportional to I_{c2}. These observations are consistent with our physical picture. Because the atomic coherence dictates the ratio of the Rabi frequencies for the coupling and revived probe fields (equation (1)), the intensity of the regenerated probe pulse is proportional to the intensity of the coupling laser when it is turned back on. Furthermore, the spatial width of the revived pulse is determined by the distribution of the atomic coherence and is thus the same as the spatial extent of the original compressed pulse. The group velocity of the probe pulse under EIT conditions is proportional to the coupling intensity[4,7]. With a larger (smaller) I_{c2}, the revived probe pulse acquires a proportionally larger (smaller) group velocity, which causes its temporal width to be inversely proportional to I_{c2}. Panel **d** shows that the intensity of the revived probe pulse can exceed that of the original input pulse, in this instance by 40%. (The observed peak-to-peak fluctuation of laser intensity is less than 10%.) The energy in the revived probe pulses is the same in all panels **a–d**, owing to the fact that the total stored amplitude of state $|2\rangle$ atoms (available to stimulate photons into the probe field) is the same in all cases. Meanings of lines and symbols as in Fig. 2.

Dissipationless pulse storage and revival processes are only possible if the ratio between the rates of dissipative and coherence-preserving events is small. When the coupling field is increased or decreased quickly compared to the duration of the probe pulse (τ) but slowly compared to $1/\Gamma$, this ratio is equal to (Z.D. and L.V.H., manuscript in preparation)

$$\frac{2\Gamma}{\Omega_c^2 + \Omega_p^2} \left(\frac{\dot{\Omega}_p}{\Omega_p} - \frac{\dot{\Omega}_c}{\Omega_c} \right) \qquad (2)$$

where Γ is the spontaneous decay rate from state $|3\rangle$. Our numerical simulations show that the probe field is constantly adjusting to match the changes in the coupling field in such a way that the terms in brackets in equation (2) nearly cancel[13,14]. Even for turn-off times faster than $1/\Gamma$, we can show that there is no decay of the coherence between states $|1\rangle$ and $|2\rangle$ as long as $\tau \gg \Gamma/(\Omega_{c_0}^2 + \Omega_{p_0}^2)$; here Ω_{c_0} and Ω_{p_0} are the Rabi frequencies before the coupling turn-off. (The adiabatic requirement introduced in ref. 12 as necessary for non-dissipative behaviour is much too strict. That requirement would inevitably break down for low coupling laser powers during turn-on or turn-off.)

We have demonstrated experimentally that coherent optical information can be stored in

570

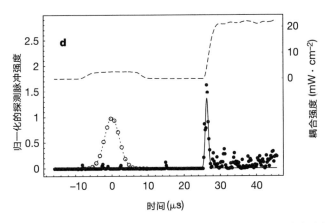

图 3. 不同的第二耦合脉冲强度(I_{c2})下再生探测脉冲的测量。第一耦合脉冲强度(I_{c1})保持恒定。**a ~ c**，记录的是分别对应 I_{c2}/I_{c1} 之比为 2，1 以及 0.5 的图形。一系列数据表明，再生脉冲的高度、时间宽度的倒数均与 I_{c2} 成正比。观察到的现象与我们的物理图像相一致。由于原子相干限定了耦合和再生探测场的拉比频率之比（等式（1）），再生探测脉冲的强度与耦合激光再次开启时的强度成正比。另外，再生脉冲的空间宽度由原子相干的分布决定，因此与最初被压缩的脉冲的空间范围相同。EIT 条件下探测脉冲的群速度与耦合强度成正比[4,7]。对于更大（或更小）的 I_{c2}，再生探测脉冲获得了按比例更大（或更小）的群速度，从而导致其时间宽度与 I_{c2} 成反比。图 **d** 显示再生的探测脉冲强度可以超过初始输入脉冲的强度，图中所示情况超过了 40%。（观察到的激光强度的峰–峰波动在 10% 以内。）再生探测脉冲的能量在所有 **a ~ d** 图中相同，这是由于在所有四种情况下，处于量子态 $|2\rangle$ 的原子（可用于将光子激发到探测场中）的总存储振幅相等。线和符号所对应的含义与图 2 一致。

无耗散的脉冲存储和再生过程只有在耗散率和相干保持事件率之比很小时才可能实现。当耦合场的增大或减小与探测脉冲的持续时间（τ）相比很迅速，而与 $1/\Gamma$ 相比变化缓慢时，这个比值等于（扎卡里·达顿和莱娜·韦斯特戈·豪，稿件准备中）

$$\frac{2\Gamma}{\Omega_c^2+\Omega_p^2}\left(\frac{\dot{\Omega}_p}{\Omega_p}-\frac{\dot{\Omega}_c}{\Omega_c}\right) \tag{2}$$

这里 Γ 为态 $|3\rangle$ 的自发衰减率。我们的数值模拟显示，探测场一直通过使等式（2）的括号内项几乎相消的方式不断调节至与耦合场变化相匹配[13,14]。甚至对于关闭时间小于 $1/\Gamma$ 的情况，我们也可以证明，只要 $\tau \gg \Gamma/(\Omega_{c_0}^2+\Omega_{p_0}^2)$，态 $|1\rangle$ 和 $|2\rangle$ 的相干就不存在衰减；这里 Ω_{c_0} 和 Ω_{p_0} 为在耦合激光关闭前的拉比频率。（文献 12 中提到的对于非耗散行为所必需的绝热要求太过严格。这个条件在低耦合激光电源开启和关闭时就不可避免地被破坏。）

本文实验论证了相干光学信息可以被存储在原子介质中，并随后在磁阱捕获的

an atomic medium and subsequently read out by using the effect of EIT in a magnetically trapped, cooled atom cloud. We have experimentally verified that the storage and read-out processes are controlled by stimulated photon transfers between two laser fields. Multiple read-outs can be achieved using a series of short coupling laser pulses. In Fig. 4a and b we show measurements of double and triple read-outs spaced by up to hundreds of microseconds. Each of the regenerated probe pulses contains part of the contents of the "atomic memory", and for the parameters chosen, the memory is depleted after the second pulse and after the third pulse.

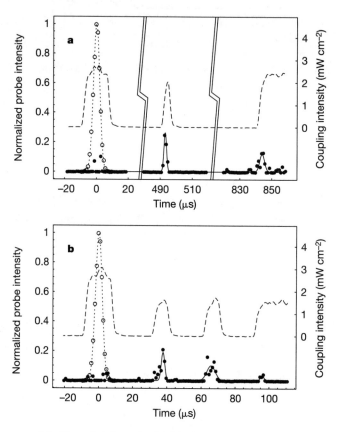

Fig. 4. Measurements of double and triple read-out of the atomic memory. To deplete the atomic memory in these cases, we use two (**a**) and three (**b**) short coupling pulses. The total energy in the two (three) revived probe pulses is measured to be the same as the energy in the single revived probe pulse obtained with a single, long coupling laser pulse (as used in Figs 2 and 3). Meanings of lines and symbols as in Fig. 2.

We believe that this system could be used for quantum information transfer; for example, to inter-convert stationary and flying qubits[15]. By injection of multiple probe pulses into a Bose–Einstein condensate—where we expect that most atomic collisions are coherence-preserving—and with use of controlled atom–atom interactions, quantum information processing may be possible during the storage time.

(**409**, 490-493; 2001)

冷原子云中利用 EIT 效应读取出来。本文通过实验验证了存储和读取过程是由两激光场之间的受激光子转移来控制的。应用一系列短耦合激光脉冲可以实现多重读取。图 4a 和 4b 显示了间隔高达几百个毫秒的二重和三重读取的测量结果。每个再生的探测脉冲包含"原子内存"的部分内容，而对于本实验所选择的参数，内存在第二和第三个脉冲之后被耗尽。

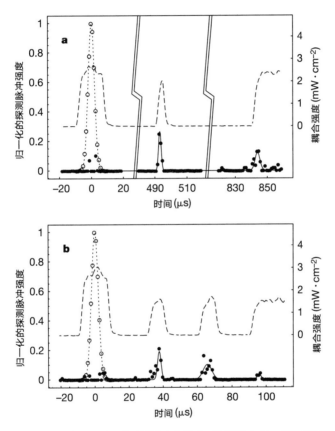

图 4. 原子内存的二重和三重读取测量。为了耗尽这些情况下的原子内存，我们使用了二重（a）和三重（b）短耦合脉冲。二（三）重再生探测脉冲的总能量与具有单一长耦合激光脉冲的单再生探测脉冲（图 2 和 3 中使用的脉冲）的能量相等。线和符号所对应的含义与图 2 一致。

我们相信该系统可以应用于量子信息传递；例如，用于定态量子比特和飞行量子比特的互相转换[15]。通过注入多重探测脉冲至玻色–爱因斯坦凝聚体（我们预计此条件下多数原子碰撞是可以相干保留的）并运用受控的原子–原子之间的相互作用，量子信息处理在存储时间之内或许是可能的。

（崔宁 翻译；石锦卫 审稿）

Chien Liu[*†], Zachary Dutton[*‡], Cyrus H. Behroozi[*†] & Lene Vestergaard Hau[*†‡]

[*] Rowland Institute for Science, 100 Edwin H. Land Boulevard, Cambridge, Massachusetts 02142, USA

[†] Division of Engineering and Applied Sciences, [‡] Department of Physics, Harvard University, Cambridge, Massachusetts 02138, USA

Received 13 October; accepted 17 November 2000.

References:

1. Harris, S. E. Electromagnetically induced transparency. *Phys. Today* **50**, 36-42 (1997).

2. Scully, M. O. & Zubairy, M. S. *Quantum Optics* (Cambridge Univ. Press, Cambridge, 1997).

3. Arimondo, E. in *Progress in Optics* (ed. Wolf, E.) 257-354 (Elsevier Science, Amsterdam, 1996).

4. Hau, L. V., Harris, S. E., Dutton, Z. & Behroozi, C. H. Light speed reduction to 17 metres per second in an ultracold atomic gas. *Nature* **397**, 594-598 (1999).

5. Kash, M. M. *et al.* Ultraslow group velocity and enhanced nonlinear optical effects in a coherently driven hot atomic gas. *Phys. Rev. Lett.* **82**, 5229-5232 (1999).

6. Budker, D., Kimball, D. F., Rochester, S. M. & Yashchuk, V. V. Nonlinear magneto-optics and reduced group velocity of light in atomic vapor with slow ground state relaxation. *Phys. Rev. Lett.* **83**, 1767-1770 (1999).

7. Harris, S. E., Field, J. E. & Kasapi, A. Dispersive properties of electromagnetically induced transparency. *Phys. Rev. A* **46**, R29-R32 (1992).

8. Grobe, R., Hioe, F. T. & Eberly, J. H. Formation of shape-preserving pulses in a nonlinear adiabatically integrable system. *Phys. Rev. Lett.* **73**, 3183-3186 (1994).

9. Xiao, M., Li, Y.-Q., Jin, S.-Z. & Gea-Banacloche, J. Measurement of dispersive properties of electromagnetically induced transparency in rubidium atoms. *Phys. Rev. Lett.* **74**, 666-669 (1995).

10. Kasapi, A., Jain, M., Yin, G. Y. & Harris, S. E. Electromagnetically induced transparency: propagation dynamics. *Phys. Rev. Lett.* **74**, 2447-2450 (1995).

11. Hau, L. V. *et al.* Near-resonant spatial images of confined Bose-Einstein condensates in a 4-Dee magnetic bottle. *Phys. Rev. A* **58**, R54-R57 (1998).

12. Fleischhauer, M. & Lukin, M. D. Dark-state polaritons in electromagnetically induced transparency. *Phys. Rev. Lett.* **84**, 5094-5097 (2000).

13. Harris, S. E. Normal modes for electromagnetically induced transparency. *Phys. Rev. Lett.* **72**, 52-55 (1994).

14. Fleischhauer, M. & Manak, A. S. Propagation of laser pulses and coherent population transfer in dissipative three-level systems: An adiabatic dressed-state picture. *Phys. Rev. A* **54**, 794-803 (1996).

15. DiVincenzo, D. P. The physical implementation of quantum computation. Preprint quant-ph/0002077 at ⟨http://xxx.lanl.gov⟩ (2000).

Acknowledgements. We thank J. Golovchenko for discussions during which the idea of the rapid turn off and on of the coupling laser first emerged. We also thank M. Burns for critical reading of the manuscript. This work was supported by the Rowland Institute for Science, the Defense Advanced Research Projects Agency, the US Airforce Office of Scientific Research, and the US Army Research Office OSD Multidisciplinary University Research Initiative Program.

Correspondence and requests for materials should be addressed to C.L. (e-mail: chien@deas.harvard.edu).

Experimental Violation of a Bell's Inequality with Efficient Detection

M. A. Rowe *et al.*

Editor's Note

By the new millennium, careful tests of John Bell's celebrated "inequalities" had indicated that quantum physics seems to be inconsistent with any local realistic interpretation—that is, with a view in which systems have definite properties independent of other parts of the universe, and no influence can travel faster than light. But argument persisted due to various "loopholes" linked to experimental imperfections, including the limited efficiency of particle detectors. Here Mary Rowe and colleagues report violation of Bell's inequalities in experiments using heavy beryllium ions, for which they were able to detect all particles and thus close one loophole. Nonetheless, not all possible local realistic interpretations of quantum theory have been excluded even now.

Local realism is the idea that objects have definite properties whether or not they are measured, and that measurements of these properties are not affected by events taking place sufficiently far away[1]. Einstein, Podolsky and Rosen[2] used these reasonable assumptions to conclude that quantum mechanics is incomplete. Starting in 1965, Bell and others constructed mathematical inequalities whereby experimental tests could distinguish between quantum mechanics and local realistic theories[1,3-5]. Many experiments[1,6-15] have since been done that are consistent with quantum mechanics and inconsistent with local realism. But these conclusions remain the subject of considerable interest and debate, and experiments are still being refined to overcome "loopholes" that might allow a local realistic interpretation. Here we have measured correlations in the classical properties of massive entangled particles ($^9Be^+$ ions): these correlations violate a form of Bell's inequality. Our measured value of the appropriate Bell's "signal" is 2.25 ± 0.03, whereas a value of 2 is the maximum allowed by local realistic theories of nature. In contrast to previous measurements with massive particles, this violation of Bell's inequality was obtained by use of a complete set of measurements. Moreover, the high detection efficiency of our apparatus eliminates the so-called "detection" loophole.

EARLY experiments to test Bell's inequalities were subject to two primary, although seemingly implausible, loopholes. The first might be termed the locality or "lightcone" loophole, in which the correlations of apparently separate events could result from unknown subluminal signals propagating between different regions of the apparatus. Aspect[16] has given a brief history of this issue, starting with the experiments of ref. 8 and highlighting the strict relativistic separation between measurements reported by the

利用高效检测来实验验证贝尔不等式的违背

罗等

编者按

新的千年到来之前，对约翰·贝尔著名的"不等式"进行的种种仔细的验证表明量子物理学似乎与任何定域实在论的解释都不相吻合，定域实在论认为系统有确定的性质，与世界的其他事物无关，且任何影响都不能超光速传播。但由于存在与实验不完美有关的各种各样的、包括粒子探测器的有效探测效率在内的"漏洞"，争论一直持续存在。本文中，玛丽·罗和其同事们报道了利用重铍离子所做的实验中贝尔不等式的违背。由于他们在实验中可以探测到所有的粒子，从而堵上了一个漏洞。然而，即使是现在，仍不能排除对量子理论的所有可能的定域实在论的解释。

定域实在论的观点认为，无论是否被测量，物体都具有确定的性质，对这些性质的测量并不受足够远发生的事件的影响[1]。爱因斯坦–波多尔斯基–罗森[2]用这些合理的假设，得出量子力学是不完备的结论。从 1965 年开始，贝尔和其他人构建了一个数学不等式，依靠实验验证能区分量子力学与定域实在论的理论[1,3-5]。已进行的很多实验[1,6-15]与量子力学理论相符，而与定域实在论不符。但这些结论仍然是大家相当感兴趣和有争议的问题，一些实验仍在不断完善以克服那些允许定域实在论解释的"漏洞"。我们现已测量了大质量的纠缠粒子（$^9Be^+$ 离子）在经典性质下的关联：这些关联违反了贝尔不等式的形式。我们得到的贝尔"信号"测量值是 2.25 ± 0.03，然而 2 是自然界定域实在论所允许的最大值。与过去的重粒子测量相比，这次背离贝尔不等式的结果是使用一组完备测量得出的。此外，我们装置的高检测效率消除了所谓的"检测"漏洞。

验证贝尔不等式的早期实验受限于两个似乎令人难以置信的初级漏洞。第一个漏洞可称为定域性的，或"光锥"漏洞，这个漏洞指的是两个明显分离的事件之间的关联有可能是某些未知的亚光速信号在装置的不同区域之间传播造成的。阿斯佩[16]给出了这个问题的简明综述，综述从文献 8 的实验开始，并着重介绍了由因斯布鲁克小组[15]报道的测量之间严格的相对论性分离。日内瓦实验[14,17]也报道了类似的结果。

Innsbruck group[15]. Similar results have also been reported for the Geneva experiment[14,17]. The second loophole is usually referred to as the detection loophole. All experiments up to now have had detection efficiencies low enough to allow the possibility that the subensemble of detected events agrees with quantum mechanics even though the entire ensemble satisfies Bell's inequalities. Therefore it must be assumed that the detected events represent the entire ensemble; a fair-sampling hypothesis. Several proposals for closing this loophole have been made[18-24]; we believe the experiment that we report here is the first to do so. Another feature of our experiment is that it uses massive particles. A previous test of Bell's inequality was carried out on protons[25], but the interpretation of the detected events relied on quantum mechanics, as symmetries valid given quantum mechanics were used to extrapolate the data to a complete set of Bell's angles. Here we do not make such assumptions.

A Bell measurement of the type suggested by Clauser, Horne, Shimony and Holt[5] (CHSH) consists of three basic ingredients (Fig. 1a). First is the preparation of a pair of particles in a repeatable starting configuration (the output of the "magic" box in Fig. 1a). Second, a variable classical manipulation is applied independently to each particle; these manipulations are labelled ϕ_1 and ϕ_2. Finally, in the detection phase, a classical property with two possible outcomes is measured for each of the particles. The correlation of these outcomes

$$q(\phi_1, \phi_2) = \frac{\mathcal{N}_{\text{same}}(\phi_1, \phi_2) - \mathcal{N}_{\text{different}}(\phi_1, \phi_2)}{\mathcal{N}_{\text{same}} + \mathcal{N}_{\text{different}}} \tag{1}$$

is measured by repeating the experiment many times. Here $\mathcal{N}_{\text{same}}$ and $\mathcal{N}_{\text{different}}$ are the number of measurements where the two results were the same and different, respectively. The CHSH form of Bell's inequalities states that the correlations resulting from local realistic theories must obey:

$$B(\alpha_1, \delta_1, \beta_2, \gamma_2) = |q(\delta_1, \gamma_2) - q(\alpha_1, \gamma_2)| + |q(\delta_1, \beta_2) + q(\alpha_1, \beta_2)| \leqslant 2 \tag{2}$$

where α_1 and δ_1 (β_2 and γ_2) are specific values of ϕ_1 (ϕ_2). For example, in a photon experiment[15], parametric down-conversion prepares a pair of photons in a singlet Einstein–Podolsky–Rosen (EPR) pair. After this, a variable rotation of the photon polarization is applied to each photon. Finally, the photons' polarization states, vertical or horizontal, are determined.

第二个漏洞通常被认为是检测漏洞。至今的所有实验检测的效率很低，以至于允许存在即使整个系综满足贝尔不等式，而检测事件的子系综与量子力学符合的可能性。因此，必须假设检测的事件代表整个系综；这是一个合理的取样假设。现已提出了堵上这个漏洞的数个建议[18-24]；我们相信，本文所述乃是首次针对堵上此漏洞所进行的实验。我们实验的另一个特点是采用大质量粒子。之前的贝尔不等式验证是利用质子进行的[25]，但检测事件的解释依据是量子力学，如量子力学中给出的对称性被用于将数据外推到贝尔角的完备集。此处我们并未采用这些假设。

由克劳塞、霍恩、希莫尼和霍尔特[5]（CHSH）提出的这类贝尔测量包括三个基本部分（图1a）。第一部分是制备在可重复的起始组态中的一对粒子（图 1 a 中"魔盒"的输出）。第二部分是对每个粒子独立地用一个可变的经典操控，这些操控标记为 ϕ_1 和 ϕ_2；最后在检测阶段，对每个粒子的经典性质进行测量，皆有两个可能的结果。这些结果的关联函数为

$$q(\phi_1, \phi_2) = \frac{N_{\text{same}}(\phi_1, \phi_2) - N_{\text{different}}(\phi_1, \phi_2)}{N_{\text{same}} + N_{\text{different}}} \tag{1}$$

式（1）是多次重复实验的测量结果。式中 N_{same} 和 $N_{\text{different}}$ 分别是两个结果相同和不同时的测量数目。贝尔不等式的 CHSH 形式表述为，由定域实在论得出的相关性必须遵从下式：

$$B(\alpha_1, \delta_1, \beta_2, \gamma_2) = |q(\delta_1, \gamma_2) - q(\alpha_1, \gamma_2)| + |q(\delta_1, \beta_2) + q(\alpha_1, \beta_2)| \leqslant 2 \tag{2}$$

式中 α_1 和 δ_1（β_2 和 γ_2）是 ϕ_1（ϕ_2）的特定值。例如，在光子实验中[15]，利用参数下转换技术把一对光子制备在爱因斯坦–波多尔斯基–罗森（EPR）单重态上，然后让每个光子通过一个角度可变的偏振片。最后确定光子的偏振状态：垂直或水平。

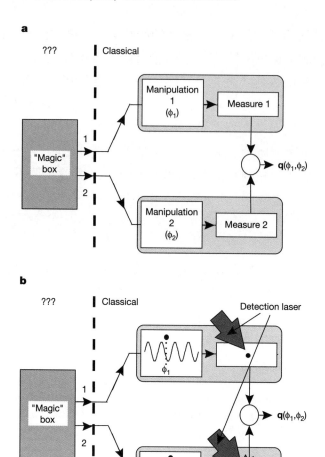

Fig. 1. Illustration of how Bell's inequality experiments work. The idea is that a "magic box" emits a pair of particles. We attempt to determine the joint properties of these particles by applying various classical manipulations to them and observing the correlations of the measurement outputs. **a**, A general CHSH type of Bell's inequality experiment. **b**, Our experiment. The manipulation is a laser wave applied with phases ϕ_1 and ϕ_2 to ion 1 and ion 2 respectively. The measurement is the detection of photons emanating from the ions upon application of a detection laser. Two possible measurement outcomes are possible, detection of few photons (as depicted for ion 1 in the figure) or the detection of many photons (as depicted for ion 2 in the figure).

Our experiment prepares a pair of two-level atomic ions in a repeatable configuration (entangled state). Next, a laser field is applied to the particles; the classical manipulation variables are the phases of this field at each ion's position. Finally, upon application of a detection laser beam, the classical property measured is the number of scattered photons emanating from the particles (which effectively measures their atomic states). Figure 1b shows how our experiment maps onto the general case. Entangled atoms produced in the context of cavity-quantum-electrodynamics[26] could similarly be used to measure Bell's inequalities.

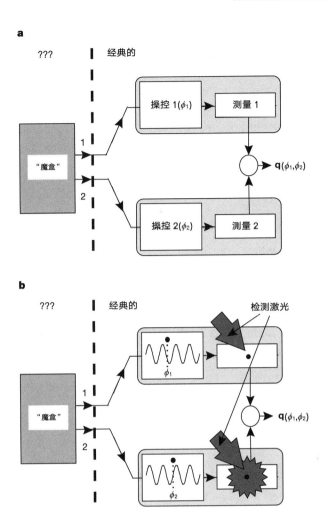

图 1. 贝尔不等式验证实验如何开展的示意图。实验思想是一个"魔盒"发射一对粒子。通过应用各种经典的操控方法，我们试图测定这些粒子的联合性质，并观测其测量输出的关联。a，一般的 CHSH 型贝尔不等式实验；b，我们的实验。操控是用相位为 ϕ_1 和 ϕ_2 的激光波分别加到离子 1 和 2 上。测量是对离子施加探测激光后，检测从离子发射出来的光子。有两个可能的测量结果，几乎检测不到光子（如图中对离子 1 的描绘）或检测到许多光子（如图中对离子 2 的描绘）。

实验中我们把一对二能级原子离子制备在可重复组态（纠缠态）上。其次，在粒子上加一激光场；经典操控的变量是每个离子所在处这个场的相位。最后，应用检测激光束之后，被测的经典性质是从粒子发射出的散射光子的数量（通过它可有效地测量粒子的原子态）。图 1b 展示了我们的实验与一般实验的对应。在腔量子电动力学意义下产生的纠缠的原子[26]能类似地用于测量贝尔不等式。

The experimental apparatus is as described in ref. 27. Two $^9\text{Be}^+$ ions are confined along the axis of a linear Paul trap with an axial centre-of-mass frequency of 5 MHz. We select two resolved levels of the $2S_{1/2}$ ground state, $|\downarrow\rangle \equiv |F=2, m_F=-2\rangle$ and $|\uparrow\rangle \equiv |F=1, m_F=-1\rangle$, where F and m_F are the quantum numbers of the total angular momentum. These states are coupled by a coherent stimulated Raman transition. The two laser beams used to drive the transition have a wavelength of 313 nm and a difference frequency near the hyperfine splitting of the states, $\omega_0 \cong 2\pi \times 1.25$ GHz. The beams are aligned perpendicular to each other, with their difference wavevector $\Delta\mathbf{k}$ along the trap axis. As described in ref. 27, it is possible in this configuration to produce the entangled state

$$|\psi_2\rangle = \frac{1}{\sqrt{2}} (|\uparrow\uparrow\rangle - |\downarrow\downarrow\rangle) \tag{3}$$

The fidelity $F=\langle\psi_2|\rho|\psi_2\rangle$, where ρ is the density matrix for the state we make, was about 88% for the data runs. In the discussion below we assume $|\psi_2\rangle$ as the starting condition for the experiment.

After making the state $|\psi_2\rangle$, we again apply Raman beams for a pulse of short duration ($\sim$400 ns) so that the state of each ion j is transformed in the interaction picture as

$$|\uparrow_j\rangle \rightarrow \frac{1}{\sqrt{2}} (|\uparrow_j\rangle - ie^{-i\phi_j}|\downarrow_j\rangle); \quad |\downarrow_j\rangle \rightarrow \frac{1}{\sqrt{2}} (|\downarrow_j\rangle - ie^{i\phi_j}|\uparrow_j\rangle) \tag{4}$$

The phase, ϕ_j, is the phase of the field driving the Raman transitions (more specifically, the phase difference between the two Raman beams) at the position of ion j and corresponds to the inputs ϕ_1 and ϕ_2 in Fig. 1. We set this phase in two ways in the experiment. First, as an ion is moved along the trap axis this phase changes by $\Delta\mathbf{k} \cdot \Delta\mathbf{x}_j$. For example, a translation of $\lambda/\sqrt{2}$ along the trap axis corresponds to a phase shift of 2π. In addition, the laser phase on both ions is changed by a common amount by varying the phase, ϕ_s, of the radio-frequency synthesizer that determines the Raman difference frequency. The phase on ion j is therefore

$$\phi_j = \phi_s + \Delta\mathbf{k} \cdot \mathbf{x}_j \tag{5}$$

In the experiment, the axial trap strength is changed so that the ions move about the centre of the trap symmetrically, giving $\Delta\mathbf{x}_1 = -\Delta\mathbf{x}_2$. Therefore the trap strength controls the differential phase, $\Delta\phi \equiv \phi_1 - \phi_2 = \Delta\mathbf{k} \cdot (\mathbf{x}_1 - \mathbf{x}_2)$, and the synthesizer controls the total phase, $\phi_{\text{tot}} \equiv \phi_1 + \phi_2 = 2\phi_s$. The calibration of these relations is discussed in the Methods.

The state of an ion, $|\downarrow\rangle$ or $|\uparrow\rangle$, is determined by probing the ion with circularly polarized light from a "detection" laser beam[27]. During this detection pulse, ions in the $|\downarrow\rangle$ or bright state scatter many photons, and on average about 64 of these are detected with a photomultiplier tube, while ions in the $|\uparrow\rangle$ or dark state scatter very few photons. For two ions, three cases can occur: zero ions bright, one ion bright, or two ions bright. In the one-ion-bright case it is not necessary to know which ion is bright because the Bell's

实验装置如文献 27 中所描述。两个 $^9Be^+$ 离子以 5 MHz 的轴向质心频率约束在沿着线性保罗阱的轴上。我们选择 $2S_{1/2}$ 基态的两个可分辨的能级，$|\downarrow\rangle \equiv |F = 2,\ m_F = -2\rangle$ 和 $|\uparrow\rangle \equiv |F = 1,\ m_F = -1\rangle$，式中 F 和 m_F 是总角动量的量子数。这些态经由相干受激拉曼跃迁耦合。用于驱动跃迁的两束激光波长为 313 nm，两者的差频接近态的超精细分裂 $\omega_0 \cong 2\pi \times 1.25\ GHz$ 的跃迁。两束激光彼此相互垂直，其波矢差 $\Delta\mathbf{k}$ 沿着阱的轴。如文献 27 中所述，在这种组态中可能产生纠缠态

$$|\psi_2\rangle = \frac{1}{\sqrt{2}}\ (|\uparrow\uparrow\rangle - |\downarrow\downarrow\rangle) \tag{3}$$

对于运行数据，保真度 $F = \langle\psi_2|\rho|\psi_2\rangle$ 约为 88%，式中 ρ 是我们所产生的态的密度矩阵。在以下讨论中，我们假定 $|\psi_2\rangle$ 是实验的起始条件。

在产生了态 $|\psi_2\rangle$ 后，我们再应用拉曼光束作为一个短脉冲(约 400 ns)，因此每个离子 j 的态在相互作用绘景中转变为

$$|\uparrow_j\rangle \rightarrow \frac{1}{\sqrt{2}}\ (|\uparrow_j\rangle - ie^{-i\phi}|\downarrow_j\rangle);\ |\downarrow_j\rangle \rightarrow \frac{1}{\sqrt{2}}\ (|\downarrow_j\rangle - ie^{i\phi}|\uparrow_j\rangle) \tag{4}$$

相位 ϕ_j 是在离子 j 的位置上驱动拉曼跃迁的场的相位(更明确地说，是两束拉曼光束之间的相位差)，对应于图 1 中的输入 ϕ_1 和 ϕ_2。我们在实验中按两种方式设定其相位。首先，在离子沿着阱轴运动时，相位变化为 $\Delta\mathbf{k} \cdot \Delta\mathbf{x}_j$。例如，沿着阱轴的平移 $\lambda/\sqrt{2}$，相当于 2π 的相移。此外，改变决定拉曼差频的射频合成器的相位 ϕ_s，使加在两个离子上的激光相位发生同量的变化。从而离子 j 的相位是

$$\phi_j = \phi_s + \Delta\mathbf{k} \cdot \mathbf{x}_j \tag{5}$$

在实验中，轴向阱的强度改变，因此离子绕阱的中心对称地运动，给出 $\Delta\mathbf{x}_1 = -\Delta\mathbf{x}_2$。因此阱的强度控制了相位差值，$\Delta\phi \equiv \phi_1 - \phi_2 = \Delta\mathbf{k} \cdot (\mathbf{x}_1 - \mathbf{x}_2)$，合成器控制了总相位 $\phi_{tot} \equiv \phi_1 + \phi_2 = 2\phi_s$。这些关系的标定在方法一节中进行讨论。

一个离子的态，$|\downarrow\rangle$ 或 $|\uparrow\rangle$，用"检测"激光束中的圆偏振光探测离子来确定[27]。在检测脉冲持续时间内，处于 $|\downarrow\rangle$ 态或明态的离子散射许多光子，用光电倍增管检测到平均约 64 个，处于 $|\uparrow\rangle$ 态或暗态的离子散射很少的光子。对两个离子而言可能出现三种情况：无离子处于明态，一个离子处于明态，两个离子处于明态。在一个

measurement requires only knowledge of whether or not the ions' states are different. Figure 2 shows histograms, each with 20,000 detection measurements. The three cases are distinguished from each other with simple discriminator levels in the number of photons collected with the phototube.

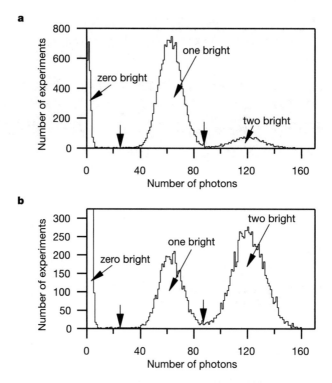

Fig. 2. Typical data histograms comprising the detection measurements of 20,000 experiments taking a total time of about 20 s. In each experiment the population in the $|\uparrow\rangle$ state is first coherently transferred to the $|F = 1, M_F = +1\rangle$ to make it even less likely to fluoresce upon application of the detection laser. The detection laser is turned on and the number of fluorescence photons detected by the phototube in 1 ms is recorded. The cut between the one bright and two bright cases is made so that the fractions of two equal distributions which extend past the cut points are equal. The vertical arrows indicate the location of the cut between the 0 (1) bright and 1 (2) bright peaks at 25 (86) counts. **a**, Data histogram with a negative correlation using $\phi_1 = 3\pi/8$ and $\phi_2 = 3\pi/8$. For these data $N_0 \cong 2,200$, $N_1 \cong 15,500$ and $N_2 \cong 2,300$. **b**, Data histogram with a positive correlation using $\phi_1 = 3\pi/8$ and $\phi_2 = -\pi/8$. For these data $N_0 \cong 7,700$, $N_1 \cong 4,400$ and $N_2 \cong 7,900$. The zero bright peak extends vertically to 2,551.

An alternative description of our experiment can be made in the language of spin-one-half magnetic moments in a magnetic field (directed in the $\hat{z}$ direction). The dynamics of the spin system are the same as for our two-level system[28]. Combining the manipulation (equation (4)) and measurement steps, we effectively measure the spin projection of each ion j in the $\hat{r}_j$ direction, where the vector $\hat{r}_j$ is in the $\hat{x}-\hat{y}$ plane at an angle ϕ_j to the $\hat{y}$ axis. Although we have used quantum-mechanical language to describe the manipulation and measurement steps, we emphasize that both are procedures completely analogous to the classical rotations of wave-plates and measurements of polarization in an optical apparatus.

离子处于明态的情况下，不必知道是哪一个离子，因为贝尔的测量仅要求知道离子态是否不同。图 2 示出了直方图，每个实验重复测量 20,000 次。用光电管收集到的光子进行计数即可区分这三种情况。

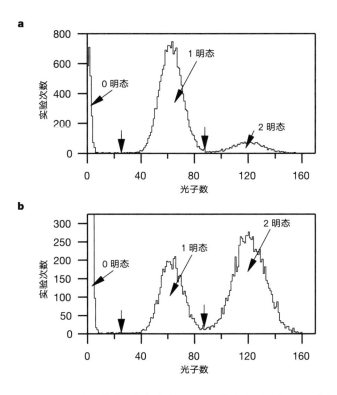

图 2. 包含 20,000 次实验的检测结果的典型数据直方图，所用的总时间约为 20 s。在每次实验中，在 $|\uparrow\rangle$ 态的粒子数，首先相干转移到 $|F=1, M_F=+1\rangle$，使得它在检测激光作用下也不发出荧光。打开检测激光，光电管在 1 ms 内记录检测到的荧光光子数。在一个离子处于明态和两个离子处于明态的情况之间进行截断，使两个相等分布扩展到截断点以外的部分所占的比例相等。垂直的箭头表示在 25(86) 计数时，在 0(1) 个离子处于明态与 1(2) 个离子处于明态峰值之间的截断。a. 用 $\phi_1=3\pi/8$ 和 $\phi_2=3\pi/8$ 的负相关数据直方图。对于这些数据，$N_0\cong 2,200$，$N_1\cong 15,500$ 和 $N_2\cong 2,300$。b. 用 $\phi_1=3\pi/8$ 和 $\phi_2=-\pi/8$ 的正相关数据直方图。对于这些数据，$N_0\cong 7,700$，$N_1\cong 4,400$ 和 $N_2\cong 7,900$。无离子处于明态的峰值将垂直扩展到 2,551。

我们的实验还能用磁场中（在 $\hat{z}$ 方向上）1/2 自旋磁矩的语言这另一种方式进行描述。自旋系统的动力学与我们的二能级系统是相同的[28]。结合操控（式(4)）和测量步骤，我们有效地测量了每个离子 j 在 $\hat{r}_j$ 方向上的自旋投影，其中矢量 $\hat{r}_j$ 在 $\hat{x}-\hat{y}$ 平面上与 $\hat{y}$ 呈角度 ϕ_j。虽然，我们已经使用量子力学的语言来描述操控和测量步骤，我们强调两者是与光学装置中波片的经典旋转和偏振的测量完全类似的步骤。

Here we calculate the quantum-mechanical prediction for the correlation function. Our manipulation step transforms the starting state, $|\psi_2\rangle$, to

$$|\psi_2'\rangle = \frac{1}{2\sqrt{2}}\left\{\left(1+e^{i(\phi_1+\phi_2)}\right)\left(|\uparrow\uparrow\rangle - e^{-i(\phi_1+\phi_2)}|\downarrow\downarrow\rangle\right) - i\left(1-e^{i(\phi_1+\phi_2)}\right)\left(e^{-i\phi_2}|\uparrow\downarrow\rangle + e^{-i\phi_1}|\downarrow\uparrow\rangle\right)\right\} \quad (6)$$

Using the measurement operators $\hat{N}_{\text{same}} = N_{\text{tot}}[|\uparrow\uparrow\rangle\langle\uparrow\uparrow| + |\downarrow\downarrow\rangle\langle\downarrow\downarrow|]$ and $\hat{N}_{\text{different}} = N_{\text{tot}}[|\uparrow\downarrow\rangle\langle\uparrow\downarrow| + |\downarrow\uparrow\rangle\langle\downarrow\uparrow|]$, the correlation function is calculated to be

$$q(\phi_1, \phi_2) = \frac{1}{8}\left[2\left|1+e^{i(\phi_1+\phi_2)}\right|^2 - 2\left|1-e^{i(\phi_1+\phi_2)}\right|^2\right] = \cos(\phi_1+\phi_2) \quad (7)$$

The CHSH inequality (equation (2)) is maximally violated by quantum mechanics at certain sets of phase angles. One such set is $\alpha_1 = -(\pi/8)$, $\delta_1 = 3\pi/8$, $\beta_2 = -(\pi/8)$ and $\gamma_2 = 3\pi/8$. With these phase angles quantum mechanics predicts

$$B\left(-\frac{\pi}{8}, \frac{3\pi}{8}, -\frac{\pi}{8}, \frac{3\pi}{8}\right) = 2\sqrt{2} \quad (8)$$

This violates the local realism condition, which requires that $B \leq 2$.

The correlation function is measured experimentally at four sets of phase angles, listed in Table 1. The experiment is repeated $N_{\text{tot}} = 20{,}000$ times at each of the four sets of phases. For each set of phases the correlation function is calculated using

$$q = \frac{(N_0 + N_2) - N_1}{N_{\text{tot}}} \quad (9)$$

Here N_0, N_1 and N_2 are the number of events with zero, one and two ions bright, respectively. The correlation values from the four sets of phase angles are combined into the Bell's signal, $B(\alpha_1, \delta_1, \beta_2, \gamma_2)$, using equation (2). The correlation values and resulting Bell's signals from five data runs are given in Table 2.

Table 1. The four sets of phase angles used for the Bell's experiment

Experiment input	ϕ_1	ϕ_2	$\Delta\phi$	ϕ_{tot}
$\alpha_1\,\beta_2$	$-\pi/8$	$-\pi/8$	0	$-\pi/4$
$\alpha_1\,\gamma_2$	$-\pi/8$	$3\pi/8$	$-\pi/2$	$+\pi/4$
$\delta_1\,\beta_2$	$3\pi/8$	$-\pi/8$	$+\pi/2$	$+\pi/4$
$\delta_1\,\gamma_2$	$3\pi/8$	$3\pi/8$	0	$+3\pi/4$

这里，我们计算了关联函数的量子力学预测结果。我们的操控步骤将起始态 $|\psi_2\rangle$ 变换为

$$|\psi_2'\rangle = \frac{1}{2\sqrt{2}}\left\{\left(1+e^{i(\phi_1+\phi_2)}\right)\left(|\uparrow\uparrow\rangle - e^{-i(\phi_1+\phi_2)}|\downarrow\downarrow\rangle\right) - i\left(1-e^{i(\phi_1+\phi_2)}\right)\left(e^{-i\phi_2}|\uparrow\downarrow\rangle + e^{-i\phi_1}|\downarrow\uparrow\rangle\right)\right\} \tag{6}$$

应用测量算符 $\hat{N}_{\text{same}} = N_{\text{tot}}[|\uparrow\uparrow\rangle\langle\uparrow\uparrow| + |\downarrow\downarrow\rangle\langle\downarrow\downarrow|]$ 和 $\hat{N}_{\text{different}} = N_{\text{tot}}[|\uparrow\downarrow\rangle\langle\uparrow\downarrow| + |\downarrow\uparrow\rangle\langle\downarrow\uparrow|]$，计算的关联函数为

$$q(\phi_1, \phi_2) = \frac{1}{8}\left[2|1+e^{i(\phi_1+\phi_2)})|^2 - 2|1-e^{i(\phi_1+\phi_2)}|^2\right] = \cos(\phi_1+\phi_2) \tag{7}$$

根据量子力学，CHSH 不等式（式(2)）在特定相角组合处达到了最大的背离。一组这样的组合为 $\alpha_1 = -(\pi/8)$，$\delta_1 = 3\pi/8$，$\beta_2 = -(\pi/8)$ 及 $\gamma_2 = 3\pi/8$。用这些相角，量子力学预测

$$B\left(-\frac{\pi}{8}, \frac{3\pi}{8}, -\frac{\pi}{8}, \frac{3\pi}{8}\right) = 2\sqrt{2} \tag{8}$$

这背离了要求 $B \leq 2$ 的定域实在论条件。

实验上用列于表 1 的四组相角测量关联函数。四组相角的每一组，实验都重复 $N_{\text{tot}} = 20,000$ 次。对每组相位，用下式计算关联函数

$$q = \frac{(N_0+N_2)-N_1}{N_{\text{tot}}} \tag{9}$$

式中 N_0, N_1, N_2 分别是以 0，1 和 2 个离子处于明态的事件数目。四组相角的相关值已用式(2)代入贝尔信号 $B(\alpha_1, \delta_1, \beta_2, \gamma_2)$ 中。表 2 给出了五次数据运行的相关值和得出的贝尔信号。

表 1. 用于贝尔实验的四组相角

实验输入	ϕ_1	ϕ_2	$\Delta\phi$	ϕ_{tot}
$\alpha_1\,\beta_2$	$-\pi/8$	$-\pi/8$	0	$-\pi/4$
$\alpha_1\,\gamma_2$	$-\pi/8$	$3\pi/8$	$-\pi/2$	$+\pi/4$
$\delta_1\,\beta_2$	$3\pi/8$	$-\pi/8$	$+\pi/2$	$+\pi/4$
$\delta_1\,\gamma_2$	$3\pi/8$	$3\pi/8$	0	$+3\pi/4$

Table 2. Correlation values and resulting Bell's signals for five experimental runs

Run number	$q(\alpha_1, \beta_2)$	$q(\alpha_1, \gamma_2)$	$q(\delta_1, \beta_2)$	$q(\delta_1, \gamma_2)$	$B(\alpha_1, \delta_1, \beta_2, \gamma_2)$
1	0.541	0.539	0.569	−0.573	2.222
2	0.575	0.570	0.530	−0.600	2.275
3	0.551	0.634	0.590	−0.487	2.262
4	0.575	0.561	0.559	−0.551	2.246
5	0.541	0.596	0.537	−0.571	2.245

The experimental angle values were $\alpha_1 = -(\pi/8)$, $\delta_1 = 3\pi/8$, $\beta_2 = -(\pi/8)$, and $\gamma_2 = 3\pi/8$. The statistical errors are 0.006 and 0.012 for the q and B values respectively. The systematic errors (see text) are 0.03 and 0.06 for the q and B values respectively.

So far we have described the experiment in terms of perfect implementation of the phase angles. In the actual experiment, however, α_1, δ_1, β_2 and γ_2 are not quite the same angles both times they occur in the Bell's inequality. In our experiment the dominant reason for this error results from the phase instability of the synthesizer, which can cause the angles to drift appreciably during four minutes, the time required to take a complete set of measurements. This random drift causes a root-mean-squared error for the correlation function of ± 0.03 on this timescale, which propagates to an error of ± 0.06 for the Bell's signal. The error for the Bell's signal from the five combined data sets is then ± 0.03, consistent with the run-to-run variation observed. Averaging the five Bell's signals from Table 2, we arrive at our experimental result, which is

$$B\left(-\frac{\pi}{8}, \frac{3\pi}{8}, -\frac{\pi}{8}, \frac{3\pi}{8}\right) = 2.25 \pm 0.03 \qquad (10)$$

If we take into account the imperfections of our experiment (imperfect state fidelity, manipulations, and detection), this value agrees with the prediction of quantum mechanics.

The result above was obtained using the outcomes of every experiment, so that no fair-sampling hypothesis is required. In this case, the issue of detection efficiency is replaced by detection accuracy. The dominant cause of inaccuracy in our state detection comes from the bright state becoming dark because of optical pumping effects. For example, imperfect circular polarization of the detection light allows an ion in the $|\downarrow\rangle$ state to be pumped to $|\uparrow\rangle$, resulting in fewer collected photons from a bright ion. Because of such errors, a bright ion is misidentified 2% of the time as being dark. This imperfect detection accuracy decreases the magnitude of the measured correlations. We estimate that our Bell's signal would be 2.37 with perfect detection accuracy.

We have thus presented experimental results of a Bell's inequality measurement where a measurement outcome was recorded for every experiment. Our detection efficiency was high enough for a Bell's inequality to be violated without requiring the assumption of fair sampling, thereby closing the detection loophole in this experiment. The ions were

表 2　五次实验运行的相关值和得出的贝尔信号

运行序号	$q(\alpha_1, \beta_2)$	$q(\alpha_1, \gamma_2)$	$q(\delta_1, \beta_2)$	$q(\delta_1, \gamma_2)$	$B(\alpha_1, \delta_1, \beta_2, \gamma_2)$
1	0.541	0.539	0.569	−0.573	2.222
2	0.575	0.570	0.530	−0.600	2.275
3	0.551	0.634	0.590	−0.487	2.262
4	0.575	0.561	0.559	−0.551	2.246
5	0.541	0.596	0.537	−0.571	2.245

实验的角度数值是 $\alpha_1 = -(\pi/8)$，$\delta_1 = 3\pi/8$，$\beta_2 = -(\pi/8)$ 及 $\gamma_2 = 3\pi/8$。对 q 和 B 的统计误差分别为 0.006 和 0.012。对 q 和 B 的系统误差（见正文）分别为 0.03 和 0.06。

在此为止，我们已用理想的相角设定描述了实验。然而，在实际实验中，两次代入贝尔不等式的角 α_1、δ_1、β_2 和 γ_2 并不是完全相同的。在我们的实验中，这个误差主要来自合成器的相位不稳定度，使相角在四分钟内产生了明显的漂移，这正是一整套测量所需要的时间。这项随机漂移对于关联函数来说，在这一时标上产生了 ±0.03 的方均根误差，它传递到贝尔信号的误差为 ±0.06。来自五个数据组的贝尔信号的误差则为 ±0.03，与观测到的每次运行变化是相符的。对表 2 中五个贝尔信号求平均，我们得到的实验结果为

$$B\left(-\frac{\pi}{8}, \frac{3\pi}{8}, -\frac{\pi}{8}, \frac{3\pi}{8} \right) = 2.25 \pm 0.03 \qquad (10)$$

如果考虑到我们实验中的不完美的方面（不完美的态保真度、操控和检测），这个结果与量子力学的预测值是一致的。

上述结果是利用每个实验的输出量得出的，因此不要求有合理取样的前提。在这种情况下，检测效率的问题由检测准确度代替。在我们实验中，态检测的不准确的主要原因来自于光抽运效应引起的明态变为暗态。例如，检测光不完美的圆偏振允许离子由 $|\downarrow\rangle$ 态抽运到 $|\uparrow\rangle$，从而处于明态的离子收集到较少的光子。由于这类误差的存在，一个离子处于明态而被错判为处于暗态的概率是 2%。这类不完美的检测准确度降低了测量相关性的量级。我们估计本文贝尔信号在完美检测准确度下将是 2.37。

我们已经给出了每次实验记录的贝尔不等式测量的实验结果。我们的检测效率高到足以不要求有合理取样的前提也可以给出贝尔不等式背离的结论，从而在这个实验中堵上了检测漏洞这个问题。离子之间相距很远，远到没有已知的相互作用可

separated by a distance large enough that no known interaction could affect the results; however, the lightcone loophole remains open here. Further details of this experiment will be published elsewhere.

Methods

Phase calibration

The experiment was run with specific phase differences of the Raman laser beam fields at each ion. In order to implement a complete set of laser phases, a calibration of the phase on each ion as a function of axial trap strength was made. We emphasize that the calibration method is classical in nature. Although quantum mechanics guided the choice of calibration method, no quantum mechanics was used to interpret the signal. General arguments are used to describe the signal resulting from a sequence of laser pulses and its dependence on the classical physical parameters of the system, the laser phase at the ion, and the ion's position.

In the calibration procedure, a Ramsey experiment was performed on two ions. The first $\pi/2$ Rabi rotation was performed identically each time. The laser phases at the ions' positions for the second $\pi/2$ Rabi rotation were varied, ϕ_1 for ion 1 and ϕ_2 for ion 2. The detection signal is the total number of photons counted during detection. With an auxiliary one-ion experiment we first established empirically that the individual signal depends only on the laser phase at an individual ion and is $C+A\cos\phi_j$. Here C and A are the offset and amplitude of the one-ion signal. We measure the detector to be linear, so that the detection signal is the sum of the two ions' individual signals. The two-ion signal is therefore

$$C+A\cos\phi_1+C+A\cos\phi_2 = 2C+2A\cos\left[\frac{1}{2}(\phi_1+\phi_2)\right]\cos\left[\frac{1}{2}(\phi_1-\phi_2)\right] \tag{11}$$

By measuring the fringe amplitude and phase as $\phi_s = (\phi_1+\phi_2)/2$ is swept, we calibrate $\phi_1-\phi_2$ as a function of trap strength and ensure that $\phi_1+\phi_2$ is independent of trap strength.

We use the phase convention that at the ion separation used for the entanglement preparation pulse the maximum of the correlation function is at $\phi_1 = \phi_2 = 0$ (or $\Delta\phi = \phi_{tot} = 0$). Our measurement procedure begins by experimentally finding this condition of $\phi_1 = \phi_2 = 0$ by keeping $\Delta\phi = 0$ and scanning the synthesizer phase to find the maximum correlation. The experiment is then adjusted to the phase angles specified above by switching the axial trap strength to set $\Delta\phi$ and incrementing the synthesizer phase to set ϕ_{tot}.

Locality issues

The ions are separated by a distance of approximately 3 µm, which is greater than 100 times the size of the wavepacket of each ion. Although the Coulomb interaction strongly couples the ions' motion, it does not affect the ions' internal states. At this distance, all known relevant interactions are expected to be small. For example, dipole–dipole interactions between the ions slightly modify the light-

以影响实验结果；但是，本文还没有解决光锥漏洞的问题。这个实验的更多细节将发表在其他地方。

方　　法

相位标定

实验是在每个离子的拉曼激光光束场的特定相位差下运行的。为了实现激光相位的完备集，每个离子上的激光相位被标定为轴向阱强度的函数。我们强调标定方法本质上是经典的。虽然，量子力学影响了标定方法的选择，但并没有使用量子力学来解释信号。一般的理论用于描述从一系列激光脉冲得出的信号，以及它与系统的经典物理参量的相关关系，经典的物理参量包括离子上的激光相位和离子的位置。

在标定过程中，对两个离子进行了拉姆齐实验。每次实验中第一个 $\pi/2$ 的拉比旋转都相同。对于第二个 $\pi/2$ 拉比旋转，离子处的激光相位发生了改变，离子 1 处为 ϕ_1，离子 2 处为 ϕ_2。检测信号是在检测期间记录的总光子数。用一个辅助的单离子实验，我们首先在实践经验上确立了单个信号仅与单个离子处的激光相位有关，信号记为 $C+A\cos\phi_j$。此处的 C 和 A 是单离子信号的偏置和振幅。我们测得检测器是线性的，因此检测信号是两个离子的单独信号之和。从而两个离子的信号为

$$C+A\cos\phi_1+C+A\cos\phi_2=2C+2A\cos\left[\frac{1}{2}(\phi_1+\phi_2)\right]\cos\left[\frac{1}{2}(\phi_1-\phi_2)\right] \tag{11}$$

当扫描 $\phi_s=(\phi_1+\phi_2)/2$ 时，通过测量条纹振幅和相位，我们标定了 $\phi_1-\phi_2$ 作为阱强度的函数，并确保 $\phi_1+\phi_2$ 与阱强度无关。

我们按照惯例，在用于纠缠制备脉冲的离子分离处，关联函数在 $\phi_1=\phi_2=0$（或 $\Delta\phi=\phi_{tot}=0$）处取得最大值。在测量的开始，我们通过保持 $\Delta\phi=0$，进而用实验方法找到 $\phi_1=\phi_2=0$ 的条件，并扫描合成器的相位来找到关联函数最大值。然后在实验中通过调整轴阱强度来设置 $\Delta\phi$，通过增加合成器相位来设置 ϕ_{tot}，把相角调整到上文给出的相应值。

定域性的争论

离子间相隔的距离约为 3 μm，这是每个离子的波包尺寸的 100 倍。虽然，库仑相互作用与离子的运动发生强烈的耦合，但它并不影响离子的内部状态。在这个距离上，所有已知的相互作用的预期都是很小的。例如，在离子之间的偶极–偶极

scattering intensity, but this effect is negligible for the ion–ion separations used[29]. Also, the detection solid angle is large enough that Young's interference fringes, if present, are averaged out[30]. Even though all known interactions would cause negligible correlations in the measurement outcomes, the ion separation is not large enough to eliminate the lightcone loophole.

We note that the experiment would be conceptually simpler if, after creating the entangled state, we separated the ions so that the input manipulations and measurements were done individually. However, unless we separated the ions by a distance large enough to overcome the lightcone loophole, this is only a matter of convenience of description and does not change the conclusions that can be drawn from the results.

(**409**, 791-794; 2001)

M. A. Rowe[*], D. Kielpinski[*], V. Meyer[*], C. A. Sackett[*], W. M. Itano[*], C. Monroe[†] & D. J. Wineland[*]

[*] Time and Frequency Division, National Institute of Standards and Technology, Boulder, Colorado 80305, USA
[†] Department of Physics, University of Michigan, Ann Arbor, Michigan 48109, USA

Received 25 October; accepted 30 November 2000.

References:

1. Clauser, J. F. & Shimony, A. Bell's theorem: experimental tests and implications. *Rep. Prog. Phys.* **41**, 1883-1927 (1978).

2. Einstein, A., Podolsky, B. & Rosen, N. Can quantum-mechanical description of reality be considered complete? *Phys. Rev.* **47**, 777-780 (1935).

3. Bell, J. S. On the Einstein-Podolsky-Rosen paradox. *Physics* **1**, 195-200 (1965).

4. Bell, J. S. in *Foundations of Quantum Mechanics* (ed. d'Espagnat, B.) 171-181 (Academic, New York, 1971).

5. Clauser, J. F., Horne, M. A., Shimony, A. & Holt, R. A. Proposed experiment to test local hidden-variable theories. *Phys. Rev. Lett.* **23**, 880-884 (1969).

6. Freedman, S. J. & Clauser, J. F. Experimental test of local hidden-variable theories. *Phys. Rev. Lett.* **28**, 938-941 (1972).

7. Fry, E. S. & Thompson, R. C. Experimental test of local hidden-variable theories. *Phys. Rev. Lett.* **37**, 465-468 (1976).

8. Aspect, A., Grangier, P. & Roger, G. Experimental realization of Einstein-Podolsky-Rosen-Bohm *Gedankenexperiment*: a new violation of Bell's inequalities. *Phys. Rev. Lett.* **49**, 91-94 (1982).

9. Aspect, A., Dalibard, J. & Roger, G. Experimental test of Bell's inequalities using time-varying analyzers. *Phys. Rev. Lett.* **49**, 1804-1807 (1982).

10. Ou, Z. Y. & Mandel, L. Violation of Bell's inequality and classical probability in a two-photon correlation experiment. *Phys. Rev. Lett.* **61**, 50-53 (1988).

11. Shih, Y. H. & Alley, C. O. New type of Einstein-Podolsky-Rosen-Bohm experiment using pairs of light quanta produced by optical parametric down conversion. *Phys. Rev. Lett.* **61**, 2921-2924 (1988).

12. Tapster, P. R., Rarity, J. G. & Owens, P. C. M. Violation of Bell's inequality over 4 km of optical fiber. *Phys. Rev. Lett.* **73**, 1923-1926 (1994).

13. Kwiat, P. G., Mattle, K., Weinfurter, H. & Zeilinger, A. New high-intensity source of polarization-entangled photon pairs. *Phys. Rev. Lett.* **75**, 4337-4341 (1995).

14. Tittel, W., Brendel, J., Zbinden, H. & Gisin, N. Violation of Bell inequalities by photons more than 10 km apart. *Phys. Rev. Lett.* **81**, 3563-3566 (1998).

15. Weihs, G. *et al.* Violation of Bell's inequality under strict Einstein locality conditions. *Phys. Rev. Lett.* **81**, 5039-5043 (1998).

16. Aspect, A. Bell's inequality test: more ideal than ever. *Nature* **398**, 189-190 (1999).

17. Gisin, N. & Zbinden, H. Bell inequality and the locality loophole: active versus passive switches. *Phys. Lett. A* **264**, 103-107 (1999).

18. Lo, T. K. & Shimony, A. Proposed molecular test of local hidden-variable theories. *Phys. Rev. A* **23**, 3003-3012 (1981).

19. Kwiat, P. G., Eberhard, P. H., Steinberg, A. M. & Chiao, R. Y. Proposal for a loophole-free Bell inequality experiment. *Phys. Rev. A* **49**, 3209-3220 (1994).

20. Huelga, S. F., Ferrero, M. & Santos, E. Loophole-free test of the Bell inequality. *Phys. Rev. A* **51**, 5008-5011 (1995).

21. Fry, E. S., Walther, T. & Li, S. Proposal for a loophole free test of the Bell inequalities. *Phys. Rev. A* **52**, 4381-4395 (1995).

22. Freyberger, M., Aravind, P. K., Horne, M. A. & Shimony, A. Proposed test of Bell's inequality without a detection loophole by using entangled Rydberg atoms. *Phys. Rev. A* **53**, 1232-1244 (1996).

23. Brif, C. & Mann, A. Testing Bell's inequality with two-level atoms via population spectroscopy. *Europhys. Lett.* **49**, 1-7 (2000).

24. Beige, A., Munro, W. J. & Knight, P. L. A Bell's inequality test with entangled atoms. *Phys. Rev. A* **62**, 052102-1–052102-9 (2000).

25. Lamehi–Rachti, M. & Mittig, W. Quantum mechanics and hidden variables: a test of Bell's inequality by the measurement of the spin correlation in low-energy proton-proton scattering. *Phys. Rev. D* **14**, 2543-2555 (1976).

26. Hagley, E. *et al.* Generation of Einstein-Podolsky-Rosen pairs of atoms. *Phys. Rev. Lett.* **79**, 1-5 (1997).

相互作用稍微改变了光散射强度，但这种影响对于所用的离子之间的距离来说是可以忽略不计的 [29]。同时，检测立体角大到足以使杨氏干涉条纹（如存在）被平均掉 [30]。即使所有已知的相互作用在测量结果中的相关性可以忽略不计，但离子之间的距离仍不会大到足以消除光锥漏洞。

我们注意到，如果在建立纠缠态后分开离子，使输入操控和测量分别进行，实验将在概念上更为简单。然而，除非我们把离子之间的距离扩大到足以克服光锥漏洞，否则，这只是描述方便与否的问题，而并不会改变从实验结果所得出的结论。

（沈乃澂 翻译；李军刚 审稿）

27. Sackett, C. A. *et al.* Experimental entanglement of four particles. *Nature* **404**, 256-259 (2000).

28. Feynman, R. P., Vernon, F. L. & Hellwarth, R. W. Geometrical representation of the Schrödinger equation for solving maser problems. *J. Appl. Phys.* **28**, 49-52 (1957).

29. Richter, T. Cooperative resonance fluorescence from two atoms experiencing different driving fields. *Optica Acta* **30**, 1769-1780 (1983).

30. Eichmann, U. *et al.* Young's interference experiment with light scattered from two atoms. *Phys. Rev. Lett.* **70**, 2359-2362 (1993).

Acknowledgements. We thank A. Ben-Kish, J. Bollinger, J. Britton, N. Gisin, P. Knight, P. Kwiat and I. Percival for useful discussions and comments on the manuscript. This work was supported by the US National Security Agency (NSA) and the Advanced Research and Development Activity (ARDA), the US Office of Naval Research, and the US Army Research Office. This paper is a contribution of the National Institute of Standards and Technology and is not subject to US copyright.

Correspondence and requests for materials should be addressed to D.J.W. (e-mail: david.wineland@boulder.nist.gov).

Guide to the Draft Human Genome

T. G. Wolfsberg *et al.*

Editor's Note

The release of the first draft of the human genome—the genetic sequence of all human chromosomes—presented the scientific community with a data set that was almost overwhelming in its implications, questions, complications and repercussions. To accompany the paper reporting the results obtained by the International Human Genome Sequencing Consortium, *Nature* here presented an analysis of how the data might be used. As the authors point out, the sequence data joined a considerable existing body of information about the structure and evolution of the human genome, ranging from detailed studies of individual genes and their protein products to linkages between particular inherited phenotypes. The paper provides some pointers for how these disparate types of data might be integrated.

There are a number of ways to investigate the structure, function and evolution of the human genome. These include examining the morphology of normal and abnormal chromosomes, constructing maps of genomic landmarks, following the genetic transmission of phenotypes and DNA sequence variations, and characterizing thousands of individual genes. To this list we can now add the elucidation of the genomic DNA sequence, albeit at "working draft" accuracy. The current challenge is to weave together these disparate types of data to produce the information infrastructure needed to support the next generation of biomedical research. Here we provide an overview of the different sources of information about the human genome and how modern information technology, in particular the internet, allows us to link them together.

THE ultimate goal of the Human Genome Project is to produce a single continuous sequence for each of the 24 human chromosomes and to delineate the positions of all genes. The working draft sequence described by the International Human Genome Sequencing Consortium was constructed by melding together sequence segments derived from over 20,000 large-insert clones[1]. All of the results of this analysis are available on a web site maintained by the University of California at Santa Cruz (http://genome.ucsc. edu). Over the next few years, draft quality sequence will be steadily replaced by more accurate data. The National Center for Biotechnology Information (NCBI) has developed a system for rapidly regenerating the genomic sequence and gene annotation as sequences of the underlying clones are revised (http://www.ncbi.nlm.nih.gov/genome/guide). Undoubtedly, others will apply a variety of approaches to large-scale annotation of genes and other features. One such project is Ensembl, a joint project of the European Bioinformatics Institute (EBI) and the Sanger Centre (http://www.ensembl.org).

596

人类基因组草图导读

沃尔夫斯伯格等

编者按

人类基因组——全人类染色体遗传序列第一份草图的发布向科学界提供了一个数据集，这个数据集所带来的启发和疑问，以及所具有的复杂性和影响力都是空前的。随着国际人类基因组测序联盟发表文章公布其成果，《自然》杂志在这里刊出了一篇关于如何使用这些数据的分析文章。正如作者所指出的，这些序列数据将大量已知的关于人类基因组结构和进化的信息连接起来，范围从单个基因及其蛋白产物的详细研究到与特定遗传表型之间的关联。本文为如何整合这些不同类型的数据提供了一些指南。

已经有大量的方法被用于研究人类基因组的结构、功能和进化。这些方法包括检测正常和异常染色体的形态，构建基因组标记图谱，追踪表型和DNA序列变异的遗传传递，以及描述数千个单基因的特征。虽然这还处于"工作草图"的准确度阶段，但我们已经可以在此列表中加上基因组DNA序列的解析。我们目前面临的挑战是将这些类型各异的数据编织在一起，进而形成能够支持下一代生物医学研究的信息学基础框架。在这里，我们将会概述这些与人类基因组相关的不同来源的信息，以及如何利用现代信息技术，特别是互联网，将它们联系在一起。

人类基因组计划的最终目标是为人类24条染色体中的每一条都构建一个单独的连续序列，并描述所有基因的位置。国际人类基因组测序联盟所绘制的工作草图序列，是通过把超过20,000个大片断插入克隆的序列合并在一起而构建得到的[1]。所有的分析结果均可在加利福尼亚大学圣克鲁兹分校维护的网站(http://genome.ucsc.edu)上获取。在未来几年间，草图质量序列会不断地被更精确的数据替代。美国国家生物技术信息中心(NCBI)已经开发了一个系统，它能在相应的克隆序列被修订后，迅速更新对应的基因组序列和基因注释(http://www.ncbi.nlm.nih.gov/genome/guide)。毋庸置疑的是其他研究小组将会采用各种各样的方法对基因以及其他一些特征进行大规模注释。其中一个项目是Ensembl，它是由欧洲生物信息研究所(EBI)和桑格中心(http://www.ensembl.org)联合建立的一个合作项目。

Pinpointing New Genes

Recent estimates have placed the number of human genes at 25,000–35,000 (refs 2, 3). More than 10,000 human genes have been catalogued in the Online *Mendelian Inheritance in Man*[4] (OMIM), which documents all inherited human diseases and their causal gene mutations. Integration of the information contained in OMIM with the working draft is facilitated by the fact that it has already been tied to reference messenger RNA sequences through a collaborative effort between OMIM, the Human Gene Nomenclature Committee and the NCBI[5,6]. As a result, the positions of many known genes have been determined by alignment of mRNAs with genomic sequences. For the remaining genes, we must currently resort to computational gene-finding methods (reviewed in refs 7, 8).

When mRNA species align differently to a genomic sequence, this indicates that alternative splicing has taken place. In the current set of full-length reference mRNAs, 11,174 transcripts have been sequenced from 10,742 distinct genes (2.4% of the genes have multiple splicing variants). Alignments of expressed sequence tag (EST) sequences to the working draft sequence, however, suggest that about 60% of human genes have multiple splicing variants, which has important implications for the complexity of human gene expression[1]. By their sheer numbers (currently over 2.5 million) we might expect ESTs to sample a larger fraction of splicing variants than would be the case for more traditional targeted approaches. For example, alignment of the mRNA of the membrane-bound metalloprotease-disintegrin ADAM23 (ref. 9) to the draft genome reveals that the gene consists of at least 23 exons. Of the many ESTs that also align to the *ADAM23* locus, one lacks the exon that encodes the transmembrane domain, which suggests an alternatively spliced, soluble protein. Although this is a biologically plausible conclusion, one should exercise caution when interpreting such results: ESTs are partial single-pass sequences that have been associated with a variety of artefacts, including sequencing errors and improper splicing[10,11].

Finding Relatives

Genes can be found through an implied relationship to something else—for example, being a putative orthologue (related to a gene in another species). To do this, it is useful to search the genomic sequence or, preferably, its mRNA sequences and protein products, using BLAST[12]. As an example, we use the mouse *Lmx1b* gene, which encodes a LIM homeobox protein that is important in pattern development[13]. When we used the protein sequence encoded by *Lmx1b* as a query in a BLAST search against the working draft human genome sequence, the best match was to a region of 9q34, and the positions of the alignments line up with the exons of the human *LMX1B* gene (Fig. 1a).

查明新基因

最近的估计结果认定人类基因的数目大概是介于 25,000 ~ 35,000 之间（参考文献 2 和 3）。已经有超过 10,000 个基因被编入在线"人类孟德尔遗传"（OMIM）数据库[4]中，这个系统记录了所有的人类遗传性疾病以及导致疾病发生的基因突变。通过 OMIM、人类基因命名委员会和 NCBI 的协作，OMIM 中包含的信息已经关联到相关的信使 RNA（mRNA）序列上，这促进了 OMIM 中的信息与工作草图的整合[5,6]。因此，很多已知基因的位置都已经通过 mRNA 与基因组序列的比对确定下来。对于剩下的基因，目前我们必须要求助于各种基于计算的基因寻找方法（参考文献 7 和 8 中的相关综述）。

当 mRNA 种类与基因组序列的比对结果不同时，意味着发生了选择性剪接。在目前的一整套全长参考 mRNA 中，有 11,174 个转录产物从 10,742 个特定基因中（2.4% 的基因有多剪接变异体的情况）测序得到。然而，将表达序列标签（EST）序列与基因组草图序列进行比对，结果却发现大概有 60% 的人类基因具有多剪接变异体，这对于解释人类基因表达的复杂性具有重要意义[1]。仅考虑其数目的话（目前超过了 250 万），我们可以期望 EST 的样本数相对于传统靶向方法具有更多的剪接变异体。举例来讲，将细胞膜结合金属蛋白酶—解整联蛋白 ADAM23（参考文献 9）所对应的 mRNA 与基因组草图进行对比发现该基因至少包含 23 个外显子。在与 *ADAM23* 位点比对的 EST 中，其中一个 EST 缺少编码跨膜区的外显子，这意味着 mRNA 发生了选择性剪接，得到的是可溶性蛋白。尽管这是生物学上合理的结论，但当解释这一类结果时我们仍然需要保持谨慎：EST 是部分单向的序列，可能会与各种各样的人为修饰相关联，包括测序错误以及不适当的剪接[10,11]。

寻找亲缘关系

通过基因与其他一些事物之间暗含的关系寻找基因——例如，推定直系同源基因（与其他物种某个基因相关）。这可以利用 BLAST[12] 搜索基因组序列，或者最好使用 mRNA 序列或者蛋白质产物来寻找基因。以小鼠的 *Lmx1b* 基因为例，这个基因编码的是一个 LIM 同源异形框蛋白，该蛋白在模式发育方面具有重要作用[13]。当我们用 *Lmx1b* 编码的蛋白质序列作为查询序列，通过 BLAST 在人类基因组草图序列数据库中进行检索时，最匹配的区域位于基因组 9p34 区域，其对应序列的位置是人类 *LMX1B* 基因外显子的区域（图 1a）。

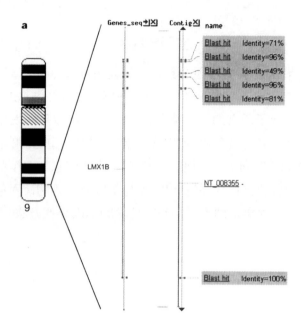

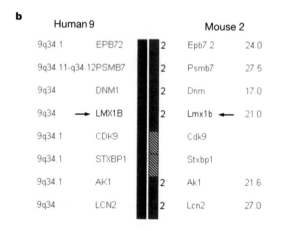

Fig. 1. *Lmx1b* encodes a transcription factor that helps to control the trajectory of motor axons during mammalian limb development. **a**, The results of a search of the six-frame translation of the draft genome sequence on the NCBI site using mouse *Lmx1b* protein NP_034855.1 as a query and using TBLASTN with standard search parameters. The best match was to a region of chromosome 9 that contains the human *LMX1B* gene. **b**, The mouse *Lmx1b* and the human *LMX1B* genes lie within a conserved syntenic block of genes, in mouse on chromosome 2 and in human on chromosome 9. This conservation of gene order supports the theory that *Lmx1b* and *LMX1B* are orthologous.

Additional support for two genes being orthologous comes from the mouse-human homology map. Despite being separated by 200 million years of evolution, mouse and human genes often fall into homologous chromosomal regions that share a conserved gene order (synteny). In fact, the working draft sequence has helped to refine the homology map and provides inferred map positions for many mouse genes[1]. The two homeobox genes fall within a conserved syntenic block between mouse chromosome 2 and human

600

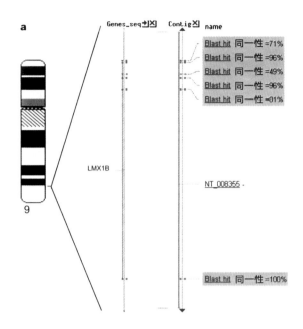

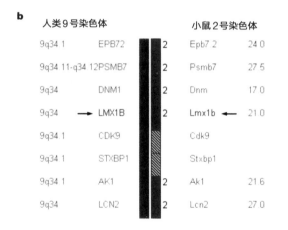

图1. *Lmx1b* 编码一个转录因子，该转录因子在哺乳动物肢体发育过程中，帮助控制运动神经元轴突的轨迹方向。**a**，在NCBI网站上，以小鼠的Lmx1b蛋白NP_034855.1作为查询序列，使用TBLASTN及标准搜索参数，在人类基因组草图序列的六种读码框翻译中进行检索的结果。最佳匹配是9号染色体上的一段包含有人类 *LMX1B* 基因的区域。**b**，小鼠的 *Lmx1b* 基因和人类的 *LMX1B* 基因都位于一个保守的线性基因区域中，小鼠的这个基因在2号染色体上，而人类的在9号染色体上。这种保守的基因排布支持 *Lmx1b* 和 *LMX1B* 是直系同源基因的推测。

　　进一步支持两个基因是直系同源的证据来自小鼠-人类同源基因图谱。尽管各自独立地进化了2亿年，小鼠和人类的基因常常位于具有保守基因序列（共线性）的同源染色体区域内。事实上，人类基因组草图已经帮助修正了小鼠和人类之间的同源基因图谱，为许多小鼠基因的定位提供了推测的图谱位置[1]。这两个同源异形框基因定位于小鼠的2号染色体和人类的9号染色体上的一个共同的保守线性区域中（图1b）。除

chromosome 9 (Fig. 1b). Furthermore, the human *LMX1B* gene has been implicated in nail patella syndrome (NPS), an autosomal recessive disorder characterized by limb and kidney defects. A mouse in which *Lmx1b* has been inactivated shows a phenotype that is strikingly similar to NPS[13]. Besides providing additional support for the conclusion of orthology, this connection may provide a useful mouse model for the human disorder. In this way, information from OMIM and mouse mutants can further define human genes.

Another way to find a gene is by looking for paralogues—family members derived by gene duplication. As an example, we used human *ADAM23*, which maps to 2q33 (ref. 14). In a BLAST search against a set of proteins predicted from the draft sequence, aside from matching itself, the best match was to a peptide from chromosome 20. No ADAM family member has previously been mapped to this chromosome. The predicted protein encoded by this gene does not begin with a methionine and appears to be incomplete at its amino terminus when aligned with other family members: this could be due to an erroneous protein prediction or a gap in the draft sequence. Computational analysis can reveal protein family domains and their relationships to three-dimensional protein structures. In this case, the putative ADAM paralogue contains both the zinc metalloprotease and disintegrin motifs characteristic of the ADAM family. Critical amino acids of the metalloprotease domain are conserved in the putative paralogue (Fig. 2b), including a trio of histidine residues in the active site (shaded yellow), which are important in complexing the zinc ion (Fig. 2a). The finding that the sequence of the new predicted ADAM member has an intact active site suggests that the predicted gene is functional, rather than being a pseudogene.

Fig. 2. The ADAM23 protein sequence NP_003803.1 was used in a BLASTP search of the Ensembl confirmed peptides produced from the 5 Sep 2000 version of the working draft sequence. As of October 2000, the best match was to the peptide ENSP00000025626 derived from chromosome 2, the predicted peptide for ADAM23. The second match was to peptide ENSP00000072108, derived from chromosome 20 and falling within GenBank Acc. No. AC055771.2. We used this predicted peptide to search a database of Pfam[24] and SMART[25] protein domains that are aligned with protein structures. This Conserved Domain Database (CDD) search at NCBI resulted in hits to reprolysin and disintegrin domains. **a**, Pfam family 01421, Reprolysin, has a structure associated with it: the zinc-dependent metalloprotease Atrolysin C (PDB: 1ATL). **b**, The query ENSP00000072108 aligns with the 1ATL protein sequence and eight other ADAMs from the Pfam Reprolysin entry. In the structure and alignment, red indicates conserved residues; grey indicates non-aligned sequences. The three histidines of the metalloprotease active site, which complex with the zinc ion, are highlighted in yellow. The structure and alignment were created using the structure viewing program Cn3D.

此之外，人类的 *LMX1B* 基因被认为与指甲髌骨综合征(NPS)的发生相关，它是一种常染色体隐性疾病，其特征是肢体和肾脏的缺陷。*Lmx1b* 基因失活的小鼠会表现出与 NPS 极为相似的症状[13]。除了可以为同源基因的结论提供更多的支持以外，这种关联性也可以为研究人类疾病提供一种有用的小鼠模型。通过这样一种方式，来自 OMIM 和小鼠的突变体的信息可以进一步定义人类的基因。

另外一种寻找基因的方法是寻找旁系同源序列——由基因复制得到的家族成员。以人类的 *ADAM23* 基因为例，它在基因组草图上对应的是 2q33(参考文献 14)。通过 BLAST 检索人类基因组草图预测的所有蛋白质发现，除了与其自身蛋白匹配外，最佳的匹配结果为 20 号染色体上的一个多肽。在此之前还没有任何一个 ADAM 蛋白质家族的成员被定位到这条染色体上。通过与其他家族成员的比对发现，推测的这个基因所编码的蛋白质并不是以甲硫氨酸起始的，它的氨基末端似乎是不完整的：这可能是由于蛋白质预测错误或者是由于在人类基因草图序列上存在缺口所引起的。通过计算分析可以揭示蛋白家族结构域以及它们与蛋白质三维结构之间的关系。在这个例子中，预测得到的 ADAM 旁系同源蛋白同时具有 ADAM 家族特征性的锌金属蛋白酶和解整联蛋白基序。金属蛋白酶结构域的关键氨基酸在推定的旁系同源蛋白中是保守的(图 2b)，这其中包括位于活性部位的组氨酸残基三联体(黄色阴影表示)，它对于络合锌离子至关重要(图 2a)。新预测的 ADAM 家族成员的序列具有一个完整的活性位点，说明这个预测的基因是有功能的，并不是一个假基因。

图 2. 在 Ensembl 中用 BLASTP 检索 ADAM23 蛋白序列 NP_003803.1，确认了 2000 年 9 月 5 日版本中人类基因草图序列得到的多肽产物。截至 2000 年 10 月，最好的匹配结果是来自 2 号染色体的多肽 ENSP00000025626，也就是预测的 ADAM23 多肽。匹配结果排在第二位的多肽是 ENSP00000072108，来自 20 号染色体，其在 GenBank 中的编号是 AC055771.2。我们将这个预测的多肽在一个数据库中检索，该数据库包含了与蛋白结构相匹配的 Pfam[24] 和 SMART[25] 蛋白结构域。在 NCBI 上搜索的保守结构域数据库(CDD)，对应的是锌金属蛋白酶和解整联蛋白结构域。**a**，Pfam 家族 01421——锌金属蛋白酶具有一个和它相关的结构：锌依赖的金属蛋白酶 Atrolysin C(PDB 代码：1ATL)。**b**，ENSP00000072108 作为查询序列，与 1ATL 蛋白序列以及 Pfam 锌金属蛋白酶条目下的另外八个 ADAMs 进行比对。在结构图和比对图中，红色代表保守性残基；灰色代表没有比对上的序列。金属蛋白酶活性位点上的三个组氨酸残基，采用黄色突出显示，它们可以与锌离子络合。图中的结构和比对结果采用结构可视化程序 Cn3D 生成。

Searching by Position

It is sometimes desirable to find genes by their position in the genome, rather than by sequence similarity. For example, when genetic or cytogenetic analysis has implicated a particular region in the aetiology of a disease, it is of interest to see what genes lie in the region. A natural way to describe positions in a sequence would be by base coordinates, but this is impractical for the working draft sequence, as the sequence is still being revised. Cytogenetic band nomenclature is more commonly used to describe positions in the genome, and many human diseases are linked to chromosomal deletions, amplifications and translocations[15]. However, to be useful in conjunction with the working draft, these designations must be related to the sequence. Towards this end, a consortium has integrated this information using fluorescence *in situ* hybridization (FISH) to localize BAC clones that also bear sequence tags that can be found in the draft sequence[16]. In addition to providing cytogenetic coordinates as entry points into the genome, they also provide mapped clone reagents that may be useful in further experimental work.

Another way to describe positions in the genome is relative to mapped sequence tagged site (STS) markers. This is particularly useful in positional cloning projects, where candidate regions are usually defined by polymorphic STSs used in genetic linkage analysis. For example, the breast cancer susceptibility locus *BRCA2* was originally localized by fine genetic mapping to a 600-kilobase (kb) interval on chromosome 13 centred around the STS marker D13S171 (ref. 17). STS markers from several genetic and physical maps have been localized in the working draft sequence using a procedure known as electronic PCR[18]. Thus, by simply looking up the position of D13S171, we can see the region around what is now known to be *BRCA2*, together with other features such as adjacent genes and markers, translocation breakpoints, and genetic variations.

Variations on a Theme

A map of DNA sequence variations will aid our understanding of complex diseases and human population dynamics. The most common class of variation is the single nucleotide polymorphism (SNP) and the total number of SNPs in the public database (dbSNP)[19] now exceeds 2.5 million, representing 1.5 million unique SNP loci. Because database entries include flanking sequence surrounding the polymorphic base(s), it is possible to localize variations within the working draft by simple sequence alignment[20].

Histone deacetylase 3 (HDAC3) is a nucleosome-remodelling enzyme that deacetylates the lysine residues of histones, affecting transcriptional repression[21]. Its genomic region contains seven mapped SNPs near the locus, one of which falls within the coding region—a G-to-C substitution that results in the nonsynonymous substitution Arg265Pro in the protein product. The three-dimensional structure of an *Aquifex aeolicus* homologue shows that this residue is at the lip of the active-site pocket[22]. Of the two classes of eukaryotic histone

基于位置的检索

有时候，通过基因在基因组上的位置来寻找基因比通过序列相似性来寻找基因更能获得令人满意的结果。举例来讲，当遗传学或者细胞遗传学分析已经暗示了某一个染色体特定区域和疾病发生密切相关时，搞清楚有什么基因位于这个区域内就显得非常有意义。描述序列位置最自然的一种方法就是采用碱基坐标，但是这种方法用于描述基因组草图序列是行不通的，因为序列还在不断的修正过程中。细胞遗传学条带系统命名法是一种更为普遍的用于描述基因组序列位置的方法，许多人类疾病都与染色体缺失、扩增以及易位相关[15]。然而，为了更好地与草图衔接使用，这些名称必须与序列相关联。为了达到这个目的，某个联盟利用荧光原位杂交技术（FISH）定位了在人类基因组草图序列中可以找到的含有序列标签的 BAC 克隆，从而整合了这些信息[16]。他们不仅可以为细胞遗传学切入基因组提供相匹配的坐标，还能提供已经比对好的克隆用于后续的实验工作。

另外一种描述基因组序列位置的方法为序列标签位点（STS）标记物定位法。这种方法特别适用于定位克隆的项目，这些候选区域通常采用遗传连锁分析中使用的多态性 STS 来定义。举例来讲，乳腺癌易感位点 BRCA2 最初就采用精细遗传图谱定位在 13 号染色体上以 STS 标记 D13S171 为中心的大约 600 kb 区域内（参考文献 17）。现在，来自多个遗传和物理图谱上的 STS 标记已经通过一种叫电子 PCR 的方法被定位在基因组草图序列上[18]。因而，简单地查询 D13S171 的位置，我们就可以看到这个区域位于 BRCA2 附近，并且还可以看到其他特征，如临近基因和标记物、易位断点以及遗传变异等。

基因组序列上的变异

DNA 序列变异图谱将帮助我们理解复杂疾病和人类种群的动态变化。最常见的一类变异是单核苷酸多态性（SNP），在公共数据库（dbSNP）[19]中存储的 SNP 总数目前已经超过了 250 万个，代表了 150 万个唯一 SNP 位点。由于数据库条目包含了多态性碱基的侧翼序列，因此采用简单的序列比对方法就可以确定这些变异在基因组草图上的位置[20]。

组蛋白脱乙酰酶 3（HDAC3）是一个核小体重构酶，可以催化组蛋白中的赖氨酸残基去乙酰化，从而影响转录抑制[21]。它所在的基因组区域包含七个确定位置的 SNP，其中一个就位于其编码区域内——一个 G 碱基到 C 碱基的替换，这会导致蛋白质产物发生非同义突变，即 265 位的精氨酸突变为脯氨酸。来自超嗜热菌的同源物的三维结构显示，这个残基位于活性位点口袋的边缘[22]。真核细胞组蛋白脱乙酰

deacetylase, one most often has Arg at this position and the other most often has Pro[23], a trait shared with the bacterial members of the histone deacetylase superfamily. This SNP might therefore occur at a functionally interesting site, and also gives pause for speculation: as bacterial members of the superfamily predominantly have a Pro in this position, perhaps Pro is the ancient residue at this site, and not Arg. Note that several high-throughput SNP discovery methods have been used to generate these data and not all SNPs have been rigorously validated.

Conclusions

The draft sequence provides us with the first comprehensive integration of diverse genomic resources. The mapping of ESTs, gene predictions, STSs and SNPs onto the draft sequence can enable identification of alternative splicing, orthologues, paralogues, map positions and coding sequence variations. Users should remember, though, that these genomic resources represent a work-in-progress, and will evolve as the genome is finished and computation methods further refined.

(**409**, 824-826; 2001)

Tyra G. Wolfsberg[*], Johanna McEntyre[†] & Gregory D. Schuler[†]

[*] Genome Technology Branch, National Human Genome Research Institute, National Institutes of Health, Bethesda, Maryland 20892, USA

[†] National Center for Biotechnology Information, National Library of Medicine, National Institutes of Health, Bethesda, Maryland 20894, USA

References:

1. International Human Genome Sequencing Consortium. Initial sequencing and analysis of the human genome. *Nature* **409**, 860-921 (2001).

2. Ewing, B. & Green, P. Analysis of expressed sequence tags indicates 35,000 human genes. *Nature Genet.* **25**, 232-234 (2000).

3. Roest Crollius, H. *et al.* Estimate of human gene number provided by genome-wide analysis using *Tetraodon nigroviridis* DNA sequence. *Nature Genet.* **25**, 235-238 (2000).

4. McKusick, V. A. *Mendelian Inheritance in Man. Catalogs of Human Genes and Genetic Disorders* (Johns Hopkins Univ. Press, Baltimore, 1998).

5. Maglott, D. R., Katz, K. S., Sicotte, H. & Pruitt, K. D. NCBI's LocusLink and RefSeq. *Nucleic Acids Res.* **28**, 126-128 (2000).

6. Pruitt, K. D., Katz, K. S., Sicotte, H. & Maglott, D. R. Introducing RefSeq and LocusLink: curated human genome resources at the NCBI. *Trends Genet.* **16**, 44-47 (2000).

7. Guigo, R., Agarwal, P., Abril, J. F., Burset, M. & Fickett, J. W. An assessment of gene prediction accuracy in large DNA sequences. *Genome Res.* **10**, 1631-1642 (2000).

8. Stormo, G. D. Gene-finding approaches for eukaryotes. *Genome Res.* **10**, 394-397 (2000).

9. Sagane, K., Ohya, Y., Hasegawa, Y. & Tanaka, I. Metalloproteinase-like, disintegrin-like, cysteine-rich proteins MDC2 and MDC3: novel human cellular disintegrins highly expressed in the brain. *Biochem. J.* **334**, 93-98 (1998).

10. Wolfsberg, T. G. & Landsman, D. A comparison of expressed sequence tags (ESTs) to human genomic sequences. *Nucleic Acids Res.* **25**, 1626-1632 (1997).

11. Wolfsberg, T. G. & Landsman, D. in *Bioinformatics: A Practical Guide to the Analysis of Genes and Proteins* (eds Baxevanis, A. D. & Ouellette, B. F. F.) (Wiley-Liss, Inc., New York, 2001).

12. Altschul, S. F. *et al.* Gapped BLAST and PSI-BLAST: a new generation of protein database search programs. *Nucleic Acids Res.* **25**, 3389-3402 (1997).

13. Chen, H. *et al.* Limb and kidney defects in Lmx1b mutant mice suggest an involvement of LMX1B in human nail patella syndrome. *Nature Genet.* **19**, 51-55 (1998).

14. Poindexter, K., Nelson, N., DuBose, R. F., Black, R. A. & Cerretti, D. P. The identification of seven metalloproteinase-disintegrin (ADAM) genes from genomic libraries. *Gene* **237**, 61-70 (1999).

15. Mitelman, F., Mertens, F. & Johansson, B. A breakpoint map of recurrent chromosomal rearrangements in human neoplasia. *Nature Genet.* **15**, 417-474 (1997).

16. The BAC Resource Consortium. Integration of cytogenetic landmarks into the draft sequence of the human genome. *Nature* **409**, 953-958 (2001).

酶分为两类，一类在这个位置上通常是精氨酸，而另一类则与细菌组蛋白脱乙酰酶超家族一样，在这个位置上通常是脯氨酸。因此这个 SNP 位点可能出现在一个有趣的功能性位点上，同时我们可以推测：鉴于细菌组蛋白脱乙酰酶超家族在这个位置上主要是脯氨酸，或许脯氨酸才是这个位置上的古老残基，而非精氨酸。需要注意的是，已经有几种高通量 SNP 发现方法被用来产生这些数据，并且不是所有的 SNP 位点都能被严格验证。

<h2 style="text-align:center">结　论</h2>

基因组草图序列使我们得以首次全面整合不同的基因组资源。将 EST、基因预测、STS 和 SNP 定位到草图序列上，可以帮助我们确认可变剪接、直系同源序列、旁系同源序列、比对位置以及编码序列的变异。但是，使用者也要注意，这些基因组资源尚在"加工"之中，会随着基因组测序的最终完成和计算方法的进一步优化而不断地修改。

（刘振明 翻译；陈捷胤 审稿）

17. Wooster, R. *et al.* Localization of a breast cancer susceptibility gene, BRCA2, to chromosome 13q12-13. *Science* **265**, 2088-2090 (1994).

18. Schuler, G. D. Sequence mapping by electronic PCR. *Genome Res.* **7**, 541-550 (1997).

19. Smigielski, E. M., Sirotkin, K., Ward, M. & Sherry, S. T. dbSNP: a database of single nucleotide polymorphisms. *Nucleic Acids Res.* **28**, 352-355 (2000).

20. The International SNP Map Working Group. A map of human genome sequence variation containing 1.42 million single nucleotide polymorphisms. *Nature* **409**, 928-933 (2001).

21. Struhl, K. Histone acetylation and transcriptional regulatory mechanisms. *Genes Dev.* **12**, 599-606 (1998).

22. Finnin, M. S. *et al.* Structures of a histone deacetylase homologue bound to the TSA and SAHA inhibitors. *Nature* **401**, 188-193 (1999).

23. Leipe, D. D. & Landsman, D. Histone deacetylases, acetoin utilization proteins and acetylpolyamine amidohydrolases are members of an ancient protein superfamily. *Nucleic Acids Res.* **25**, 3693-3697 (1997).

24. Bateman, A. *et al.* The Pfam protein families database. *Nucleic Acids Res.* **28**, 263-266 (2000).

25. Schultz, J., Copley, R. R., Doerks, T., Ponting, C. P. & Bork, P. SMART: a web-based tool for the study of genetically mobile domains. *Nucleic Acids Res.* **28**, 231-234 (2000).

Acknowledgements. We thank G. Marth, S. Sherry, D. Landsman, D. Church and D. Lipman for suggestions and review of the manuscript.

Correspondence should be addressed to G.D.S. (e-mail: schuler@ncbi.nlm.nih.gov).

Mining the Draft Human Genome

E. Birney *et al.*

Editor's Note

The first draft of the human genome provided a wealth of information of potential interest to researchers in many fields, ranging from medicine to palaeontology. The sequence was made publicly available, but making productive use of it required some knowledge both of the nature of the information it contained and the tools provided for navigating these data. Here some researchers from the team involved in the genome project supply a quick guide to the possibilities of such data mining.

Now that the draft human genome sequence is available, everyone wants to be able to use it. However, we have perhaps become complacent about our ability to turn new genomes into lists of genes. The higher volume of data associated with a larger genome is accompanied by a much greater increase in complexity. We need to appreciate both the scale of the challenge of vertebrate genome analysis and the limitations of current gene prediction methods and understanding.

IN this issue, accompanying the description of the sequence[1], there are nine data-mining papers that interrogate the genome from distinct biological perspectives. These range from broad topics—cancer[2], addiction[3], gene expression[4], immunology[5] and evolutionary genomics[6]—to the more focused: membrane trafficking[7], cytoskeleton[8], cell cycle[9] and circadian clock[10]. The findings reported by these authors are likely to be indicative of many people's experiences with the draft human genome: frustrating and rewarding in equal measures.

The Current Data Set

The human genome—the first vertebrate genome sequence to be determined—seems likely to be quite representative of what we will find in other vertebrate genomes. It is around 30 times larger than the recently sequenced worm and fly genomes, and 250 times larger than that of yeast, the first eukaryotic genome to be sequenced[11]. Despite its size, it seems likely to have only two or three times as many genes as the fly and worm genomes, with the coding regions of genes accounting for only 3% of the DNA. Repeat sequences form a large proportion of the remaining DNA, around 46%. These repeats may or may not have a function, but they are certainly characteristic of large vertebrate genomes. The rest of the sequence contains promoters, transcriptional regulatory sequences and other

610

挖掘人类基因组草图

人类基因组的第一份草图为从医学到古生物学等众多领域的研究者提供了大量有用的信息。序列已经可以公开获取，但是要高效利用这些序列既需要了解这些序列所包含信息的本质，也需要了解浏览这些数据的工具。在这里，一些参与该基因组项目的研究者为挖掘这些数据的可能性提供了快速指南。

既然已经有了人类基因组草图序列，每个人都希望能够使用它。然而，我们可能已经为我们拥有将新基因组变成一个个基因列表的能力而沾沾自喜。基因组越大，包含的数据越多，同时还伴随着复杂程度大大增加。我们需要意识到对脊椎动物基因组进行分析所面临挑战的规模，以及目前在基因预测方法上和认识上的局限性。

在这一期中，除了对序列的描述[1]，还有九篇数据挖掘方面的研究论文从截然不同的生物学视角对基因组进行审视。这些研究的范围从宽泛的主题——癌症[2]、成瘾性[3]、基因表达[4]、免疫学[5]和进化基因组学[6]，到更具体的主题——膜转运[7]、细胞骨架[8]、细胞周期[9]和生物钟[10]。这些作者报道的发现可能代表了许多人类基因组草图研究者的经验：挫折和收获一样多。

目前的数据集

人类基因组——作为第一个被测序的脊椎动物基因组——在我们将要发现的其他脊椎动物基因组中，似乎很具有代表性。它大约是目前已经测序的蠕虫和苍蝇基因组的30倍，是第一个测序的真核生物——酵母的基因组的250倍[11]。尽管很大，但是它所包含的基因数目看起来只有蠕虫和苍蝇基因组的两到三倍，因为它的基因编码区仅占整个DNA长度的3%。重复序列构成了其余DNA的一大部分，大概占到46%。这些重复序列可能有功能，也可能没有功能，但是它们确实是大型脊椎动物基因组的典型特征。剩下的序列包含启动子、转录调控序列和其他一些尚未知晓

features, as yet unknown.

The International Human Genome Sequencing Consortium has been sequencing the genome in fragments of about 100–200 kilobases (kb). These fragments exist as bacterial artificial chromosome (BAC) clones, which are derived from sequences whose chromosomal location is known. Each newly generated sequence is deposited in the high-throughput genome sequence (HTGS) division of the International Nucleotide Database (GenBank/EMBL/DDBJ) within 24 hours of being assembled and is assigned a unique identifier (its accession number). For the working draft, about 75% of clones are "unfinished": each still consists of about 10–20 unassembled sequence fragments. Sequencing centres are continuously reading new sequence data from these clones until all the gaps are eliminated, at which point the sequence is declared "finished". As the HTGS entries are updated they retain the same accession numbers, but their version numbers increase.

There is a great deal of overlap between BAC clones, so it is typically more convenient to view a cleaned up version of the raw data, in which the sequences of the clones are correctly ordered and overlapped to remove redundancy and create a contiguous DNA sequence for each chromosome. These virtual chromosome sequences change continuously as gaps are closed and fragment ordering is refined.

Finding Genes

With over 30 genomes sequenced, the casual observer could be forgiven for thinking that gene prediction, or annotation, was a problem filed neatly under "solved". Unfortunately this is far from true. The large size of the genome makes finding the genes much more difficult. The protein-coding parts of human genes, called exons, are split into pieces in the genome and these pieces are separated by non-coding sequence called introns. Nearly all of the increase in gene size in human compared with fly or worm is due to the introns becoming much longer (about 50 kb versus 5 kb). The protein-encoding exons, on the other hand, are roughly the same size. This decrease in signal (exon) to noise (intron) ratio in the human genome leads to misprediction by computational gene-finding strategies.

Many methods for predicting genes are based on compositional signals that are found in the DNA sequence. These methods detect characteristics that are expected to be associated with genes, such as splice sites and coding regions, and then piece this information together to determine the complete or partial sequence of a gene. Unfortunately, these *ab initio* methods tend to produce false positives, leading to overestimates of gene numbers, which means that we cannot confidently use them for annotation. They also do not work well with unfinished sequence that has gaps and errors, which may give rise to frameshifts, when the reading frame of the gene is disrupted by the addition or removal of bases.

Thankfully, there is a wealth of data that we can use to produce more reliable gene predictions. Information on expressed sequences (expressed sequence tags (ESTs) and

612

的特征性序列。

国际人类基因组测序联盟一直在对长约 100～200 kb 的基因组片段进行测序。这些片段以细菌人工染色体（BAC）克隆的形式存在，这些克隆来自于那些染色体位置已知的序列。每一条新测序的序列都会在组装完成后的 24 小时内存入国际核酸序列数据库（GenBank/EMBL/DDBJ）中的高通量测序基因组序列（HTGS）子库，并且被分配一个唯一的标识符（它的编号）。对于整个工作草图而言，还有大概 75% 的克隆处于"未完成"的状态：它们中的每一个大约包含 10～20 个没有完成组装的序列片段。测序中心正在不断地从这些克隆中读取新的序列数据，直到序列上所有的空缺都被消除为止，直到此时，整个序列的测序工作才能宣告"完成"。在 HTGS 条目更新的过程中，原有编号会保留下来，但是它们的版本号会增加。

细菌人工染色体克隆之间会有大量重叠，因此通常查看原始数据处理后的版本会比较方便，在整洁版中，这些克隆的序列按照正确的顺序进行排列、相互重叠时移除冗余部分，为每一条染色体构造一个连续的 DNA 序列。伴随着序列内空缺的闭合和片段排列顺序的修正，这些实际的染色体序列会不断地改变。

寻 找 基 因

面对 30 多个已经测序的基因组，对旁观者而言，认为"基因的预测或注释领域的问题已经解决"也情有可原。但不幸的是，事实远非如此。庞大的基因组让寻找基因的工作变得更加困难。人类基因中蛋白编码的部分称为外显子，在基因组中被非编码序列的内含子分隔成为片段。与苍蝇和蠕虫基因相比，几乎所有人类基因大小的增加都是由内含子变长造成的（约为 50 kb 比 5 kb）。而另一方面，编码蛋白的外显子，长度则大致相同。人类基因组中的这种信（外显子）噪（内含子）比降低的情况，会导致通过计算寻找基因的策略给出错误的预测。

许多预测基因的方法都是基于在 DNA 序列中找到的组成信号。这些方法首先探测某些期望与基因相关的特征，例如剪接位点和编码区，然后把这些信息拼接在一起，从而确定基因的完整序列或部分序列。然而不幸的是，这些从头预测的方法倾向于给出假阳性的结果，导致基因数目的高估，这就意味着我们不能自信地应用它们来进行基因注释。此外，这些方法也不能很好地处理含有缺口和错误的未完成序列，因为如果增加或者移除的碱基破坏了基因的阅读框，就可能会产生移码。

万幸的是，已经有丰富的数据可以用来进行更为可靠的基因预测。来自人和其他生物的表达序列（表达序列标签（EST）和互补 DNA）和蛋白质信息，为从浩瀚的

complementary DNAs) and proteins from humans and other organisms provide a more accurate resource for resolving gene structures against the vast genomic background. The most effective algorithms integrate gene-prediction methods with similarity comparisons. Such algorithms are integral to software programs such as GeneWise[12], Genomescan[13] and Genie[14], which provide accurate, automatic predictions, whereas BLAST or FASTA programs typically require considerable manual effort to determine the complete structure of a single gene.

The most powerful tool for finding genes may be other vertebrate genomes. Comparing conserved sequence regions between two closely related organisms will enable us to find genes and other important regions in both genomes with no previous knowledge of the gene content of either. The next couple of years should see the sequencing of the mouse, zebrafish and *Tetraodon* genomes. The preliminary sequence of *Tetraodon* has already proved useful in estimating gene numbers[15], and shows much promise for the use of comparative genomics in gene prediction.

Resources Available to the User

There are a number of resources currently available for perusing the human genome. "Human Genome Central"[16] attempts to gather together the most useful web sites (see http://www.ensembl.org/genome/central/ or http://www.ncbi.nlm.nih.gov/genome/central). The best starting point for the uninitiated will be a site such as those of NCBI, Ensembl or the University of California Santa Cruz (UCSC). These sites offer a mixture of genomic viewers and web-searchable datasets, and allow analysis of the human genome sequence without the need to run complex software locally.

For more involved analysis, it might be necessary to download some of the data locally. Useful downloadable sequence-oriented datasets include protein datasets (available from Ensembl and NCBI) and the assembled DNA sequence for regions of the genome, available at UCSC. Other genomic datasets are also available, such as the global physical map from The Genome Sequencing Center in St Louis and the single nucleotide polymorphism (SNP) database from NCBI. Raw sequence data is available from the International Nucleotide Database (GenBank/EMBL/DDBJ), but this data is generally more difficult to handle because it is very fragmentary, can contain contaminating non-human DNA and may include misleading information such as incorrect map assignment.

This loose network of sites will probably coalesce into a more coordinated network of sites offering informative web pages and resources. NCBI, Ensembl and UCSC are developing new, more accessible resources that will become available within the next year.

614

基因组背景中解析出基因结构提供了更为精确的数据源。最有效的算法是把基因预测方法和相似性比较整合在一起。这类算法已经被整合进一系列的软件程序中，例如 GeneWise[12]、Genomescan[13] 以及 Genie[14]，它们可以提供准确的自动化预测。与之相比，使用 BLAST 和 FASTA 程序通常需要相当多的人工工作来确定一个基因的完整结构。

寻找基因最为有效的工具也许就是其他脊椎动物的基因组。比较两个近缘生物的序列保守区可以使我们不必知道两个基因组中任何一个基因的内容就能发现这两个基因组中的基因和其他重要区域。未来几年应该会看到对小鼠、斑马鱼和河鲀鱼基因组的测序。河鲀鱼基因组的初步测序结果在预测基因数目方面的作用已经得到证实[15]，同时显示出比较基因组学在基因预测方面的应用很有前景。

用户可以利用的资源

目前，已经有了一些可用于详尽分析人类基因组的资源。"人类基因组中心"[16] 正试图将最有用的网站（参见 http://www.ensembl.org/genome/central/ 或者 http://www.ncbi.nlm.nih.gov/genome/central）集合在一起。对于缺少经验的人而言，最好的起点就是类似于 NCBI、Ensembl 或者加州大学圣克鲁兹分校（UCSC）这样的一些站点。这些站点都提供了基因组浏览器以及可通过网络搜索的数据库，这样可以允许使用者对人类基因组序列进行分析，而无需在本地运行复杂的软件。

对于更进一步的分析，可能就需要将一些数据下载到本地了。有用的可供下载的定位于序列的数据库，包括蛋白质数据库（可以在 Ensembl 和 NCBI 上获得）以及基因组区域组装好的 DNA 序列，可以在 UCSC 上获得。其他一些基因组数据库也是可以利用的，比如来自圣路易斯的基因组测序中心的完整物理图谱，以及 NCBI 提供的单核苷酸多态性（SNP）数据库。原始的序列数据可以从国际核酸数据库（GenBank/EMBL/DDBJ）获得，但是这类数据通常更难处理，因为它们的片段化非常严重，可能包含受污染的非人类 DNA 序列，还可能包含误导性信息，比如错误的图谱排列。

这些松散的网络站点有可能合并成为一个更为协同的网络，可以提供信息丰富的网页以及资源。NCBI、Ensembl 以及 UCSC 等正在开发新的、更容易使用的资源，这些资源在接下来的一年内将可以供人使用。

How to Use the Resources

There are two main ways to use the human genome sequence. First, we can look for a homologue of a protein that is known from another organism. For example, Clayton et al.[10] looked for relatives of the *Drosophila* period clock protein and found the three known relatives and a possible fourth cousin on chromosome 7. Or we can try and find all of the proteins belonging to a particular family—in ref. 4, Tupler et al. catalogue all homeobox domains[4]. The easiest way to approach these problems is to use a protein set. This sidesteps the frustration of predicting genes, but makes the researcher reliant on the quality of the predictions being provided. For most of the accompanying reports, a single protein set was the most useful resource provided. For example, Nestler et al. searched for G-protein receptor kinases[3] using PSI-BLAST, which searches only protein datasets.

What are the potential pitfalls of the data? Human genes are hard to predict and are often fragmented. If each end of a query protein matches to a different predicted protein, we should suspect that the query sequence may in fact be two parts of a fragmented gene. The two matched human genes should be in the same or adjacent genomic locations. Pollard[8] discovered that fragmentation complicated the analysis of myosin genes. In addition, the unfinished human genomic DNA may contain contamination, particularly from bacteria but also from other sources. Contaminating DNA is routinely removed from finished sequence, but some is still present in unfinished sequence. If the predicted gene matches a bacterial gene more closely than any vertebrate gene then it will almost always be a contaminant. Futreal et al.[2] were led up a blind alley for a week before they discovered that cDNA contamination in draft genomic sequences was giving the false impression of multiple p53 proteins in the genome.

During the assembly of unfinished human genomic data it is possible to create artificial duplications, which can result in artefacts in the subsequent analysis. Very similar gene sequences found within the same clone may represent duplicate genes, but could also be the result of an assembly error. This also means that predicted protein sets may contain artificial duplications, leading to overestimation of the number of members in a family.

What does this analysis tell us? For Bock et al.[7], the draft genome revealed a list of the molecular players involved in membrane trafficking, providing a platform for experiments that may complete our understanding of this area of biology. In contrast, Murray and Marks[9] found no new cyclin-dependent kinases, indicating that they were all found by traditional experimental techniques. Futreal et al. had a similar experience for known cancer genes, but suggest that with new techniques the genome will provide new avenues of cancer research[2].

The interpretation of unfinished draft genomic data may seem like hard work. But it is something to become accustomed to, because we expect future vertebrate genomes to be

如何使用这些资源

使用人类基因组序列有两种主要方式。首先，我们可以寻找与其他生物中已知蛋白质同源的蛋白质。例如，克莱顿等人[10]在人类基因组序列中寻找与果蝇周期性时钟蛋白具有亲缘关系的蛋白，结果在 7 号染色体上发现了三个已知的同源蛋白以及可能的第四个远亲蛋白。或者，我们可以尝试寻找属于一个特定家族的所有蛋白质——在参考文献 4 中，图普勒等人整理了所有的同源异型框结构域[4]。解决这些问题最简单的途径是使用一个蛋白质数据库。这样就规避了预测基因的困难，但这使研究者要依赖提供的数据质量。对于大多数研究产生的报告，单一的蛋白质数据库是所提供的最为有效的数据源。举例来讲，内斯特勒等人使用 PSI-BLAST 程序搜索 G 蛋白受体激酶[3]，这个过程只搜索了蛋白质数据库。

这些数据潜在的缺陷是什么？人类基因难以预测，并且通常是片段化的。如果所查询蛋白的每一端匹配上了不同的预测蛋白，我们应该怀疑查询序列实际上可能是一个片段化基因的两个部分。两个匹配上的人类基因应当在基因组上处于相同或者邻近的位置。波拉德[8]发现基因的片段化使得对肌球蛋白基因的分析变得复杂。除此之外，没有完成的人类基因组 DNA 可能会含有污染，特别是来自细菌的污染，当然，也有可能是其他来源的污染。按照规程，污染 DNA 会从完成的序列中移除，但是仍然会有一些留存在尚未完成的序列中。如果预测的基因与细菌基因的相似性超过了任何脊椎动物的基因，那么几乎可以肯定就是污染物。富特雷亚尔等人[2]被带进了死胡同里，他们花了将近一个星期的时间，才发现基因组草图序列中的 cDNA 污染造成了人类基因组中含有多个 p53 蛋白这一错误印象。

在对未完成的人类基因组数据进行组装的过程中，有可能会造成人为的重复，从而导致后续序列分析中的假象。在同一个克隆中找到的非常相似的基因序列可能代表基因复制，但也可能是组装错误的结果。这也意味着预测得到的蛋白质数据库中可能包含人为引入的重复，从而导致高估某个家族中所包含成员的数目。

这个分析能告诉我们什么？对于博克等人而言[7]，人类基因组草图揭示了一系列参与细胞膜转运过程的分子成员，为实验提供了一个平台，可能完善我们对生物学该领域的理解。与之相反，默里和马克斯[9]没有找到新的周期蛋白依赖性激酶，说明它们都已经被传统的实验技术发现了。弗特利尔等人对已知的癌症基因也有相似的经验，但他们的研究说明，随着新技术的使用，基因组将会为癌症的研究提供新的途径[2]。

对尚未完成的基因组草图数据的解析看起来是一项困难的工作。但是它会成为一件平常的事情，因为我们预计未来脊椎动物基因组最先会以草图的形式发布出来。

released initially in draft form. The database providers must develop better ways of viewing the data; and researchers need to be educated in how to use them. That said, there are many undiscovered treasures in the current data set waiting to be found by intuition, hard work and experimental verification. Good luck, and happy hunting!

(409, 827-828; 2001)

Ewan Birney[*], **Alex Bateman**[†], **Michele E. Clamp**[†] & **Tim J. Hubbard**[†]

[*] The European Bioinformatics Institute, Wellcome Trust Genome Campus, Hinxton, Cambridge, CB10 1SA, UK
[†] The Sanger Centre, Wellcome Trust Genome Campus, Hinxton, Cambridge CB10 1SA, UK

References:

1. International Human Genome Sequencing Consortium. Initial sequencing and analysis of the human genome. *Nature* **409**, 860-921 (2001).

2. Futreal, A., Wooster, R., Kasprzyk, A., Birney, E. & Stratton, S. Cancer and genomics. *Nature* **409**, 850- 852 (2001).

3. Nestler, E. J. & Landsman, E. Learning about addiction from the genome. *Nature* **409**, 834-835 (2001).

4. Tupler, R., Perini, G. & Green, M. R. Expressing the human genome. *Nature* **409**, 832-833 (2001).

5. Fahrer, A. M., Bazan, J. F., Papathanasiou, P., Nelms, K. A. & Goodnow, C. C. A genomic view of immunology. *Nature* **409**, 836-838 (2001).

6. Li, W-H., Gu, Z., Wang, H. & Nekrutenko, A. Evolutionary analyses of the human genome. *Nature* **409**, 847-849 (2001).

7. Bock, J. B., Matern, H. T., Peden, A. A. & Scheller, R. H. A genomic perspective on membrane compartment organization. *Nature* **409**, 839-841 (2001).

8. Pollard, T. D. Genomics, the cytoskeleton and motility. *Nature* **409**, 842-843 (2001).

9. Murray, A. W. & Marks, D. Can sequencing shed light on cell cycling? *Nature* **409**, 844-846 (2001).

10. Clayton, J. D., Kyriacou, C. P. & Reppert, S. M. Keeping time with the human genome. *Nature* **409**, 829-831 (2001).

11. Goffeau, A. *et al.* The Yeast Genome Directory. *Nature* **387** (suppl.), 1-105 (1997).

12. Birney, E. & Durbin, R. Using GeneWise in the *Drosophila* annotation experiment. *Genome Res.* **10**, 547-548 (2000).

13. Burge *et al. Nature Genet.* (submitted).

14. Reese, M. G., Kulp, D., Tammana, H., Haussler, D. Genie—gene finding in *Drosophila melanogaster. Genome Res.* **10**, 529-538 (2000).

15. Crollius, H. R. *et al.* Characterization and repeat analysis of the compact genome of the freshwater pufferfish *Tetraodon nigroviridis. Genome Res.* **10**, 939-949 (2000).

16. Genome website set up to help with sequence analysis. *Nature* **406**, 929 (2000).

Correspondence should be addressed to E.B. (e-mail: birney@ebi.ac.uk).

数据库的提供者必须开发更好的方式来查看数据；研究者则需要接受培训以运用这些方法。也就是说，在现有的数据库中，还有许多未知的宝藏等待我们通过感知、努力工作以及实验验证去发现。祝大家好运，并开始快乐的探索旅程！

（刘振明 翻译；解彬彬 审稿）

Initial Sequencing and Analysis of the Human Genome*

International Human Genome Sequencing Consortium

Editor's Note

This paper details the first draft sequence of the human genome. The data, which were published at the same time as Celera Genomics' privately-funded human genome sequence, were the results of an international collaboration between 20 sequencing centres in 6 different countries. This draft sequence, which covers about 94% of the human genome, suggests a presence of some 30,000 to 40,000 protein-coding genes. The final count, revealed with the publication of the complete genome sequence two years later, was down-graded to around 25,000. But the paper remains important because it was the largest extensively sequenced genome of its time, the first vertebrate genome to be extensively sequenced, and uniquely, a first glimpse at the genome of our own species, holding clues to human development, physiology, medicine and evolution.

The human genome holds an extraordinary trove of information about human development, physiology, medicine and evolution. Here we report the results of an international collaboration to produce and make freely available a draft sequence of the human genome. We also present an initial analysis of the data, describing some of the insights that can be gleaned from the sequence.

THE rediscovery of Mendel's laws of heredity in the opening weeks of the 20th century[1-3] sparked a scientific quest to understand the nature and content of genetic information that has propelled biology for the last hundred years. The scientific progress made falls naturally into four main phases, corresponding roughly to the four quarters of the century. The first established the cellular basis of heredity: the chromosomes. The second defined the molecular basis of heredity: the DNA double helix. The third unlocked the informational basis of heredity, with the discovery of the biological mechanism by which cells read the information contained in genes and with the invention of the recombinant DNA technologies of cloning and sequencing by which scientists can do the same.

---

*This is a shortened version of the original paper. Some, but by no means all, of the omissions have been indicated in the text. Lists of authors and affiliations are given in the original paper. Citations are numbered herein as they are in the original paper; only those references that are cited in this version are listed in the "References" section.

人类基因组的初步测序与分析[*]

国际人类基因组测序联盟

编者按

本文详细解析了人类基因组的第一个草图序列。本文的数据是 6 个国家的 20 个测序中心的国际合作的结果，与塞莱拉基因组公司资助的人类基因组序列同时发表。草图序列覆盖了约 94% 的人类基因组，提示可能存在 30,000 ~ 40,000 个蛋白质编码基因。随着两年后完整基因组序列的发表，最终蛋白质编码基因计数结果降至 25,000 个。但是，本文仍旧占据重要地位，因为它是其所处时代最大规模的测序基因组，也是第一个经大规模测序的脊椎动物的基因组，而且非常独特的是，这也是对我们自身物种基因组的第一次探索，获得了关于人类发育、生理学、医药和进化的诸多研究线索。

人类基因组蕴含着与人类发育、生理学、医药和进化相关的海量信息。本文报道了国际合作组织的测序结果，该组织完成了人类基因组草图序列，该信息可以免费获取。我们也对这些数据进行了初步分析，描述了从序列中获得的启示。

---

20 世纪初，孟德尔遗传定律的重新发现[1-3]激发了人们对遗传信息的本质与内涵的探索，推动了生物学近一百年多年的发展。这些科学进展自然地分成了四个主要的阶段，大体上对应了 20 世纪的四个二十五年。第一个阶段建立了遗传的细胞学基础：染色体。第二个阶段定义了遗传的分子基础：DNA 双螺旋。第三个阶段解密了遗传的信息学基础，发现了细胞读取包含在基因中的遗传信息的生物学机制，并发明了克隆和测序的 DNA 重组技术，使得科学家可以进行 DNA 重组。

---

[*] 这是原文的缩略版。有一些删减在文本中标示出来了，但并非所有的删减都进行了标注。作者及单位名单请参见原文。本文中引用参考文献处的编号与原文一致；文末"References"部分仅保留了本文中有所引用的文献。

621

The last quarter of a century has been marked by a relentless drive to decipher first genes and then entire genomes, spawning the field of genomics. The fruits of this work already include the genome sequences of 599 viruses and viroids, 205 naturally occurring plasmids, 185 organelles, 31 eubacteria, seven archaea, one fungus, two animals and one plant.

Here we report the results of a collaboration involving 20 groups from the United States, the United Kingdom, Japan, France, Germany and China to produce a draft sequence of the human genome. The draft genome sequence was generated from a physical map covering more than 96% of the euchromatic part of the human genome and, together with additional sequence in public databases, it covers about 94% of the human genome. The sequence was produced over a relatively short period, with coverage rising from about 10% to more than 90% over roughly fifteen months. The sequence data have been made available without restriction and updated daily throughout the project. The task ahead is to produce a finished sequence, by closing all gaps and resolving all ambiguities. Already about one billion bases are in final form and the task of bringing the vast majority of the sequence to this standard is now straightforward and should proceed rapidly.

The sequence of the human genome is of interest in several respects. It is the largest genome to be extensively sequenced so far, being 25 times as large as any previously sequenced genome and eight times as large as the sum of all such genomes. It is the first vertebrate genome to be extensively sequenced. And, uniquely, it is the genome of our own species.

Much work remains to be done to produce a complete finished sequence, but the vast trove of information that has become available through this collaborative effort allows a global perspective on the human genome. Although the details will change as the sequence is finished, many points are already clear.

- The genomic landscape shows marked variation in the distribution of a number of features, including genes, transposable elements, GC content, CpG islands and recombination rate. This gives us important clues about function. For example, the developmentally important HOX gene clusters are the most repeat-poor regions of the human genome, probably reflecting the very complex coordinate regulation of the genes in the clusters.

- There appear to be about 30,000–40,000 protein-coding genes in the human genome— only about twice as many as in worm or fly. However, the genes are more complex, with more alternative splicing generating a larger number of protein products.

- The full set of proteins (the "proteome") encoded by the human genome is more complex than those of invertebrates. This is due in part to the presence of vertebrate-specific protein domains and motifs (an estimated 7% of the total), but more to the fact that vertebrates appear to have arranged pre-existing components into a richer collection of domain architectures.

20 世纪的最后二十五年，科学家们通过不懈努力破译了第一组基因，随后进一步破译了完整的基因组，从而开创了新的研究领域——基因组学。这项工作的研究成果包括了 599 个病毒和类病毒、205 个天然质粒、185 个细胞器、31 株真细菌、7 株古菌、1 株真菌、2 种动物和 1 种植物的基因组序列。

本文报道了来自美国、英国、日本、法国、德国和中国的 20 多个团队的合作研究成果，他们合作完成了人类基因组草图的构建。基因组草图序列由物理图谱产生，该图谱覆盖了基因组常染色质 96% 的序列，加上公共数据库中的其他序列，大约覆盖了人类基因组 94% 的序列。序列的构建所耗费的时间相对较短，序列覆盖度从大约 10% 达到 90% 以上只用了大约 15 个月。序列数据可开放获取，没有任何限制，且在项目进行过程中每天更新。项目下一步的任务是构建基因组完成图，修补所有序列缺口，并解决所有序列歧义。人类基因组中已经有 10 亿个碱基得到了确认，将大部分序列按上述标准落实是项目的首要任务，需要迅速完成。

人类基因组序列从多个方面来看都十分有趣。它是目前被大规模测序的基因组中最大的一个，是之前测序的任意一个基因组的 25 倍，也是之前测序的基因组总和的 8 倍。它是第一个被大规模测序的脊椎动物的基因组。而且，特别的是，它还是我们人类自己的基因组。

为了构建基因组完成图，仍有大量工作需要做，但是在合作者的共同努力下，我们已经获得了海量有价值的信息，可以从全球视角认识人类基因组。尽管当序列全部完成时一些细节会改变，但是很多要点已经理清。

- 基因组全景显示，基因组的大量特征在分布上存在显著差异，包括基因、转座元件、GC 含量、CpG 岛和重组率，这为我们提供了重要的功能线索。例如，对发育非常重要的 HOX 基因簇是基因组上重复序列最贫瘠的区域，这可能反映了基因簇中复杂的基因间协同调控。

- 人类基因组上约有 30,000 ~ 40,000 个蛋白质编码基因，仅约为线虫或果蝇的两倍。然而，人类基因更为复杂，具有更多的可变剪接，可产生更多的蛋白质产物。

- 人类基因组所编码的全部蛋白质（"蛋白质组"）比无脊椎动物更为复杂。这一现象部分是由于脊椎动物特异性蛋白质结构域和基序的出现（约占总量的 7%），但更重要的是，脊椎动物似乎已经将先前存在的组分组装为更丰富的结构域集合体系。

- Hundreds of human genes appear likely to have resulted from horizontal transfer from bacteria at some point in the vertebrate lineage. Dozens of genes appear to have been derived from transposable elements.

- Although about half of the human genome derives from transposable elements, there has been a marked decline in the overall activity of such elements in the hominid lineage. DNA transposons appear to have become completely inactive and long-terminal repeat (LTR) retroposons may also have done so.

- The pericentromeric and subtelomeric regions of chromosomes are filled with large recent segmental duplications of sequence from elsewhere in the genome. Segmental duplication is much more frequent in humans than in yeast, fly or worm.

- Analysis of the organization of Alu elements explains the longstanding mystery of their surprising genomic distribution, and suggests that there may be strong selection in favour of preferential retention of Alu elements in GC-rich regions and that these "selfish" elements may benefit their human hosts.

- The mutation rate is about twice as high in male as in female meiosis, showing that most mutation occurs in males.

- Cytogenetic analysis of the sequenced clones confirms suggestions that large GC-poor regions are strongly correlated with "dark G-bands" in karyotypes.

- Recombination rates tend to be much higher in distal regions (around 20 megabases (Mb)) of chromosomes and on shorter chromosome arms in general, in a pattern that promotes the occurrence of at least one crossover per chromosome arm in each meiosis.

- More than 1.4 million single nucleotide polymorphisms (SNPs) in the human genome have been identified. This collection should allow the initiation of genome-wide linkage disequilibrium mapping of the genes in the human population.

In this paper, we start by presenting background information on the project and describing the generation, assembly and evaluation of the draft genome sequence. We then focus on an initial analysis of the sequence itself: the broad chromosomal landscape; the repeat elements and the rich palaeontological record of evolutionary and biological processes that they provide; the human genes and proteins and their differences and similarities with those of other organisms; and the history of genomic segments. (Comparisons are drawn throughout with the genomes of the budding yeast *Saccharomyces cerevisiae*, the nematode worm *Caenorhabditis elegans*, the fruitfly *Drosophila melanogaster* and the mustard weed *Arabidopsis thaliana*; we refer to these for convenience simply as yeast, worm, fly and mustard weed.) Finally, we discuss applications of the sequence to biology and medicine and describe next steps in the project. A full description of the methods is provided as Supplementary Information on *Nature*'s web site (http://www. nature.com).

- 几百个人类基因似乎来自细菌的横向转移，这可能发生在脊椎动物世系的某个点。很多基因似乎起源于转座元件。

- 尽管人类基因组约半数起源于转座元件，类人猿世系中的这一类元件的整体活性出现了显著的降低。DNA 转座子似乎已完全失活，长末端重复序列（LTR）反转录转座子可能也已失活。

- 染色体的近着丝粒和亚端粒区域被近期发生的大片段的重复序列所填补，这些序列来自基因组的其他区域。在人类基因组中，片段重复的发生概率远高于酵母、果蝇或线虫。

- 对 Alu 元件结构组织的分析解释了长期困扰我们的谜团，即它们惊人的全基因组范围的分布，提示了可能存在强烈的选择，使 Alu 元件偏好性地保留在GC 富集区。这些"自私"元件可能使其人类宿主收益。

- 男性减数分裂的突变率约为女性的两倍，显示多数突变发生在男性中。

- 对已测序的克隆的细胞遗传学分析确认了一种假设，即大片段的 GC 匮乏区与染色体核型中的"暗 G 带"存在强烈关联。

- 染色体远端（约 20 兆碱基（Mb））的重组率显著升高，在染色体短臂上则为平均值，这种模式促使每条染色体臂在每次减数分裂时至少发生一次交换。

- 在人类基因组中识别出了超过 140 万个单核苷酸多态性（SNP），这使我们可以在人类种群中建立基因组范围的连锁不平衡图谱。

在本文中，我们从展示人类基因组计划的背景信息开始，描述了基因组草图序列的产生、组装和评估。随后，我们聚焦于序列本身的初步分析：宽泛的染色体全景；重复序列元件，以及丰富的关于进化的古生物学证据以及它们提供的生物学过程；人类基因和蛋白质与其他生物的基因和蛋白质的差异和相似性；基因组片段的历史。（这些比较是通过出芽生殖的酿酒酵母、线虫类的秀丽隐杆线虫、黑腹果蝇、芥草拟南芥的基因组提取的；方便起见，我们将这四种生物简称为酵母、线虫、果蝇和芥草。）最后，我们讨论了这些序列在生物学和医药方面的应用，并描述了人类基因组计划的下一步工作。《自然》杂志网站（http://www.nature.com）上的补充信息提供了关于研究方法的完整描述。

We recognize that it is impossible to provide a comprehensive analysis of this vast dataset, and thus our goal is to illustrate the range of insights that can be gleaned from the human genome and thereby to sketch a research agenda for the future.

Background to the Human Genome Project

The Human Genome Project arose from two key insights that emerged in the early 1980s: that the ability to take global views of genomes could greatly accelerate biomedical research, by allowing researchers to attack problems in a comprehensive and unbiased fashion; and that the creation of such global views would require a communal effort in infrastructure building, unlike anything previously attempted in biomedical research.

The idea of sequencing the entire human genome was first proposed in discussions at scientific meetings organized by the US Department of Energy and others from 1984 to 1986 (refs 21, 22). A committee appointed by the US National Research Council endorsed the concept in its 1988 report[23], but recommended a broader programme, to include: the creation of genetic, physical and sequence maps of the human genome; parallel efforts in key model organisms such as bacteria, yeast, worms, flies and mice; the development of technology in support of these objectives; and research into the ethical, legal and social issues raised by human genome research. The programme was launched in the US as a joint effort of the Department of Energy and the National Institutes of Health. In other countries, the UK Medical Research Council and the Wellcome Trust supported genomic research in Britain; the Centre d'Etude du Polymorphisme Humain and the French Muscular Dystrophy Association launched mapping efforts in France; government agencies, including the Science and Technology Agency and the Ministry of Education, Science, Sports and Culture supported genomic research efforts in Japan; and the European Community helped to launch several international efforts, notably the programme to sequence the yeast genome. By late 1990, the Human Genome Project had been launched, with the creation of genome centres in these countries. Additional participants subsequently joined the effort, notably in Germany and China. In addition, the Human Genome Organization (HUGO) was founded to provide a forum for international coordination of genomic research. Several books[24-26] provide a more comprehensive discussion of the genesis of the Human Genome Project.

Through 1995, work progressed rapidly on two fronts. The first was construction of genetic and physical maps of the human and mouse genomes[27-31], providing key tools for identification of disease genes and anchoring points for genomic sequence. The second was sequencing of the yeast[32] and worm[33] genomes, as well as targeted regions of mammalian genomes[34-37]. These projects showed that large-scale sequencing was feasible and developed the two-phase paradigm for genome sequencing. In the first, "shotgun", phase, the genome is divided into appropriately sized segments and each segment is covered to a high degree of redundancy (typically, eight- to tenfold) through the sequencing of randomly selected subfragments. The second is a "finishing" phase, in which sequence gaps are closed and

我们认识到本文无法提供关于这些海量数据的全面分析，因此我们的目标是，举例说明从人类基因组中所得到的具有普遍意义的新发现，从而建立起未来的研究框架。

人类基因组计划的背景

人类基因组计划起源于 20 世纪 80 年代初期的两个重要的远见：通过让研究者以深刻而无偏见的方式挑战难题，全面把握基因组的能力可以极大地促进生物医药的研究；这一全局视野的产生需要全世界在科研的基础设施建设方面共同努力，而不像之前试图进行的任何生物医药研究。

对人类全基因组进行测序的想法是在一次学术会议的讨论中产生的，该会议由美国能源部和其他机构在 1984～1986 年组织（参考文献 21 和 22）。美国国家研究委员会在其 1988 年的报告中签署同意了这一概念[23]，但是他们建议将项目的目标拓宽，包括：建立人类基因组的遗传图谱、物理图谱和序列图谱；同时建立关键模式生物如细菌、酵母、线虫、果蝇和小鼠的上述三种图谱；发展支持上述目标的技术；研究因人类基因组研究而产生的伦理、法律和社会议题。该项目由美国能源部和美国国立卫生研究院联合启动。在其他国家，英国医学研究理事会和维康信托基金会支持了英国的基因组研究；法国人类多态性研究中心和肌肉萎缩症协会落实了法国的测序工作；政府机构，包括日本科学振兴机构、文部科学省支持了日本的基因组研究；欧洲共同体发起了若干国际项目，尤其是酵母基因组测序项目。至 1990年末，人类基因组计划已经展开，并在这些国家建立了基因组研究中心。随后有更多的国家加入到这一项目，主要是德国和中国。而且还成立了国际人类基因组组织（HUGO），为基因组研究的国际协作提供论坛。一些著作[24-26]提供了关于人类基因组计划产生的更加深入的讨论。

1995 年，研究工作在以下两个前沿快速进展。第一，建立了人类和小鼠基因组的遗传和物理图谱[27-31]，提供了识别疾病基因和基因组序列锚定位点的关键工具。第二，测定了酵母[32]和线虫[33]的基因组以及哺乳动物基因组的靶区域[34-37]。这些项目表明了大规模测序是灵活可变的，并建立了基因组测序的两步法范例。在第一步即"鸟枪"阶段，基因组被打断成合适大小的片段，通过随机挑取亚片段测序，每个片段都被覆盖且高度冗余（通常来说，8～10 倍）。第二步是"完成"阶段，在这一阶段，通过直接分析，填补序列的缺口，纠正序列歧义。上述结果也显示，完整

remaining ambiguities are resolved through directed analysis. The results also showed that complete genomic sequence provided information about genes, regulatory regions and chromosome structure that was not readily obtainable from cDNA studies alone.

The human genome sequencing effort moved into full-scale production in March 1999. The idea of first producing a draft genome sequence was revived at this time, both because the ability to finish such a sequence was no longer in doubt and because there was great hunger in the scientific community for human sequence data. In addition, some scientists favoured prioritizing the production of a draft genome sequence over regional finished sequence because of concerns about commercial plans to generate proprietary databases of human sequence that might be subject to undesirable restrictions on use[42-44].

The consortium focused on an initial goal of producing, in a first production phase lasting until June 2000, a draft genome sequence covering most of the genome. Such a draft genome sequence, although not completely finished, would rapidly allow investigators to begin to extract most of the information in the human sequence. Experiments showed that sequencing clones covering about 90% of the human genome to a redundancy of about four- to fivefold ("half-shotgun" coverage) would accomplish this[45,46]. The draft genome sequence goal has been achieved, as described below.

The second sequence production phase is now under way. Its aims are to achieve full-shotgun coverage of the existing clones during 2001, to obtain clones to fill the remaining gaps in the physical map, and to produce a finished sequence (apart from regions that cannot be cloned or sequenced with currently available techniques) no later than 2003.

Strategic Issues

Hierarchical shotgun sequencing

Soon after the invention of DNA sequencing methods[47,48], the shotgun sequencing strategy was introduced[49-51]; it has remained the fundamental method for large-scale genome sequencing[52-54] for the past 20 years. The approach has been refined and extended to make it more efficient. For example, improved protocols for fragmenting and cloning DNA allowed construction of shotgun libraries with more uniform representation. The practice of sequencing from both ends of double-stranded clones ("double-barrelled" shotgun sequencing) was introduced by Ansorge and others[37] in 1990, allowing the use of "linking information" between sequence fragments.

Practical difficulties arise because of repeated sequences and cloning bias. The human genome is filled (> 50%) with repeated sequences, including interspersed repeats derived from transposable elements, and long genomic regions that have been duplicated in tandem, palindromic or dispersed fashion. Such features complicate the assembly of a correct and finished genome sequence.

的基因组序列提供了关于基因、调控区和染色体结构的信息，这些信息是无法仅从cDNA 研究中获得的。

1999 年 3 月，人类基因组测序经过努力，进入了全序列产生阶段。首先生成一个基因组草图序列的想法在这一时期重新流行起来，这一方面是因为完成这样的序列的能力已经不再是问题，另一方面也是因为科学共同体对人类序列数据有着强大的需求。而且，一些科学家支持优先产生基因组草图序列，而不是局部完成序列。因为一些商业计划企图对人类基因组序列数据库申请专利，这可能会对数据的使用造成不良限制[42-44]。

科学共同体聚焦于产生数据这一首要目标，即第一阶段持续到 2000 年 6 月，在此之前使基因组草图序列覆盖基因组的大部分区域。这一基因组草图序列尽管没有彻底完成，但也使得研究者们能够尽快地提取到人类基因组中的大部分信息。实验显示，测序克隆覆盖约 90% 的人类基因组，约 4 ~ 5 倍冗余（"半鸟枪"覆盖率）[45,46]。如下所示，人类基因组草图序列目标已完成。

第二阶段正在进行中，旨在于 2001 年获得目前所有克隆的全鸟枪覆盖，并获得能够填补物理图谱缺口的克隆，以便在 2003 年之前生成基因组序列完成图（除了现有技术无法克隆和测序的基因组区域之外）。

策略问题

层次鸟枪法测序

DNA 测序方法[47,48]发明以后，很快就出现了鸟枪法测序[49-51]的策略；在过去的二十多年里，鸟枪法仍然是大规模基因组测序[52-54]的基础方法。鸟枪法也在不断地补充和改进，以便更加高效。例如，DNA 片段化和克隆的方法改进后，鸟枪法构建的文库更加均一。1990 年，安佐格等人[37]提出了双链克隆的两端测序法（"双管"鸟枪法测序），使测序片段之间"连接信息"的使用成为可能。

由于重复序列和克隆偏好性的存在，实际操作存在不少困难。人类基因组含有大量（ > 50%）重复序列，包括起源于转座元件的散在重复序列，以及以串联、回文或散在等重复形式存在的长基因组区域。这些情况使一个正确完整的基因序列的组装工作更加复杂。

There are two approaches for sequencing large repeat-rich genomes. The first is a whole-genome shotgun sequencing approach, as has been used for the repeat-poor genomes of viruses, bacteria and flies, using linking information and computational analysis to attempt to avoid misassemblies. The second is the "hierarchical shotgun sequencing" approach (Fig. 2), also referred to as "map-based", "BAC-based" or "clone-by-clone". This approach involves generating and organizing a set of large-insert clones (typically 100–200 kb each) covering the genome and separately performing shotgun sequencing on appropriately chosen clones. Because the sequence information is local, the issue of long-range misassembly is eliminated and the risk of short-range misassembly is reduced.

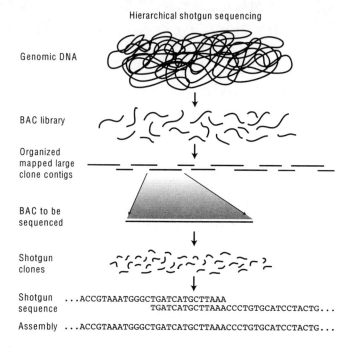

Fig 2. Idealized representation of the hierarchical shotgun sequencing strategy. A library is constructed by fragmenting the target genome and cloning it into a large-fragment cloning vector; here, BAC vectors are shown. The genomic DNA fragments represented in the library are then organized into a physical map and individual BAC clones are selected and sequenced by the random shotgun strategy. Finally, the clone sequences are assembled to reconstruct the sequence of the genome.

A biotechnology company, Celera Genomics, has chosen to incorporate the whole-genome shotgun approach into its own efforts to sequence the human genome. Their plan[60,61] uses a mixed strategy, involving combining some coverage with whole-genome shotgun data generated by the company together with the publicly available hierarchical shotgun data generated by the International Human Genome Sequencing Consortium. If the raw sequence reads from the whole-genome shotgun component are made available, it may be possible to evaluate the extent to which the sequence of the human genome can be assembled without the need for clone-based information. Such analysis may help to refine sequencing strategies for other large genomes.

对富含重复序列的大型基因组进行测序，目前有两种途径。一种是全基因组鸟枪测序法，这种方法用于完成病毒、细菌、果蝇等重复较少的基因组测序，使用片段链接的信息和电脑分析来避免错误组装。第二种是"层次鸟枪法测序"（图 2），也称为"基于图谱""基于 BAC"或"连续克隆"法。这种方法包括生成并组装一组能覆盖整个基因组的大片段克隆（通常每个 100~200 kb），选择其中合适的克隆进行鸟枪法测序。由于所得的序列信息都是局部的，这样就不存在长程错误组装的问题，也降低了短程错误组装的风险。

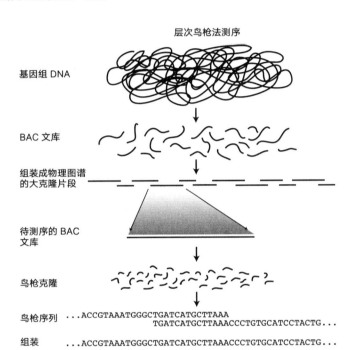

图 2. 理想的层次鸟枪法测序策略。通过目的基因组片段化并克隆到大片段载体上构建文库；此处显示的是 BAC 文库。库中的基因组 DNA 片段组建成物理图谱，挑选出单个的 BAC 克隆利用随机鸟枪法测序。最后，对克隆序列进行组装并重建基因组序列。

一家生物技术公司——塞莱拉基因选择在他们的方法中融入全基因组鸟枪法来完成人类基因组测序。他们计划[60,61]使用混合式策略，将公司全基因组鸟枪法测序得到的数据与国际人类基因组测序联盟公布的层次鸟枪法数据联合。如果可获得全基因组鸟枪法组分的原始数据，就有可能估算出在不依赖克隆信息的前提下，多大程度上可对人类基因组进行组装。这样的分析会帮助修正今后其他大型基因组测序的策略。

Coordination and public data sharing

The Human Genome Project adopted two important principles with regard to human sequencing. The first was that the collaboration would be open to centres from any nation. Although potentially less efficient, in a narrow economic sense, than a centralized approach involving a few large factories, the inclusive approach was strongly favoured because we felt that the human genome sequence is the common heritage of all humanity and the work should transcend national boundaries, and we believed that scientific progress was best assured by a diversity of approaches. The collaboration was coordinated through periodic international meetings (referred to as "Bermuda meetings" after the venue of the first three gatherings) and regular telephone conferences. Work was shared flexibly among the centres, with some groups focusing on particular chromosomes and others contributing in a genome-wide fashion.

The second principle was rapid and unrestricted data release. The centres adopted a policy that all genomic sequence data should be made publicly available without restriction within 24 hours of assembly[79,80]. Pre-publication data releases had been pioneered in mapping projects in the worm[11] and mouse genomes[30,81] and were prominently adopted in the sequencing of the worm, providing a direct model for the human sequencing efforts. We believed that scientific progress would be most rapidly advanced by immediate and free availability of the human genome sequence. The explosion of scientific work based on the publicly available sequence data in both academia and industry has confirmed this judgement.

Generating the Draft Genome Sequence

Generating a draft sequence of the human genome involved three steps: selecting the BAC clones to be sequenced, sequencing them and assembling the individual sequenced clones into an overall draft genome sequence.

The draft genome sequence is a dynamic product, which is regularly updated as additional data accumulate en route to the ultimate goal of a completely finished sequence. The results below are based on the map and sequence data available on 7 October 2000, except as otherwise noted. At the end of this section, we provide a brief update of key data.

Clone selection

The hierarchical shotgun method involves the sequencing of overlapping large-insert clones spanning the genome. For the Human Genome Project, clones were largely chosen from eight large-insert libraries containing BAC or P1-derived artificial chromosome (PAC) clones (refs 82–88). The libraries were made by partial digestion of genomic DNA with restriction

项目协调与公共数据共享

关于人类基因组全序列，人类基因组计划采用了两项重要的原则。一是此项计划的合作面向全世界所有国家的研究中心。尽管从狭隘的经济学角度看，集中于几个大的中心的工作效率会更高，但我们强烈赞成更具包容性的合作，因为我们认为，人类基因组序列是全人类共同的财富，此项工作应跨越国家的界限，而且我们相信，多样化的途径是获得科学进展的最好保证。合作中的协调工作是通过定期的国际会议（百慕大会议，由前三次的会议地点命名）和日常电话会议完成。某些小组集中于特定的染色体，而另一些则着眼于全基因组的范围，各中心间的进展分享非常灵活。

第二点原则是快速的不受限的数据发布。所有的测序中心都遵循一个政策：在基因组序列完成组装的 24 小时内，要将其不受限地公布于众[79,80]。在线虫[11]和小鼠[30,81]基因组的图谱工作中，公开发表前的数据发布已经为人类基因组测序做了很好的榜样。我们确信，伴随人类基因组数据的及时和无偿使用，科学研究将获得迅速的进展。在学术界和产业界，基于公共可获取序列的科学研究成果的激增充分肯定了我们的判断。

生成基因组草图序列

生成人类基因组草图序列包括三步：选择要测序的 BAC 克隆，测序，将测序完成的单个克隆组装为整个基因组草图序列。

基因组草图序列是动态产物，草图序列会因数据的不断积累而经常更新，最终目标是完整的完成序列。除非特别注明，下文的结论都是基于 2000 年 10 月 7 日的图谱与序列数据。在本节的末尾，我们会提供关键数据的简要更新。

克隆选择

层次鸟枪法的测序包括遍布整个基因组的相互重叠的大片段插入克隆。对人类基因组计划来说，克隆大部分选自 8 个大片段插入文库，包括 BAC 或 P1 人工染色体（PAC）克隆（参考文献 82～88）。文库是由限制性内切酶部分消化基因组 DNA 后

enzymes. Together, they represent around 65-fold coverage (redundant sampling) of the genome. Libraries based on other vectors, such as cosmids, were also used in early stages of the project.

The libraries were prepared from DNA obtained from anonymous human donors in accordance with US Federal Regulations for the Protection of Human Subjects in Research (45CFR46) and following full review by an Institutional Review Board. Briefly the opportunity to donate DNA for this purpose was broadly advertised near the two laboratories engaged in library construction. Volunteers of diverse backgrounds were accepted on a first-come, first-taken basis. Samples were obtained after discussion with a genetic counsellor and written informed consent. The samples were made anonymous as follows: the sampling laboratory stripped all identifiers from the samples, applied random numeric labels, and transferred them to the processing laboratory, which then removed all labels and relabelled the samples. All records of the labelling were destroyed. The processing laboratory chose samples at random from which to prepare DNA and immortalized cell lines.

Because the sequencing project was shared among twenty centres in six countries, it was important to coordinate selection of clones across the centres. Most centres focused on particular chromosomes or, in some cases, larger regions of the genome. We also maintained a clone registry to track selected clones and their progress. In later phases, the global map provided an integrated view of the data from all centres, facilitating the distribution of effort to maximize coverage of the genome. Before performing extensive sequencing on a clone, several centres routinely examined an initial sample of 96 raw sequence reads from each subclone library to evaluate possible overlap with previously sequenced clones.

Sequencing

The selected clones were subjected to shotgun sequencing. Detailed protocols are available on the web sites of many of the individual centres.

The overall sequencing output rose sharply during production (Fig. 4). Following installation of new sequence detectors beginning in June 1999, sequencing capacity and output rose approximately eightfold in eight months to nearly 7 million samples processed per month, with little or no drop in success rate (ratio of useable reads to attempted reads). By June 2000, the centres were producing raw sequence at a rate equivalent to onefold coverage of the entire human genome in less than six weeks. This corresponded to a continuous throughput exceeding 1,000 nucleotides per second, 24 hours per day, seven days per week. This scale-up resulted in a concomitant increase in the sequence available in the public databases (Fig. 4).

构建的。总的来说，样品覆盖 65 倍的基因组（样品冗余）。在计划早期，我们也应用了其他的一些载体，比如黏粒等。

制备文库所使用的 DNA 来自匿名的捐献者，符合美国联邦法律对于保护研究中的人类受试对象的规定（45CFR46），并且接受机构审查委员会的监督。简单地说就是构建文库的两个实验室在其附近广泛宣传了此次捐献 DNA 的机会。不同背景的志愿者的接收采取先到先取的方式。在与遗传顾问讨论并签署知情同意书后，样品被采集。样品的匿名程序如下：采样实验室去掉所有样品的标签，用随机数字进行编号，将样品移交给处理实验室，后者再次去掉标签并重新标记。所有的标记记录都被销毁。处理实验室会随机选取制备 DNA 的样品，并且将细胞系永生化。

由于测序计划是在 6 个国家的 20 个中心共同完成的，协调各中心的克隆选择是非常重要的。大部分中心的测序集中于特定染色体，另一些则着眼于全基因组范围。我们维护了一个克隆注册表，用以跟踪选出的克隆及后续进展。在后期，全局图谱提供了来自所有研究中心的数据整合视图，这样有助于在各中心间分配工作从而获得基因组的最大覆盖度。在对某个克隆进行详细测序以前，几个中心通常会对最初样品的每一个亚克隆文库的 96 个原始读序进行检验，评估其与已完成测序的克隆出现重叠的可能性。

测序

选出的克隆随后被送去做鸟枪法测序。具体的操作方法可在各中心的网站上获得。

在计划完成过程中，测序结果的输出量在急剧增加（图 4）。1999 年 6 月开始安装新的测序仪后，测序能力和输出量在 8 个月内增长至原来的 8 倍，每月可处理近 700 万样品，而成功率（可用读序与尝试读序的比率）几乎没有下降。到 2000 年 6 月，所有中心获得原始序列的速度相当于在不到 6 周时间内将整个人类基因组覆盖一遍。这相当于每周七天、每天工作 24 小时、每秒测出的核苷酸超过 1,000 个的连续通量。这种扩容也同步增加了公共数据库中的可用序列（图 4）。

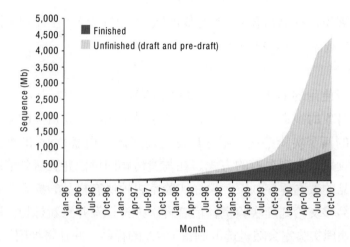

Fig 4. Total amount of human sequence in the High Throughput Genome Sequence (HTGS) division of GenBank. The total is the sum of finished sequence (red) and unfinished (draft plus predraft) sequence (yellow).

Although the main emphasis was on producing a draft genome sequence, the centres also maintained sequence finishing activities during this period, leading to a twofold increase in finished sequence from June 1999 to June 2000 (Fig. 4). The total amount of human sequence in this final form stood at more than 835 Mb on 7 October 2000, or more than 25% of the human genome. This includes the finished sequences of chromosomes 21 and 22 (refs 93, 94). As centres have begun to shift from draft to finished sequencing in the last quarter of 2000, the production of finished sequence has increased to an annualized rate of 1 Gb per year and is continuing to rise.

Assembly of the draft genome sequence

We then set out to assemble the sequences from the individual large-insert clones into an integrated draft sequence of the human genome. The assembly process had to resolve problems arising from the draft nature of much of the sequence, from the variety of clone sources, and from the high fraction of repeated sequences in the human genome. (Editorial note: several details of the assembly procedure have been omitted in this shortened version.)

The result of the assembly process is an integrated draft sequence of the human genome. (Editorial note: details of sequence quality assessment have been omitted.)

The contiguity of the draft genome sequence at each level is an important feature.

Genome coverage

We assessed the nature of the gaps within the draft genome sequence, and attempted to

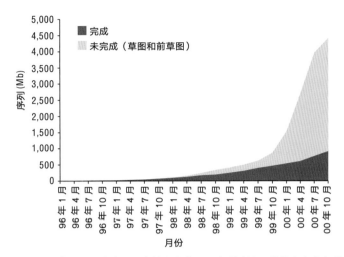

图 4. GenBank 中经高通量基因组测序 (HTGS) 的人类基因组序列总量。总数为完成序列 (红色) 和未完成 (草图和前草图) 序列 (黄色) 的总量。

尽管工作重点在于生成基因组草图序列，但各中心在此期间也一直在做最终序列的整理，使 1999 年 6 月到 2000 年 6 月之间完成序列的数量增加了一倍 (图 4)。2000 年 10 月 7 日，最终形式的人类基因组序列总量超过 835 Mb，大于人类基因组的 25%。这其中包含了测序完成的第 21 和 22 号染色体 (参考文献 93 和 94)。2000 年最后一个季度开始，各中心的工作从草图测序转向完成序列，完成序列的产量每年增加 1 Gb，并持续上升。

基因组草图序列的组装

随后，我们着手将单独的大片段插入克隆的序列组装成人类基因组草图序列。装配过程必须解决大部分序列以草图形式存在产生的问题、不同克隆来源产生的问题和人类基因组中高比例的重复序列产生的问题。(编者注：在这篇删减版中组装过程的细节被删减了。)

组装过程的结果是产生一个整合的人类基因组草图序列。(编者注：序列质量的测试细节被删减了。)

基因组草图序列在各层次上的连续性是一个重要特征。

基因组覆盖度

我们对基因组草图序列中间隙的性质进行了评估，并尝试估计人类基因组未表

estimate the fraction of the human genome not represented within the current version.

Our results indicate that about 88% of the human genome is represented in the draft genome sequence and about 94% in the combined publicly available sequence databases. The figure of 88% agrees well with our independent estimates above that about 3%, 5% and 4% of the genome reside in the three types of gap in the draft genome sequence.

We arrived at a total human genome size estimate of around 3,200 Mb, which compares favourably with previous estimates based on DNA content.

We also independently estimated the size of the euchromatic portion of the genome by determining the fraction of the 5,615 random raw sequences that matched the finished portion of the human genome (whose total length is known with greater precision). Twenty-nine per cent of these raw sequences found a match among 835 Mb of nonredundant finished sequence. This leads to an estimate of the euchromatic genome size of 2.9 Gb. This agrees reasonably with a prediction based on the length of the draft genome sequence.

Broad Genomic Landscape

What biological insights can be gleaned from the draft sequence? In this section, we consider very large-scale features of the draft genome sequence: the distribution of GC content, CpG islands and recombination rates, and the repeat content and gene content of the human genome. The draft genome sequence makes it possible to integrate these features and others at scales ranging from individual nucleotides to collections of chromosomes. Unless noted, all analyses were conducted on the assembled draft genome sequence described above.

Figure 9 (Editorial note: since the Figure is oversized and cannot fit here, it is provided on *Nature*'s web site) provides a high-level view of the contents of the draft genome sequence, at a scale of about 3.8 Mb per centimetre. Of course, navigating information spanning nearly ten orders of magnitude requires computational tools to extract the full value. We have created and made freely available various "Genome Browsers". These web-based computer tools allow users to view an annotated display of the draft genome sequence, with the ability to scroll along the chromosomes and zoom in or out to different scales.

Fig. 9. Overview of features of draft human genome. The Figure shows the occurrences of twelve important types of feature across the human genome. Large grey blocks represent centromeres and centromeric heterochromatin (size not precisely to scale). Each of the feature types is depicted in a track, from top to bottom as follows. (1) Chromosome position in Mb. (2) The approximate positions of Giemsa-stained chromosome bands at the 800 band resolution. (3) Level of coverage in the draft genome sequence. Red, areas covered by finished clones; yellow, areas covered by predraft sequence. Regions covered by draft sequenced clones are in orange, with darker shades reflecting increasing shotgun sequence coverage. (4) GC content. Percentage of bases in a 20,000 base window that are C or G. (5) Repeat density. Red line, density of SINE class repeats in a 100,000-base window; blue

示在当前版本中的部分。

我们的结果显示，约 88% 的人类基因组存在于草图序列中，94% 在公开数据库中可获取。我们在前文的独立评估显示，约 3%、5% 和 4% 的基因组存在于基因组草图序列的三种间隙中，88% 的数字与这一结果良好吻合。

我们估计人类基因组约 3,200 Mb，与先前基于 DNA 含量的估计值接近。

我们还通过测算 5,615 个随机原始序列与人类基因组完成序列（精度更高的整体长度）的匹配度，独立估算了基因组中常染色质部分的体量。这些原始序列中的 29% 在 835 Mb 的无冗余完成序列中找到了匹配，由此估计常染色质体量为 2.9 Gb，这与依据基因组草图序列长度所做的预测结果相当。

宏观基因组蓝图

从草图序列中我们可以获得哪些生物学启示呢？在这一部分，我们将考虑那些基因组草图序列体现出的非常大范围上的特征：GC 含量分布，CpG 岛和重组率，人类基因组的重复含量和基因含量。基因组草图序列使得从单核苷酸到染色体集合等各个尺度范围上的特征的整合成为可能。除特殊说明外，以下所有分析都是基于前面所描述的组装后的基因组草图序列。

图 9（编者注：图片过大，此处无法呈现，请在《自然》杂志网站获取）提供了基因组草图序列的高级视图，每厘米约对应 3.8 Mb。当然，浏览的内容跨越近 10 个数量级，需要计算工具来提取出完整的信息。我们已经创建并免费提供各种"基因组浏览器"。这些基于网页的电脑工具使使用者能看到基因组草图序列的注释，也能沿着染色体滚动并放大或缩小到不同的比例。

图 9. 人类基因组草图特征概览。本图展示了人类基因组上 12 种重要的特征序列。大灰色方块代表着丝粒和着丝粒异染色质（大小与实际并非准确相关）。每一种特征类型都标注在基因组上，从上到下依次如下文所述。(1) 染色体位置，单位为 Mb。(2) 吉姆萨染色法染出的染色体条带位置，分辨率为 800 个条带。(3) 基因组草图序列的覆盖程度。红色，该区域被最终克隆覆盖；黄色，该区域被预组装序列覆盖。被草图序列覆盖的区域用橙色表示，深色阴影表示不断增长的鸟枪测序覆盖率。(4) GC 含量。在 20,000 碱基窗口中碱基 C 或 G 的百分比。(5) 重复序列密度。红线，SINE 类重复序列在 100,000 碱基窗口的密度；蓝线，LINE 类重复序列在 100,000 碱基窗口的密度。(6) SNP 在 50,000 碱基窗口的密

line, density of LINE class repeats in a 100,000-base window. (6) Density of SNPs in a 50,000-base window. The SNPs were detected by sequencing and alignments of random genomic reads. Some of the heterogeneity in SNP density reflects the methods used for SNP discovery. Rigorous analysis of SNP density requires comparing the number of SNPs identified to the precise number of bases surveyed. (7) Non-coding RNA genes. Brown, functional RNA genes such as tRNAs, snoRNAs and rRNAs; light orange, RNA pseudogenes. (8) CpG islands. Green ticks represent regions of ~200 bases with CpG levels significantly higher than in the genome as a whole, and GC ratios of at least 50%. (9) Exofish ecores. Regions of homology with the pufferfish *T. nigroviridis*[292] are blue. (10) ESTs with at least one intron when aligned against genomic DNA are shown as black tick marks. (11) The starts of genes predicted by Genie or Ensembl are shown as red ticks. The starts of known genes from the RefSeq database[110] are shown in blue. (12) The names of genes that have been uniquely located in the draft genome sequence, characterized and named by the HGM Nomenclature Committee. Known disease genes from the OMIM database are red, other genes blue. This Figure is based on an earlier version of the draft genome sequence than analysed in the text, owing to production constraints. We are aware of various errors in the Figure, including omissions of some known genes and misplacements of others. Some genes are mapped to more than one location, owing to errors in assembly, close paralogues or pseudogenes. Manual review was performed to select the most likely location in these cases and to correct other regions. For updated information, see http://genome.ucsc.edu/ and http://www.ensembl.org/.

In addition to using the Genome Browsers, one can download from these sites the entire draft genome sequence together with the annotations in a computer-readable format. The sequences of the underlying sequenced clones are all available through the public sequence databases. URLs for these and other genome websites are listed in Box 2.

Box 2. Sources of publicly available sequence data and other relevant genomic information

http://genome.ucsc.edu/

University of California at Santa Cruz

Contains the assembly of the draft genome sequence used in this paper and updates

http://genome.wustl.edu/gsc/human/Mapping/

Washington University

Contains links to clone and accession maps of the human genome

http://www.ensembl.org

EBI/Sanger Centre

Allows access to DNA and protein sequences with automatic baseline annotation

http://www.ncbi.nlm.nih.gov/genome/guide/

NCBI

Views of chromosomes and maps and loci with links to other NCBI resources

http://www.ncbi.nlm.nih.gov/genemap99/

Gene map 99: contains data and viewers for radiation hybrid maps of EST-based STSs

---

度。SNP 是通过读取随机基因组测序和比对检测到的。SNP 密度的异质性反映了 SNP 检测方法的不同。SNP 密度的精确分析需要将识别出的 SNP 数量与所研究的精确的碱基数量作比较。(7) 非编码 RNA 基因。棕色，功能 RNA 基因如 tRNA、snoRNA 和 rRNA；浅橙色，RNA 假基因。(8) CpG 岛。绿色钩代表约 200 个碱基的区域，该区域的 CpG 水平显著高于全基因组，且 GC 含量至少为 50%。(9) 鱼类外显子。与河豚序列[292]同源的区域为蓝色。(10) 与基因组序列比对，带有至少一个内含子的 EST 用黑色钩表示。(11) 通过 Genie 或 Ensembl 预测的基因起始位点用红色钩表示。从 RefSeq 数据库[110]中得到的已知基因的起始位点用蓝色表示。(12) 每一个基因特有的名字已经定位在基因组草图序列中，经 HGM 命名委员会表征并命名。来自 OMIM 数据库的已知疾病基因用红色表示，其他基因用蓝色表示。限于输出限制，本图是基于基因组草图序列的早期版本而不是本文分析的内容绘制的。我们意识到图中有几处错误，包括遗漏了一些已知基因以及搞错了其他一些基因的位置。由于组装、相近的同源序列或假基因方面的错误，一些基因被标注在多个位点。通过人工检查我们挑出了上述情况中最有可能的位置，并将其更正。关于更新的信息，请见 http://genome.ucsc.edu/ 和 http://ensembl.org/。

除了使用基因组浏览器之外，使用者还可以从这些网站下载计算机可读的基因组草图序列和相应的注释。那些基础测序的克隆的序列都可以从公共序列数据库获得。上述网站和其他基因组网站的 URL 列表在框 2 中。

框 2. 可公开获取的序列数据及其他相关基因组信息的来源

http://genome.ucsc.edu/
加州大学圣克鲁兹分校
包含本文及后续更新中使用的基因组草图序列的组装

http://genome.wustl.edu/gsc/human/Mapping/
华盛顿大学
包含人类基因组的克隆和接入图谱的链接

http://www.ensembl.org
EBI/桑格中心
可以访问带有自动基线注释的 DNA 和蛋白质序列

http://www.ncbi.nlm.nih.gov/genome/guide/
NCBI
可浏览染色体、图谱和基因座，并带有其他 NCBI 资源的链接

http://www.ncbi.nlm.nih.gov/genemap99/
Gene map 99：包含基于 EST 的 STS 辐射杂交图谱的数据和阅读器

http://compbio.ornl.gov/channel/index.html

Oak Ridge National Laboratory

Java viewers for human genome data

http://hgrep.ims.u-tokyo.ac.jp/

RIKEN and the University of Tokyo

Gives an overview of the entire human genome structure

http://snp.cshl.org/

The SNP Consortium

Includes a variety of ways to query for SNPs in the human genome

http://www.ncbi.nlm.nih.gov/Omim/

Online *Mendelian Inheritance in Man*

Contain information about human genes and disease

http://www.nhgri.nih.gov/ELSI/ and http://www.ornl.gov/hgmis/elsi/elsi.html

NHGRI and DOE

Contains information, links and articles on a wide range of social, ethical and legal issues

Long-range variation in GC content

The existence of GC-rich and GC-poor regions in the human genome was first revealed by experimental studies involving density gradient separation, which indicated substantial variation in average GC content among large fragments. Subsequent studies have indicated that these GC-rich and GC-poor regions may have different biological properties, such as gene density, composition of repeat sequences, correspondence with cytogenetic bands and recombination rate[112-117]. Many of these studies were indirect, owing to the lack of sufficient sequence data.

The draft genome sequence makes it possible to explore the variation in GC content in a direct and global manner. Visual inspection (Fig. 9) confirms that local GC content undergoes substantial long-range excursions from its genome-wide average of 41%.

There are huge regions (> 10 Mb) with GC content far from the average.

Long-range variation in GC content is evident not just from extreme outliers, but throughout the genome.

The correlation between GC content domains and various biological properties is of great interest, and this is likely to be the most fruitful route to understanding the basis of variation

http://compbio.ornl.gov/channel/index.html
橡树岭国家实验室
用于人类基因组数据的 Java 阅读器

http://hgrep.ims.u-tokyo.ac.jp/
RIKEN 及东京大学
给出整个人类基因组结构的总览

http://snp.cshl.org/
SNP 联盟
包含查询人类基因组中 SNP 的多种方式

http://www.ncbi.nlm.nih.gov/Omim/
在线人类孟德尔遗传数据库
包含关于人类基因和疾病的信息

http://www.nhgri.nih.gov/ELSI/and http://www.ornl.gov/hgmis/elsi/elsi.html
NHGRI 及 DOE
包含大量与社会、伦理和法律问题相关的信息、链接及文章

GC 含量的远距离变化

人类基因组中存在 GC 富集区和 GC 匮乏区，这一特征最初是通过密度梯度离心的实验发现的，揭示了大片段之间平均 GC 含量存在显著变化。随后的研究表明，GC 富集区和 GC 匮乏区可能具有不同的生物学特性，如基因密度、重复序列组成以及细胞遗传学带和重组率的对应关系[112-117]。由于缺乏足够的序列数据，许多研究是间接的。

基因组草图序列的完成使直接而全面地研究 GC 含量变化成为可能。直接观察（图 9）证实，全基因组范围内的平均 GC 含量是 41%，而局部 GC 含量与之相比则有大量的远程偏移。

有很大的区域（＞10 Mb）的 GC 含量远高于平均值。

GC 含量的远程变化是很明显的，不仅来自极端的异常点，而是存在于整个基因组。

区域的 GC 含量与各种生物学性质之间的相关性引起了人们极大的兴趣，这可

in GC content. We confirm the existence of strong correlations with both repeat content and gene density.

CpG islands

A related topic is the distribution of so-called CpG islands across the genome. The dinucleotide CpG is notable because it is greatly under-represented in human DNA, occurring at only about one-fifth of the roughly 4% frequency that would be expected by simply multiplying the typical fraction of Cs and Gs (0.21×0.21). The deficit occurs because most CpG dinucleotides are methylated on the cytosine base, and spontaneous deamination of methyl-C residues gives rise to T residues. (Spontaneous deamination of ordinary cytosine residues gives rise to uracil residues that are readily recognized and repaired by the cell.) As a result, methyl-CpG dinucleotides steadily mutate to TpG dinucleotides. However, the genome contains many "CpG islands" in which CpG dinucleotides are not methylated and occur at a frequency closer to that predicted by the local GC content. CpG islands are of particular interest because many are associated with the 5′ ends of genes[122-127].

We searched the draft genome sequence for CpG islands.

The density of CpG islands varies substantially among some of the chromosomes. Most chromosomes have 5–15 islands per Mb, with a mean of 10.5 islands per Mb. However, chromosome Y has an unusually low 2.9 islands per Mb, and chromosomes 16, 17 and 22 have 19–22 islands per Mb. The extreme outlier is chromosome 19, with 43 islands per Mb. Similar trends are seen when considering the percentage of bases contained in CpG islands. The relative density of CpG islands correlates reasonably well with estimates of relative gene density on these chromosomes, based both on previous mapping studies involving ESTs and on the distribution of gene predictions discussed below.

Comparison of genetic and physical distance

The draft genome sequence makes it possible to compare genetic and physical distances and thereby to explore variation in the rate of recombination across the human chromosomes. We focus here on large-scale variation. Finer variation is examined in an accompanying paper[131].

Two striking features emerge from analysis of these data. First, the average recombination rate increases as the length of the chromosome arm decreases. A similar trend has been seen in the yeast genome[132,133], despite the fact that the physical scale is nearly 200 times as small.

The second observation is that the recombination rate tends to be suppressed near the centromeres and higher in the distal portions of most chromosomes, with the increase largely in the terminal 20–35 Mb. The increase is most pronounced in the male meiotic map.

644

能是了解 GC 含量变化本质的最有效途径。我们确定重复序列与基因密度存在强相关性。

CpG 岛

一个相关的话题是所谓的 CpG 岛在基因组内的分布。CpG 二核苷酸引人注意是因为它在人类 DNA 中所占比例非常少，如果按 CpG 出现的理论概率计算，简单将 C 和 G 的占比相乘（0.21×0.21）即大约 4%，而 CpG 的出现率仅为 4% 的五分之一。发生缺失的原因是大多数 CpG 二核苷酸的胞嘧啶发生甲基化，甲基化的 C 残基自发脱氨基而使 T 残基增加（普通胞嘧啶的自发脱氨基使尿嘧啶残基增加，尿嘧啶很容易被细胞识别和修复）。结果，甲基化 CpG 二核苷酸稳步突变为 TpG 二核苷酸。然而，基因组含有许多 "CpG 岛"，岛内 CpG 二核苷酸没有被甲基化，发生频率与局部 GC 含量预测频率接近。CpG 岛非常有趣，因为许多 CpG 岛与基因的 5′ 端相关[122-127]。

我们在基因组草图序列中搜寻 CpG 岛。

CpG 岛的密度在不同染色体之间差异很大。大多数染色体每 Mb 有 5~15 个岛，平均每 Mb 有 10.5 个岛。不过，Y 染色体的 CpG 岛密度异常低，每 Mb 仅有 2.9 个，而 16 号、17 号和 22 号染色体每 Mb 则有 19~22 个岛。最为异常的是 19 号染色体，每 Mb 约有 43 个 CpG 岛。考虑 CpG 岛所含碱基的百分比时能看到类似的趋势。基于之前包含 EST 的图谱研究和下面讨论的基因预测的分布，CpG 岛的相对密度与这些染色体上估算的相对基因密度之间相关性良好。

遗传距离与物理距离的比较

基因组草图序列的完成，使遗传距离和物理距离的比较成为可能，进而可以探索人类染色体间重组率的变化。此处，我们将重点放在大范围内的变化上，更细微的变化在附带的论文中讨论[131]。

数据分析显示出两个明显特征：一是平均重组率随着染色体臂长的减少而增加。尽管酵母基因组的物理大小约为人类的 200 分之一，但相似的趋势仍然存在[132,133]。

第二个观察结果是，在大多数染色体上，接近着丝粒的重组率倾向于被抑制，而重组率在远端部分更高，在末端 20~35 Mb 更是大幅增加。这种重组率的增长在男性减数分裂图谱中更加显著。

Why is recombination higher on smaller chromosome arms? A higher rate would increase the likelihood of at least one crossover during meiosis on each chromosome arm, as is generally observed in human chiasmata counts[135]. Crossovers are believed to be necessary for normal meiotic disjunction of homologous chromosome pairs in eukaryotes.

Mechanistically, the increased rate of recombination on shorter chromosome arms could be explained if, once an initial recombination event occurs, additional nearby events are blocked by positive crossover interference on each arm. Evidence from yeast mutants in which interference is abolished shows that interference plays a key role in distributing a limited number of crossovers among the various chromosome arms in yeast[136]. An alternative possibility is that a checkpoint mechanism scans for and enforces the presence of at least one crossover on each chromosome arm.

Variation in recombination rates along chromosomes and between the sexes is likely to reflect variation in the initiation of meiosis-induced double-strand breaks (DSBs) that initiate recombination. DSBs in yeast have been associated with open chromatin[137,138], rather than with specific DNA sequence motifs. With the availability of the draft genome sequence, it should be possible to explore in an analogous manner whether variation in human recombination rates reflects systematic differences in chromosome accessibility during meiosis.

Repeat Content of the Human Genome

A puzzling observation in the early days of molecular biology was that genome size does not correlate well with organismal complexity. For example, *Homo sapiens* has a genome that is 200 times as large as that of the yeast *S. cerevisiae*, but 200 times as small as that of *Amoeba dubia*[139,140]. This mystery (the C-value paradox) was largely resolved with the recognition that genomes can contain a large quantity of repetitive sequence, far in excess of that devoted to protein-coding genes (reviewed in refs 140, 141).

In the human, coding sequences comprise less than 5% of the genome (see below), whereas repeat sequences account for at least 50% and probably much more. Broadly, the repeats fall into five classes: (1) transposon-derived repeats, often referred to as interspersed repeats; (2) inactive (partially) retroposed copies of cellular genes (including protein-coding genes and small structural RNAs), usually referred to as processed pseudogenes; (3) simple sequence repeats, consisting of direct repetitions of relatively short k-mers such as $(A)_n$, $(CA)_n$ or $(CGG)_n$; (4) segmental duplications, consisting of blocks of around 10–300 kb that have been copied from one region of the genome into another region; and (5) blocks of tandemly repeated sequences, such as at centromeres, telomeres, the short arms of acrocentric chromosomes and ribosomal gene clusters. (These regions are intentionally under-represented in the draft genome sequence and are not discussed here.)

Repeats are often described as "junk" and dismissed as uninteresting. However, they actually represent an extraordinary trove of information about biological processes. The repeats

为什么在较小的染色体臂上重组率较高？通常在人类交叉计数中观察到，更高的重组率会增加每条染色体臂在减数分裂过程中至少进行一次交换的可能性[135]。交换被认为是真核生物中同源染色体配对正常减数分裂分离所必需的。

从机制上来看，较短染色体臂上重组率的增加可以这样解释：一旦发生第一次重组事件，每条染色体臂上的交换都会被干扰，进而阻止附近区域重组事件的发生。对干扰作用被破坏的酵母突变体的研究表明，在酵母各染色体臂之间分配有限的交换次数时，干扰作用发挥了关键作用[136]。另一种可能性是检查点机制扫描基因组并促成每个染色体臂上至少存在一个交换。

染色体上以及性别之间的重组率的差异可能反映了减数分裂导致的双链DNA断裂（DSB）起始过程的差异，该过程是重组的起始阶段。酵母中的DSB与开放的染色质而不是特定的DNA序列基序有关[137,138]。如果得到了基因组草图序列，就能用相似的方法探索人类重组率的差异是否反映了减数分裂过程中染色体可接近性的系统性差异。

人类基因组重复序列含量

早期分子生物学领域令人困惑的一个观察结果是基因组大小与生物复杂性的相关性不强。例如，智人的基因组大小是酿酒酵母的200倍，但也是变形虫的两百分之一[139,140]。这个谜团（C值悖论）在很大程度上得到了解决，是由于认识到基因组可以包含大量的重复序列，远远超过了蛋白质编码基因的序列（参考文献140和141中有综述）。

人类的编码序列在基因组占比小于5%（见下文），而重复序列至少占50%，甚至更多。概括地说，重复序列分为五类：（1）转座子起源的重复，通常称为散在重复；（2）基因（包括蛋白质编码基因和小的结构RNA）的退化失活（或部分失活）拷贝，通常称为加工过的假基因；（3）简单重复序列，由相对比较短的核苷酸如$(A)_n$、$(CA)_n$或$(CGG)_n$正向重复组成；（4）片段重复，由$10 \sim 300$ kb的大片段组成，这些片段是从基因组的一个区域复制到另一区域形成；（5）串联重复序列，如着丝粒、端粒、近端着丝粒染色体的短臂和核糖体基因簇等区域。（这些区域在基因组草图序列中代表性不足，本文不讨论。）

重复序列通常被描述为"垃圾"，并且被认为是没有意义的。但实际上它们代表了一个关于生物学过程的特殊信息库。这些重复构成了丰富的古生物学记录，包含

constitute a rich palaeontological record, holding crucial clues about evolutionary events and forces. As passive markers, they provide assays for studying processes of mutation and selection. It is possible to recognize cohorts of repeats "born" at the same time and to follow their fates in different regions of the genome or in different species. As active agents, repeats have reshaped the genome by causing ectopic rearrangements, creating entirely new genes, modifying and reshuffling existing genes, and modulating overall GC content. They also shed light on chromosome structure and dynamics, and provide tools for medical genetic and population genetic studies.

The human is the first repeat-rich genome to be sequenced, and so we investigated what information could be gleaned from this majority component of the human genome. Although some of the general observations about repeats were suggested by previous studies, the draft genome sequence provides the first comprehensive view, allowing some questions to be resolved and new mysteries to emerge.

Transposon-derived repeats

Most human repeat sequence is derived from transposable elements[142,143]. We can currently recognize about 45% of the genome as belonging to this class. Much of the remaining "unique" DNA must also be derived from ancient transposable element copies that have diverged too far to be recognized as such.

In mammals, almost all transposable elements fall into one of four types, of which three transpose through RNA intermediates and one transposes directly as DNA. These are long interspersed elements (LINEs, an ancient and common innovation in eukaryotes), short interspersed elements (SINEs, which are freeloaders on LINEs), LTR retrotransposons and DNA transposons.

Transposable elements employ different strategies to ensure their evolutionary survival. LINEs and SINEs rely almost exclusively on vertical transmission within the host genome[154] (but see refs 148, 155). DNA transposons are more promiscuous, requiring relatively frequent horizontal transfer. LTR retroposons use both strategies, with some being long-term active residents of the human genome (such as members of the ERVL family) and others having only short residence times.

Currently recognized SINEs, LINEs, LTR retroposons and DNA transposon copies comprise 13%, 20%, 8% and 3% of the sequence, respectively. We expect these densities to grow as more repeat families are recognized, among which will be lower copy number LTR elements and DNA transposons, and possibly high copy number ancient (highly diverged) repeats.

Age distribution. The age distribution of the repeats in the human genome provides a rich "fossil record" stretching over several hundred million years. The ancestry and

648

关于进化事件和进化驱动力的重要线索。作为一种被动标记，它们提供了用于研究突变过程和选择过程的材料。我们可以识别在同一时间"出生"的一群重复序列，追踪它们在基因组的不同区域或不同物种中的命运。作为活跃分子，重复序列通过引起异位重排、创建全新基因、修饰和重组现有基因及调整总体 GC 含量等重塑了基因组。它们还揭示了染色体结构和动力学，并为医学遗传学和群体遗传学研究提供了工具。

人类是第一个经过测序的富含重复序列的基因组，因此我们研究了从人类基因组大多数组分中可以收集到哪些信息。虽然以前的研究也提出了一些关于重复的普遍观察结果，但基因组草图序列第一次提供了序列的全貌，在解决某些问题的同时产生了新的谜团。

转座子起源的重复

大多数人类重复序列来源于转座元件[142,143]。目前，我们可以认定约 45% 的基因组属于这一类。剩下的"独特"DNA 中，许多应该也来自远古转座元件的拷贝，这些拷贝已经因分化得太远而无法识别。

在哺乳动物中，几乎所有的转座元件都可归入四个大类，其中三类是由 RNA 介导，一类是 DNA 直接转座完成。这四类分别是：长散在重复序列(LINE，在真核生物中古老而常见的创新序列)，短散在重复序列(SINE，LINE 中的自由插入序列)、LTR 逆转座子和 DNA 转座子。

转座元件采用不同的策略来确保它们的进化生存。LINE 和 SINE 几乎完全依赖宿主基因组内的垂直传递[154](请参见参考文献 148 和 155)。DNA 转座子更加杂乱，需要相对频繁的水平转移。LTR 逆转座子则使用上述两种策略，一些在人类基因组长期活跃驻扎(例如 ERVL 家族的成员)，其他则是短暂停留。

目前公认的 SINE、LINE、LTR 逆转座子和 DNA 转座子的拷贝数分别占序列的 13%、20%、8% 和 3%。我们预计上述数字会随着更多重复家族的识别而增加，其中包括较低拷贝数的 LTR 元件和 DNA 转座子，可能还有高拷贝数的远古(高度分化)重复。

年代分布　人类基因组中重复序列的年龄分布提供了绵延数亿年的丰富的"化

approximate age of each fossil can be inferred by exploiting the fact that each copy is derived from, and therefore initially carried the sequence of, a then-active transposon and, being generally under no functional constraint, has accumulated mutations randomly and independently of other copies. We can infer the sequence of the ancestral active elements by clustering the modern derivatives into phylogenetic trees and building a consensus based on the multiple sequence alignment of a cluster of copies. Using available consensus sequences for known repeat subfamilies, we calculated the per cent divergence from the inferred ancestral active transposon for each of three million interspersed repeats in the draft genome sequence.

The percentage of sequence divergence can be converted into an approximate age in millions of years (Myr) on the basis of evolutionary information.

Several facts are apparent from analysis.

First, most interspersed repeats in the human genome predate the eutherian radiation. This is a testament to the extremely slow rate with which nonfunctional sequences are cleared from vertebrate genomes.

Second, LINE and SINE elements have extremely long lives. The monophyletic LINE1 and Alu lineages are at least 150 and 80 Myr old, respectively. In earlier times, the reigning transposons were LINE2 and MIR[148,158]. The SINE MIR was perfectly adapted for reverse transcription by LINE2, as it carried the same 50-base sequence at its 3' end. When LINE2 became extinct 80–100 Myr ago, it spelled the doom of MIR.

Third, there were two major peaks of DNA transposon activity. The first involved Charlie elements and occurred long before the eutherian radiation; the second involved Tigger elements and occurred after this radiation. Because DNA transposons can produce large-scale chromosome rearrangements[159-162], it is possible that widespread activity could be involved in speciation events.

Fourth, there is no evidence for DNA transposon activity in the past 50 Myr in the human genome. The youngest two DNA transposon families that we can identify in the draft genome sequence (MER75 and MER85) show 6–7% divergence from their respective consensus sequences representing the ancestral element, indicating that they were active before the divergence of humans and new world monkeys. Moreover, these elements were relatively unsuccessful, together contributing just 125 kb to the draft genome sequence.

Finally, LTR retroposons appear to be teetering on the brink of extinction, if they have not already succumbed. For example, the most prolific elements (ERVL and MaLRs) flourished for more than 100 Myr but appear to have died out about 40 Myr ago[163,164].

More generally, the overall activity of all transposons has declined markedly over the past

石记录"。每个"化石"的祖先和大致年龄可以通过探究以下事实来推断：每个拷贝来源于一个当时活跃的转座子并因此在最初携带其序列，同时在没有功能选择压力的情况下独立于其他拷贝之外随机累积突变。我们可以通过将现代衍生物聚类为系统发生树并基于拷贝簇的多重序列比对建立共识，来推断祖先活性元件的序列。根据已知的重复亚家族的共有序列，我们计算了基因组草图序列中三百万个散在重复序列各自与推测的祖先活跃转座子的百分比差异。

依据进化的信息，序列差异的百分比可近似转化为以百万年计（Myr）的时间。

分析后有几点明显的特征：

首先，人类基因组中大多数散在重复序列早于真兽亚纲的扩张，这证明了非功能性序列从脊椎动物基因组中被清除的速率非常缓慢。

其次，LINE 元件和 SINE 元件具有很长的寿命。单源 LINE1 和 Alu 谱系至少有 1.5 亿年和 8,000 万年的历史。早期，居于统治地位的转座子是 LINE2 和 MIR[148,158]。SINE MIR 的 3′ 端携带与 LINE2 相同的 50 个碱基序列，因而完全适用于 LINE2 的逆转录。当 LINE2 在 8,000 万 ~ 1 亿年前灭绝时，MIR 也难逃厄运。

第三，DNA 转座子的活性曾有两次峰值。第一次涉及 Charlie 元件，发生在真兽亚纲扩张之前很久；第二次涉及 Tigger 元件，发生在扩张后。因为 DNA 转座子可以产生大规模的染色体重排[159-162]，所以转座子的广泛活动可能涉及形成物种的事件。

第四，在人类基因组中，过去 5,000 万年里没有 DNA 转座子活性的证据。我们在基因组草图序列中鉴定出的最年轻的两个 DNA 转座子家族（MER75 和 MER85）显示出与各自的代表祖先元件的共有序列有 6% ~ 7%的差异，这表明它们在人类与新世界猴分化之前是有活性的。而且，这些元件没那么成功，合在一起仅占基因组草图序列 125 kb 的长度。

最后，即使 LTR 逆转座子还没有屈服，但似乎已在灭绝的边缘了。例如，最丰富的元件（ERVL 和 MaLRs）兴盛了 1 亿年，但它们已经在 4,000 万年前消失了[163,164]。

更普遍的是，在过去的 3,500 万 ~ 5,000 万年间，所有转座子的整体活性显著下

35–50 Myr, with the possible exception of LINE1. Indeed, apart from an exceptional burst of activity of Alus peaking around 40 Myr ago, there would appear to have been a fairly steady decline in activity in the hominid lineage since the mammalian radiation. The extent of the decline must be even greater than it appears because old repeats are gradually removed by random deletion and because old repeat families are harder to recognize and likely to be under-represented in the repeat databases.

What explains the decline in transposon activity in the lineage leading to humans? There is no similar decline in the mouse genome.

Comparison with other organisms. In contrast to their possible extinction in humans, LTR retroposons are alive and well in the mouse. These evolutionary findings are consistent with the empirical observations that new spontaneous mutations are 30 times more likely to be caused by LINE insertions in mouse than in human (~3% versus 0.1%)[170] and 60 times more likely to be caused by transposable elements in general. It is estimated that around 1 in 600 mutations in human are due to transpositions, whereas 10% of mutations in mouse are due to transpositions (mostly IAP insertions).

The contrast between human and mouse suggests that the explanation for the decline of transposon activity in humans may lie in some fundamental difference between hominids and rodents. Population structure and dynamics would seem to be likely suspects. Rodents tend to have large populations, whereas hominid populations tend to be small and may undergo frequent bottlenecks. Evolutionary forces affected by such factors include inbreeding and genetic drift, which might affect the persistence of active transposable elements[171]. Studies in additional mammalian lineages may shed light on the forces responsible for the differences in the activity of transposable elements[172]. (Editorial note: a detailed discussion of inter-species comparisons has been omitted.)

Distribution by GC content. We next focused on the correlation between the nature of the transposons in a region and its GC content. We calculated the density of each repeat type as a function of the GC content in 50-kb windows. As has been reported[142,173-176], LINE sequences occur at much higher density in AT-rich regions (roughly fourfold enriched), whereas SINEs (MIR, Alu) show the opposite trend (for Alu, up to fivefold lower in AT-rich DNA). LTR retroposons and DNA transposons show a more uniform distribution, dipping only in the most GC-rich regions.

The preference of LINEs for AT-rich DNA seems like a reasonable way for a genomic parasite to accommodate its host, by targeting gene-poor AT-rich DNA and thereby imposing a lower mutational burden.

The contrary behaviour of SINEs, however, is baffling. How do SINEs accumulate in GC-rich DNA, particularly if they depend on the LINE transposition machinery[178]?

降，也许只有LINE1是个例外。事实上，除了4,000万年前Alu的活性爆发并达到顶峰之外，自哺乳动物扩张以来，类人猿世系的转座子活性似乎已经相当稳定地下降了。由于古老的重复序列逐渐地随机删除，同时古老重复序列家族难以识别且可能在数据库中代表不足，实际下降幅度应该更大。

什么能够解释人类谱系中的转座子活性下降呢？在小鼠的基因组中并未观察到同样的现象。

与其他生物的比较　与在人类中可能灭绝的情况相反，LTR逆转座子在小鼠中存活得很好。这些进化上的发现与经验观察值相吻合，在小鼠中，由LINE插入产生新的自发突变的可能性是人类中的30倍（约3%相对于0.1%）[170]，由转座元件插入产生的可能性是人类的60倍。据估计，人类突变中约六百分之一由转座引起，在小鼠中由转座引起的突变占到10%（大多数为IAP插入）。

人类与小鼠的对比表明，人类转座子活性的下降可能基于类人猿世系和啮齿类动物的某些根本差异。种群结构和种群动态似乎是个疑点。啮齿动物往往拥有大种群，而人类种群倾向于变小，并经常遇到瓶颈。进化动力受近亲繁殖和遗传漂变等因素的影响，可能会影响活跃转座元件的持续[171]。其他哺乳动物谱系的研究也许能揭示转座元件活性差异的原因[172]。（编者注：关于种群间比较的详细讨论被删减了。）

GC含量的分布　下面我们将重点关注区域内转座子的性质和GC含量之间的关系。以50 kb为窗口，我们将每类重复序列的密度与GC含量关联起来。正如已报道的[142,173-176]，LINE序列分布在AT富集区（大约4倍富集），而SINE（MIR、Alu）呈现相反的趋势（Alu在富含AT的DNA区域密度低至五分之一）。逆转录子LTR和DNA转座子呈现更均匀的分布，仅处于GC富集区。

LINE序列对AT富集DNA的偏好似乎是适应基因组寄生生活的合理途径，AT富集DNA通常基因分布较少，插入后引起的突变负担更小。

然而，SINE的相反行为令人困惑。SINE如何在GC富集DNA中积累，特别是如果它们依赖于LINE转座机制[178]？

We used the draft genome sequence to investigate this mystery by comparing the proclivities of young, adolescent, middle-aged and old Alus. Strikingly, recent Alus show a preference for AT-rich DNA resembling that of LINEs, whereas progressively older Alus show a progressively stronger bias towards GC-rich DNA. These results indicate that the GC bias must result from strong pressure: there is a 13-fold enrichment of Alus in GC-rich DNA within the last 30 Myr, and possibly more recently.

These observations indicate that there may be some force acting particularly on Alus. This could be a higher rate of random loss of Alus in AT-rich DNA, negative selection against Alus in AT-rich DNA or positive selection in favour of Alus in GC-rich DNA. The first two possibilities seem unlikely because AT-rich DNA is gene-poor and tolerates the accumulation of other transposable elements. The third seems more feasible, in that it involves selecting in favour of the minority of Alus in GC-rich regions rather than against the majority that lie in AT-rich regions. But positive selection for Alus in GC-rich regions would imply that they benefit the organism.

Our results may support the controversial idea that SINEs actually earn their keep in the genome. Clearly, much additional work will be needed to prove or disprove the hypothesis that SINEs are genomic symbionts.

Fast living on chromosome Y. The pattern of interspersed repeats can be used to shed light on the unusual evolutionary history of chromosome Y. Our analysis shows that the genetic material on chromosome Y is unusually young, probably owing to a high tolerance for gain of new material by insertion and loss of old material by deletion. Overall, chromosome Y seems to maintain a youthful appearance by rapid turnover.

Interspersed repeats on chromosome Y can also be used to estimate the relative mutation rates, α_m and α_f, in the male and female germlines. Chromosome Y always resides in males, whereas chromosome X resides in females twice as often as in males. The substitution rates, μ_Y and μ_X, on these two chromosomes should thus be in the ratio $\mu_Y:\mu_X = (\alpha_m):(\alpha_m+2\alpha_f)/3$, provided that one considers equivalent neutral sequences. Page and colleagues[192] obtained an estimate of $\mu_Y:\mu_X = 1.36$, corresponding to $\alpha_m:\alpha_f = 1.7$.

Our estimate is in reasonable agreement with that of Page *et al.*, although it is based on much more total sequence (360 kb on Y, 1.6 Mb on X) and a much longer time period. Various theories have been proposed for the higher mutation rate in the male germline, including the greater number of cell divisions in the formation of sperm than eggs and different repair mechanisms in sperm and eggs. (Editorial note: a detailed discussion of transposon dynamics and simple sequence repeats has been omitted.)

Segmental duplications

A remarkable feature of the human genome is the segmental duplication of portions of

我们使用基因组草图序列，通过比较"青年"、"青少年"、"中年"和"老年"Alu序列的倾向性来调查这个谜团。引人注目的是，较年轻的Alu展示了一种类似于LINE对AT富集DNA优先选择的倾向性，而逐渐变老的Alu对富含GC的DNA的偏好逐渐增强。这些结果表明，GC偏好性应该是强大的压力造成的。Alu序列在GC富集DNA中约13倍富集，这发生在距今3,000万年内或更近。

这些观察结果表明，可能会有一些力量在Alu上发挥独特的作用。这可能是富含AT的DNA随机丢失Alu的概率较高，AT富集DNA对Alu的负选择或者GC富集DNA对Alu的正选择。前两种似乎可能性不大，因为AT富集DNA缺乏基因，并且容忍其他转座元件的积累。第三种看起来更可行，因为它意味着在GC富集区选择偏好少数Alu，而不是在AT富集区排斥多数Alu。但是GC富集区正向选择Alu意味着生物体将因此获益。

我们的研究结果也许支持了这样一个有争议的观点，那就是SINE在基因组中争得了一席之地。显然，若想证实或否定SINE是基因组共生体的假设，需要做更多的工作。

在Y染色体上快速生存　散在重复的模式可用于阐明Y染色体不寻常的进化历史。我们的分析表明，Y染色体上的遗传物质异常年轻，可能对于是对插入新物质和删除旧物质具有较高耐受性。总的来说，Y染色体似乎通过快速更新而保持年轻的外貌。

Y染色体上的散布重复也可用于估算男性和女性生殖系中的相对突变率 α_m 和 α_f。染色体Y总是存在于雄性中，而染色体X存在于女性中的频率是男性的两倍。这两种染色体上的替代率 μ_Y 和 μ_X 应符合比值 $\mu_Y : \mu_X = (\alpha_m) : (\alpha_m + 2\alpha_f)/3$，前提是考虑相等的中性序列。佩奇及其同事[192]估算 $\mu_Y : \mu_X = 1.36$，相应的 $\alpha_m : \alpha_f = 1.7$。

尽管我们的估计基于更多的全序列（Y染色体360 kb，X染色体1.6 Mb）和更长的时间周期，但是与佩奇等人的估计相当吻合。对于男性生殖系中的较高突变率，目前已经提出了不同理论，包括精子形成中细胞分裂的数量多于卵子以及精子和卵子中修复机制不同。（编者注：关于转座子动力学和简单序列重复的详细讨论被删除了。）

片段重复

人类基因组的一个显著特征是基因组序列的部分片段重复[215-217]。这种片段重复

genomic sequence[215-217]. Such duplications involve the transfer of 1–200-kb blocks of genomic sequence to one or more locations in the genome. The locations of both donor and recipient regions of the genome are often not tandemly arranged, suggesting mechanisms other than unequal crossing-over for their origin. They are relatively recent, inasmuch as strong sequence identity is seen in both exons and introns (in contrast to regions that are considered to show evidence of ancient duplications, characterized by similarities only in coding regions). Indeed, many such duplications appear to have arisen in very recent evolutionary time, as judged by high sequence identity and by their absence in closely related species.

Segmental duplications can be divided into two categories. First, interchromosomal duplications are defined as segments that are duplicated among nonhomologous chromosomes.

The second category is intrachromosomal duplications, which occur within a particular chromosome or chromosomal arm.

Until now, the identification and characterization of segmental duplications have been based on anecdotal reports. The availability of the entire genomic sequence will make it possible to explore the nature of segmental duplications more systematically.

We performed a global genome-wide analysis to characterize the amount of segmental duplication in the genome.

The finished sequence consists of at least 3.3% segmental duplication. Interchromosomal duplication accounts for about 1.5% and intrachromosomal duplication for about 2%, with some overlap (0.2%) between these categories. We analysed the lengths and divergence of the segmental duplications. The duplications tend to be large (10–50 kb) and highly homologous, especially for the interchromosomal segments. The sequence divergence for the interchromosomal duplications appears to peak between 96.5% and 97.5%. This may indicate that interchromosomal duplications occurred in a punctuated manner. It will be intriguing to investigate whether such genomic upheaval has a role in speciation events.

We compared the entire human draft genome sequence (finished and unfinished) with itself to identify duplications with 90–98% sequence identity. The draft genome sequence contains at least 3.6% segmental duplication. The actual proportion will be significantly higher, because we excluded many true matches with more than 98% sequence identity (at least 1.1% of the finished sequence). Although exact measurement must await a finished sequence, the human genome seems likely to contain about 5% segmental duplication, with most of this sequence in large blocks (> 10 kb). Such a high proportion of large duplications clearly distinguishes the human genome from other sequenced genomes, such as the fly and worm.

涉及 1 ~ 200 kb 的序列转移至基因组一个或多个位置。基因组的供体和受体区域的位置通常不是串联排列的，这提示可能存在与其起源的不等交换不同的其他机制。这类重复是相对较晚出现的，因为不管是外显子还是内含子都能观察到较强的序列一致性（与此相反，那些被认为古老的重复序列仅在编码区域中具有相似性）。事实上，许多这样的重复似乎是在最近的进化年代中出现的，它们不仅序列一致性高，而且不会出现在关系很近的物种之间。

片段重复可以分为两类。第一类，染色体间的重复被定义为非同源染色体间的重复片段。

第二类是染色体内重复，发生在特定的染色体或染色体臂内。

到目前为止，对片段重复的确定和描述基本来自坊间的报道。全基因组序列的可用性使得人们可以更系统地探索片段重复的性质。

我们在全基因组范围内进行分析，识别片段重复的数量。

完成序列包含至少 3.3% 的片段重复。染色体间重复约占 1.5%，染色体内重复约占 2%，两类之间有部分重叠（0.2%）。我们分析了片段重复的长度和差异性。重复的片段往往很大（10 ~ 50 kb）并且高度同源，尤其是染色体间重复。染色体间重复的序列差异的峰值似乎在 96.5% 和 97.5% 之间。这可能表明染色体间重复以间断的形式发生。调查此类基因组的巨变在物种形成中是否发挥作用将是非常有趣的。

我们将人类全基因组草图序列（完成和未完成）与其自身进行比较，用以鉴定具有 90% ~ 98% 序列一致性的重复序列。基因组草图序列至少包含 3.6% 的片段重复，我们排除了许多序列一致性在 98% 以上（至少 1.1% 的完成序列）的真实匹配，因此实际比例将更高。精确值必须等到基因组序列完成后才可测量，但人类基因组很可能含有约 5% 的片段重复，其中大部分重复序列是大片段的（> 10 kb）。人类基因组内大段重复的比例如此之高是其与其他已测序基因组（如果蝇和线虫）的明显区别。

Gene Content of the Human Genome

Genes (or at least their coding regions) comprise only a tiny fraction of human DNA, but they represent the major biological function of the genome and the main focus of interest by biologists. They are also the most challenging feature to identify in the human genome sequence.

The ultimate goal is to compile a complete list of all human genes and their encoded proteins, to serve as a "periodic table" for biomedical research[243]. But this is a difficult task. In organisms with small genomes, it is straightforward to identify most genes by the presence of long open reading frames (ORFs). In contrast, human genes tend to have small exons (encoding an average of only 50 codons) separated by long introns (some exceeding 10 kb). This creates a signal-to-noise problem, with the result that computer programs for direct gene prediction have only limited accuracy. Instead, computational prediction of human genes must rely largely on the availability of cDNA sequences or on sequence conservation with genes and proteins from other organisms. This approach is adequate for strongly conserved genes (such as histones or ubiquitin), but may be less sensitive to rapidly evolving genes (including many crucial to speciation, sex determination and fertilization).

Noncoding RNAs

Although biologists often speak of a tight coupling between "genes and their encoded protein products", it is important to remember that thousands of human genes produce noncoding RNAs (ncRNAs) as their ultimate product[244]. There are several major classes of ncRNA. (1) Transfer RNAs (tRNAs) are the adapters that translate the triplet nucleic acid code of RNA into the amino-acid sequence of proteins; (2) ribosomal RNAs (rRNAs) are also central to the translational machinery, and recent X-ray crystallography results strongly indicate that peptide bond formation is catalysed by rRNA, not protein[245,246]; (3) small nucleolar RNAs (snoRNAs) are required for rRNA processing and base modification in the nucleolus[247,248]; and (4) small nuclear RNAs (snRNAs) are critical components of spliceosomes, the large ribonucleoprotein (RNP) complexes that splice introns out of pre-mRNAs in the nucleus.

We can identify genomic sequences that are homologous to known ncRNA genes, using BLASTN or, in some cases, more specialized methods.

It is sometimes difficult to tell whether such homologous genes are orthologues, paralogues or closely related pseudogenes (because inactivating mutations are much less obvious than for protein-coding genes). For tRNA, there is sufficiently detailed information about the cloverleaf secondary structure to allow true genes and pseudogenes to be distinguished with high sensitivity. For many other ncRNAs, there is much less structural information and so we employ an operational criterion of high sequence similarity (> 95% sequence identity

658

人类基因组的基因含量

基因(至少是它们的编码区)只占人类 DNA 的很小一部分,但体现了基因组的主要生物学功能,是生物学家研究的焦点。它们也是人类基因组序列中识别起来最富挑战性的一项特征。

我们的终极目标是编制一张用于生物医学研究的"元素周期表",罗列所有人类基因及其编码的蛋白质[243]。但是这是一项很困难的工作。对于基因组较小的物种来说,通过长的开放阅读框(ORF)可以直接确定大部分基因。与此相反,人类基因倾向于含有被大的内含子(有些超过 10 kb)分隔开的小的外显子(仅平均编码 50 个密码子)。这就产生了信噪比问题,因此利用电脑程序直接预测基因的准确性很有限。因此,人类基因的电脑预测要极大地依赖于 cDNA 序列或来自其他物种的基因和蛋白质的保守序列。这个方法对高度保守的基因(如组蛋白或泛素)足以胜任,但是对于快速进化的基因(包括很多对物种形成、性别决定和受精非常重要的基因)就没那么敏感了。

非编码 RNA

虽然生物学家经常说"基因与其编码的蛋白质产物"连接紧密,但请务必记得还有数以千计的人类基因产生非编码 RNA(ncRNA)作为其最终产物[244]。ncRNA 主要分以下几大类:(1)转运 RNA(tRNA),负责将 RNA 的三联体密码翻译为蛋白质的氨基酸序列;(2)核糖体 RNA(rRNA),是转录机器的核心,最近的 X 射线晶体学结果显示肽键的形成是由 rRNA 而非蛋白质催化[245,246];(3)小核仁 RNA(snoRNA),参与核仁中的 rRNA 编辑和碱基修饰[247,248];(4)小核 RNA(snRNA),剪切体的关键组分,剪切体是大型核糖核蛋白(RNP)复合物,负责在细胞核中将前 mRNA 的内含子切除。

我们可以利用 BLASTN 或其他更专业的方法识别与已知的 ncRNA 基因同源的基因组序列。

有时,很难讲清这些同源关系是直系同源、旁系同源还是密切相关的假基因(因为与编码蛋白质的基因相比,ncRNA 基因的失活突变更不明显)。对 tRNA 来说,三叶草的二级结构提供了足够多的信息,用来判定基因和假基因的灵敏度很高。但是对于其他 ncRNA 来说,结构信息要少得多,因此我们引入了一条序列高度相似(大于 95% 的一致性和大于 95% 的全长)的操作标准来区别基因和假基因。这些工

and > 95% full length) to distinguish true genes from pseudogenes. These assignments will eventually need to be reconciled with experimental data.

Transfer RNA genes.   In the draft genome sequence, we find only 497 human tRNA genes. This appears to include most of the known human tRNA species. The draft genome sequence contains 37 of 38 human tRNA species listed in a tRNA database[253], allowing for up to one mismatch.

The results indicate that the human has fewer tRNA genes than the worm, but more than the fly. This may seem surprising, but tRNA gene number in metazoans is thought to be related not to organismal complexity, but more to idiosyncrasies of the demand for tRNA abundance in certain tissues or stages of embryonic development. For example, the frog *Xenopus laevis*, which must load each oocyte with a remarkable 40 ng of tRNA, has thousands of tRNA genes[254].

The tRNA genes are dispersed throughout the human genome. However, this dispersal is nonrandom. tRNA genes have sometimes been seen in clusters at small scales[262,263] but we can now see striking clustering on a genome-wide scale. More than 25% of the tRNA genes (140) are found in a region of only about 4 Mb on chromosome 6. This small region, only about 0.1% of the genome, contains an almost sufficient set of tRNA genes all by itself. The 140 tRNA genes contain a representative for 36 of the 49 anticodons found in the complete set; and of the 21 isoacceptor types, only tRNAs to decode Asn, Cys, Glu and selenocysteine are missing. Many of these tRNA genes, meanwhile, are clustered elsewhere; 18 of the 30 Cys tRNAs are found in a 0.5-Mb stretch of chromosome 7 and many of the Asn and Glu tRNA genes are loosely clustered on chromosome 1. More than half of the tRNA genes (280 out of 497) reside on either chromosome 1 or chromosome 6. Chromosomes 3, 4, 8, 9, 10, 12, 18, 20, 21 and X appear to have fewer than 10 tRNA genes each; and chromosomes 22 and Y have none at all (each has a single pseudogene). (Editorial note: much of the discussion of RNA genes has been omitted.)

Our observations confirm the striking proliferation of ncRNA-derived pseudogenes. There are hundreds or thousands of sequences in the draft genome sequence related to some of the ncRNA genes. The most prolific pseudogene counts generally come from RNA genes transcribed by RNA polymerase III promoters, including U6, the hY RNAs and SRP-RNA. These ncRNA pseudogenes presumably arise through reverse transcription. The frequency of such events gives insight into how ncRNA genes can evolve into SINE retroposons, such as the tRNA-derived SINEs found in many vertebrates and the SRP-RNA-derived Alu elements found in humans.

Protein-coding genes

(Editorial note: this section has been significantly shortened.) Identifying the protein-coding genes in the human genome is one of the most important applications of the

660

作最终需要实验数据来协调。

转运 RNA 基因 在基因组草图序列中，我们仅找到了 497 个人类 tRNA 基因。这包含了目前已知的大多数 tRNA 种类。在 tRNA 数据库[253]列举的 38 种人类 tRNA 中，基因组草图中出现了 37 种，允许最多出现一次错配。

研究结果显示，人类的 tRNA 基因少于线虫，但多于果蝇。虽然出人意料，但在多细胞生物中，tRNA 基因的数量被认为与生物体的复杂程度无关，而是与某些组织的对 tRNA 需求或胚胎发育的阶段等特征有关。如非洲爪蟾，它的卵母细胞会携带多达 40 ng 的 tRNA，而它的 tRNA 基因也多达上千种[254]。

tRNA 基因分散在整个人类基因组中。但并非随机分布。曾经也发现过 tRNA 基因小规模聚集成簇[262,263]，但是我们现在看到，tRNA 基因在基因组范围内有非常显著的基因簇。超过 25% 的 tRNA(140 个)位于 6 号染色体的长约 4 Mb 的一个区域内。这个小区域仅约占人类基因组总长的 0.1%，但却包含了几乎一整套必需的 tRNA 基因。这 140 个 tRNA 基因包含 49 个反密码子中典型的 36 个；而 21 个同工型中只缺少天冬酰胺、半胱氨酸、谷氨酸和硒代半胱氨酸的解码 tRNA。这些 tRNA 基因大多成簇存在于其他位置，30 个半胱氨酸 tRNA 中的 18 个在 7 号染色体的 0.5 Mb 区域内，很多天冬酰胺和谷氨酸 tRNA 基因则松散地成簇出现在 1 号染色体上。超过一半的 tRNA 基因(497 个中的 280 个)位于 1 号或 6 号染色体上，第 3、4、8、9、10、12、18、20、21 号染色体和 X 染色体上各自分布不到 10 个 tRNA 基因，22 号和 Y 染色体则完全没有 tRNA 基因(各自含有一个假基因)。(编者注：关于 RNA 基因的大部分讨论被删减了。)

我们的观察证实了起源于 ncRNA 的假基因的显著增殖过程。基因组草图中，成千上万的序列与某些 ncRNA 基因有关。大量的假基因一般来自 RNA 聚合酶Ⅲ启动子转录的 RNA 基因，包括 U6、hY RNA 和 SRP-RNA。这些 ncRNA 假基因很可能起源于逆转录。这种事件发生的频率也让我们有机会了解，ncRNA 基因是怎样进化为 SINE 逆转座子的，就像 tRNA 起源的 SINE 存在于很多脊椎动物基因组中，SRP-RNA 起源的 Alu 元件存在于人类基因组中。

蛋白质编码基因

(编者注：这部分有大量的删减。)在人类基因组中识别出蛋白质编码基因，是

sequence data, but also one of the most difficult challenges.

Towards a complete index of human genes. We focused on creating an initial index of human genes and proteins. This index is quite incomplete, owing to the difficulty of gene identification in human DNA and the imperfect state of the draft genome sequence. Nonetheless, it is valuable for experimental studies and provides important insights into the nature of human genes and proteins.

Gene identification is difficult in human DNA. The signal-to-noise ratio is low: coding sequences comprise only a few per cent of the genome and an average of about 5% of each gene; internal exons are smaller than in worms; and genes appear to have more alternative splicing. The challenge is underscored by the work on human chromosomes 21 and 22. Even with the availability of finished sequence and intensive experimental work, the gene content remains uncertain, with upper and lower estimates differing by as much as 30%. The initial report of the finished sequence of chromosome 22 (ref. 94) identified 247 previously known genes, 298 predicted genes confirmed by sequence homology or ESTs and 325 *ab initio* predictions without additional support. Many of the confirmed predictions represented partial genes. In the past year, 440 additional exons (10%) have been added to existing gene annotations by the chromosome 22 annotation group, although the number of confirmed genes has increased by only 17 and some previously identified gene predictions have been merged[286].

Creating an initial gene index. We set out to create an initial integrated gene index (IGI) and an associated integrated protein index (IPI) for the human genome.

Evaluation of IGI/IPI. We used several approaches to evaluate the sensitivity, specificity and fragmentation of the IGI/IPI set.

Comparison with "new" known genes. One approach was to examine newly discovered genes arising from independent work that were not used in our gene prediction effort. We identified 31 such genes: 22 recent entries to RefSeq and 9 from the Sanger Centre's gene identification program on chromosome X. Of these, 28 were contained in the draft genome sequence and 19 were represented in the IGI/IPI. This suggests that the gene prediction process has a sensitivity of about 68% (19/28) for the detection of novel genes in the draft genome sequence and that the current IGI contains about 61% (19/31) of novel genes in the human genome.

Comparison with RIKEN mouse cDNAs. In a less direct but larger-scale approach, we compared the IGI gene set to a set of mouse cDNAs sequenced by the Genome Exploration Group of the RIKEN Genomic Sciences Center[309]. Around 81% of the genes in the RIKEN mouse set showed sequence similarity to the human genome sequence, whereas 69% showed sequence similarity to the IGI/IPI. This suggests a sensitivity of 85% (69/81).

662

序列数据最重要的应用之一，也是最艰巨的挑战之一。

生成人类基因的完整索引　我们聚焦于产生一个人类基因与蛋白质的初始索引。由于在人类 DNA 中识别基因的困难，以及基因组草图序列的未完成状态，该索引离完成之日尚早。尽管如此，它对实验性研究仍有价值，并可从中洞察人类基因与蛋白质的本质。

在人类 DNA 中识别基因是很困难的。有信噪比很低的问题：编码序列只构成了基因组很小的一部分，平均只占每个基因的 5%；基因内部的外显子比线虫的还小；基因可能存在更多的可变剪接。对人类 21 和 22 号染色体进行测序所面临的挑战被低估了。即使能够得到最终序列和深入的实验研究，基因含量仍旧是不确定的，估计值上限和下限的差异可达 30%。22 号染色体完成序列的初始报告（参考文献 94）确认了 247 个已知基因、298 个根据序列同源性或表达序列标签 EST 预测的基因以及 325 个没有其他证据支持的从头预测的基因。许多预测基因不完整。在过去的一年里，在 22 号染色体的注释数据组中，尽管已确定基因的数量只增长了 17 个，一些之前识别出来的基因也有所重叠，我们在现有基因注释的基础上又发现了 440 个外显子（10%）[286]。

生成初始基因索引　我们着手对人类基因组生成一个初始的整合基因索引（IGI），和一个相关的整合蛋白质索引（IPI）。

IGI/IPI 的评估　我们用几种方法评估 IPI/IGI 数据组的灵敏性、特异性和片段化。

与"新的"已知基因的比较　一种方法是检查那些新发现的基因，它们来自我们的基因预测中尚未使用的独立工作。我们发现了 31 个这样的基因：有 22 个最近刚加入 RefSeq 数据库，有 9 个来自桑格中心的 X 染色体基因识别项目。在这些基因中，有 28 个包含在基因组草图序列中，有 19 个出现在 IGI/IPI 数据组中。这提示，基因预测过程中，从基因组草图序列中预测新基因的灵敏性约为 68%（19/28），而目前的 IGI 数据组包含了人类基因组中 61%（19/31）的新基因。

与 RIKEN 小鼠 cDNA 比较　通过一个较为间接而规模较大的方法，我们将 IGI 基因数据组与小鼠的 cDNA 数据组进行比较，该小鼠的 cDNA 序列是由日本理化研究所（RIKEN）基因组科学中心的基因组探索团队测序的[309]。在 RIKEN 小鼠数据组中，有 81% 的基因与人类基因组存在序列相似性，而有 69% 的基因与 IGI/IPI 数据组存在序列相似性。这提示了 85%（69/81）的灵敏度。

Chromosomal distribution. Finally, we examined the chromosomal distribution of the IGI gene set. The average density of gene predictions is 11.1 per Mb across the genome, with the extremes being chromosome 19 at 26.8 per Mb and chromosome Y at 6.4 per Mb. It is likely that a significant number of the predictions on chromosome Y are pseudogenes (this chromosome is known to be rich in pseudogenes) and thus that the density for chromosome Y is an overestimate.

Summary. We are clearly still some way from having a complete set of human genes. The current IGI contains significant numbers of partial genes, fragmented and fused genes, pseudogenes and spurious predictions, and it also lacks significant numbers of true genes. This reflects the current state of gene prediction methods in vertebrates even in finished sequence, as well as the additional challenges related to the current state of the draft genome sequence. Nonetheless, the gene predictions provide a valuable starting point for a wide range of biological studies and will be rapidly refined in the coming year.

The analysis above allows us to estimate the number of distinct genes in the IGI, as well as the number of genes in the human genome. The IGI set contains about 15,000 known genes and about 17,000 gene predictions. Assuming that the gene predictions are subject to a rate of overprediction (spurious predictions and pseudogenes) of 20% and a rate of fragmentation of 1.4, the IGI would be estimated to contain about 24,500 actual human genes. Assuming that the gene predictions contain about 60% of previously unknown human genes, the total number of genes in the human genome would be estimated to be about 31,000. This is consistent with most recent estimates based on sampling, which suggest a gene number of 30,000–35,000.

Comparative proteome analysis

Knowledge of the human proteome will provide unprecedented opportunities for studies of human gene function. Often clues will be provided by sequence similarity with proteins of known function in model organisms. Such initial observations must then be followed up by detailed studies to establish the actual function of these molecules in humans.

For example, 35 proteins are known to be involved in the vacuolar protein-sorting machinery in yeast. Human genes encoding homologues can be found in the draft human sequence for 34 of these yeast proteins, but precise relationships are not always clear. In nine cases there appears to be a single clear human orthologue (a gene that arose as a consequence of speciation); in 12 cases there are matches to a family of human paralogues (genes that arose owing to intra-genome duplication); and in 13 cases there are matches to specific protein domains[311-314]. Hundreds of similar stories emerge from the draft sequence, but each merits a detailed interpretation in context. To treat these subjects properly, there will be many following studies, the first of which appear in accompanying papers[315-323].

Genes shared with fly, worm and yeast. IPI.1 contains apparent homologues of 61%

染色体分布 最后，我们检验了 IGI 数据组的染色体分布。基因预测在全基因组范围内的平均密度是每 Mb 分布 11.1 个基因，极限是 19 号染色体上每 Mb 分布 26.8 个基因，Y 染色体上每 Mb 分布 6.4 个基因。可能 Y 染色体上大量预测出来的是假基因（已知这条染色体富含假基因），因此 Y 染色体的密度被高估了。

小结 我们显然还未得到完整的人类基因数据组。目前的 IGI 包含了大量的不完整基因、片段化和融合基因、假基因以及错误的预测；它也同样缺少相当数量的真基因。这反映了在脊椎动物完成序列中基因预测方法的当前状态，以及与基因组草图序列当前状态相关的更多挑战。尽管如此，基因预测还是为大范围的生物学研究提供了宝贵的起点，并将在未来快速优化。

上述分析使我们能够估计 IGI 中不同基因的数量以及人类基因组中的基因数量。IGI 数据组中包含约 15,000 个已知基因和约 17,000 个预测基因。假设赋予基因预测 20% 的高估率（错误预测和假基因）以及 1.4 倍的片段化率，则 IGI 数据组估计应含 24,500 个实际人类基因。假设基因预测包含 60% 未知人类基因，则人类基因组中的基因总数估算为约 31,000 个。这与最近基于采样的估计结果一致，该结果为 30,000 至 35,000 个。

比较蛋白质组学分析

人类蛋白质组学的知识将为研究人类基因功能提供前所未有的重大机遇。这些线索常常由模式生物中已知功能的蛋白质的序列相似性所提供。必须进行详细的研究跟进这些初始的观察，从而发现这些分子在人类中的实际功能。

例如，已知酵母的液泡蛋白分选机制涉及 35 个蛋白质。在人类基因组草图序列中，可以找到 34 个酵母此类蛋白质的人类基因编码的同源物，但它们之间精细的关系还不清楚。有 9 个可能是人类的直系同源物（由于物种形成产生的基因）；有 12 个可能与人类旁系同源物（基因组内的序列倍增产生的基因）家族匹配；有 13 个与特异性的蛋白质结构域匹配[311-314]。在草图序列中产生了几百个相似的例子，但每一个都值得在文中详细解析。为妥善处理这些研究对象，我们需要许多后续研究，最先开始的工作就出现在附随的文章中[315-323]。

与果蝇、线虫和酵母共有的基因 IPI.1 包含了果蝇蛋白质组中 61% 的显著同源

of the fly proteome, 43% of the worm proteome and 46% of the yeast proteome. We next considered the groups of proteins containing likely orthologues and paralogues (genes that arose from intragenome duplication) in human, fly, worm and yeast.

We identified 1,308 groups of proteins, each containing at least one predicted orthologue in each species and many containing additional paralogues. The 1,308 groups contained 3,129 human proteins, 1,445 fly proteins, 1,503 worm proteins and 1,441 yeast proteins. These 1,308 groups represent a conserved core of proteins that are mostly responsible for the basic "housekeeping" functions of the cell, including metabolism, DNA replication and repair, and translation.

Most proteins do not show simple 1-1-1 orthologous relationships across the three animals.

New architectures from old domains.   Whereas there appears to be only modest invention at the level of new vertebrate protein domains, there appears to be substantial innovation in the creation of new vertebrate proteins. This innovation is evident at the level of domain architecture, defined as the linear arrangement of domains within a polypeptide. New architectures can be created by shuffling, adding or deleting domains, resulting in new proteins from old parts.

We quantified the number of distinct protein architectures found in yeast, worm, fly and human by using the SMART annotation resource[339]. The human proteome set contained 1.8 times as many protein architectures as worm or fly and 5.8 times as many as yeast. This difference is most prominent in the recent evolution of novel extracellular and transmembrane architectures in the human lineage. Human extracellular proteins show the greatest innovation: the human has 2.3 times as many extracellular architectures as fly and 2.0 times as many as worm.

Conclusion.   Five lines of evidence point to an increase in the complexity of the proteome from the single-celled yeast to the multicellular invertebrates and to vertebrates such as the human. Specifically, the human contains greater numbers of genes, domain and protein families, paralogues, multidomain proteins with multiple functions, and domain architectures. According to these measures, the relatively greater complexity of the human proteome is a consequence not simply of its larger size, but also of large-scale protein innovation.

An important question is the extent to which the greater phenotypic complexity of vertebrates can be explained simply by two- or threefold increases in proteome complexity. The real explanation may lie in combinatorial amplification of these modest differences, by mechanisms that include alternative splicing, post-translational modification and cellular regulatory networks. The potential numbers of different proteins and protein–protein interactions are vast, and their actual numbers cannot readily be discerned from the genome sequence. Elucidating such system-level properties presents one of the great challenges for modern biology. (Editorial note: a discussion of genome segmental history and comparison of conserved segments between human and mouse genomes has been omitted.)

666

物、线虫蛋白质组中 43% 的显著同源物及酵母蛋白质组中 46% 的显著同源物。我们接下来考虑的蛋白质类群包括人类、果蝇、线虫和酵母中可能的直系同源物和旁系同源物(来自基因组内的基因加倍)。

我们识别出 1,308 个蛋白质类群,每个类群至少包含每个物种中一个预测的直系同源物,有许多类群包含更多的旁系同源物。1,308 个类群中包含了 3,129 个人类蛋白质、1,445 个果蝇蛋白质、1,503 个线虫蛋白质和 1,441 个酵母蛋白质。这 1,308 个类群代表了一组保守的核心蛋白质,它们大多数主要负责细胞中最基本的"管家"功能,包括代谢、DNA 复制和修复以及翻译。

在这三种动物中,大多数蛋白质未显示出 1-1-1 的直系同源关系。

来自旧结构域的新型组织结构   尽管看起来在脊椎动物新的蛋白质结构域水平只有少量发现,但在产生脊椎动物新的蛋白质方面有实质性新发现。这一发现在结构域组织结构水平十分显著,即在多肽中结构域的线性排列。新的组织结构可通过变换结构域位置、添加或删除结构域来实现,使旧的组件中产生新的蛋白质。

通过 SMART 注释资源[339],我们对酵母、线虫、果蝇和人类中找到的不同蛋白质组织结构进行计数。人类蛋白质组包含的蛋白质组织结构数量是线虫或果蝇的 1.8 倍,是酵母的 5.8 倍。在人类谱系新的细胞外和跨膜组织结构的近期演化中,这些差异最为显著。人类细胞外蛋白质中的新发现是最多的:人类细胞外组织结构数量是果蝇的 2.3 倍、线虫的 2.0 倍。

结论   从单细胞的酵母到多细胞的无脊椎动物再到脊椎动物,例如人类,蛋白质组的复杂性逐渐升高,有 5 条证据均指向这一点。具体地说,人类包含更多的基因、结构域和蛋白质家族、旁系同源物、具有多功能的多结构域蛋白质以及结构域组织结构。通过这些衡量可以发现,人类蛋白质组相对高的复杂性不仅仅是由于它的基因组有更大的尺寸,也是由于大规模的蛋白质新组合。

一个重要的问题是,脊椎动物表型的复杂程度可简单地用蛋白质组复杂性的两倍或者三倍的增长来解释。真正的解释可能存在于这些基本差异的组合扩增,可通过包括可变剪接、翻译后修饰和细胞调控网络在内的机制实现。不同的蛋白质和蛋白质–蛋白质相互作用的潜在数量是巨大的,也无法从基因组序列中可靠地辨明实际数量。对这些系统水平的性质的阐释,代表了现代生物学面临的巨大挑战之一。(编者注:对基因组片段历史以及人类和小鼠基因组保守片段的比较被删除了。)

Applications to Medicine and Biology

In most research papers, the authors can only speculate about future applications of the work. Because the genome sequence has been released on a daily basis over the past four years, however, we can already cite many direct applications. We focus on a handful of applications chosen primarily from medical research.

Disease genes

A key application of human genome research has been the ability to find disease genes of unknown biochemical function by positional cloning[388]. This method involves mapping the chromosomal region containing the gene by linkage analysis in affected families and then scouring the region to find the gene itself. Positional cloning is powerful, but it has also been extremely tedious. When the approach was first proposed in the early 1980s[9], a researcher wishing to perform positional cloning had to generate genetic markers to trace inheritance; perform chromosomal walking to obtain genomic DNA covering the region; and analyse a region of around 1 Mb by either direct sequencing or indirect gene identification methods. The first two barriers were eliminated with the development in the mid-1990s of comprehensive genetic and physical maps of the human chromosomes, under the auspices of the Human Genome Project. The remaining barrier, however, has continued to be formidable.

All that is changing with the availability of the human draft genome sequence. The human genomic sequence in public databases allows rapid identification *in silico* of candidate genes, followed by mutation screening of relevant candidates, aided by information on gene structure. For a mendelian disorder, a gene search can now often be carried out in a matter of months with only a modestly sized team.

At least 30 disease genes[55,389-422] have been positionally cloned in research efforts that depended directly on the publicly available genome sequence. As most of the human sequence has only arrived in the past twelve months, it is likely that many similar discoveries are not yet published. In addition, there are many cases in which the genome sequence played a supporting role, such as providing candidate microsatellite markers for finer genetic linkage analysis.

The genome sequence has also helped to reveal the mechanisms leading to some common chromosomal deletion syndromes. In several instances, recurrent deletions have been found to result from homologous recombination and unequal crossing over between large, nearly identical intrachromosomal duplications. Examples include the DiGeorge/velocardiofacial syndrome region on chromosome 22 (ref. 238) and the Williams-Beuren syndrome recurrent deletion on chromosome 7 (ref. 239).

在医学与生物学上的应用

对基因组序列在未来的应用，大多数研究性论文的作者仅能做出预测。然而，由于基因组序列的数据在过去的四年中每天都在增加，现在我们已经可以引用很多直接的应用实例了。我们在这里精选了几个医学上的研究应用。

疾病基因

目前，人类基因组研究一项最关键的应用是通过定位克隆来找到生化功能未知的致病基因[388]。这种方法涉及通过家系的连锁分析将包括该基因的染色体进行定位，再从该区域搜寻基因。定位克隆是有效的，但也极度冗余。当此方法在 20 世纪 80 年代早期被首次提出时[9]，研究者要做定位克隆首先要生成一个遗传标记跟踪该基因；然后通过染色体步移获得覆盖整个区域的基因组 DNA；再通过直接测序或间接鉴定基因的方法来分析 1 Mb 左右的区域。在 20 世纪 90 年代中期，在人类基因组计划的助力下，人类染色体遗传图谱和物理图谱的广泛绘制使前两步障碍已经扫除。即便如此，逾越剩下的障碍仍然非常困难。

当人类基因组草图完成后，一切都为之改观。我们可以利用储存在公共数据库中的人类基因组序列，通过生物信息学方法对目标基因进行快速鉴定，随后还可根据基因结构信息对目标基因的突变进行筛选。对某种孟德尔遗传疾病来说，一个中等规模的团队现在大约只需数月就能完成寻找特定基因的工作。

直接基于已发表的基因组序列，至少 30 个致病基因[55,389-422]已经完成定位克隆。由于大多数的人类基因组序列是在过去 12 个月内进入数据库的，可能还有很多此类研究尚未发表。此外，基因组序列还在很多研究中起到辅助作用，比如为精细的遗传连锁分析提供可用的微卫星标记。

基因组序列还帮助揭示了某些常见染色体缺失综合征的致病机制。在一些疾病中，我们发现频发性缺失是由大段的、几乎完全相同的染色体内部重复序列的同源重组和不等交换导致的。比如 22 号染色体上区域缺失引起的迪格奥尔格 /腭心面综合征（参考文献 238）和 7 号染色体上频发性缺失引起的威廉姆斯综合征（参考文献 239）。

Drug targets

Over the past century, the pharmaceutical industry has largely depended upon a limited set of drug targets to develop new therapies. A recent compendium[426,427] lists 483 drug targets as accounting for virtually all drugs on the market. Knowing the complete set of human genes and proteins will greatly expand the search for suitable drug targets. Although only a minority of human genes may be drug targets, it has been predicted that the number will exceed several thousand, and this prospect has led to a massive expansion of genomic research in pharmaceutical research and development. (Editorial note: the original text includes discussion of some specific applications of this new genomic information to drug development.)

Basic biology

Response to certain bitter tastes.   Recently, investigators mapped this trait in both humans and mice and then searched the relevant region of the human draft genome sequence for G-protein coupled receptors. These studies led, in quick succession, to the discovery of a new family of such proteins, the demonstration that they are expressed almost exclusively in taste buds, and the experimental confirmation that the receptors in cultured cells respond to specific bitter substances[433-435].

The Next Steps

Considerable progress has been made in human sequencing, but much remains to be done to produce a finished sequence. Even more work will be required to extract the full information contained in the sequence. Many of the key next steps are already underway.

Developing the IGI and IPI

A high priority will be to refine the IGI and IPI to the point where they accurately reflect every gene and every alternatively spliced form. Several steps are needed to reach this ambitious goal.

Finishing the human sequence will assist in this effort, but the experiences gained on chromosomes 21 and 22 show that sequence alone is not enough to allow complete gene identification. One powerful approach is cross-species sequence comparison with related organisms at suitable evolutionary distances. The sequence coverage from the pufferfish *T. nigroviridis* has already proven valuable in identifying potential exons[292]; this work is expected to continue from its current state of onefold coverage to reach at least fivefold coverage later this year. The genome sequence of the laboratory mouse will provide a

670

药物靶点

在过去的一个世纪里，制药产业极大地依赖有限的药物靶点来寻找新的治疗方法。最近发布的一项纲要总结了目前市场上所有药物的 483 个靶点[426,427]。而对人类的所有基因和蛋白质的揭示将极大地扩展对适当药物靶点的寻找。尽管目前只有少数人类基因可作为药物靶点，但可预见的是，这个数量将超过几千种，而此预期已经带来了药物研发领域中基因组研究的大发展。（编者注：原文包括新的基因信息在药物研发中的特定应用的讨论。）

基础生物学

对某些苦味的反应　最近，研究者们将这一特征定位在人类和小鼠的基因组上，并在人类基因组草图序列的相关区域中寻找 G 蛋白偶联受体。这些连贯的研究导致了这类蛋白质新家族的发现，证明了这些基因只在味蕾中表达，并实验验证了在体外培养的细胞中这些受体对苦味底物有信号反应[433-435]。

后　　续

尽管人类基因组序列已经取得了相当的进展，但要获得完整的最终序列，还有很多工作要做。而挖掘序列中包含的所有信息需要更多的工作。后续计划中的很多关键步骤已在进行中。

发展 IGI 和 IPI

高度优先的任务是将 IGI 和 IPI 精细到点，以精确地反映每一个基因和每一个可变剪接形式。这个宏伟的目标还需要几个步骤。

人类基因组序列的完成对这项工作有很大的帮助，但是从 21 号和 22 号染色体上获取的经验表明，仅有序列信息还不足以完成全部基因的识别。与演化距离适当的物种进行跨物种序列比对是一个有力的工具。河豚鱼基因组序列的覆盖率对识别潜在的外显子很有价值[292]，这项工作有望在今年从一倍覆盖率达到至少五倍覆盖率。实验室小鼠的基因组序列将为外显子识别提供相当有力的工具，序列相似性有

particularly powerful tool for exon identification, as sequence similarity is expected to identify 95–97% of the exons, as well as a significant number of regulatory domains[436-438].

Another important step is to obtain a comprehensive collection of full-length human cDNAs, both as sequences and as actual clones. The Mammalian Gene Collection project has been underway for a year[18] and expects to produce 10,000–15,000 human full-length cDNAs over the coming year, which will be available without restrictions on use. The Genome Exploration Group of the RIKEN Genomic Sciences Center is similarly developing a collection of cDNA clones from mouse[309], which is a valuable complement because of the availability of tissues from all developmental time points.

Large-scale identification of regulatory regions

The one-dimensional script of the human genome, shared by essentially all cells in all tissues, contains sufficient information to provide for differentiation of hundreds of different cell types, and the ability to respond to a vast array of internal and external influences. Much of this plasticity results from the carefully orchestrated symphony of transcriptional regulation. Although much has been learned about the *cis*-acting regulatory motifs of some specific genes, the regulatory signals for most genes remain uncharacterized. It will also be of considerable interest to study epigenetic modifications such as cytosine methylation on a genome-wide scale, and to determine their biological consequences[446,447]. Towards this end, a pilot Human Epigenome Project has been launched[448,449].

Sequencing of additional large genomes

More generally, comparative genomics allows biologists to peruse evolution's laboratory notebook—to identify conserved functional features and recognize new innovations in specific lineages. Determination of the genome sequence of many organisms is very desirable. Already, projects are underway to sequence the genomes of the mouse, rat, zebrafish and the pufferfishes *T. nigroviridis* and *Takifugu rubripes*. Plans are also under consideration for sequencing additional primates and other organisms that will help define key developments along the vertebrate and nonvertebrate lineages.

To realize the full promise of comparative genomics, however, it needs to become simple and inexpensive to sequence the genome of any organism. Sequencing costs have dropped 100-fold over the last 10 years, corresponding to a roughly twofold decrease every 18 months. This rate is similar to "Moore's law" concerning improvements in semiconductor manufacture.

望识别 95% ~ 97% 的外显子以及大量的调控结构域[436-438]。

另一个重要的工作是获得人类基因组的全长 cDNA 库，既有序列信息，又有实际使用的克隆。哺乳动物基因库计划已经执行了一年[18]，预计将在明年产生 10,000 ~ 15,000 个人类全长 cDNA，并向公众开放使用。同样，日本理化研究所基因组科学中心的基因组探索团队也用小鼠建立了 cDNA 克隆库[309]，该库是极有价值的补充，因为它含有所有发育时间点的组织。

调控区域的大规模识别

人类基因组的一维文本信息在所有组织中为所有细胞共享，为几百种不同细胞类型的分化提供充足的信息，并对大量内部和外部的影响做出反应。这种可塑性多数来自精细安排的转录调控。尽管已经对一些特定基因的顺式作用调控模体做了很多研究，大多数基因的调控信号仍然未知。人们对研究表观遗传学修饰，例如基因组范围内的胞嘧啶甲基化，以及确定它们的生物学效果也有很大的兴趣[446,447]。一个领航性的人类表观基因组计划已经为实现此目标而发起[448,449]。

更多的大基因组测序

通常来说，比较基因组学使得生物学家能够利用演化的实验室数据识别保守的功能特征以及特定谱系中的新序列。我们需要确定大量物种的基因组序列。已经有一些项目在测定小鼠、大鼠、斑马鱼、河豚鱼和红旗东方鲀的基因组。人们也在酝酿测定更多灵长类动物和其他生物的基因组，这将帮助解释脊椎动物与非脊椎动物的关键发育过程。

要建立完善的比较基因组学，就需要简单而廉价地测定生物体的基因组。在过去的 10 年中，测序费用已经下降为最初的 100 分之一，大约每 18 个月下降一半。这一速率与半导体制造业发展的"摩尔定律"相符。

Completing the catalogue of human variation

The human draft genome sequence has already allowed the identification of more than 1.4 million SNPs, comprising a substantial proportion of all common human variation. This program should be extended to obtain a nearly complete catalogue of common variants and to identify the common ancestral haplotypes present in the population. In principle, these genetic tools should make it possible to perform association studies and linkage disequilibrium studies[376] to identify the genes that confer even relatively modest risk for common diseases. Launching such an intense era of human molecular epidemiology will also require major advances in the cost efficiency of genotyping technology, in the collection of carefully phenotyped patient cohorts and in statistical methods for relating large-scale SNP data to disease phenotype.

From sequence to function

The scientific program outlined above focuses on how the genome sequence can be mined for biological information. In addition, the sequence will serve as a foundation for a broad range of functional genomic tools to help biologists to probe function in a more systematic manner. These will need to include improved techniques and databases for the global analysis of: RNA and protein expression, protein localization, protein–protein interactions and chemical inhibition of pathways.

Concluding Thoughts

The Human Genome Project is but the latest increment in a remarkable scientific program whose origins stretch back a hundred years to the rediscovery of Mendel's laws and whose end is nowhere in sight. In a sense, it provides a capstone for efforts in the past century to discover genetic information and a foundation for efforts in the coming century to understand it.

We find it humbling to gaze upon the human sequence now coming into focus. In principle, the string of genetic bits holds long-sought secrets of human development, physiology and medicine. In practice, our ability to transform such information into understanding remains woefully inadequate.

The scientific work will have profound long-term consequences for medicine, leading to the elucidation of the underlying molecular mechanisms of disease and thereby facilitating the design in many cases of rational diagnostics and therapeutics targeted at those mechanisms. But the science is only part of the challenge. We must also involve society at large in the work ahead. We must set realistic expectations that the most important benefits will not be reaped overnight. Moreover, understanding and wisdom will be required to ensure that

674

完成人类变异目录

在人类基因组草图中已经识别出超过 1,400,000 个 SNP，它们构成了人类基因组序列变异的大部分。这一项目应扩展以建立基本包括所有常见变异的目录，并识别出种群中常见的祖先单倍型。原则上，这些遗传学工具应当使我们能够开展相关性研究和连锁不平衡研究[376]以辨别出即便是仅有中度风险的常见疾病基因。要开启这样一个人类分子流行病学的时代，也需要基因分型技术的成本效率、精细分型的患者群体的收集以及将大规模 SNP 数据与疾病分型关联起来的统计学方法等方面的重大进展。

从序列到功能

科学项目勾勒出了以上研究热点，关注如何从基因组数据中挖掘生物学信息。而且，序列数据还是大规模功能基因组学工具的基础，能帮助生物学家以更加系统化的方式验证功能。这些都需要优化技术和数据库，用于如下全局性分析：RNA 和蛋白质表达、蛋白质定位、蛋白质–蛋白质相互作用和通路的化学抑制。

结　　语

人类基因组计划只是一个举世瞩目的科学研究计划中的最新增长点，它的起源可回溯至一百年前孟德尔遗传定律的重新发现，而它的终点则延伸至未知的未来。从某种意义上讲，它是过去一个世纪以来发现遗传学信息科学研究的顶峰，并为下一个世纪解析这些数据提供了基础。

只关注目前聚焦的人类基因组序列令人感到惭愧。理论上，这些遗传信息蕴含着关于人类发育、生理和医学中我们长期探求的秘密。而实际上，我们将这些信息转化理解的能力严重不足。

科学研究工作蕴含着深刻而长远的医学成果，能够阐释疾病的分子机制，也因此方便根据这些机制设计合理的诊断学和治疗学手段。但是，科学只是挑战的一部分。在开展工作之前，我们也必须关注到全社会。我们必须正确面对现实的期待，不能急于求成。而且，我们需要理解和智慧来确保这些好处能够公平合理地普惠大

these benefits are implemented broadly and equitably. To that end, serious attention must be paid to the many ethical, legal and social implications (ELSI) raised by the accelerated pace of genetic discovery. This paper has focused on the scientific achievements of the human genome sequencing efforts. This is not the place to engage in a lengthy discussion of the ELSI issues, which have also been a major research focus of the Human Genome Project, but these issues are of comparable importance and could appropriately fill a paper of equal length.

Finally, it is has not escaped our notice that the more we learn about the human genome, the more there is to explore.

"We shall not cease from exploration. And the end of all our exploring will be to arrive where we started, and know the place for the first time. "—T. S. Eliot[450]

(**409**, 860-921; 2001)

Received 7 December 2000; accepted 9 January 2001.

References:

1. Correns, C. Untersuchungen u ber die Xenien bei Zea mays. *Berichte der Deutsche Botanische Gesellschaft*, **17**, 410-418 (1899).

2. De Vries, H. Sur la loi de disjonction des hybrides. *Comptes Rendue Hebdemodaires, Acad. Sci. Paris* **130**, 845-847 (1900).

3. von Tschermack, E. Uber Künstliche Kreuzung bei Pisum sativum. *Berichte der Deutsche Botanische Gesellschaft*, **18**, 232-239 (1900).

9. Botstein, D., White, R. L., Skolnick, M. & Davis, R. W. Construction of a genetic linkage map in man using restriction fragment length polymorphisms. *Am. J. Hum. Genet.* **32**, 314-331 (1980).

11. Coulson, A., Sulston, J., Brenner, S. & Karn, J. Toward a physical map of the genome of the nematode *Caenorhabditis elegans*. *Proc. Natl Acad. Sci. USA* **83**, 7821-7825 (1986).

18. Strausberg, R. L., Feingold, E. A., Klausner, R. D. & Collins, F. S. The mammalian gene collection. *Science* **286**, 455-457 (1999).

21. Sinsheimer, R. L. The Santa Cruz Workshop-May 1985. *Genomics* **5**, 954-956 (1989).

22. Palca, J. Human genome-Department of Energy on the map. *Nature* **321**, 371 (1986).

23. National Research Council *Mapping and Sequencing the Human Genome* (National Academy Press, Washington DC, 1988).

24. Bishop, J. E. & Waldholz, M. *Genome* (Simon and Schuster, New York, 1990).

25. Kevles, D. J. & Hood, L. (eds) *The Code of Codes: Scientific and Social Issues in the Human Genome Project* (Harvard Univ. Press, Cambridge, Massachusetts, 1992).

26. Cook-Deegan, R. *The Gene Wars: Science, Politics, and the Human Genome* (W. W. Norton & Co., New York, London, 1994).

27. Donis-Keller, H. *et al.* A genetic linkage map of the human genome. *Cell* **51**, 319-337 (1987).

28. Gyapay, G. *et al.* The 1993-94 Genethon human genetic linkage map. *Nature Genet.* **7**, 246-339 (1994).

29. Hudson, T. J. *et al.* An STS-based map of the human genome. *Science* **270**, 1945-1954 (1995).

30. Dietrich, W. F. *et al.* A comprehensive genetic map of the mouse genome. *Nature* **380**, 149-152 (1996).

31. Nusbaum, C. *et al.* A YAC-based physical map of the mouse genome. *Nature Genet.* **22**, 388-393 (1999).

32. Oliver, S. G. *et al.* The complete DNA sequence of yeast chromosome III. *Nature* **357**, 38-46 (1992).

33. Wilson, R. *et al.* 2.2 Mb of contiguous nucleotide sequence from chromosome III of *C. elegans*. *Nature* **368**, 32-38 (1994).

34. Chen, E. Y. *et al.* The human growth hormone locus: nucleotide sequence, biology, and evolution. *Genomics* **4**, 479-497 (1989).

35. McCombie, W. R. *et al.* Expressed genes, Alu repeats and polymorphisms in cosmids sequenced from chromosome 4p16.3. *Nature Genet.* **1**, 348-353 (1992).

36. Martin-Gallardo, A. *et al.* Automated DNA sequencing and analysis of 106 kilobases from human chromosome 19q13.3. *Nature Genet.* **1**, 34-39 (1992).

37. Edwards, A. *et al.* Automated DNA sequencing of the human HPRT locus. *Genomics* **6**, 593-608 (1990).

42. Marshall, E. A second private genome project. *Science* **281**, 1121 (1998).

43. Marshall, E. NIH to produce a `working draft' of the genome by 2001. *Science* **281**, 1774-1775 (1998).

44. Pennisi, E. Academic sequencers challenge Celera in a sprint to the finish. *Science* **283**, 1822-1823 (1999).

45. Bouck, J., Miller, W., Gorrell, J. H., Muzny, D. & Gibbs, R. A. Analysis of the quality and utility of random shotgun sequencing at low redundancies. *Genome Res.* **8**, 1074-1084 (1998).

46. Collins, F. S. *et al.* New goals for the U. S. Human Genome Project: 1998-2003. *Science* **282**, 682-689 (1998).

众。为此，我们应当严肃地关注加速发展的遗传学发现导致的伦理、法律和社会问题（ELSI）。本文聚焦于人类基因组测序产生的科学成就。这里并未就 ELSI 问题进行长篇大论，但这些问题已经成为人类基因组研究的主要热点之一，这些问题也同等重要，可以写一篇同样长的论文。

最终我们意识到，对人类基因组了解得越多，就有越多未知需要去探索。

"我们不会停止求索，万般求索终将抵达最初的起点，此时才把这个地方看个透彻。"——艾略特[450]

（王海纳 任奕 翻译；于军 审稿）

47. Sanger, F. & Coulson, A. R. A rapid method for determining sequences in DNA by primed synthesis with DNA polymerase. *J. Mol. Biol.* **94**, 441-448 (1975).

48. Maxam, A. M. & Gilbert, W. A new method for sequencing DNA. *Proc. Natl Acad. Sci. USA* **74**, 560- 564 (1977).

49. Anderson, S. Shotgun DNA sequencing using cloned DNase I-generated fragments. *Nucleic Acids Res.* **9**, 3015-3027 (1981).

50. Gardner, R. C. *et al.* The complete nucleotide sequence of an infectious clone of cauliflower mosaic virus by M13mp7 shotgun sequencing. *Nucleic Acids Res.* **9**, 2871-2888 (1981).

51. Deininger, P. L. Random subcloning of sonicated DNA: application to shotgun DNA sequence analysis. *Anal. Biochem.* **129**, 216-223 (1983).

52. Chissoe, S. L. *et al.* Sequence and analysis of the human ABL gene, the BCR gene, and regions involved in the Philadelphia chromosomal translocation. *Genomics* **27**, 67-82 (1995).

53. Rowen, L., Koop, B. F. & Hood, L. The complete 685-kilobase DNA sequence of the human beta T cell receptor locus. *Science* **272**, 1755-1762 (1996).

54. Koop, B. F. *et al.* Organization, structure, and function of 95 kb of DNA spanning the murine T-cell receptor C alpha/C delta region. *Genomics* **13**, 1209-1230 (1992).

55. Wooster, R. *et al.* Identification of the breast cancer susceptibility gene BRCA2. *Nature* **378**, 789-792 (1995).

60. Venter, J. C. *et al.* Shotgun sequencing of the human genome. *Science* **280**, 1540-1542 (1998).

61. Venter, J. C. *et al.* The sequence of the human genome. *Science* **291**, 1304-1351 (2001).

79. Bentley, D. R. Genomic sequence information should be released immediately and freely in the public domain. *Science* **274**, 533-534 (1996).

80. Guyer, M. Statement on the rapid release of genomic DNA sequence. *Genome Res.* **8**, 413 (1998).

81. Dietrich, W. *et al.* A genetic map of the mouse suitable for typing intraspecific crosses. *Genetics* **131**, 423-447 (1992).

82. Kim, U. J. *et al.* Construction and characterization of a human bacterial artificial chromosome library. *Genomics* **34**, 213-218 (1996).

83. Osoegawa, K. *et al.* Bacterial artificial chromosome libraries for mouse sequencing and functional analysis. *Genome Res.* **10**, 116-128 (2000).

84. Marra, M. A. *et al.* High throughput fingerprint analysis of large-insert clones. *Genome Res.* **7**, 1072- 1084 (1997).

85. Marra, M. *et al.* A map for sequence analysis of the *Arabidopsis thaliana* genome. *Nature Genet.* **22**, 265-270 (1999).

86. The International Human Genome Mapping Consortium. A physical map of the human genome. *Nature* **409**, 934-941 (2001).

87. Zhao, S. *et al.* Human BAC ends quality assessment and sequence analyses. *Genomics* **63**, 321-332 (2000).

88. Mahairas, G. G. *et al.* Sequence-tagged connectors: A sequence approach to mapping and scanning the human genome. *Proc. Natl Acad. Sci. USA* **96**, 9739-9744 (1999).

93. Hattori, M. *et al.* The DNA sequence of human chromosome 21. *Nature* **405**, 311-319 (2000).

94. Dunham, I. *et al.* The DNA sequence of human chromosome 22. *Nature* **402**, 489-495 (1999).

110. Pruit, K. D. & Maglott, D. R. RefSeq and LocusLink: NCBI gene-centered resources. *Nucleic Acids Res.* **29**, 137-140 (2001).

112. Hurst, L. D. & Eyre-Walker, A. Evolutionary genomics: reading the bands. *Bioessays* **22**, 105-107 (2000).

113. Saccone, S. *et al.* Correlations between isochores and chromosomal bands in the human genome. *Proc. Natl Acad. Sci. USA* **90**, 11929-11933 (1993).

114. Zoubak, S., Clay, O. & Bernardi, G. The gene distribution of the human genome. *Gene* **174**, 95-102 (1996).

115. Gardiner, K. Base composition and gene distribution: critical patterns in mammalian genome organization. *Trends Genet.* **12**, 519-524 (1996).

116. Duret, L., Mouchiroud, D. & Gautier, C. Statistical analysis of vertebrate sequences reveals that long genes are scarce in GC-rich isochores. *J. Mol. Evol.* **40**, 308-317 (1995).

117. Saccone, S., De Sario, A., Della Valle, G. & Bernardi, G. The highest gene concentrations in the human genome are in telomeric bands of metaphase chromosomes. *Proc. Natl Acad. Sci. USA* **89**, 4913-4917 (1992).

122. Bird, A., Taggart, M., Frommer, M., Miller, O. J. & Macleod, D. A fraction of the mouse genome that is derived from islands of nonmethylated, CpG-rich DNA. *Cell* **40**, 91-99 (1985).

123. Bird, A. P. CpG islands as gene markers in the vertebrate nucleus. *Trends Genet.* **3**, 342-347 (1987).

124. Chan, M. F., Liang, G. & Jones, P. A. Relationship between transcription and DNA methylation. *Curr. Top. Microbiol. Immunol.* **249**, 75-86 (2000).

125. Holliday, R. & Pugh, J. E. DNA modification mechanisms and gene activity during development. *Science* **187**, 226-232 (1975).

126. Larsen, F., Gundersen, G., Lopez, R. & Prydz, H. CpG islands as gene markers in the human genome. *Genomics* **13**, 1095-1107 (1992).

127. Tazi, J. & Bird, A. Alternative chromatin structure at CpG islands. *Cell* **60**, 909-920 (1990).

131. Yu, A. Comparison of human genetic and sequence-based physical maps. *Nature* **409**, 951-953 (2001).

132. Kaback, D. B., Guacci, V., Barber, D. & Mahon, J. W. Chromosome size-dependent control of meiotic recombination. *Science* **256**, 228-232 (1992).

133. Riles, L. *et al.* Physical maps of the 6 smallest chromosomes of *Saccharomyces cerevisiae* at a resolution of 2.6-kilobase pairs. *Genetics* **134**, 81-150 (1993).

135. Laurie, D. A. & Hulten, M. A. Further studies on bivalent chiasma frequency in human males with normal karyotypes. *Ann. Hum. Genet.* **49**, 189-201 (1985).

136. Roeder, G. S. Meiotic chromosomes: it takes two to tango. *Genes Dev.* **11**, 2600-2621 (1997).

137. Wu, T.-C. & Lichten, M. Meiosis-induced double-strand break sites determined by yeast chromatin structure. *Science* **263**, 515-518 (1994).

138. Gerton, J. L. *et al.* Global mapping of meiotic recombination hotspots and coldspots in the yeast *Saccharomyces cerevisiae*. *Proc. Natl Acad. Sci. USA* **97**, 11383-11390 (2000).

139. Li, W. -H. *Molecular Evolution* (Sinauer, Sunderland, Massachusetts, 1997).

140. Gregory, T. R. & Hebert, P. D. The modulation of DNA content: proximate causes and ultimate consequences. *Genome Res.* **9**, 317-324 (1999).

141. Hartl, D. L. Molecular melodies in high and low C. *Nature Rev. Genet.* **1**, 145-149 (2000).

142. Smit, A. F. Interspersed repeats and other mementos of transposable elements in mammalian genomes. *Curr. Opin. Genet. Dev.* **9**, 657-663 (1999).

143. Prak, E. L. & Haig, H. K. Jr Mobile elements and the human genome. *Nature Rev. Genet.* **1**, 134-144 (2000).

148. Smit, A. F. The origin of interspersed repeats in the human genome. *Curr. Opin. Genet. Dev.* **6**, 743- 748 (1996).

154. Malik, H. S., Burke, W. D. & Eickbush, T. H. The age and evolution of non-LTR retrotransposable elements. *Mol. Biol. Evol.* **16**, 793-805 (1999).

155. Kordis, D. & Gubensek, F. Bov-B long interspersed repeated DNA (LINE) sequences are present in *Vipera ammodytes* phospholipase A2 genes and in genomes of Viperidae snakes. *Eur. J. Biochem.* **246**, 772-779 (1997).

158. Smit, A. F., Toth, G., Riggs, A. D., & Jurka, J. Ancestral, mammalian-wide subfamilies of LINE-1 repetitive sequences. *J. Mol. Biol.* **246**, 401-417 (1995).

159. Lim, J. K. & Simmons, M. J. Gross chromosome rearrangements mediated by transposable elements In *Drosophila melanogaster*. *Bioessays* **16**, 269-275 (1994).

160. Caceres, M., Ranz, J. M., Barbadilla, A., Long, M. & Ruiz, A. Generation of a widespread Drosophila inversion by a transposable element. *Science* **285**, 415-418 (1999).

161. Gray, Y. H. It takes two transposons to tango: transposable-element-mediated chromosomal rearrangements. *Trends Genet.* **16**, 461-468 (2000).

162. Zhang, J. & Peterson, T. Genome rearrangements by nonlinear transposons in maize. *Genetics* **153**, 1403-1410 (1999).

163. Smit, A. F. Identification of a new, abundant superfamily of mammalian LTR-transposons. *Nucleic Acids Res.* **21**, 1863-1872 (1993).

164. Cordonnier, A., Casella, J. F. & Heidmann, T. Isolation of novel human endogenous retrovirus-like elements with foamy virus-related pol sequence. *J. Virol.* **69**, 5890-5897 (1995).

170. Kazazian, H. H. Jr & Moran, J. V. The impact of L1 retrotransposons on the human genome. *Nature Genet.* **19**, 19-24 (1998).

171. Malik, H. S. & Eickbush, T. H. NeSL-1, an ancient lineage of site-specific non-LTR retrotransposons from *Caenorhabditis elegans*. *Genetics* **154**, 193-203 (2000).

172. Casavant, N. C. *et al.* The end of the LINE?: lack of recent L1 activity in a group of South American rodents. *Genetics* **154**, 1809-1817 (2000).

173. Meunier-Rotival, M., Soriano, P., Cuny, G., Strauss, F. & Bernardi, G. Sequence organization and genomic distribution of the major family of interspersed repeats of mouse DNA. *Proc. Natl Acad. Sci. USA* **79**, 355-359 (1982).

174. Soriano, P., Meunier-Rotival, M. & Bernardi, G. The distribution of interspersed repeats is nonuniform and conserved in the mouse and human genomes. *Proc. Natl Acad. Sci. USA* **80**, 1816- 1820 (1983).

175. Goldman, M. A., Holmquist, G. P., Gray, M. C., Caston, L. A. & Nag, A. Replication timing of genes and middle repetitive sequences. *Science* **224**, 686-692 (1984).

176. Manuelidis, L. & Ward, D. C. Chromosomal and nuclear distribution of the *Hind*III 1.9-kb human DNA repeat segment. *Chromosoma* **91**, 28-38 (1984).

178. Jurka, J. Sequence patterns indicate an enzymatic involvement in integration of mammalian retroposons. *Proc. Natl Acad. Sci. USA* **94**, 1872-1877 (1997).

192. Bohossian, H. B., Skaletsky, H. & Page, D. C. Unexpectedly similar rates of nucleotide substitution found in male and female hominids. *Nature* **406**, 622-625 (2000).

215. Ji, Y., Eichler, E. E., Schwartz, S. & Nicholls, R. D. Structure of chromosomal duplicons and their role in mediating human genomic disorders. *Genome Res.* **10**, 597-610 (2000).

216. Eichler, E. E. Masquerading repeats: paralogous pitfalls of the human genome. *Genome Res.* **8**, 758- 762 (1998).

217. Mazzarella, R. & D. Schlessinger, D. Pathological consequences of sequence duplications in the human genome. *Genome Res.* **8**, 1007-1021 (1998).

238. Shaikh, T. H. *et al.* Chromosome 22-specific low copy repeats and the 22q11.2 deletion syndrome: genomic organization and deletion endpoint analysis. *Hum. Mol. Genet.* **9**, 489-501 (2000).

239. Francke, U. Williams-Beuren syndrome: genes and mechanisms. *Hum. Mol. Genet.* **8**, 1947-1954 (1999).

243. Lander, E. S. The new genomics: Global views of biology. *Science* **274**, 536-539 (1996).

244. Eddy, S. R. Noncoding RNA genes. *Curr. Op. Genet. Dev.* **9**, 695-699 (1999).

245. Ban, N., Nissen, P., Hansen, J., Moore, P. B. & Steitz, T. A. The complete atomic structure of the large ribosomal subunit at 2.4 angstrom resolution. *Science* **289**, 905-920 (2000).

246. Nissen, P., Hansen, J., Ban, N., Moore, P. B. & Steitz, T. A. The structural basis of ribosome activity in peptide bond synthesis. *Science* **289**, 920-930 (2000).

247. Weinstein, L. B. & Steitz, J. A. Guided tours: from precursor snoRNA to functional snoRNP. *Curr. Opin. Cell Biol.* **11**, 378-384 (1999).

248. Bachellerie, J.-P. & Cavaille, J. in *Modification and Editing of RNA* (ed. Benne, H. G. a. R.) 255-272 (ASM, Washington DC, 1998).

253. Sprinzl, M., Horn, C., Brown, M., Ioudovitch, A. & Steinberg, S. Compilation of tRNA sequences and sequences of tRNA genes. *Nucleic Acids Res.* **26**, 148-153 (1998).

254. Long, E. O. & Dawid, I. B. Repeated genes in eukaryotes. *Annu. Rev. Biochem.* **49**, 727-764 (1980).

262. Buckland, R. A. A primate transfer-tRNA gene cluster and the evolution of human chromosome 1. *Cytogenet. Cell Genet.* **61**, 1-4 (1992).

263 Gonos, E. S. & Goddard, J. P. Human tRNA-Glu genes: their copy number and organization. *FEBS Lett.* **276**, 138-142 (1990).

286. Dunham, I. The gene guessing game. *Yeast* **17**, 218-224 (2000).

292. Roest Crollius, H. *et al.* Estimate of human gene number provided by genome-wide analysis using *Tetraodon nigroviridis* DNA sequence. *Nature Genet.* **25**, 235-238 (2000).

309. The RIKEN Genome Exploration Research Group Phase II Team and the FANTOM Consortium. Functional annotation of a full-length mouse cDNA collection. *Nature* **409**, 685-690 (2001).

311. Janin, J. & Chothia, C. Domains in proteins: definitions, location, and structural principles. *Methods Enzymol.* **115**, 420-430 (1985).

312. Ponting, C. P., Schultz, J., Copley, R. R., Andrade, M. A. & Bork, P. Evolution of domain families. *Adv. Protein Chem.* **54**, 185-244 (2000).

313. Doolittle, R. F. The multiplicity of domains in proteins. *Annu. Rev. Biochem.* **64**, 287-314 (1995).

314. Bateman, A. & Birney, E. Searching databases to find protein domain organization. *Adv. Protein Chem.* **54**, 137-157 (2000).

315. Futreal, P. A. *et al.* Cancer and genomics. *Nature* **409**, 850-852 (2001).

679

316. Nestler, E. J. & Landsman, D. Learning about addiction from the human draft genome. *Nature* **409**, 834-835 (2001).

317. Tupler, R., Perini, G. & Green, M. R. Expressing the human genome. *Nature* **409**, 832-835 (2001).

318. Fahrer, A. M., Bazan, J. F., Papathanasiou, P., Nelms, K. A. & Goodnow, C. C. A genomic view of immunology. *Nature* **409**, 836-838 (2001).

319. Li, W. -H., Gu, Z., Wang, H. & Nekrutenko, A. Evolutionary analyses of the human genome. *Nature* **409**, 847-849 (2001).

320. Bock, J. B., Matern, H. T., Peden, A. A. & Scheller, R. H. A genomic perspective on membrane compartment organization. *Nature* **409**, 839-841 (2001).

321. Pollard, T. D. Genomics, the cytoskeleton and motility. *Nature* **409**, 842-843 (2001).

322. Murray, A. W. & Marks, D. Can sequencing shed light on cell cycling? *Nature* **409**, 844-846 (2001).

323. Clayton, J. D., Kyriacou, C. P. & Reppert, S. M. Keeping time with the human genome. *Nature* **409**, 829-831 (2001).

339. Schultz, J., Copley, R. R., Doerks, T., Ponting, C. P. & Bork, P. SMART: a web-based tool for the study of genetically mobile domains. *Nucleic Acids Res.* **28**, 231-234 (2000).

376. Lander, E. S. & Schork, N. J. Genetic dissection of complex traits. *Science* **265**, 2037-2048 (1994).

388. Collins, F. S. Positional cloning moves from perdition to traditional. *Nature Genet.* **9**, 347-350 (1995).

389. Nagamine, K. *et al.* Positional cloning of the APECED gene. *Nature Genet.* **17**, 393-398 (1997).

390. Reuber, B. E. *et al.* Mutations in PEX1 are the most common cause of peroxisome biogenesis disorders. *Nature Genet.* **17**, 445-448 (1997).

391. Portsteffen, H. *et al.* Human PEX1 is mutated in complementation group 1 of the peroxisome biogenesis disorders. *Nature Genet.* **17**, 449-452 (1997).

392. Everett, L. A. *et al.* Pendred syndrome is caused by mutations in a putative sulphate transporter gene (PDS). *Nature Genet.* **17**, 411-422 (1997).

393. Coffey, A. J. *et al.* Host response to EBV infection in X-linked lymphoproliferative disease results from mutations in an SH2-domain encoding gene. *Nature Genet.* **20**, 129-135 (1998).

394. Van Laer, L. *et al.* Nonsyndromic hearing impairment is associated with a mutation in DFNA5. *Nature Genet.* **20**, 194-197 (1998).

395. Sakuntabhai, A. *et al.* Mutations in ATP2A2, encoding a Ca2+pump, cause Darier disease. *Nature Genet.* **21**, 271-277 (1999).

396. Gedeon, A. K. *et al.* Identification of the gene (SEDL) causing X-linked spondyloepiphyseal dysplasia tarda. *Nature Genet.* **22**, 400-404 (1999).

397. Hurvitz, J. R. *et al.* Mutations in the CCN gene family member WISP3 cause progressive pseudorheumatoid dysplasia. *Nature Genet.* **23**, 94-98 (1999).

398. Laberge-le Couteulx, S. *et al.* Truncating mutations in CCM1, encoding KRIT1, cause hereditary cavernous angiomas. *Nature Genet.* **23**, 189-193 (1999).

399. Sahoo, T. *et al.* Mutations in the gene encoding KRIT1, a Krev-1/rap1a binding protein, cause cerebral cavernous malformations (CCM1). *Hum. Mol. Genet.* **8**, 2325-2333 (1999).

400. McGuirt, W. T. *et al.* Mutations in COL11A2 cause non-syndromic hearing loss (DFNA13). *Nature Genet.* **23**, 413-419 (1999).

401. Moreira, E. S. *et al.* Limb-girdle muscular dystrophy type 2G is caused by mutations in the gene encoding the sarcomeric protein telethonin. *Nature Genet.* **24**, 163-166 (2000).

402. Ruiz-Perez, V. L. *et al.* Mutations in a new gene in Ellis-van Creveld syndrome and Weyers acrodental dysostosis. *Nature Genet.* **24**, 283-286 (2000).

403. Kaplan, J. M. *et al.* Mutations in ACTN4, encoding alpha-actinin-4, cause familial focal segmental glomerulosclerosis. *Nature Genet.* **24**, 251-256 (2000).

404. Escayg, A. *et al.* Mutations of SCN1A, encoding a neuronal sodium channel, in two families with GEFS+2. *Nature Genet.* **24**, 343-345 (2000).

405. Sacksteder, K. A. *et al.* Identification of the alpha-aminoadipic semialdehyde synthase gene, which is defective in familial hyperlysinemia. *Am. J. Hum. Genet.* **66**, 1736-1743 (2000).

406. Kalaydjieva, L. *et al.* N-myc downstream-regulated gene 1 is mutated in hereditary motor and sensory neuropathy-Lom. *Am. J. Hum. Genet.* **67**, 47-58 (2000).

407. Sundin, O. H. *et al.* Genetic basis of total colourblindness among the Pingelapese islanders. *Nature Genet.* **25**, 289-293 (2000).

408. Kohl, S. *et al.* Mutations in the CNGB3 gene encoding the beta-subunit of the cone photoreceptor cGMP-gated channel are responsible for achromatopsia (ACHM3) linked to chromosome 8q21. *Hum. Mol. Genet.* **9**, 2107-2116 (2000).

409. Avela, K. *et al.* Gene encoding a new RING-B-box-coiled-coil protein is mutated in mulibrey nanism. *Nature Genet.* **25**, 298-301 (2000).

410. Verpy, E. *et al.* A defect in harmonin, a PDZ domain-containing protein expressed in the inner ear sensory hair cells, underlies usher syndrome type 1C. *Nature Genet.* **26**, 51-55 (2000).

411. Bitner-Glindzicz, M. *et al.* A recessive contiguous gene deletion causing infantile hyperinsulinism, enteropathy and deafness identifies the usher type 1C gene. *Nature Genet.* **26**, 56-60 (2000).

412. The May-Hegglin/Fechtner Syndrome Consortium. Mutations in MYH9 result in the May-Hegglin anomaly, and Fechtner and Sebastian syndromes. *Nature Genet.* **26**, 103-105 (2000).

413. Kelley, M. J., Jawien, W., Ortel, T. L. & Korczak, J. F. Mutation of MYH9, encoding non-muscle myosin heavy chain A, in May-Hegglin anomaly. *Nature Genet.* **26**, 106-108 (2000).

414. Kirschner, L. S. *et al.* Mutations of the gene encoding the protein kinase A type I-a regulatory subunit in patients with the Carney complex. *Nature Genet.* **26**, 89-92 (2000).

415. Lalwani, A. K. *et al.* Human nonsyndromic hereditary deafness DFNA17 is due to a mutation in non-muscle myosin MYH9. *Am. J. Hum. Genet.* **67**, 1121-1128 (2000).

416. Matsuura, T. *et al.* Large expansion of the ATTCT pentanucleotide repeat in spinocerebellar ataxia type 10. *Nature Genet.* **26**, 191-194 (2000).

417. Delettre, C. *et al.* Nuclear gene OPA1, encoding a mitochondrial dynamin-related protein, is mutated in dominant optic atrophy. *Nature Genet.* **26**, 207-210 (2000).

418. Pusch, C. M. *et al.* The complete form of X-linked congenital stationary night blindness is caused by mutations in a gene encoding a leucine-rich repeat protein. *Nature Genet.* **26**, 324-327 (2000).

419. The ADHR Consortium. Autosomal dominant hypophosphataemic rickets is associated with mutations in FGF23. *Nature Genet.* **26**, 345-348 (2000).

420. Bomont, P. *et al.* The gene encoding gigaxonin, a new member of the cytoskeletal BTB/kelch repeat family, is mutated in giant axonal neuropathy. *Nature Genet.* **26**, 370-374 (2000).

421. Tullio-Pelet, A. *et al.* Mutant WD-repeat protein in triple-A syndrome. *Nature Genet.* **26**, 332-335 (2000).

422. Nicole, S. *et al.* Perlecan, the major proteoglycan of basement membranes, is altered in patients with Schwartz-Jampel syndrome (chondrodystrophic myotonia). *Nature Genet.* **26**, 480-483 (2000).

426. Drews, J. Research & development. Basic science and pharmaceutical innovation. *Nature Biotechnol.* **17**, 406 (1999).

427. Drews, J. Drug discovery: a historical perspective. *Science* **287**, 1960-1964 (2000).

433. Matsunami, H., Montmayeur, J. P. & Buck, L. B. A family of candidate taste receptors in human and mouse. *Nature* **404**, 601-604 (2000).

434. Adler, E. *et al.* A novel family of mammalian taste receptors. *Cell* **100**, 693-702 (2000).

435. Chandrashekar, J. *et al.* T2Rs function as bitter taste receptors. *Cell* **100**, 703-711 (2000).

436. Hardison, R. C. Conserved non-coding sequences are reliable guides to regulatory elements. *Trends Genet.* **16**, 369-372 (2000).

437. Onyango, P. *et al.* Sequence and comparative analysis of the mouse 1-megabase region orthologous to the human 11p15 imprinted domain. *Genome Res.* **10**, 1697-1710 (2000).

438. Bouck, J. B., Metzker, M. L. & Gibbs, R. A. Shotgun sample sequence comparisons between mouse and human genomes. *Nature Genet.* **25**, 31-33 (2000).

446. Feil, R. & Khosla, S. Genomic imprinting in mammals: an interplay between chromatin and DNA methylation? *Trends Genet.* **15**, 431-434 (1999).

447. Robertson, K. D. & Wolffe, A. P. DNA methylation in health and disease. *Nature Rev. Genet.* **1**, 11-19 (2000).

448. Beck, S., Olek, A. & Walter, J. From genomics to epigenomics: a loftier view of life. *Nature Biotechnol.* **17**, 1144-1144 (1999).

449. Hagmann, M. Mapping a subtext in our genetic book. *Science* **288**, 945-946 (2000).

450. Eliot, T. S. in *T. S. Eliot. Collected Poems 1909-1962* (Harcourt Brace, New York, 1963).

Supplementary Information is available on *Nature*'s World-Wide Web site (http://www.nature.com) or as paper copy from the London editorial office of *Nature*.

Correspondence and requests for materials should be addressed to E. S. Lander (e-mail: lander@genome.wi.mit.edu), R. H. Waterston (e-mail: bwaterst@watson.wustl.edu), J. Sulston (e-mail: jes@sanger.ac.uk) or F. S. Collins (e-mail: fc23a@nih.gov).

DNA sequence databases

GenBank, National Center for Biotechnology Information, National Library of Medicine, National Institutes of Health, Bldg. 38A, 8600 Rockville Pike, Bethesda, Maryland 20894, USA

EMBL, European Bioinformatics Institute, Wellcome Trust Genome Campus, Hinxton, Cambridge CB10 1SD, UK

DNA Data Bank of Japan, Center for Information Biology, National Institute of Genetics, 1111 Yata, Mishima-shi, Shizuoka-ken 411-8540, Japan

A Map of Human Genome Sequence Variation Containing 1.42 Million Single Nucleotide Polymorphisms

The International SNP Map Working Group

Editor's Note

This study reveals the first detailed map of the 1.42 million single nucleotide polymorphisms (SNPs) found across the human genome. The data, amassed by The International SNP Map Working Group, provide interesting first glimpses into the patterns of variation found across the human genome. Although the vast majority of human DNA sequences are the same, single base pair changes in the code (SNPs) can affect our response to disease, drugs and environmental factors such as toxins and viruses. This makes SNPs a valuable tool for biomedical research and drug design. SNPs also change little between generations, making them useful for studies of evolutionary history. This map integrates all of the publicly available SNPs with known genes and other genomic features, and was made possible by and was published with the first draft sequence of the human genome.

We describe a map of 1.42 million single nucleotide polymorphisms (SNPs) distributed throughout the human genome, providing an average density on available sequence of one SNP every 1.9 kilobases. These SNPs were primarily discovered by two projects: The SNP Consortium and the analysis of clone overlaps by the International Human Genome Sequencing Consortium. The map integrates all publicly available SNPs with described genes and other genomic features. We estimate that 60,000 SNPs fall within exon (coding and untranslated regions), and 85% of exons are within 5 kb of the nearest SNP. Nucleotide diversity varies greatly across the genome, in a manner broadly consistent with a standard population genetic model of human history. This high-density SNP map provides a public resource for defining haplotype variation across the genome, and should help to identify biomedically important genes for diagnosis and therapy.

INHERITED differences in DNA sequence contribute to phenotypic variation, influencing an individual's anthropometric characteristics, risk of disease and response to the environment. A central goal of genetics is to pinpoint the DNA variants that contribute most significantly to population variation in each trait. Genome-wide linkage analysis and positional cloning have identified hundreds of genes for human diseases[1] (http://ncbi.nlm.nih.gov/OMIM), but nearly all are rare conditions in which mutation of a single gene is necessary and sufficient to cause disease. For common diseases, genome-wide linkage studies have had limited success, consistent with a more complex genetic

682

包含 142 万个单核苷酸多态性的
人类基因组序列变异图谱

国际 SNP 图谱研究组

编者按

这项研究首次展示了人类基因组中已发现的 142 万个单核苷酸多态性(SNP)的详细图谱。这份由国际 SNP 图谱研究组积累的数据让人们首次目睹了人类基因组的变异图谱。尽管绝大多数的人类 DNA 序列是一致的,但是序列中单个碱基对的改变(SNP)就能够影响我们对疾病、药物和环境因素(例如毒素和病毒)的反应。这使得SNP 成为生物医药研究和药物设计行业的有力工具。SNP 在不同世代中变化不大,这对于进化史的研究是很有帮助的。这份图谱整合了所有可公开获取的 SNP 与已知基因及其他基因组特征,依据人类基因组第一个草图序列得以完成,并与后者共同发表。

我们描绘了一张遍及人类基因组的包含 142 万个单核苷酸多态性(SNP)的图谱,这张图谱给我们提供了可用序列中 SNP 的平均密度:每 1,900 个碱基一个 SNP。这些 SNP 最初是在两个项目中发现的:SNP 国际联合会和国际人类基因组测序协会的克隆重叠分析。该图谱整合了所有可公开获取的 SNP 与已知基因和其他基因组特征。我们估计有 60,000 个 SNP 在外显子(编码和非翻译区)中,85% 的外显子距离最近的 SNP 5kb 以内。不同基因组间的核苷酸多样性变化很大,但变化方式大致符合人类历史标准群体遗传学模型。高密度的 SNP 图谱可以为定义整个基因组的单体型变异提供公开资源,帮助识别生物医学相关的重要基因,为诊断与治疗提供帮助。

可遗传的 DNA 序列差异导致了表型上的变异,这影响了个体的人体测量特征、疾病的风险和环境应激反应。遗传学的一个核心目标是精确找到在各个性状中对种群变异贡献最显著的 DNA 变异。全基因组连锁分析和定位克隆已经鉴定出人类疾病的数百个基因[1](http://ncbi.nlm.nih.gov/OMIM),但单个基因的突变足以引起疾病的情况十分罕见。对于常见疾病,全基因连锁研究的成果有限,这与基因结构的复

architecture. If each locus contributes modestly to disease aetiology, more powerful methods will be required.

One promising approach is systematically to explore the limited set of common gene variants for association with disease[2-4]. In the human population most variant sites are rare, but the small number of common polymorphisms explain the bulk of heterozygosity[3] (see also refs 5–11). Moreover, human genetic diversity appears to be limited not only at the level of individual polymorphisms, but also in the specific combinations of alleles (haplotypes) observed at closely linked sites[8,11-14]. As these common variants are responsible for most heterozygosity in the population, it will be important to assess their potential impact on phenotypic trait variation.

If limited haplotype diversity is general, it should be practical to define common haplotypes using a dense set of polymorphic markers, and to evaluate each haplotype for association with disease. Such haplotype-based association studies offer a significant advantage: genomic regions can be tested for association without requiring the discovery of the functional variants. The required density of markers will depend on the complexity of the local haplotype structure, and the distance over which these haplotypes extend, neither of which is yet well defined.

Current estimates (refs 13–17) indicate that a very dense marker map (30,000–1,000,000 variants) would be required to perform haplotype-based association studies. Most human sequence variation is attributable to SNPs, with the rest attributable to insertions or deletions of one or more bases, repeat length polymorphisms and rearrangements. SNPs occur (on average) every 1,000–2,000 bases when two human chromosomes are compared[5,6,9,18-20], and are thus present at sufficient density for comprehensive haplotype analysis. SNPs are binary, and thus well suited to automated, high-throughput genotyping. Finally, in contrast to more mutable markers, such as microsatellites[21], SNPs have a low rate of recurrent mutation, making them stable indicators of human history. We have constructed a SNP map of the human genome with sufficient density to study human haplotype structure, enabling future study of human medical and population genetics.

Identification and Characteristics of SNPs

The map contains all SNPs that were publicly available in November 2000. Over 95% were discovered by The SNP Consortium (TSC) and the public Human Genome Project (HGP). TSC contributed 1,023,950 candidate SNPs (http://snp.cshl.org) identified by shotgun sequencing of genomic fragments drawn from a complete (45% of data) or reduced (55% of data) representation of the human genome[18,22]. Individual contributions were: Whitehead Institute, 589,209 SNPs from 2.57 million (M) passing reads; Sanger Centre, 262,279 SNPs from 1.16M passing reads; Washington University, 172,462 SNPs from 1.69M passing reads. TSC SNPs were discovered using a publicly available panel of 24 ethnically diverse individuals[23]. Reads were aligned to one another and to the available genome

杂性有关。如果每个突变位点的发现对疾病病因学贡献不大，我们就需要更有力的方法。

一个可行的方法是系统地探索出一组数目有限的常规基因变异来研究相关疾病[2-4]。人类群体中大多数变异位点是罕见的，但是少数的常见多态性解释了大量杂合性的存在[3]（也可见于参考文献 5~11）。此外，对人类基因多样性的研究不该受限于个体多态性水平，而且要研究相邻连锁位点观察到的等位基因（单体型）的特异组合[8,11-14]。正因为这些常见的变异是引起种群中大多数杂合性的原因，因此，评价它们在表型性状变异中的潜在影响就十分重要。

如果有限的单体型多态性普遍存在，那么用多态性标记物的密集组合来定义常见的单体型，评估疾病相关的每个单体型，应该会变得切实可行。这些基于单体型的相关研究提供给我们一个重要的优势：无需发现功能型变异，基因组区域可用来测试相关性。标记所需的密度将取决于局部单体型结构的复杂程度，以及这些单体型延伸的范围，这两者均未被很好地界定。

现有的估计（参考文献 13~17）指出，做基于单体型的关联研究需要一个十分密集的标记图谱（30,000~1,000,000 个突变体）。大部分人类序列变异可归因于 SNP，而剩下的则归因于一个或更多碱基的插入或缺失、重复长度多态性和重排。当将两个人类染色体进行比较时，（平均）每 1,000~2,000 个碱基出现一个 SNP[5,6,9,18-20]。对于复杂的单体型分析，SNP 的这种出现频率足够了。SNP 是二等位型的，因此很适合自动化的、高通量基因分型。最后，与更可变的标记如微卫星[21]相比，SNP 有较低的频发突变，这使得它们可以作为人类历史的稳定指标。我们已经绘制了一个具有足够密度的人类基因组 SNP 图谱，用来研究人类单体型结构，从而使未来人类医学和群体遗传学的研究成为可能。

SNP 的鉴定及特性

这张图谱包含 2000 年 11 月公开可获取的所有 SNP。超过 95% 的 SNP 是由 SNP 国际联合会（TSC）和公共人类基因组计划（HGP）发现的。TSC 贡献了 1,023,950 个候选 SNP（http://snp.cshl.org），它们是通过鸟枪测序法测定来自完整的（45% 的数据）或者缩短的（55% 的数据）人类基因组碎片确定的[18,22]。各机构的贡献如下：怀特黑德研究所，从 257 万合格读长中得到 589,209 个 SNP 位点；桑格中心，从 116 万合格读长中得到 262,279 个 SNP 位点；华盛顿大学，从 169 万合格读长中得到 172,462 个 SNP 位点。TSC 的 SNP 位点是从 24 个种族不同个体可公开获取的基因组检测

sequence, followed by detection of single base differences using one of two validated algorithms: Polybayes[24] and the neighbourhood quality standard (NQS[18,22]).

An additional 971,077 candidate SNPs were identified as sequence differences in regions of overlap between large-insert clones (bacterial artificial chromosomes (BACs) or P1-derived artificial chromosomes (PACs)) sequenced by the HGP. Two groups (NCBI/Washington University (556,694 SNPs): G.B., P.Y.K. and S.S.; and The Sanger Centre (630,147 SNPs): J.C.M. and D.R.B.) independently analysed these overlaps using the two detection algorithms. This approach contributes dense clusters of SNPs throughout the genome. The remaining 5% of SNPs were discovered in gene-based studies, either by automated detection of single base differences in clusters of overlapping expressed sequence tags[24-28] or by targeted resequencing efforts (see ftp://ncbi.nlm.nih.gov/snp/human/submit_format/*/*publicat.rep.gz).

It is critical that candidate SNPs have a high likelihood of representing true polymorphisms when examined in population studies. Although many methods and contributors are represented on the map (see above), most SNPs (> 95%) were contributed by two large-scale efforts that uniformly applied automated methods. Random samples of these SNPs have been evaluated by confirmation in the original DNA samples (where possible) to rule out false positives, and in independent population samples to determine allele frequency. The TSC centres and two outside laboratories (Orchid and Cold Spring Harbor Laboratory) successfully genotyped 1,585 TSC SNPs in the 24 DNA samples used for discovery (http://snp.cshl.org); having surveyed all chromosomes in which each SNP could have been identified, any non-polymorphic candidates must represent false positives. In these tests, 1,500 SNPs (95%) were polymorphic, 67 (4%) non-polymorphic (false positives) and 18 (1%) uniformly heterozygous (previously unrecognized repeats). These high validation rates were observed separately for subsets of SNPs discovered by reduced representation shotgun and genomic alignment, and for subsets identified with Polybayes and the NQS. Thus, these algorithms appear to generate few false positive SNPs. The small number (1%) of uniformly "heterozygous" candidate SNPs show that the methods also exclude nearly all low-copy repeats.

The allele frequencies of a set of SNPs have been evaluated[29] in independent populations using pooled resequencing. Samples of TSC ($n = 502$) and overlap SNPs ($n = 774$) were studied in population samples of European, African American and Chinese descent, revealing 82% to be polymorphic in at least one ethnic group at frequencies above the detection threshold of pooled resequencing ($\sim 10\%$). The remaining 18% presumably represent SNPs with a frequency less than 10% in the populations surveyed and false positives. Furthermore, 77% of SNPs had a minor allele frequency of more than 20% in at least one population, and 27% had an allele frequency higher than 20% in all three ethnic groups. TSC and overlap SNPs had similar distributions across the populations, showing that they are comparable in quality and frequency. The high proportion of SNPs with significant population frequency is expected after SNP discovery in two or a few chromosomes, given standard assumptions about human population history[18,29,30].

686

区域发现的[23]。序列进行拼接并与已有基因组序列进行比对，接着用 Polybayes[24]
和周边质量标准（NQS[18,22]）两个有效算法之一进行单个碱基差异性的检测。

额外的 971,077 个候选 SNP 是在 HGP 测序得到的大片段插入克隆（细菌人工染色体（BAC）或者 P1 衍生人工染色体（PAC））的重叠区中发现的序列差异。两个机构（美国国家生物技术信息中心/华盛顿大学（556,694 个 SNP 位点）：邦尼、郭沛恩和谢里；桑格中心（630,147 个 SNP 位点）：马利金和本特利）用两种检测算法对这些重叠区域进行单独的分析。这种方法有助于形成遍及基因组的密集的 SNP 簇。剩下 5% 的 SNP 则是在基于基因的研究中发现的，要么是通过成簇的重叠表达序列标签中单碱基差异性的自动检测[74-78]，要么是通过靶向重测序（见于 ftp://ncbi.nlm.nih.gov/snp/human/submit_format/*/*publicat.rep.gz）。

至关重要的是，检测群体研究发现，候选 SNP 极有可能代表真正的多态性。尽管图谱来源于许多方法以及贡献者（如上所述），但大多数 SNP（＞95%）都是由运用同样自动化方法的两大机构发现的。这些 SNP 的随机样本都经过了原始 DNA 样本（可获得的）的评估来排除假阳性，并且在独立群体样本中检测等位基因的频率。TSC 中心和两个外部实验室（兰花和冷泉港实验室）从 24 个用于研究的样本中，成功地对 1,585 个 TSC SNP 位点进行了基因分型（http://snp.cshl.org）；对所有鉴定到 SNP 的染色体都进行了调研，任何非多态性候选位点都被认定为是假阳性。在这些测试中，1,500 个 SNP（95%）是多态性的，67 个 SNP（4%）是非多态性的（假阳性），而 18 个 SNP（1%）是一致的杂合子（先前无法识别的重复）。这些高验证率分别是在简化鸟枪和基因组比对发现的 SNP 子集以及 Polybayes 和 NQS 鉴定的子集中观察到的。因此，这些算法几乎很少出现假阳性的 SNP。少量（1%）一致的"杂合子"候选 SNP 表明这些方法也几乎排除了所有的低拷贝重复序列。

采用混合重测序的方法，对一组 SNP 在独立群体中的等位基因频率进行了评估[29]。在欧洲人、非洲裔美国人和中国人的人群样本中，研究了 TSC（$n = 502$）和重叠 SNP（$n = 774$）的样本，发现在至少一个族群中，82% 的样本具有多态性，其频率高于混合重测序检测阈值（约 10%）。剩下的 18% 可能代表 SNP 低于 10% 的受访群体以及假阳性。此外，在至少一个群体中，77% 的 SNP 等位基因频率超过 20%，在三个种族群体中，27% 的 SNP 等位基因频率高于 20%。TSC 和重叠 SNP 在各群体间有相似的分布，这表明他们在性质和频率上相似。考虑到关于人类群体历史的标准假设，在两个或少数染色体中发现 SNP 后，预计有高比例的 SNP 伴随显著的群体频率[18,29,30]。

Description of the SNP Map

We mapped the sequence flanking each SNP by alignment to the genomic sequence of large-insert clones in GenBank. These alignments were converted into chromosomal coordinates according to the publicly available genome assemblies of July and September 2000 (http://genome.ucsc.edu). Candidate SNPs were included in the final map only if they mapped to a single location in the genome assembly. Integrated displays of SNPs, genes and other features are available at the ENSEMBL (http://www.ensembl.org), NCBI (National Center for Biotechnology Information; http://www.ncbi.nlm.nih.gov), UCSC (University of California at Santa Cruz; http://genome.ucsc.edu) and TSC (http://snp.cshl.org) websites.

The nonredundant SNP total of 1,433,393 is fewer than the sum of individual submissions (2,067,476) because some SNPs (mainly in regions of BAC overlap) were discovered by more than one effort. Of these, 1,419,190 mapped to unique locations in the 2.7 gigabases (Gb) of assembled human genome sequence, providing an average density of one SNP every 1.91 kb. TSC SNPs, which are more evenly distributed than those from clone overlaps, were found on average every 3.05 kb. SNP density (Table 1) is relatively constant across the autosomes. To characterize the distribution of SNPs, we examined 366,192 SNPs that fell within finished sequence. Most of the genome contains SNPs at high density (Fig. 1): 90% of contiguous 20-kb windows contain one or more SNPs, as do 63% of 5-kb windows and 28% of 1-kb windows. Only 4% of genome sequence falls in gaps between SNPs of > 80 kb, and some of these gaps are covered by SNPs that are discovered but not yet mapped owing to gaps in the genome assembly.

To evaluate the density of SNPs in regions within and surrounding genes, we used the September 2000 release of RefSeq[31]. In total, 14,534 SNPs map to within these 7,000 carefully annotated, non-redundant messenger RNAs, equivalent to about two exonic SNPs per gene (coding and untranslated regions). Extrapolating two exonic SNPs per gene to the approximately 30,000 human genes[32], we estimate there to be 60,000 exonic SNPs in this collection. The density of SNPs in exons (one SNP per 1.08 kb; Table 1) is higher than in the genome as a whole, owing to the contribution of efforts targeted to exonic regions.

We also assessed the distribution of SNPs in the genomic locus surrounding each of the RefSeq mRNAs. We assigned the RefSeq exons to their genomic locations, restricting analysis to the 2,960 RefSeq mRNAs mapping onto finished sequence. As we cannot define the extent of the noncoding (regulatory) regions of each gene, we arbitrarily defined each "gene locus" as extending from 10 kb upstream of the start of the first exon to the end of the last exon. By this definition, 93% of gene loci contain at least one SNP, and 98% are within 5 kb of the nearest SNP; also, 59% of gene loci contained five or more SNPs, and 39% ten or more. Of 24,953 exons, 85% were within 5 kb of the nearest SNP. Thus, most exons should be close enough to at least one SNP for haplotype-based association studies, where the functional variant may be some distance from the SNPs used in the study.

SNP 图谱的描述

我们将每个 SNP 的侧翼序列与 GenBank 中的大片段插入克隆基因组序列进行比对。依照 2000 年 7 月和 9 月发布的可公开获取的基因组组装图(http://genome.ucsc.edu),将比对转换成染色体坐标。候选 SNP 只有在基因组组装中被比对到单一位置才能包含在最终图谱中。SNP、基因和其他特征的综合信息可在如下网站获得:ENSEMBL(http://www.ensembl.org)、NCBI(美国国家生物技术中心;http://www.ncbi.nlm.nih.gov)、UCSC(加州大学圣克鲁兹分校;http://genome.ucsc.edu)和 TSC(http://snp.cshl.org)。

非冗余的 SNP 总计有 1,433,393 个,少于各机构提交的总数(2,067,476),这是因为一些 SNP(主要在 BAC 重叠区)被发现的次数多于一次。其中的 1,419,190 个 SNP 在人类基因组 2.7 千兆碱基(Gb)的序列中坐标唯一,平均密度为每 1.91kb 一个 SNP。TSC SNP 比克隆重叠中得到的 SNP 分布更加均匀,其平均密度为每 3.05kb 一个。SNP 密度(表 1)在常染色体间相对恒定。为了描绘 SNP 的分布,我们检查了位于已完成测序区域的 366,192 个 SNP。多数基因组包含高密度 SNP(图 1):90% 的相邻 20 kb 阅读窗内含有一个或多个 SNP,63% 的 5 kb 阅读窗和 28% 的 1 kb 阅读窗同样如此。只有 4% 的基因组序列位于相距大于 80 kb 的两个 SNP 之间的缺口中,有些缺口被已发现的 SNP 覆盖,但由于基因组组装上的缺陷还没有被绘制出来。

为了评估基因区域内和周围的 SNP 密度,我们采用了 2000 年 9 月发布的参考序列[31]。总体说来,14,534 个 SNP 比对到了 7,000 个精确注释、非冗余的信使 RNA 上,这等同于每个基因约有两个外显子 SNP(编码区和非翻译区)。以此推算到大约 30,000 个人类基因中[32],我们估计会有 60,000 个外显子 SNP。外显子中 SNP 的密度(每 1.08 kb 一个 SNP;表 1)比全部 SNP 在基因组中的密度要高,这是由把外显子区作为测序靶标导致的。

我们也估算了每个参考序列 mRNA 周围基因座上 SNP 的分布。我们将参考序列外显子分配到它们的基因组位置,将分析限制在已完成序列图谱比对的 2,960 个 mRNA 参考序列上。由于不能界定每个基因非编码(调控)区的大小,我们人为地定义每个"基因座"延伸范围为从第一个外显子起始位置上游 10 kb 到最后一个外显子。通过此定义,93% 的基因座至少包含一个 SNP,而 98% 的基因座在 5 kb 内就有一个 SNP;此外,59% 的基因座包含 5 个或者更多的 SNP,而 39% 拥有 10 个及以上的 SNP。24,953 个外显子中,85% 的外显子在 5 kb 以内有一个 SNP。因此,多数外显子应该至少靠近一个基于单体型关联研究的 SNP,然而功能性变异可能与研究中使用的 SNP 有一定的距离。

Table 1. SNP distribution by chromosome

Chromosome	Length (bp)	All SNPs		TSC SNPs	
		SNPs	kb per SNP	SNPs	kb per SNP
1	214,066,000	129,931	1.65	75,166	2.85
2	222,889,000	103,664	2.15	76,985	2.90
3	186,938,000	93,140	2.01	63,669	2.94
4	169,035,000	84,426	2.00	65,719	2.57
5	170,954,000	117,882	1.45	63,545	2.69
6	165,022,000	96,317	1.71	53,797	3.07
7	149,414,000	71,752	2.08	42,327	3.53
8	125,148,000	57,834	2.16	42,653	2.93
9	107,440,000	62,013	1.73	43,020	2.50
10	127,894,000	61,298	2.09	42,466	3.01
11	129,193,000	84,663	1.53	47,621	2.71
12	125,198,000	59,245	2.11	38,136	3.28
13	93,711,000	53,093	1.77	35,745	2.62
14	89,344,000	44,112	2.03	29,746	3.00
15	73,467,000	37,814	1.94	26,524	2.77
16	74,037,000	38,735	1.91	23,328	3.17
17	73,367,000	34,621	2.12	19,396	3.78
18	73,078,000	45,135	1.62	27,028	2.70
19	56,044,000	25,676	2.18	11,185	5.01
20	63,317,000	29,478	2.15	17,051	3.71
21	33,824,000	20,916	1.62	9,103	3.72
22	33,786,000	28,410	1.19	11,056	3.06
X	131,245,000	34,842	3.77	20,400	6.43
Y	21,753,000	4,193	5.19	1,784	12.19
RefSeq	15,696,674	14,534	1.08		
Totals	2,710,164,000	1,419,190	1.91	887,450	3.05

Length (bp) is from the public Genome Assembly of 5 September 2000. Density of SNPs on each chromosome is influenced by the amount of available genome sequence included in the Genome Assembly, depth of overlap coverage from TSC reads and clone overlaps, and the underlying heterozygosity (Table 2). Data are presented for the entire dataset (All SNPs) and for those from the SNP consortium (TSC SNPs), as the latter are more evenly spaced than those from clone overlaps.

表 1. SNP 在染色体上的分布

染色体	长度（bp）	所有 SNP		TSC SNP	
		SNP	SNP 间的距离 kb	SNP	SNP 间的距离 kb
1	214,066,000	129,931	1.65	75,166	2.85
2	222,889,000	103,664	2.15	76,985	2.90
3	186,938,000	93,140	2.01	63,669	2.94
4	169,035,000	84,426	2.00	65,719	2.57
5	170,954,000	117,882	1.45	63,545	2.69
6	165,022,000	96,317	1.71	53,797	3.07
7	149,414,000	71,752	2.08	42,327	3.53
8	125,148,000	57,834	2.16	42,653	2.93
9	107,440,000	62,013	1.73	43,020	2.50
10	127,894,000	61,298	2.09	42,466	3.01
11	129,193,000	84,663	1.53	47,621	2.71
12	125,198,000	59,245	2.11	38,136	3.28
13	93,711,000	53,093	1.77	35,745	2.62
14	89,344,000	44,112	2.03	29,746	3.00
15	73,467,000	37,814	1.94	26,524	2.77
16	74,037,000	38,735	1.91	23,328	3.17
17	73,367,000	34,621	2.12	19,396	3.78
18	73,078,000	45,135	1.62	27,028	2.70
19	56,044,000	25,676	2.18	11,185	5.01
20	63,317,000	29,478	2.15	17,051	3.71
21	33,824,000	20,916	1.62	9,103	3.72
22	33,786,000	28,410	1.19	11,056	3.06
X	131,245,000	34,842	3.77	20,400	6.43
Y	21,753,000	4,193	5.19	1,784	12.19
参考序列	15,696,674	14,534	1.08		
总计	2,710,164,000	1,419,190	1.91	887,450	3.05

长度（bp）来自 2000 年 9 月 5 日公布的基因组组装信息。每条染色体上 SNP 的密度受基因组组装收录的可用基因组序列的数量、TSC 读长和克隆重叠的覆盖度的深度，以及潜在的杂合性影响（表 2）。表中数据来自整个数据集（所有 SNP）和 SNP 国际联合会（TSC SNP），后者比来自克隆重叠的 SNP 分布更均匀。

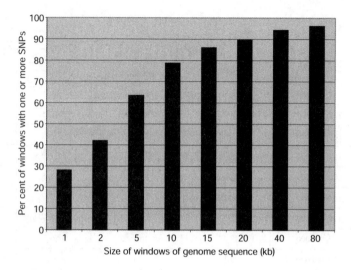

Fig. 1. Distribution of SNP coverage across intervals of finished sequence. Windows of defined size (in chromosome coordinates) were examined for whether they contained one or more SNPs. Analysis was restricted to the 900 Mb of available finished sequence.

The density of SNPs obtained at any given location depends upon the methods of SNP discovery contributing at each position (TSC, BAC overlap or targeted), the availability of genome sequence for SNP discovery and mapping, and the rate of nucleotide diversity. Of these, only nucleotide diversity is a fundamental characteristic of the region and population studied. To chart the landscape of human genome sequence polymorphism, we performed a genome-wide analysis of nucleotide diversity.

Analysis of Nucleotide Diversity

Describing the underlying pattern of nucleotide diversity required a polymorphism survey performed at high density, in a single, defined population sample, and analysed with a uniform set of tools. We reanalysed 4.5M passing sequence reads generated by TSC using genomic alignment using the NQS (see Methods). This set contained 1.2 billion aligned bases and 920,752 heterozygous positions. We measured nucleotide sequence variation using the normalized measure of heterozygosity (π), representing the likelihood that a nucleotide position will be heterozygous when compared across two chromosomes selected randomly from a population. π also estimates the population genetic parameter $\Theta = 4N_e\mu$ in a model in which sites evolve neutrally, with mutation rate μ, in a constant-sized population of effective size N_e. For the human genome, π was 7.51×10^{-4}, or one SNP for every 1,331 bp surveyed in two chromosomes drawn from the NIH diversity panel. This value agrees with smaller surveys of human genome variation[18-20].

We next examined the heterozygosity of individual chromosomes (Table 2). The autosomes were quite similar to one another, with 20 of 22 within 10% of the genome-wide average for autosomes (7.65×10^{-4}). Two had more extreme values: chromosome 21 ($\pi = 5.19 \times 10^{-4}$) and

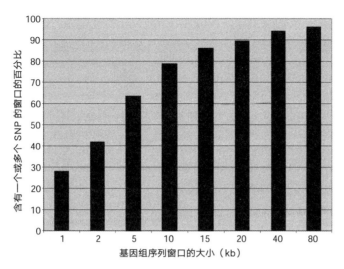

图 1. 已完成序列区间里 SNP 的覆盖分布。检查已定义大小的窗口 (在染色体坐标内) 是否包含一个或多个 SNP。分析限于 900 Mb 可用的已完成序列。

任何给定位置的 SNP 密度取决于各位点发现 SNP 时所用的方法 (TSC、BAC 重叠或者靶向)、基因组序列对于 SNP 发现和定位的可利用性以及核苷酸多样性的比率。这其中，只有核苷酸多样性是区域和群体研究的基本特性。为了绘制人类基因组序列的多态性图谱，我们进行了一个全基因组范围内的核苷酸多样性分析。

核苷酸多样性分析

描述核苷酸多样性的潜在模式需要在高密度、单一的群体样本中进行多态性调查，并且使用一组统一的工具进行分析。我们使用 NQS 方法 (见方法) 对 TSC 产生的 4.5M 数据进行了重新分析 (见方法)。这一组包含了 12 亿对碱基和 920,752 个杂合位点。我们用测量杂合性的标准方法 (π) 测量核苷酸序列的差异，这代表着当随机从一个群体中选出两条染色体比对时一个核苷酸位置是杂合子的可能性。π 也估算了群体遗传参数 $\Theta = 4N_e\mu$，在模型中其位点进化中立，突变率为 μ，群体有效大小为常量 N_e。对于人类基因组，π 是 7.51×10^{-4}，或者来自 NIH 多样性模型中两个染色体调查得到每 1,331 bp 一个 SNP。这个估值与人类基因组变异的小型调查一致[18-20]。

我们接下来检测了单个染色体的杂合性 (表 2)。常染色体彼此十分相似，22 对常染色体中有 20 对在全基因组平均值的 10% 以内 (7.65×10^{-4})。另外两对的值则非常极端：21 号染色体 (π = 5.19×10^{-4})，15 号染色体 (π = 8.79×10^{-4})。这些观测结

chromosome 15 ($\pi = 8.79 \times 10^{-4}$). Whether these observations are due to statistical fluctuations or methodological issues, or are biologically meaningful, will require investigation. The most striking difference in heterozygosity is the lower diversity of the sex chromosomes. The lower rate of polymorphism on the X chromosome may be explained by both a lower effective population size (N_e) and lower mutation rate (μ) in $\Theta = 4N_e\mu$. Because the X chromosome is hemizygous in males, the effective population size is three-quarters of that of the autosomes. In addition, μ is higher in male than in female meiosis, with $\mu_{male}/\mu_{female} \approx 1.7/1.0$ (ref. 33). As the X chromosome undergoes male meiosis only 1/3 of the time, the overall rate of mutation in the X chromosome is expected to be 91% that of the autosomes ($\mu_X = 1.23/1.35 = 0.91$). Thus, the diversity of the X chromosome is predicted to be 69% that of the autosomes. The observed heterozygosity of the X chromosome was 4.69×10^{-4}, or 61% of the average value of the autosomes. Thus, the population genetic considerations described above could largely explain the lower heterozygosity on the X chromosome. It is possible that strong selection on the X chromosome (owing to hemizygosity in males) or other factors might partially explain this observation.

Table 2. Nucleotide diversity by chromosome

Chromosome	Heterozygous positions	High-quality bp examined	$\pi(\times 10^{-4})$
1	71,483	92,639,616	7.72
2	81,860	111,060,861	7.37
3	61,190	81,359,748	7.52
4	59,922	74,162,156	8.08
5	56,344	77,924,663	7.23
6	53,864	72,380,717	7.44
7	52,010	68,527,550	7.59
8	44,477	57,476,056	7.74
9	41,329	50,834,047	8.13
10	43,040	52,184,561	8.25
11	47,477	56,680,783	8.38
12	38,607	51,160,578	7.55
13	35,250	43,915,606	8.03
14	35,083	47,425,180	7.40
15	27,847	31,682,199	8.79
16	22,994	27,736,356	8.29
17	21,247	27,124,496	7.83
18	24,711	30,357,102	8.14
19	11,499	15,060,544	7.64
20	22,726	31,795,754	7.15
21	26,160	50,367,158	5.19

果无论是由统计学波动或方法论的问题造成的，还是具有生物学意义，都需要进行研究。杂合性方面最显著的差异是性染色体中较低的多样性。X 染色体的低多态性也许可以用 $\Theta = 4N_e\mu$ 公式中较小的有效群体大小 (N_e) 和较低的突变率 (μ) 来解释。因为 X 染色体在男性中是半合子，其有效群体大小是常染色体的四分之三。此外，男性减数分裂比女性减数分裂时的 μ 更高，$\mu_{male}/\mu_{female} \approx 1.7/1.0$ (参考文献 33)。由于 X 染色体在男性减数分裂中只占 1/3 的时间，那么 X 染色体的总体突变率应该是常染色体的 91%($\mu_X = 1.23/1.35 = 0.91$)。因此，X 染色体的多样性预计是常染色体的 69%。观测到的 X 染色体的杂合性为 4.69×10^{-4}，或者说是常染色体平均值的 61%。因此，上述关于群体遗传学的考虑很大程度上可以解释 X 染色体的低杂合性。X 染色体上的强烈选择(由于在男性中是半合子)，或者其他因素可能会部分地解释这项观察结果。

表 2. 染色体的核苷酸多样性

染色体	杂合性位置	检查到的高质量 bp	$\pi(\times 10^{-4})$
1	71,483	92,639,616	7.72
2	81,860	111,060,861	7.37
3	61,190	81,359,748	7.52
4	59,922	74,162,156	8.08
5	56,344	77,924,663	7.23
6	53,864	72,380,717	7.44
7	52,010	68,527,550	7.59
8	44,477	57,476,056	7.74
9	41,329	50,834,047	8.13
10	43,040	52,184,561	8.25
11	47,477	56,680,783	8.38
12	38,607	51,160,578	7.55
13	35,250	43,915,606	8.03
14	35,083	47,425,180	7.40
15	27,847	31,682,199	8.79
16	22,994	27,736,356	8.29
17	21,247	27,124,496	7.83
18	24,711	30,357,102	8.14
19	11,499	15,060,544	7.64
20	22,726	31,795,754	7.15
21	26,160	50,367,158	5.19

Continued

Chromosome	Heterozygous positions	High-quality bp examined	$\pi(\times 10^{-4})$
22	17,469	20,478,378	8.53
X	23,818	50,809,568	4.69
Y	348	2,304,916	1.51
Total	920,752	1,225,448,590	7.51

Heterozygosity (π) of each chromosome. The data were filtered to remove repetitive sequences and heterozygosity calculated as described in the methods. Heterozygous positions and high-quality bases examined were counted separately for each pairwise comparison of read to genome, and then summed over each chromosome.

The Y chromosome has the lowest observed heterozygosity of any chromosome. It is divided into two regions: a pseudoautosomal region at either telomeric end that recombines with the X chromosome and is highly heterozygous[34], and the non-recombining Y (NRY). The genome assembly used for this analysis contains only the NRY, which shows very little diversity: 348 SNPs in 2,304,916 bases ($\pi = 1.51 \times 10^{-4}$). These values agree reasonably with previous estimates for NRY[35,36]. The lower diversity of NRY is influenced by a smaller effective population size (20% that of the autosomes), counterbalanced by the higher mutation rate of male meiosis ($\mu_Y = 1.7/1.35 = 1.26 \times$ that of the autosomes). These factors predict that the Y chromosome would have a diversity 31% that of the autosomes, as compared to the observed 20%. Other influences might include selection against deleterious alleles, patterns of male dispersal[35] and a correlation of diversity with recombination rate[19].

To look at diversity on a finer scale, we divided each chromosome into contiguous 200,000-bp bins according to the public Genome Assembly of 5 September 2000. The distribution of heterozygosity among these bins ranges from zero (12 bins, each with zero SNPs over an average of 24,720 bp examined) to 60×10^{-4} (357 SNPs in a bin surveying 58,755 bp). Although 95% of bins display nucleotide diversity values between 2.0×10^{-4} and 15.8×10^{-4}, the pattern is variable (Fig. 2a, b; see also Supplementary Information). One measure of the spread in the data is the coefficient of variation (CV), the ratio of the standard deviation (σ) to the mean (μ) of the heterozygosity π of each individual read. For the observed data, the CV($\sigma_{observed}/\mu_{observed}$) was 1.93, considerably larger than would be expected if every base had uniform diversity, corresponding to a Poisson sampling process ($\sigma_{Poisson}/\mu_{Poisson} = 1.73$). It was expected that the observed distribution would be much more variable than a Poisson process, because both biochemical and evolutionary forces cause diversity to be nonuniform across the genome. Biological factors may include rates of mutation and recombination at each locus. For example, heterozygosity is correlated with the GC content for each read (Fig. 2c), reflecting, at least in part, the high frequency of CpG to TpG mutations arising from deamination of methylated 5-methylcytosine. Population genetic forces are likely to be even more important: each locus has its own history, with samples at some loci tracing back to a recent common ancestor, and other loci describing more ancient genealogies. The time to the most recent common ancestor at a particular stretch of DNA is variable, and represents the opportunity for sequence divergence; thus, the expected pattern of heterozygosity is more heterogeneous than if every locus shared the same history[37,38].

染色体	杂合性位置	检查到的高质量 bp	$\pi(\times 10^{-4})$
22	17,469	20,478,378	8.53
X	23,818	50,809,568	4.69
Y	348	2,304,916	1.51
总计	920,752	1,225,448,590	7.51

每条染色体的杂合性(π)。数据过滤掉了重复序列，杂合性按照方法里描述的进行计算。杂合位置和高质量碱基的检查是由读长与基因组两两比较分别计算得到的，然后将每条染色体进行合计。

Y 染色体观察到的杂合性比任何一个染色体都低。它分成了两个区：一个是假常染色体区，在与 X 染色体结合的两个端粒末端之一，是高度杂合的[34]；另一个是 Y 染色体非重组区(NRY)。本研究对基因组进行组装的分析只包含 NRY 区，该区域有较低的多样性：2,304,916 个碱基中只有 348 个 SNP 位点($\pi = 1.51 \times 10^{-4}$)。这些值与之前对 NRY 的估计一致[35,36]。NRY 的低多样性受较小的有效群体大小的影响（常染色体的 20%），这抵消了男性减数分裂时的高突变率($\mu_Y = 1.7/1.35 = 1.26 \times$ 常染色体的 μ)。这些因素预示与观察到的 20% 相比，Y 染色体将有常染色体 31% 的多样性。其他的影响可能包括对有害等位基因的选择，男性的传播模式[35]以及多样性与重组率的相关性[19]。

为了在更精细的尺度上研究多样性，我们根据 2000 年 9 月 5 日公布的基因组组装结果，把每条染色体分成了连续 200,000 bp 的分箱。这些分箱的杂合性分布变化从 0(12 个分箱平均长度是 24,720 bp，每个分箱有 0 个 SNP)到 60×10^{-4}(该分箱长度是 58,755 bp，有 357 个 SNP)。尽管 95% 的分箱呈现的核苷酸多样性值在 2.0×10^{-4} 到 15.8×10^{-4} 之间，但模式是不同的(图 2a 和 2b；也可见于补充信息)。衡量数据分散性的一种方法是差异系数(CV)，即每个个体序列杂合性(π)的标准差(σ)与平均值(μ)之比。对于观察数据，CV 值($\sigma_{观测}/\mu_{观测}$)是 1.93，比如果每个碱基具有统一的多样性所预期的值要大得多，相当于泊松抽样过程($\sigma_{泊松}/\mu_{泊松} = 1.73$)。预期观测到的分布将会比泊松过程变化多得多，因为生化和进化力都会导致整个基因组的多样性不均匀。生物因素可能包括每个基因座的突变和重组率。例如，每个序列的杂合性与其 GC 含量相关(图 2c)，至少在一定程度上能反映由于甲基化 5-甲基胞嘧啶脱氨引起的 CpG 到 TpG 突变高频发生的原因。群体遗传学力量也许更重要：每个基因座有它自己的历史，一些基因座可以追溯至一个最近的共同祖先，而其他的基因座可能有更多的古老家系。在特定的 DNA 分支上追溯到最近共同祖先的时间是不同的，这代表了序列分化的可能性；因此，相比如果每个基因座都具有相同的历史情况，杂合性的期望模式会更多样[37,38]。

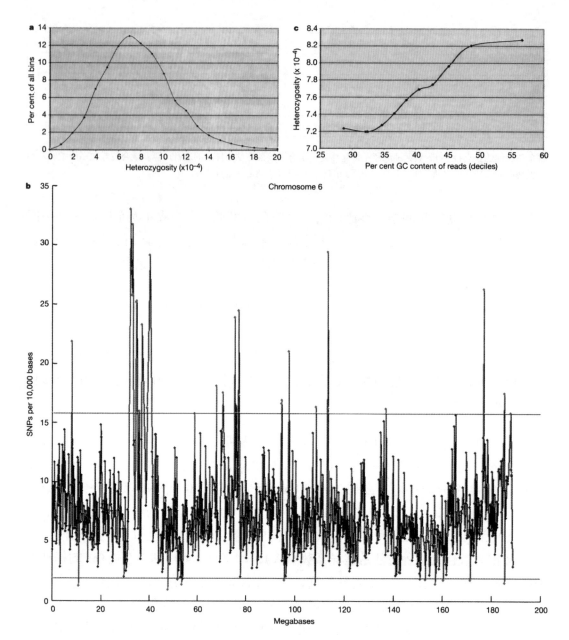

Fig. 2. Distribution of heterozygosity. **a**, The genome was divided into contiguous bins of 200,000 bp based on chromosome coordinates, and the number of high-quality bases examined and heterozygosity calculated for each. A histogram was generated of the distribution of heterozygosity values across all such bins. **b**, Heterozygosity was calculated across contiguous 200,000-bp bins on Chromosome 6. The blue lines represent the values within which 95% of regions fall: 2.0×10^{-4}–15.8×10^{-4}. Red, bins falling outside this range. The extended region of unusually high heterozygosity centred at 34 Mb corresponds to the HLA. **c**, Correlation of nucleotide diversity with GC content of each read (autosomes only). The GC content and heterozygosity of reads from the heterozygosity analysis was calculated after sorting of reads by GC content and separation into 10 bins of equal size. Each bin contains ~150 Mb of aligned, high-quality sequence.

698

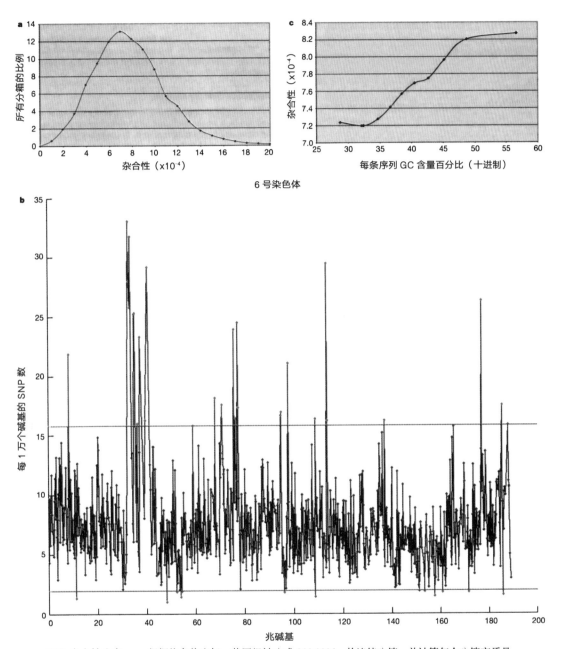

图 2. 杂合性分布。**a**，根据染色体坐标，基因组被分成 200,000 bp 的连续分箱，并计算每个分箱高质量碱基的数目和杂合性。所有这些分箱的杂合性值的分布形成一个直方图。**b**，6 号染色体上所有连续的 200,000 bp 分箱的杂合性。蓝线代表的区间为 $2.0 \times 10^{-4} \sim 15.8 \times 10^{-4}$，95% 的值落在这之间。红色代表落在该范围外面的分箱。相对于 HLA，显著高杂合性的扩展区集中在 34 Mb。**c**，每条序列（只有常染色体）的核苷酸多样性与 GC 含量的相关性。首先将序列按 GC 含量进行排序，然后平均分成 10 个分箱再进行杂合性分析，包括 GC 含量和序列杂合性计算。每个分箱包含约 150 Mb 比对上的高质量的序列。

To assess whether gene history would account for the observed variation in heterozygosity, we compared the observed CV to that expected under a standard coalescent population genetic model. For each read, we adjusted μ on the basis of its per cent GC and length, and simulated genealogical histories under the assumption of a constant-sized population with $N_e = 10,000$. The CV determined under this model ($\sigma_{constant-size}/\mu_{constant-size} = 1.96$) is a close match to the observed data. To estimate standard deviations around these estimates of the CV, it was necessary to consider that tightly linked regions may display correlated histories, and thus are nonindependent. We sampled subsets of the data chosen to minimize correlation among reads (see Methods), providing estimates of the mean and standard deviation of CV for the observed and simulated data (Table 3). These results indicate that the observed pattern of genome-wide heterozygosity is broadly consistent with predictions of this standard population genetic model (for comparison, see an analysis of variation in heterozygosity in the mouse genome)[39]. However, much work will be required to assess additional factors that could influence this distribution: biological factors such as variation in mutation and recombination rates, historical forces such as bottlenecks[40,41], expansions or admixture of differentiated populations, evolutionary selection, and methodological artefacts.

Table 3. Coefficients of variation for the observed data and the Poisson and coalescent models

SNPs per read	Observed	Poisson	Coalescent
0	8,796 ± 43	8,256 ± 52	8,767 ± 50
1	2,247 ± 44	3,040 ± 49	2,332 ± 46
2	668 ± 24	617 ± 24	663 ± 26
3	214 ± 14	99 ± 9	200 ± 15
4	102 ± 10	16 ± 4	66 ± 9
σ/μ	1.94 ± 0.02	1.72 ± 0.02	1.96 ± 0.03

Observed distribution of heterozygosity and comparison to expectation under Poisson and coalescent population genetic models. The autosomes were divided into 200,000-bp bins according to chromosome coordinates and one read randomly selected from each bin. This procedure was chosen to minimize the correlation in gene history of nearby regions, under the simplifying assumption that reads 200,000 bp apart and selected from unrelated individuals will have uncorrelated genealogies. Correlation of gene history does not influence the expected mean value of the CV, but does effect its variance. The random selection of reads and generation of expected distributions were repeated 100 times: presented are the mean and standard deviation of the number of reads in which 0, 1, 2, 3, or 4 SNPs were observed or predicted under each scenario. The Poisson model reports the number of such reads expected to display 0–4 SNPs under Poisson sampling of each read with a heterozygosity adjusted for length and GC content (Fig. 2c). Even in this reduced data set, the Poisson model can be rejected at $P < 10^{-99}$. The coalescent simulation[38] assumed a constant-sized population of effective size 10,000 and free recombination among reads. For each read, μ was scaled according to its length and GC content (Fig. 2c). Each sampled read was assigned a coalescent history from a simulated distribution and the number of SNPs predicted. The coefficient of variation of the estimate of heterozygosity is presented, with the mean and standard deviation of the 100 sampling runs shown.

Regions of low diversity were more prevalent on the sex chromosomes. Whereas only 2.5% of 200,000-bp bins across the genome had $\pi < 2.0 \times 10^{-4}$, 15% of bins on the X chromosome[42] and 89% on the Y chromosome (NRY) had these levels of diversity. Regions

为了评估基因历史能否解释观察到的杂合性的变异，我们将观测到的 CV 与标准的联合群体遗传模型下的预期值进行了对比。对每条序列来说，我们依据 GC 的百分比和长度来调整 μ，并在恒定种群大小 N_e = 10,000 的假定条件下模拟系谱历史。这个模式（$\sigma_{恒定大小}$/$\mu_{恒定大小}$ = 1.96）下得到的 CV 与观测数据高度匹配。为了评估这些 CV 估值的标准差，必须要考虑紧密相连区域可能表现出相关联的历史，这些区域是非独立的。为了使关联性降到最小，我们抽取了这些数据的子集样本（见方法），并提供了观测和模拟数据 CV 的平均值和标准差（表 3）。这些结果表明观测到的全基因组杂合性的模式与该标准群体遗传学模型的预测大体一致（为了进行比较，请参阅小鼠基因组杂合性变异的分析）[39]。然而，我们还需要做许多工作来评价可能影响分布的其他因素：生物因素如突变和重组率的变异，历史力比如瓶颈[40,41]，不同群体的扩张或混合，进化选择以及方法学误差。

表 3. 观测数据、泊松模型以及联合模型的差异系数

每个读长的 SNP	观测值	泊松模型	联合模型
0	8,796 ± 43	8,256 ± 52	8,767 ± 50
1	2,247 ± 44	3,040 ± 49	2,332 ± 46
2	668 ± 24	617 ± 24	663 ± 26
3	214 ± 14	99 ± 9	200 ± 15
4	102 ± 10	16 ± 4	66 ± 9
σ/μ	1.94 ± 0.02	1.72 ± 0.02	1.96 ± 0.03

观测到的杂合性分布，以及在泊松模型和联合群体遗传模型下的比较结果。依据染色体坐标，常染色体被分成 200,000 bp 的分箱，并随机从每个分箱中选择一条序列。该过程是为了使相邻区域基因座的相关性降到最小，在一种简化思维的假设下，每条序列间隔是 200,000 bp，并且选自不相关的个体，具有不相关的家族系谱。基因史的关联性并不影响 CV 预期的平均值，但会影响它的方差。为了达到期望的分布情况，对序列和世代进行随机选择并重复 100 次；表中呈现的是能观测到的或预测的包含 0、1、2、3 或者 4 个 SNP 位点的序列数量的平均值和标准差。泊松模型报告了这样的序列数量，期望在每次利用泊松抽样抽取序列时，预期能显示 0~4 个 SNP，并且依据长度和 GC 含量对杂合性进行调整（图 2c）。即使在这个简化的数据集中，在 $P < 10^{-99}$ 时，泊松模型仍然可以被拒绝。联合模拟[38]假定了一个具有 10,000 个有效个体的恒定大小的群体，序列之间可以进行自由重组。依据每条序列的长度和 GC 含量来衡量其 μ 值（图 2c）。每个抽样的序列从模拟分布和预测 SNP 的数目中分配一个联合历史。表中呈现的是杂合性估算的差异系数以及 100 次运行的平均值和标准差。

在性染色体上，低多样性区域更普遍。尽管整个基因组中只有 2.5% 的 200,000 bp 分箱的 π 值小于 2.0×10^{-4}，但 X 染色体上 15% 的分箱[42]和 Y 染色体（NRY）上 89%

of low diversity may be explained by the smaller effective population size of the sex chromosomes and the variable underlying distribution of heterozygosity. Strong selection acting on the sex chromosomes in males might also have a role, but this hypothesis requires further testing. Regions of high heterozygosity were also observed. One was found on chromosome 6 (Fig. 2b, centred on 34 Mb), and was confirmed to represent the HLA locus, which has high nucleotide diversity owing to balancing selection[43]. Other regions of varying size were observed on this and other chromosomes (Fig. 2c and Supplementary Information). Some of these highly diverse regions might have also experienced balancing selection, but there are other possible explanations: for example, sampling fluctuations of the coalescent distribution, regions with high rates of mutation and/or recombination, unrecognized duplications in the human genome and sequencing of a rare haplotype by the HGP (to which the TSC reads were compared).

Given the unfinished state of publicly available sequence data and genome assembly, it will be important to reevaluate these estimates as more complete genome sequence becomes available.

Implications for Medical and Population Genetics

We describe a map of publicly available SNPs (as of November 2000), fully integrated with the sequence, physical and genetic maps of the human genome. We anticipate immediate application to studies of human population genetics, candidate-gene studies for disease association, and eventually unbiased, genome-wide association scans. First, the map provides an unprecedented tool for studying the character of human sequence variation. We use these data to describe the first genome-wide view of how human DNA sequence varies in the population, and the public availability of these data should fuel future research into biological and population genetic influences on human genetic diversity.

Second, insights into human evolutionary history will be obtained by using SNPs from the map to characterize haplotype diversity throughout the genome. Human haplotype structure remains largely unexplored, and this map makes it possible to define the extent and variation of haplotype identity, the number and frequencies of common haplotypes, and their distribution among and within existing ethnic groups.

Most practically, where a gene has been implicated in causing disease (by chromosomal position relative to linkage peaks, known biological function or expression pattern), it is desirable exhaustively to survey allelic variation for any association to disease. Using the SNP map, it should be possible to evaluate the extent to which common haplotypes contribute to disease risk. As the speed and efficiency of SNP genotyping increases, such studies will fuel increasingly comprehensive tests of the hypothesis that common variants contribute significantly to the risk of common diseases. To the extent that such studies are successful, they should profoundly affect our understanding of disease, methods of diagnosis, and ultimately the development of new and more effective therapies.

的分箱具有这些多样性水平。这些低多样性的区域可能由性染色体有效种群大小较小以及潜在的杂合性分布不同所致。作用在男性性染色体上的强选择可能也起到一定作用，但这个假说需要进一步的验证。此外还发现了一些高度杂合性的区域。其中一个在 6 号染色体上（图 2b，集中于 34 Mb），并被证实是 HLA 的基因座；由于平衡选择[13]，该区域具有高的核苷酸多样性。在 6 号和其他染色体上观察到大小不一的其他区域（图 2c 及补充信息）。一些高度多样化区域可能经历了平衡选择，但是也有一些其他可能的解释：比如联合模型分布抽样的波动性，带有高突变率和（或）重组率的区域，人类基因组未识别到的重复序列，以及由 HGP 进行的罕见单体型测序（以 TSC 读长作为对比）。

考虑到可公开获取序列数据和基因组组装的未完成状态，当更完整的基因组序列可以获取之时，重新评价这些估算显得尤为重要。

对医学和群体遗传学的影响

我们充分地整合了人类基因组的序列、物理图谱以及遗传图谱，描绘了一张可公开获取的 SNP 图谱（截止 2000 年 11 月）。我们预计该图谱可直接应用于人类群体基因组学研究和疾病相关的候选基因研究，并最终应用于无偏移的全基因组关联扫描。首先，图谱为人类序列差异特性的研究提供了空前的工具。我们用这些数据来描述关于群体中人类 DNA 序列变化的第一个全基因组视图，而这些数据的可公开获取性将促进未来关于生物和群体遗传学对人类遗传多样性的影响的研究。

其次，我们将利用图谱中的 SNP 来洞悉人类进化史，从而描述整个基因组的单体型多样性。人类单体型结构仍然有大部分未被探索，而这张图谱使得对单体型特征的范围和变异，常见单体型的数量和频率，以及它们在现有种群内的分布等的界定变得可能。

最实际的应用就是，当一个基因引起疾病时（通过与连锁峰相关的染色体位置，已知生物学功能或表达模式），可以详尽地调查与疾病有任何相关的等位基因的变异。利用 SNP 图谱，我们可以评估常见单体型促成疾病风险的程度。随着 SNP 基因分型的速度和效率的增加，这样的研究有助于更加全面地检验常见突变显著增加常见疾病风险这一假说。从某种程度上说这些研究是成功的，它们会深刻影响我们对疾病的理解和诊断方法，并最终开发新的更有效的治疗方法。

Methods

SNP identification

Candidate SNPs were identified by detection of high-confidence base differences in aligned sequences. For TSC, sequence reads were filtered to exclude low quality reads and those containing predominantly known repetitive sequence. Sequences were aligned to each other using the reduced representation shotgun (RRS) method, and by genomic alignment (GA) as described[18,22]. For GA of TSC data, reads were compared to available large-insert clones (finished and draft with available PHRAP quality scores) in GenBank. For the analysis of clone overlaps, all available finished and unfinished genomic sequence accessions were aligned. Two methods were used to detect SNPs. The NQS relies upon the sequence trace quality surrounding the SNP base to increase base-calling confidence[18,22]; most data discovered using the NQS was processed using SsahaSNP, an ultrafast, hash-based implementation of the algorithm (Z.N., A. Cox and J.C.M, manuscript in preparation). The second method calculates confidence scores on the basis of a Bayesian analysis of confidence scores[24]. A variety of methods were used to find SNPs in expressed sequence tag (EST) overlaps[24,25,27] and for targeted resequencing; details of the remaining SNPs can be found in the individual dbSNP entries (www.ncbi.nlm.nih.gov/SNP/).

Mapping of SNPs and features

MEGABLAST[44] was used to align TSC SNP flanking sequences to the genomic sequence accessions. A SNP was considered mapped if a high-quality match (99% identity or greater) was found across the available flanking sequence of no less than 270 bp. SNPs that matched more than three accessions with identity > 98% were judged to be possible repetitive regions and set aside. SNP coordinates were generated relative to the OO18 build of the genome assembly (5 September 2000) and the OO15 build (15 July 2000), using the AGP format files provided by D. Haussler (http://genome.ucsc.edu).

The NCBI RefSeq mRNA transcripts[31] were aligned to the Genome Assembly using the NCBI SPIDEY alignment tool. Alignment required > 97% sequence similarity between mRNA and genome sequence; alignments were refined by taking into account the donor/acceptor sites. In cases where CDS annotations were available in the GenBank record, exons of the CDS were aligned within the confines of the mRNA alignment. Regions of known human repeats were annotated directly using RepeatMasker (A. Smit, unpublished).

Nucleotide diversity analysis

To characterize nucleotide diversity, we required a data set in which all data could be analysed both for the number of high-quality bases meeting quality standards for SNP detection, and for

方　法

SNP 的鉴定

利用比对序列中高置信度碱基的差异进行候选 SNP 的鉴定。对 TSC 来说，将序列读长进行过滤处理，从而排除低质量读长以及那些包含显著已知重复序列的读长。利用简化代表性鸟枪法(RRS)和描述的基因组比对(GA)[18,22] 方法进行序列比对。对 TSC 数据的 GA 而言，序列与 GenBank 中可用的长的插入克隆(包含 PHRAP 质量值的已完成和未完成序列)进行比较。为了分析克隆重叠群，所有可用的完成和未完成基因组序列(含有基因库中编号)均被进行比对。两种方法用来对 SNP 进行鉴定。一种是 NQS 方法，该方法依赖围绕在 SNP 碱基周围序列的质量追踪来增加碱基识别的置信度[18,22]；多数数据使用 NQS 方法都是通过使用 SsahaSNP 软件来进行分析的，SsahaSNP 是一种超快的基于哈希算法实现的 SNP 鉴定工具(宁泽民、考克斯和马利金，稿件准备中)。第二种方法在基于置信评分的贝叶斯分析基础上计算置信评分[24]。我们使用多种方法在表达序列标签(EST)中鉴定了 SNP[24,25,27] 并进行靶向测序；剩余 SNP 的信息可以在单独的 dbSNP 条目中找到(www.ncbi.nlm.nih.gov/SNP/)。

SNP 图谱及其特征

利用 MEGABLAST 软件[44] 将 TSC SNP 侧翼序列与基因组含有基因库编号的序列进行比对。对于任何一个 SNP 位点，如果发现在可用侧翼序列中不少于 270 bp 的高质量碱基与库中序列相匹配(99% 的一致性或更高)，则认为该 SNP 被比对上。如果一个 SNP 匹配三个以上的基因组序列(98% 相似性)，那么这个 SNP 定义为重复区，将被去掉。基于 OO18 (2000 年 9 月 5 日)和 OO15(2000 年 7 月 15 日)构建的基因组组装结果，使用豪斯勒提供的 AGP 文件格式(http://genome.ucsc.edu)生成 SNP 坐标。

利用 NCBI SPIDEY 比对工具将 NCBI 参考序列中的 mRNA 转录本[31] 与基因组组装结果进行比对。比对要求 mRNA 与基因组序列间有 >97% 的序列相似性；考虑供 / 受体位置对结果进行调整。当 CDS 在 GenBank 中有注释信息时，CDS 的外显子区域在信使 RNA 的范围内进行比对。人类基因组中已知的重复片段区域直接使用 RepeatMasker 软件进行注释(斯米特，未发表)。

核苷酸多样性分析

为了描述核苷酸的多样性，我们需要一个数据集，该数据集里所有的数据都满足 SNP 检测的质量标准中高质量碱基的数量以及 SNP 数量。为了确保分析的均一性，我们只分

the number of SNPs. To ensure homogeneity of analysis, we performed a single analysis of 4.5 million high-quality TSC reads from the Sanger Centre, Washington University in St. Louis and the Whitehead Center for Genome Research. The GC content of these reads was 41%, the same as the genome as a whole[32], and the distribution of read GC content across deciles of the genome (sorted by GC content) was within 10% of the expected value for all bins. The read coverage was well distributed: 88% of contiguous 200,000-bp windows contained over 10,000 aligned bases (5%) surveyed for SNPs (see below). Using a single analytic tool (SsahaSNP, an implementation of the NQS; Z.N., A. Cox and J.C.M, in preparation), these reads were aligned to the available genome sequence (finished and draft with quality scores) and the number of high-quality bases (meeting NQS) and SNPs counted. We limited the analysis to SNPs found by genomic alignment so that the cluster depth of each comparison would be exactly two chromosomes. We precisely measured the target size for SNP discovery by counting the number of positions meeting the NQS. This is desirable because alignments contain positions of both high and low quality, but only those meeting the NQS are candidates for SNP discovery. Where a single TSC read aligned to multiple (overlapping) BACs from the HGP, we averaged the number of SNPs and aligned bp for all pairwise alignments of that read; this weighted evenly those reads mapping to a single BAC and those aligning to a region of overlap. Reads representing repeat loci were excluded using validated criteria[18,22]: alignments of reads to genome were excluded if they were less than 99% identical. The genome was then divided into contiguous bins of 200,000 bp (based on chromosome-relative coordinates). Individual reads were filtered for repeats: any that aligned to more than one bin in the genome assembly were rejected. Finally, heterozygous positions and bases meeting the NQS were counted. As a final filter for regions containing a high proportion of repeats, we reject any bin for which more than 10% of the reads mapping to that bin also mapped to another chromosome. Finally, to avoid statistical fluctuation due to inadequate sampling, we examined only the 88% of bins in which at least 10,000 aligned bases met the NQS and thus could be examined for SNPs.

Coalescent modelling was performed by simulation[38], and assumed a constant-sized population of 10,000 individuals and a mutation rate adjusted for each read on the basis of its GC content (Fig. 2c) and length. To assess the standard deviation around this estimate, the simulation was repeated 100 times. For the observed data, calculating a standard deviation around the CV is difficult owing to the correlation of gene history for closely linked sites. In expectation, this correlation should not alter the mean of the observed coefficient of variation, but does influence its variance. To estimate the variance around the CV for the observed data, we selected 100 reduced data sets, each containing one randomly chosen read from each 200,000-bp bin along the autosomes. In using this approach, we assume that these reads, 200,000 bp apart and sampled from unrelated individuals, have independent genealogies. This random sampling procedure was repeated 100 times to estimate the mean and variance of the observed CV.

The data for the heterozygosity analysis, including the coordinates of each bin, the number of bases examined and number of SNPs identified, is available as Supplementary Information.

(**409**, 928-933; 2001)

析了来自桑格中心、圣路易斯华盛顿大学以及怀特黑德基因组研究中心的 450 万个高质量 TSC 序列。这些序列的 GC 含量是 41%，与整体基因组一致[32]，序列 GC 含量跨基因组十分位数的分布（按 GC 含量分类）在所有分箱预期值的 10% 以内。序列覆盖分布良好：对连续 200,000 bp 窗口区域进行 SNP 分析发现，有 10,000 多个碱基（5%）分布在 88% 的区域中（见下文）。只利用一种分析软件（SsahaSNP，NQS 的一种实现方法；宁泽民、考克斯和马利金，稿件准备中）将这些序列与可用的基因组序列（含有质量值的完成的或未完成的序列）进行比对，计算高质量碱基（符合 NQS）和 SNP 的数目。我们对由基因组比对发现的 SNP 设置限制条件，使每个比较的重叠群深度都正好是两条染色体。通过计算符合 NQS 的位置数量，可以精确计算 SNP 发现区域的目标大小。因为比对信息中包括高质量和低质量碱基的位置，只有那些符合 NQS 的才能作为候选 SNP 位点，这一点是可取的。当单个 TSC 序列比对到来自 HGP 的多个（重叠）BAC 上时，我们对该序列上 SNP 的数目和比对的长度取平均数，这对那些映射到图谱的单一 BAC 和那些与一个重叠区域匹配的序列进行了加权平均。重复基因座上的序列使用如下确认准则进行排除[18,22]：如果与基因组比对的相似性小于 99%，那么该序列被去掉。然后，基因组被分成连续 200,000 bp 的分箱（基于染色体相对坐标）。对每条序列都进行如下重复序列过滤：如果匹配到基因组中多于一个分箱，序列被去掉。最终，计算符合 NQS 的杂合子的位置和碱基数目。最后一步过滤是针对高比例重复区的筛选，当某一分箱中 10% 以上的序列既比对到该分箱中，同时又比对到其他染色体上，那么该分箱被去掉。最终，为了避免由于采样不足引起的统计波动，我们只选择了其中 88% 的分箱进行 SNP 鉴定，这些分箱中至少有 10,000 个比对的碱基符合 NQS。

我们模拟了联合模型[38]，该模型假定了恒定大小为 10,000 个体的群体，并根据 GC 含量（图 2c）和长度对每个序列的突变率进行了调整。将模拟重复 100 次以评估该估算附近的标准差。对于观测到的数据来说，由于紧密相连位点的基因史具有相关性，根据 CV 计算标准差很困难。在预期中，这种相关性应该不会改变观测到的变异系数的均值，但是会影响其方差。为了依据 CV 对观测数据的方差进行估算，我们选定 100 个简化数据集，每个数据集均包含从每 200,000 bp 的常染色体分箱中随机选择的一个序列。运用这种方法，我们假定这些序列相距 200,000 bp 并且从来自不相关的个体中抽样，有独立的家系图谱。该随机抽样过程重复 100 次，以估算观测到的 CV 的平均值和方差。

杂合性分析的数据在补充信息中，这些信息包括每个分箱的坐标、被检查的碱基数目和被鉴定的 SNP 数目。

（杨晶 翻译；胡松年 审稿）

The International SNP Map Working Group[*]
[*] A full list of authors appears at the end of this paper.

Received 28 November; accepted 27 December 2000.

References:

1. Collins, F. S. Of needles and haystacks: finding human disease genes by positional cloning. *Clin. Res.* **39**, 615-623 (1991).

2. Collins, F. S., Guyer, M. S. & Charkravarti, A. Variations on a theme: cataloging human DNA sequence variation. *Science* **278**, 1580-1581 (1997).

3. Lander, E. S. The new genomics: global views of biology. *Science* **274**, 536-539 (1996).

4. Risch, N. & Merikangas, K. The future of genetic studies of complex human diseases. *Science* **273**, 1516-1517 (1996).

5. Li, W. H. & Sadler, L. A. Low nucleotide diversity in man. *Genetics* **129**, 513-523 (1991).

6. Cargill, M. *et al.* Characterization of single-nucleotide polymorphisms in coding regions of human genes [published erratum appears in *Nature Genet.* **23**, 373 (1999)]. *Nature Genet.* **22**, 231-238(1999).

7. Cambien, F. *et al.* Sequence diversity in 36 candidate genes for cardiovascular disorders. *Am. J. Hum. Genet.* **65**, 183-191 (1999).

8. Fullerton, S. M. *et al.* Apolipoprotein E variation at the sequence haplotype level: implications for the origin and maintenance of a major human polymorphism. *Am. J. Hum. Genet.* **67**, 881-900 (2000).

9. Halushka, M. K. *et al.* Patterns of single-nucleotide polymorphisms in candidate genes for blood-pressure homeostasis. *Nature Genet.* **22**, 239-247 (1999).

10. Nickerson, D. A. *et al.* DNA sequence diversity in a 9.7-kb region of the human lipoprotein lipase gene. *Nature Genet.* **19**, 233-240 (1998).

11. Rieder, M. J., Taylor, S. L., Clark, A. G. & Nickerson, D. A. Sequence variation in the human angiotensin converting enzyme. *Nature Genet.* **22**, 59-62 (1999).

12. Templeton, A. R., Weiss, K. M., Nickerson, D. A., Boerwinkle, E. & Sing, C. F. Cladistic structure within the human lipoprotein lipase gene and its implications for phenotypic association studies. *Genetics* **156**, 1259-1275 (2000).

13. Eaves, I. A. *et al.* The genetically isolated populations of Finland and sardinia may not be a panacea for linkage disequilibrium mapping of common disease genes. *Nature Genet.* **25**, 320-323 (2000).

14. Taillon-Miller, P. *et al.* Juxtaposed regions of extensive and minimal linkage disequilibrium in human Xq25 and Xq28. *Nature Genet.* **25**, 324-328 (2000).

15. Kruglyak, L. Prospects for whole-genome linkage disequilibrium mapping of common disease genes. *Nature Genet.* **22**, 139-144 (1999).

16. Collins, A., Lonjou, C. & Morton, N. E. Genetic epidemiology of single-nucleotide polymorphisms. *Proc. Natl Acad. Sci. USA* **96**, 15173-15177 (1999).

17. Reich, D. E. *et al.* Linkage disequilibrium in the human genome. *Nature* (submitted).

18. Altshuler, D. *et al.* An SNP map of the human genome generated by reduced representation shotgun sequencing. *Nature* **407**, 513-516 (2000).

19. Nachman, M. W., Bauer, V. L., Crowell, S. L. & Aquadro, C. F. DNA variability and recombination rates at X-linked loci in humans. *Genetics* **150**, 1133-1141 (1998).

20. Wang, D. G. *et al.* Large-scale identification, mapping, and genotyping of single-nucleotide polymorphisms in the human genome. *Science* **280**, 1077-1082 (1998).

21. Jorde, L. B. Linkage disequilibrium and the search for complex disease genes. *Genome Res.* **10**, 1435- 1444 (2000).

22. Mullikin, J. C. *et al.* An SNP map of human chromosome 22. *Nature* **407**, 516-520 (2000).

23. Collins, F. S., Brooks, L. D. & Chakravarti, A. A DNA polymorphism discovery resource for research on human genetic variation [published erratum appears in *Genome Res.* **9**, 210 (1999)]. *Genome Res.* **8**, 1229-1231 (1998).

24. Marth, G. T. *et al.* A general approach to single-nucleotide polymorphism discovery. *Nature Genet.* **23**, 452-456 (1999).

25. Buetow, K. H., Edmonson, M. N. & Cassidy, A. B. Reliable identification of large numbers of candidate SNPs from public EST data. *Nature Genet.* **21**, 323-325 (1999).

26. Gu, Z., Hillier, L. & Kwok, P. Y. Single nucleotide polymorphism hunting in cyberspace. *Hum. Mutat.* **12**, 221-225 (1998).

27. Irizarry, K. *et al.* Genome-wide analysis of single-nucleotide polymorphisms in human expressed sequences. *Nature Genet.* **26**, 233-236 (2000).

28. Picoult-Newberg, L. *et al.* Mining SNPs from EST databases. *Genome Res.* **9**, 167-174 (1999).

29. Marth, G. T. *et al.* Single nucleotide polymorphisms in the public database: how useful are they? *Nature Genet.* (submitted).

30. Yang, Z. *et al.* Sampling SNPs. *Nature Genet.* **26**, 13-14 (2000).

31. Pruitt, K. D., Katz, K. S., Sicotte, H. & Maglott, D. R. Introducing RefSeq and LocusLink: curated human genome resources at the NCBI. *Trends Genet.* **16**, 44-47 (2000).

32. International Human Genome Sequencing Consortium. Initial sequencing and analysis of the human genome. *Nature* **409**, 860-921 (2001).

33. Bohossian, H. B., Skaletsky, H. & Page, D. C. Unexpectedly similar rates of nucleotide substitution found in male and female hominids. *Nature* **406**, 622-625 (2000).

34. Cooke, H. J., Brown, W. R. & Rappold, G. A. Hypervariable telomeric sequences from the human sex chromosomes are pseudoautosomal. *Nature* **317**, 687-692 (1985).

35. Shen, P. *et al.* Population genetic implications from sequence variation in four Y chromosome genes. *Proc. Natl Acad. Sci. USA* **97**, 7354-7359 (2000).

36. Underhill, P. A. *et al.* Detection of numerous Y chromosome biallelic polymorphisms by denaturing high-performance liquid chromatography. *Genome Res.* **7**, 996-1005 (1997).

37. Tajima, F. Evolutionary relationship of DNA sequences in finite populations. *Genetics* **105**, 437-460 (1983).

38. Hudson, R. R. in *Oxford Surveys in Evolutionary Biology* (eds Futuyma, D. & Antonovics, J.) 1-44 (Oxford Univ. Press, Oxford, 1991).

39. Lindblad-Toh, K. *et al.* Large-scale discovery and genotyping of single-nucleotide polymorphisms in the mouse. *Nature Genet.* **24**, 381-386 (2000).

708

40. Kimmel, M. *et al.* Signatures of population expansion in microsatellite repeat data. *Genetics* **148**, 1921-1930 (1998).

41. Reich, D. E. & Goldstein, D. B. Genetic evidence for a Paleolithic human population expansion in Africa [published erratum appears in *Proc. Natl Acad. Sci. USA* **95**, 11026 (1998)]. *Proc. Natl Acad. Sci. USA* **95**, 8119-8123 (1998).

42. Miller, R. D., Taillon-Miller, P. & Kwok, P. Y. Regions of low single-nucleotide polymorphism (SNP) incidence in human and orangutan Xq: deserts and recent coalescences. *Genomics* (in the press).

43. Horton, R. *et al.* Large-scale sequence comparisons reveal unusually high levels of variation in the HLA-DQB1 locus in the class II region of the human MHC. *J. Mol. Biol.* **282**, 71-97 (1998).

44. Zhang, Z., Schwartz, S., Wagner, L. & Miller, W. A greedy algorithm for aligning DNA sequences. *J. Comput. Biol.* **7**, 203-214 (2000).

Supplementary Information is available on *Nature*'s World-Wide Web site (http://www.nature.com) or as paper copy from the London editorial office of *Nature*.

Acknowledgements. The SNP Consortium, the Wellcome Trust and the National Human Genome Research Institute funded SNP discovery and data management at Cold Spring Harbor Laboratories, The Sanger Centre, Washington University in St. Louis, and the Whitehead/MIT Center for Genome Research. Work in P.Y.K.'s laboratory is supported in part by grants from the SNP Consortium and the National Human Genome Research Institute. P.Y.K. thanks Q. Li, M. Minton, R. Donaldson and S. Duan for technical assistance. D.M.A. was supported during a phase of this work under a Postdoctoral Fellowship for Physicians from the Howard Hughes Medical Institute. For full list of contributors to TSC programme, see www.snp.cshl.org.

Correspondence and requests for materials should be addressed to D.A. (e-mail: altshul@genome.wi.mit.edu) or D.B. (e-mail: drb@sanger.ac.uk).

* **The International SNP Map Working Group** (contributing institutions are listed alphabetically).

Cold Spring Harbor Laboratories: Ravi Sachidanandam[1], David Weissman[1], Steven C. Schmidt[1], Jerzy M. Kakol[1] & Lincoln D. Stein[1]

National Center for Biotechnology Information: Gabor Marth[2] & Steve Sherry[2]

The Sanger Centre: James C. Mullikin[3], Beverley J. Mortimore[3], David L. Willey[3], Sarah E. Hunt[3], Charlotte G. Cole[3], Penny C. Coggill[3], Catherine M. Rice[3], Zemin Ning[3], Jane Rogers[3], David R. Bentley[3]

Washington University in St. Louis: Pui-Yan Kwok[4], Elaine R. Mardis[4], Raymond T. Yeh[4], Brian Schultz[4], Lisa Cook[4], Ruth Davenport[4], Michael Dante[4], Lucinda Fulton[4], LaDeana Hillier[4], Robert H. Waterston[4] & John D. McPherson[4]

Whitehead/MIT Center for Genome Research: Brian Gilman[5], Stephen Schaffner[5], William J. Van Etten[5,6], David Reich[5], John Higgins[5], Mark J. Daly[5], Brendan Blumenstiel[5], Jennifer Baldwin[5], Nicole Stange-Thomann[5], Michael C. Zody[5], Lauren Linton[5], Eric S. Lander[5,7] & David Altshuler[5,8]

1, Cold Spring Harbor, New York 11724, USA; 2, Building 38A, 8600 Rockville Pike, Bethesda, Maryland 20894, USA; 3, Wellcome Trust Genome Campus, Hinxton, Cambridge, CB10 1SA, UK; 4, 660 S. Euclid Ave, St. Louis, Missouri 63110, USA; 5, 9 Cambridge Center, Cambridge, Massachusetts 02139, USA; 6, Present address: Blackstone Technology Group, Boston, Massachusetts 02110, USA; 7, Department of Biology, Massachusetts Institute of Technology, Cambridge, Massachusetts 02142, USA; 8, Departments of Genetics and Medicine, Harvard Medical School; Department of Molecular Biology and Diabetes Unit, Massachusetts General Hospital, Boston, Massachusetts 02114, USA.

Superconductivity at 39 K in Magnesium Diboride

J. Nagamatsu *et al.*

Editor's Note

By 2001, a variety of materials based on copper oxides that superconduct at temperatures over 130 K had been observed. Here physicist Jun Nagamatsu and colleagues reported their observation of superconductivity at a temperature of "just" 39 K, but in a very different material: the metallic compound magnesium diboride. This work established a new record for the superconducting transition temperature in metallic compounds, for which the previous high had been, since 1973, only 23 K. Later work would show that the behaviour of this compound fits the standard Bardeen–Cooper–Schrieffer theory of superconductivity, although with some peculiarities. Due to the low cost of its constituent elements and fabrication, magnesium diboride has become a widely used superconductor for practical applications.

In the light of the tremendous progress that has been made in raising the transition temperature of the copper oxide superconductors (for a review, see ref. 1), it is natural to wonder how high the transition temperature, T_c, can be pushed in other classes of materials. At present, the highest reported values of T_c for non-copper-oxide bulk superconductivity are 33 K in electron-doped $Cs_xRb_yC_{60}$ (ref. 2), and 30 K in $Ba_{1-x}K_xBiO_3$ (ref. 3). (Hole-doped C_{60} was recently found[4] to be superconducting with a T_c as high as 52 K, although the nature of the experiment meant that the supercurrents were confined to the surface of the C_{60} crystal, rather than probing the bulk.) Here we report the discovery of bulk superconductivity in magnesium diboride, MgB_2. Magnetization and resistivity measurements establish a transition temperature of 39 K, which we believe to be the highest yet determined for a non-copper-oxide bulk superconductor.

THE samples were prepared from powdered magnesium (Mg; 99.9%) and powdered amorphous boron (B; 99%) in a dry box. The powders were mixed in an appropriate ratio (Mg:B = 1:2), ground and pressed into pellets. The pellets were heated at 973 K under a high argon pressure, 196 MPa, using a hot isostatic pressing (HIP) furnace (O₂Dr.HIP, Kobelco) for 10 hours. Powder X-ray diffraction was performed by a conventional X-ray spectrometer with a graphite monochromator (RINT-2000, Rigaku). Intensity data were collected with CuKα radiation over a 2θ range from 5° to 80° at a step width of 0.02°.

Figure 1 shows a typical X-ray diffraction pattern of MgB_2 taken at room temperature. All the intense peaks can be indexed assuming an hexagonal unit cell, with $a = 3.086$ Å and $c = 3.524$ Å. Figure 2 shows the crystal structure of MgB_2 (ref. 5), of which the space

二硼化镁在 39 K 时的超导性

永松纯等

编者按

到 2001 年为止，在温度超过 130 K 的多种基于铜氧化物的材料中已观测到超导性。本文中物理学家永松纯及其同事们报道了在温度"仅仅为"39 K 时观测到的超导性，但这是在一种非常不一样的材料中观测到的：金属化合物二硼化镁。这项工作创造了金属化合物超导转变温度的一个新纪录，因为从 1973 年到此之前，这种转变温度最高只有 23 K。后续的工作表明，尽管存在一些特殊性，这种化合物的性质符合标准的 BCS（巴丁–库珀–施里弗）超导理论。由于原料和制造成本低，二硼化镁已被广泛地用于超导体的实际应用中。

鉴于在提高铜氧化物的超导转变温度上所取得的巨大成就（见参考文献 1 中的综述），其他类型材料的超导转变温度 T_c 可以被提高到何种程度自然而然地引起了人们的关注。现有报道中，非铜氧化物体超导性的 T_c 最高值，在电子掺杂的 $Cs_xRb_yC_{60}$ 中为 33 K（参考文献 2），在 $Ba_{1-x}K_xBiO_3$ 中为 30 K（参考文献 3）。（近期发现空穴掺杂的 C_{60} 超导转变温度 T_c 高达 52 K[4]，尽管那次实验的原本目的是为了揭示超导电流仅局限于 C_{60} 晶体表面，而非检测整体。）本文中报告了二硼化镁（MgB_2）中体超导的发现。磁化和电阻测量表明超导转变温度为 39 K，我们相信这是非铜氧化物体超导的至今为止所确定的最高转变温度。

样品是用粉末状镁（Mg; 99.9%）和粉末状的非晶硼（B; 99%）在干燥箱中制备的。两种粉末以适当的比例（Mg：B = 1：2）混合，研磨并挤压成圆片。将这些圆片置于热等静压（HIP）炉（O_2Dr.HIP，日本神钢集团）中，在 196 Mpa 的高氩压强和 973 K 的温度下加热 10 小时。粉末 X 射线衍射通过带有石墨单色仪的常规 X 射线光谱仪（RINT-2000，日本理学公司）来进行。强度数据利用 Cu 靶的 Kα 辐射采集，2θ 范围从 5° 到 80°，步长宽度为 0.02°。

图 1 为一种典型的在室温下获取的 MgB_2 的 X 射线衍射图。假定一种六方晶胞，其 a = 3.086 Å，c = 3.524 Å，可以将所有高强度的峰进行标记。图 2 展示了 MgB_2

group is $P6/mmm$ (no.191). As shown in Fig. 2, the boron atoms are arranged in layers, with layers of Mg interleaved between them. The structure of each boron layer is the same as that of a layer in the graphite structure: each boron atom is here equidistant from three other boron atoms. Therefore, MgB_2 is composed of two layers of boron and magnesium along the c axis in the hexagonal lattice.

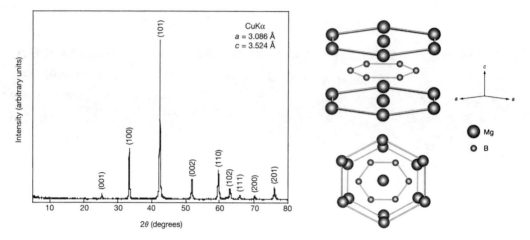

Fig. 1. X-ray diffraction pattern of MgB_2 at room temperature.   Fig. 2. Crystal structure of MgB_2.

Magnetization measurements were also performed with a SQUID magnetometer (MPMSR2, Quantum Design). Figure 3 shows the magnetic susceptibility ($\chi = M/H$, where M is magnetization and H is magnetic field) of MgB_2 as a function of temperature, under conditions of zero field cooling (ZFC) and field cooling (FC) at 10 Oe. The existence of the superconducting phase was then confirmed unambiguously by measuring the Meissner effect on cooling in a magnetic field. The onset of a well-defined Meissner effect was observed at 39 K. A superconducting volume fraction of 49% under a magnetic field of 10 Oe was obtained at 5 K, indicating that the superconductivity is bulk in nature. The standard four-probe technique was used for resistivity measurements.

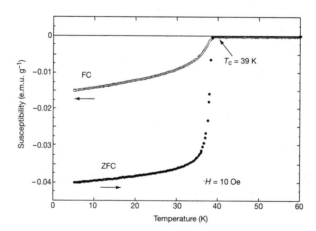

Fig. 3. Magnetic susceptibility χ of MgB_2 as a function of temperature. Data are shown for measurements under conditions of zero field cooling (ZFC) and field cooling (FC) at 10 Oe.

的晶体结构(参考文献 5),空间群为 *P6/mmm*(编号 191)。如图 2 所示,硼原子是按层排列的,镁层交错插入硼层。每一层的硼原子结构与石墨每层的结构相同:其中每个硼原子与其他三个硼原子是等距的。因此,MgB_2 是由两层硼以及一层镁沿六方晶格 *c* 轴构成的。

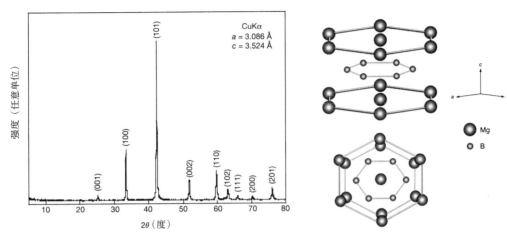

图 1. 室温下 MgB_2 的典型 X 射线衍射图   图 2. MgB_2 的晶体结构

磁化强度测量同样是通过超导量子干涉磁强计(SQUID)完成(MPMSR2, Quantum Design)。图 3 显示了 MgB_2 的磁化率($\chi = M/H$,其中 M 为磁化强度,H 为磁场)在零场冷却(ZFC)以及在 10 Oe 的场冷却(FC)条件下随温度变化的曲线。通过测量在磁场中冷却状态下的迈斯纳效应,明确地证实了超导相的存在。在温度为 39 K 时,开始观测到明确的迈斯纳效应。在温度为 5 K 时,10 Oe 磁场下超导体积分数为 49%,这就表明本质上是体超导性。在电阻测量中我们使用的是常规的四探针技术。

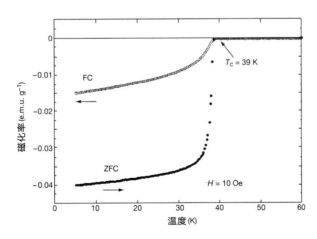

图 3. MgB_2 的磁化率 χ 随温度的变化。数据分别表示在零场冷却(ZFC)以及在 10 Oe 的场冷却(FC)条件下的测量结果。

Figure 4 shows the temperature dependence of the resistivity of MgB_2 under zero magnetic field. The onset and end-point transition temperatures are 39 K and 38 K, respectively, indicating that the superconductivity was truly realized in this system.

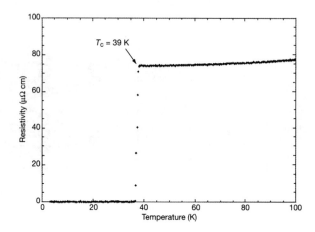

Fig. 4. Temperature dependence of the resistivity of MgB_2 under zero magnetic field.

(**410**, 63-64; 2001)

Jun Nagamatsu[*], Norimasa Nakagawa[*], Takahiro Muranaka[*], Yuji Zenitani[*] & Jun Akimitsu[*†]

[*] Department of Physics, Aoyama-Gakuin University, Chitosedai, Setagaya-ku, Tokyo 157-8572, Japan

[†] CREST, Japan Science and Technology Corporation, Kawaguchi, Saitama 332-0012, Japan

Received 24 January; accepted 5 February 2001.

References:

1. Takagi, H. in *Proc. Int. Conf. on Materials and Mechanisms of Superconductivity, High Temperature Superconductors VI. Physica C* **341-348**, 3-7 (2000).

2. Tanigaki, K. *et al.* Superconductivity at 33 K in $Cs_xRb_yC_{60}$. *Nature* **352**, 222-223 (1991).

3. Cava, R. V. *et al.* Superconductivity near 30 K without copper: the $Ba_{0.6}K_{0.4}BiO_3$ perovskite. *Nature* **332**, 814-816 (1988).

4. Schön, J. H., Kloc, Ch. & Batlogg, B. Superconductivity at 52 K in hole-doped C_{60}. *Nature* **408**, 549-552 (2000).

5. Jones, M. & Marsh, R. The preparation and structure of magnesium boride, MgB_2. *J. Am. Chem. Soc.* **76**, 1434-1436 (1954).

Acknowledgements. This work was partially supported by a Grant-in-Aid for Science Research from the Ministry of Education, Science, Sports and Culture, Japan and by a grant from CREST.

Correspondence and requests for materials should be addressed to J.A. (e-mail: jun@soliton.phys.aoyama.ac.jp).

图 4 显示了 MgB₂ 在零磁场下电阻率随温度的变化关系。转变温度的起止点分别出现在 39 K 和 38 K，这就表明在该系统中真正实现了超导性。

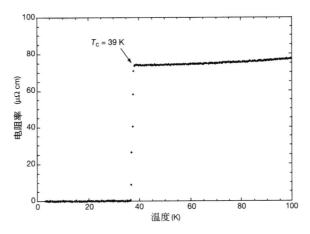

图 4. 零磁场下 MgB₂ 的电阻率随温度的变化关系

（崔宁 翻译；韩汝珊 郭建栋 审稿）

New Hominin Genus from Eastern Africa Shows Diverse Middle Pliocene Lineages

M. G. Leakey *et al.*

Editor's Note

The story of human evolution is not a linear progress of forms, each passing the baton to its successor. Instead, it is more like a bush, with several coexistent lineages. However, most interpretations of human evolution tend to converge on a single lineage between four and three million years ago, represented by *Australopithecus afarensis*. This model was exploded at the turn of the millennium in this paper by Meave Leakey and colleagues, who describe a 3.5-million-year-old cranium from West Turkana that looked radically different from *A. afarensis*. It had a much flatter and more human-like face, a signal that different lineages ran very deep. The status of this form, *Kenyanthropus platyops*, was soon debated, given the difficulties of reconstructing a skull from many fragments.

Most interpretations of early hominin phylogeny recognize a single early to middle Pliocene ancestral lineage, best represented by *Australopithecus afarensis*, which gave rise to a radiation of taxa in the late Pliocene. Here we report on new fossils discovered west of Lake Turkana, Kenya, which differ markedly from those of contemporary *A. afarensis*, indicating that hominin taxonomic diversity extended back, well into the middle Pliocene. A 3.5 Myr-old cranium, showing a unique combination of derived facial and primitive neurocranial features, is assigned to a new genus of hominin. These findings point to an early diet-driven adaptive radiation, provide new insight on the association of hominin craniodental features, and have implications for our understanding of Plio–Pleistocene hominin phylogeny.

THE eastern African hominin record between 4 and 3 Myr is represented exclusively by a single species, *A. afarensis*, and its possible ancestor, *Australopithecus anamensis*, which are commonly thought to belong to the lineage ancestral to all later hominins[1,2]. This apparent lack of diversity in the middle Pliocene contrasts markedly with the increasingly bushy phylogeny evident in the later hominin fossil record. To study further the time interval between 4 and 3 Myr, fieldwork in 1998 and 1999 focused on sites of this age at Lomekwi in the Nachukui Formation, west of Lake Turkana. New hominin discoveries from Lomekwi, as well as two mandibles and isolated molars recovered previously[3] (Table 1), indicate that multiple species existed between 3.5 and 3.0 Myr. The new finds include a well-preserved temporal bone, two partial maxillae, isolated teeth, and most importantly a largely complete, although distorted, cranium. We assign the latter specimen to a new hominin genus on the basis of its unique combination of primitive and derived features.

东非发现的古人类新属表现出
中上新世多样的人类谱系

利基等

编者按

人类演化的故事并不是形式逐个传递的线性变化过程。相反，人类演化是一种灌木式的发展，存在许多并存的谱系。但是许多人试图将人类演化集中到 300 万～400 万年前的一个谱系上去，即南方古猿阿法种。在千禧年之际，米芙·利基及其同事描述了图尔卡纳湖西岸发现的一件 350 万年前的颅骨，它看上去与南方古猿阿法种形态迥异，这件标本使得这一理论发生了翻天覆地的变革。这件标本有着更扁平且更加像人的脸，这是不同谱系各自深入演化的证据。由于从许多块碎片复原头骨的难度较大，所以肯尼亚扁脸人的地位很快就引起了争论。

大多数关于早期古人类系统发育的解释认为，从早上新世到中上新世，人类只有一个祖先谱系，以南方古猿阿法种最具代表性，它在晚上新世产生了辐射演化。本文中我们报道了在肯尼亚图尔卡纳湖西部发现的新化石，它与同时代的南方古猿阿法种化石显著不同，暗示着古人类的分类学多样性可以上溯到中上新世时期。一具 350 万年的颅骨显示出衍生的面部特征和原始的脑颅特征镶嵌的独特的综合特征，我们将其划分到了一个古人类的新属中。这些发现指向一种早期的受饮食驱动的适应性辐射，为古人类的颅牙特征的相关性提供了新的视角，对我们理解上新世–更新世时期的古人类的系统发育也具有提示性。

在 400 万年前到 300 万年前之间，东非的古人类记录只有一个物种，那就是南方古猿阿法种以及它可能的祖先——南方古猿湖畔种，它们通常被认为是后来所有古人类的祖先[1,2]。中上新世这种明显的多样性缺乏，与后来的古人类化石记录中显著的灌木式系统发育形成了鲜明对比。为了进一步研究 400 万年前到 300 万年前这段时间间隔，于 1998 年和 1999 年进行的野外考察工作主要集中在图尔卡纳湖西部纳丘库伊组的洛梅奎的这一年代的遗址上。在洛梅奎取得的新的古人类发现以及之前发掘的两个下颌骨和游离的臼齿[3]（表 1）暗示着在 350 万年前到 300 万年前之间存在着多个物种。这些新发现包括一个保存完好的颞骨、两块上颌骨部分、游离的牙齿和一块虽然变形但是大部分完整的非常重要的颅骨。由于后一标本独特地结合了原始和衍生的特征，所以我们将其划分到了一个古人类新属中。

717

Table 1. Hominin specimens from the lower Lomekwi and Kataboi Members

KNM-WT	Description	Year	Discoverer	Locality	Measurements (mm)
8556	Mandible fragment: symphysis, right body with RP_3–RM_1, isolated partial RM_2, RM_3, LP_3	1982	N. Mutiwa	LO-5	RP_3, 9.8, 12.4; RP_4, 11.3, 12.6; RM_1, 13.7, 12.9; RM_2, NA, NA; RM_3, (17.5), (14.1); LP_3, 9.8, 12.5
8557	$LM_{1/2}$	1982	N. Mutiwa	LO-4	NA, (11.5)
16003	RM^3	1985	M. Kyeva	LO-5	13.3, 14.6
16006	Left mandible fragment with M_2 fragment and M_3	1985	N. Mutiwa	LO-4E	M_2, NA, NA; M_3, 15.3, 13.1
38332	Partial RM^3 crown	1999	M. Eregae	LO-4E	NA, 14.8
38333	$LM_{1/2}$ crown	1999	M. Eregae	LO-4E	13.1, 12.1
38334	$LM_{1/2}$	1999	M. Eregae	LO-4W	12.1, 11.5
38335	$RM_{1/2}$ crown fragment	1999	M. Eregae	LO-4E	NA
38337	$RM^{1/2}$	1999	R. Moru	LO-4E	11.5, 12.3
38338	Partial $RM^{1/2}$ crown	1999	N. Mutiwa	LO-4E	NA
38339	$LM_{1/2}$ crown	1999	J. Erus	LO-4W	12.8, 12.7
38341	Partial $LM_{2/3}$	1999	G. Ekalale	LO-4E	NA
38342	$LM_{1/2}$ crown	1999	J. Erus	LO-4E	12.8, (11.3)
38343	Right maxilla fragment with I^2 and P^3 roots and partial C; mandible fragment with partial P_4 and M_1 roots	1998	J. Erus	LO-4W	NA
38344	$RM_{1/2}$ crown	1998	M. Eregae	LO-9	12.8, 12.2
38346	Partial $RM^{1/2}$	1998	M. Mutiwa	LO-5	NA
38347	LdM_2 crown	1998	R. Moru	LO-5	11.7, 9.6
38349	$RM_{1/2}$ crown	1998	W. Mangao	LO-5	13.5, 12.6
38350	Left maxilla fragment with P^3 and P^4 roots and partial M^1	1998	B. Onyango	LO-5	LM^1: (10.5), (12.0)
38352	Partial $RM_{1/2}$	1998	W. Mangao	LO-5	NA, 11.5
38355	Partial $RM^{1/2}$ crown	1998	M. Eregae	LO-9	NA
38356	Partial $RM^{1/2}$ crown	1998	M. Eregae & J. Kaatho	LO-9	12.8, NA
38357	$RM_{1/2}$	1998	G. Ekalale	LO-5	12.8, 11.8
38358	Associated RI^2, LM_2 fragment, LM_3 RM^3 fragment, four crown fragments	1998	G. Ekalale	LO-5	RI^2, 7.5, 7.5, 9.1; LM_3, 15.3, 13.2
38359	Associated RM_1, RM_2	1998	M. Eregae	LO-5	RM_1, 12.7, 11.6; RM_2, 13.9, 12.2
38361	Associated (partial) germs of I^1, LI^2, RC, LRP^3, LRP^4	1998	R. Moru	LO-5	I^1, NA, (8.0), (11.5); LI^2, 7.6, > 5.9, 8.3; LP^3, (9.3), (12.0)
38362	Associated partial $LM^{1/2}$, $RM^{1/2}$	1998	R. Moru	LO-5	$RM^{1/2}$, 12.9, 14.3
39949	Partial LP_4	1998	R. Moru	LO-5	NA
39950	RM_3	1998	R. Moru	LO-5	16.0, 14.5
39951	$RM_{1/2}$ fragment	1998	R. Moru	LO-5	NA

表1. 从下洛梅奎段和卡塔博伊段发掘的古人类标本

KNM-WT	描述	年份	发现者	产地	测量（毫米）
8556	下颌骨碎片：联合部、右下颌体带有 RP$_3$~RM$_1$、游离的部分 RM$_2$、RM$_3$、LP$_3$	1982	N. 穆蒂瓦	LO-5	RP$_3$, 9.8, 12.4; RP$_4$, 11.3, 12.6; RM$_1$, 13.7, 12.9; RM$_2$, NA, NA; RM$_3$, (17.5), (14.1); LP$_3$, 9.8, 12.5
8557	LM$_{1/2}$	1982	N. 穆蒂瓦	LO-4	NA, (11.5)
16003	RM^3	1985	基耶瓦	LO-5	13.3, 14.6
16006	带有 M$_2$ 碎片与 M$_3$ 的左下颌骨碎片	1985	N. 穆蒂瓦	LO-4E	M$_2$, NA, NA; M$_3$, 15.3, 13.1
38332	部分 RM^3 齿冠	1999	埃雷加埃	LO-4E	NA, 14.8
38333	LM$_{1/2}$ 齿冠	1999	埃雷加埃	LO-4E	13.1, 12.1
38334	LM$_{1/2}$	1999	埃雷加埃	LO-4W	12.1, 11.5
38335	RM$_{1/2}$ 齿冠碎片	1999	埃雷加埃	LO-4E	NA
38337	RM$^{1/2}$	1999	莫鲁	LO-4E	11.5, 12.3
38338	部分 RM$^{1/2}$ 齿冠	1999	N. 穆蒂瓦	LO-4E	NA
38339	LM$_{1/2}$ 齿冠	1999	埃鲁斯	LO-4W	12.8, 12.7
38341	部分 LM$_{2/3}$	1999	埃卡拉莱	LO-4E	NA
38342	LM$_{1/2}$ 齿冠	1999	埃鲁斯	LO-4E	12.8, (11.3)
38343	有 I^2 和 P^3 齿根与部分 C 的右上颌骨；有部分 P$_4$ 与 M$_1$ 齿根的下颌骨碎片	1998	埃鲁斯	LO-4W	NA
38344	RM$_{1/2}$ 齿冠	1998	埃雷加埃	LO-9	12.8, 12.2
38346	部分 RM$^{1/2}$	1998	M. 穆蒂瓦	LO-5	NA
38347	LdM$_2$ 齿冠	1998	莫鲁	LO-5	11.7, 9.6
38349	RM$_{1/2}$ 齿冠	1998	曼加奥	LO-5	13.5, 12.6
38350	有 P^3 和 P^4 齿根和部分 M^1 的左上颌骨碎片	1998	翁扬戈	LO-5	LM^1: (10.5), (12.0)
38352	部分 RM$_{1/2}$	1998	曼加奥	LO-5	NA, 11.5
38355	部分 RM$^{1/2}$ 齿冠	1998	埃雷加埃	LO-9	NA
38356	部分 RM$^{1/2}$ 齿冠	1998	埃雷加埃 & 卡阿索	LO-9	12.8, NA
38357	RM$_{1/2}$	1998	埃卡拉莱	LO-5	12.8, 11.8
38358	齿列 RI^2、LM$_2$ 碎片、LM$_3$、RM^3 碎片、4 个齿冠碎片	1998	埃卡拉莱	LO-5	RI^2, 7.5, 7.5, 9.1; LM$_3$, 15.3, 13.2
38359	齿列 RM$_1$、RM$_2$	1998	埃雷加埃	LO-5	RM$_1$, 12.7, 11.6; RM$_2$, 13.9, 12.2
38361	I^1、LI^2、RC、LRP^3、LRP^4 齿列（部分）牙胚	1998	莫鲁	LO-5	I^1, NA, (8.0), (11.5); LI^2, 7.6, > 5.9, 8.3; LP^3, (9.3), (12.0)
38362	齿列 LM$^{1/2}$ 部分、RM$^{1/2}$	1998	莫鲁	LO-5	RM$^{1/2}$, 12.9, 14.3
39949	部分 LP$_4$	1998	莫鲁	LO-5	NA
39950	RM$_3$	1998	莫鲁	LO-5	16.0, 14.5
39951	LM$_{1/2}$ 碎片	1998	莫鲁	LO-5	NA

Continued

KNM-WT	Description	Year	Discoverer	Locality	Measurements (mm)
39952	LM$_{1/2}$	1998	R. Moru	LO-5	NA
39953	LM$_{1/2}$ fragment	1998	R. Moru	LO-5	NA
39954	Two tooth fragments	1998	R. Moru	LO-5	NA
39955	L$_{\text{C}}$ fragment	1998	R. Moru	LO-5	NA
40000	Cranium	1999	J. Erus	LO-6N	RM^2, 11.4, 12.4
40001	Right temporal bone	1998	P. Gathogo	LO-5	NA

Dental measurements taken as in ref. 34. Mesiodistal crown diameter followed by buccolingual or labiolingual diameter, and for incisors and canines, labial crown height. Values in parentheses are estimates. NA, Not available. L or R in the "Description" column indicates the left or right side. C̲, upper canine; d, deciduous.

Description of *Kenyanthropus platyops*

Order Primates LINNAEUS 1758
Suborder Anthropoidea MIVART 1864
Superfamily Hominoidea GRAY 1825
Kenyanthropus gen. nov.

Etymology. In recognition of Kenya's contribution to the understanding of human evolution through the many specimens recovered from its fossil sites.

Generic diagnosis. A hominin genus characterized by the following morphology: transverse facial contour flat at a level just below the nasal bones; tall malar region; zygomaticoalveolar crest low and curved; anterior surface of the maxillary zygomatic process positioned over premolars and more vertically orientated than the nasal aperture and nasoalveolar clivus; nasoalveolar clivus long and both transversely and sagittally flat, without marked juga; moderate subnasal prognathism; incisor alveoli parallel with, and only just anterior to, the bicanine line; nasal cavity entrance stepped; palate roof thin and flexed inferiorly anterior to the incisive foramen; upper incisor (I^1 and I^2) roots near equal in size; upper premolars (P^3, P^4) mostly three-rooted; upper first and second molars (M^1 and M^2) small with thick enamel; tympanic element mediolaterally long and lacking a petrous crest; external acoustic porus small. *Kenyanthropus* can be distinguished from *Ardipithecus ramidus* by its buccolingually narrow M^2, thick molar enamel, and a temporal bone with a more cylindrical articular eminence and deeper mandibular fossa. It differs from *A. anamensis*, *A. afarensis*, *A. africanus* and *A. garhi* in the derived morphology of the lower face, particularly the moderate subnasal prognathism, sagittally and transversely flat nasoalveolar clivus, anteriorly positioned maxillary zygomatic process, similarly sized I^1 and I^2 roots, and small M^1 and M^2 crowns. From *A. afarensis* it also differs by a transversely flat midface, a small, external acoustic porus, and the absence of an occipital/marginal venous sinus system, and from *A. africanus* by a tall malar region, a low and curved zygomaticoalveolar crest, a narrow nasal aperture, the absence of anterior facial pillars,

720

续表

KNM-WT	描述	年份	发现者	产地	测量（毫米）
39952	LM$_{1/2}$	1998	莫鲁	LO-5	NA
39953	LM$_{1/2}$ 碎片	1998	莫鲁	LO-5	NA
39954	2 个牙齿碎片	1998	莫鲁	LO-5	NA
39955	L$_C$ 碎片	1998	莫鲁	LO-5	NA
40000	颅骨	1999	埃鲁斯	LO-6N	RM^2, 11.4, 12.4
40001	右颞骨	1998	加索戈	LO-5	NA

牙齿的尺寸测量方法见参考文献 34。近中远中端齿冠的直径之后是颊舌侧或唇舌侧直径，门齿和犬齿的直径之后是唇侧齿冠高度。圆括号中的数值是估计值。NA，不可用。"描述"栏中的 L 或 R 表示左侧或右侧。C，上犬齿；d，乳齿。

肯尼亚扁脸人的描述

灵长目 Primates LINNAEUS 1758

类人猿亚目 Anthropoidea MIVART 1864

人超科 Hominoidea GRAY 1825

肯尼亚人（新属）*Kenyanthropus* gen. nov.

词源学 考虑了肯尼亚通过其化石遗址发掘出来的许多标本对于理解人类演化的贡献。

属的鉴别特征 这是一个具有以下形态特征的人属：横向的面部轮廓平坦，恰好与鼻骨下方相平；高的颧骨区；颧骨齿槽脊低而弯曲；上颌骨颧突的前表面位于前臼齿之上，比鼻孔和鼻齿槽斜坡的方向更加垂直；鼻齿槽斜坡长，水平方向和矢状方向都很平，没有明显的隆起；适中的鼻下凸颌；门齿槽与双犬齿线平行并且刚刚位于其前面；鼻腔开口呈梯形；颚顶薄，在门齿孔前面向下弯曲；上门齿（I^1 和 I^2）齿根大小几乎相等；上前臼齿（P^3、P^4）大部分是三齿根的；第一和第二上臼齿（M^1 和 M^2）小，牙釉质厚；鼓室部分中间外侧方向上很长，无岩脊；外耳门小。可以通过肯尼亚人的颊舌侧窄的 M^2、厚的臼齿釉质以及具有更接近圆柱体的关节隆起和更深的下颌窝的颞骨将其与地猿原始种区别开来。在下面部的衍生形态方面，它与南方古猿湖畔种、南方古猿阿法种、南方古猿非洲种和南方古猿惊奇种都不同，尤其是适中的鼻下凸颌、矢状方向和横向都很平的鼻齿槽斜坡、位于前面的上颌骨颧突、相似大小的 I^1 和 I^2 齿根、小的 M^1 和 M^2 齿冠等。其与南方古猿阿法种的不同之处还有：横向扁平的中面部、小的外耳门以及缺少枕窦/边缘窦系统等；与南方古猿非洲种的不同之处包括：高的颧骨区、低而弯曲的颧骨齿槽脊、狭窄的鼻孔、缺少前面柱、管状的长而无脊的鼓室部分以及小的外耳门等。肯尼亚人缺少在傍人埃

a tubular, long and crestless tympanic element, and a small, external acoustic porus. *Kenyanthropus* lacks the suite of derived dental and cranial features found in *Paranthropus aethiopicus*, *P. boisei* and *P. robustus* (Table 2), and the derived cranial features of species indisputably assigned to *Homo* (For example, *H. erectus s.l.* and *H. sapiens*, but not *H. rudolfensis* and *H. habilis*)[4].

Table 2. Derived cranial features of *Paranthropus*, and their character state in *K. platyops* and *H. rudolfensis*

	Paranthropus aethiopicus	Paranthropus boisei	Paranthropus robustus	Kenyanthropus platyops	Homo rudolfensis
Upper molar size	Large	Large	Moderate	Small	Moderate
Enamel thickness	Hyperthick	Hyperthick	Hyperthick	Thick	Thick
Palatal thickness	Thick	Thick	Thick	Thin	Thin
Incisor alveoli close to bicanine line*	Present	Present	Present	Present	Present
Nasoalveolar clivus	Gutter	Gutter	Gutter	Flat	Flat
Midline subnasal prognathism	Strong	Moderate	Moderate	Weak	Weak
Upper I^2 root to lateral nasal aperture	Medial	Medial	Medial	Lateral	Lateral
Nasal cavity entrance	Smooth	Smooth	Smooth	Stepped	Stepped
Zygomaticoalveolar crest	Straight, high	Straight, high	Straight, high	Curved, low	Curved, low
Anteriorly positioned zygomatic process of maxilla*	Present	Present	Present	Present	Present
Midface transverse contour	Concave, dished	Concave, dished	Concave, dished	Flat	Flat
Malar region	Wide	Wide	Wide	Tall	Tall
Malar orientation to lateral nasal margin	Aligned	Aligned	Aligned	More vertical	More vertical
Facial hafting, frontal trigone	High, present	High, present	High, present	Low, absent	Low, absent
Postorbital constriction	Marked	Marked	Marked	Moderate	Moderate
Initial supraorbital course of temporal lines	Medial	Medial	Medial	Posteromedial	Posteromedial
Tympanic vertically deep and plate-like	Present	Present	Present	Absent	Absent
Position external acoustic porus	Lateral	Lateral	Lateral	Medial	Medial
Mandibular fossa depth	Shallow	Deep	Deep	Moderate	Moderate
Foramen magnum heart shaped	Present	Present	Absent	Absent	Absent
Occipitomarginal sinus	Unknown	Present	Present	Absent	Absent

Hypodigm of *H. rudolfensis* as in ref. 35. See refs 1, 8, 11, 36–40 for detailed discussions of the features.
* Character states shared by *Paranthropus* and *K. platyops*.

塞俄比亚种、傍人鲍氏种和傍人粗壮种（表2）中观察到的一系列牙齿和颅骨衍生特征，还缺少那些无可置疑地被划分到人属的物种（例如广义直立人和智人，但不包括鲁道夫人和能人）所具有的颅骨衍生特征[4]。

表 2. 傍人颅骨的衍生特征及其在肯尼亚扁脸人和鲁道夫人中的特征状态

	傍人埃塞俄比亚种	傍人鲍氏种	傍人粗壮种	肯尼亚扁脸人	鲁道夫人
上臼齿尺寸	大	大	中	小	中
牙釉质厚度	超厚	超厚	超厚	厚	厚
颚厚度	厚	厚	厚	薄	薄
门齿槽靠近双犬齿线 *	存在	存在	存在	存在	存在
鼻齿槽斜坡	沟状	沟状	沟状	平坦	平坦
中线鼻下凸颌	强	中	中	弱	弱
上 I^2 齿根相对梨状孔位置	中间	中间	中间	侧向	侧向
鼻腔开口	平滑	平滑	平滑	梯形	梯形
颧骨齿槽脊	直而高	直而高	直而高	弯而低	弯而低
位于前方的上颌骨颧骨突 *	存在	存在	存在	存在	存在
中面部横向轮廓	中凹碟形	中凹碟形	中凹碟形	平坦	平坦
颧骨区	宽	宽	宽	高	高
颧骨相对于鼻侧边缘	齐平	齐平	齐平	更垂直	更垂直
面部，额三角	高，存在	高，存在	高，存在	低，缺失	低，缺失
眶后缩狭	明显	明显	明显	中	中
颞骨线的初始上眶路径	中间	中间	中间	中后	中后
鼓室垂直方向深，呈板状	存在	存在	存在	缺失	缺失
外耳门位置	侧向	侧向	侧向	中间	中间
下颌窝深度	浅	深	深	适中	适中
心形枕骨大孔	存在	存在	缺失	缺失	缺失
枕边缘窦	未知	存在	存在	缺失	缺失

鲁道夫人的种型群见参考文献35。对特征的详细讨论见参考文献1、8、11、36～40。

* 傍人和肯尼亚扁脸人共有的特征状态。

Type species. *Kenyanthropus platyops* sp. nov.

Etymology. From the Greek *platus*, meaning flat, and *opsis*, meaning face; thus referring to the characteristically flat face of this species.

Specific diagnosis. Same as for genus.

Types. The holotype is KNM-WT 40000 (Fig. 1a–d), a largely complete cranium found by J. Erus in August 1999. The paratype is KNM-WT 38350 (Fig. 1e), a partial left maxilla found by B. Onyango in August 1998. The repository is the National Museums of Kenya, Nairobi.

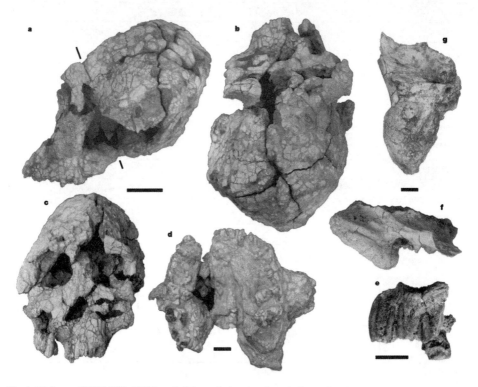

Fig. 1. Holotype KNM-WT 40000. **a**, Left lateral view (markers indicate the plane separating the distorted neurocranium and the well-preserved face). **b**, Superior view. **c**, Anterior view. **d**, Occlusal view of palate. Paratype KNM-WT 38350. **e**, Lateral view. KNM-WT 40001. **f**, Lateral view. **g**, Inferior view. Scale bars: **a–c**, 3 cm; **d–g**, 1 cm.

Localities. Lomekwi localities are situated in the Lomekwi and Topernawi river drainages in Turkana district, northern Kenya (Fig. 2). The type locality LO-6N is at 03° 54.03′ north latitude, 035° 44.40′ east longitude.

Horizon. The type specimen is from the Kataboi Member, 8 m below the Tulu Bor Tuff and 12 m above the Lokochot Tuff, giving an estimated age of 3.5 Myr. The paratype is from the lower Lomekwi Member, 17 m above the Tulu Bor Tuff, with an estimated age of 3.3 Myr.

724

模式种　肯尼亚扁脸人新种。

词源　希腊语中 *platus* 意思是扁平的，*opsis* 意思是面部；因此是指该物种特征性的扁平面部。

种的鉴别特征　与属的特征是一样的。

类型　正模标本是 KNM-WT 40000(图 1a～1d)，是由埃鲁斯于 1999 年 8 月发现的一个大部分完整的颅骨。副模标本是 KNM-WT 38350(图 1e)，是由翁扬戈于 1998 年 8 月发现的部分左上颌骨。它们都被保存在位于内罗毕的肯尼亚国家博物馆中。

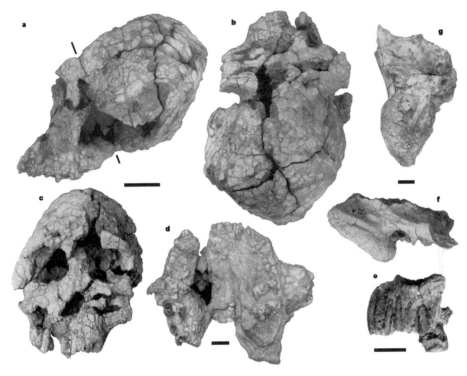

图 1. 正模标本 KNM-WT 40000。**a**，左侧面观(标记表示将变形的脑颅与保存完好的面部分离开来的平面)。**b**，上面观。**c**，前面观。**d**，颚的咬合面观。副模标本 KNM-WT 38350。**e**，侧面观。KNM-WT 40001。**f**，侧面观。**g**，下面观。比例尺：**a**～**c**，3 厘米；**d**～**g**，1 厘米。

产地　洛梅奎遗址位于肯尼亚北部的图尔卡纳地区的洛梅奎和托佩尔纳维河流域(图 2)。LO-6N 典型地点位于北纬 03°54.03′，东经 035°44.40′。

地层　该类型标本是从卡塔博伊段发掘出来的，位于图卢博尔凝灰岩之下 8 米、洛科乔特凝灰岩之上 12 米处，估计其年代为 350 万年。副模标本是在下洛梅奎段发掘出来的，位于图卢博尔凝灰岩之上 17 米处，估计其年代为 330 万年。

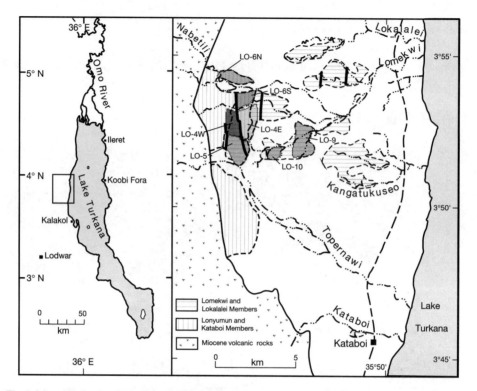

Fig. 2. Map showing localities of fossil collection in upper Lomekwi and simplified geology. The boundary between the Kataboi and Lomekwi Members is the base of the Tulu Bor Tuff, indicated as a dashed line through LO-4E and LO-4W. Faults are shown as thick lines; minor faults are omitted. LO-4E and LO-4W are of different shades to distinguish them from each other.

Cranial Description and Comparisons

The overall size of the KNM-WT 40000 cranium falls within the range of *A. afarensis* and *A. africanus*. It is preserved in two main parts, the neurocranium with the superior and lateral orbital margins, but lacking most of the cranial base; and the face, lacking the premolar and anterior tooth crowns and the right incisor roots. Most of the vault is heavily distorted, both through post-mortem diploic expansion and compression from an inferoposterior direction (Fig. 1a, b). The better preserved facial part shows some lateral skewing of the nasal area, anterior displacement of the right canine, and some expansion of the alveolar and zygomatic processes (Fig. 1c–d), but allows for reliable assessment of its morphology.

Only the right M^2 crown is sufficiently preserved to allow reliable metric dental comparisons. It is particularly small, falling below the known ranges of other early hominin species (Fig. 3a). Likewise, the estimated M^1 crown size of KNM-WT 38350 (Table 1) corresponds to minima for *A. anamensis*, *A. afarensis* and *H. habilis*, and is below the ranges for other African early hominins[5-7]. Molar enamel thickness in both specimens is comparable to that in *A. anamensis* and *A. afarensis*. CT scans show that both P^3 and P^4 of KNM-WT 40000 have a lingual root and two well-separated buccal roots. This morphology, thought

726

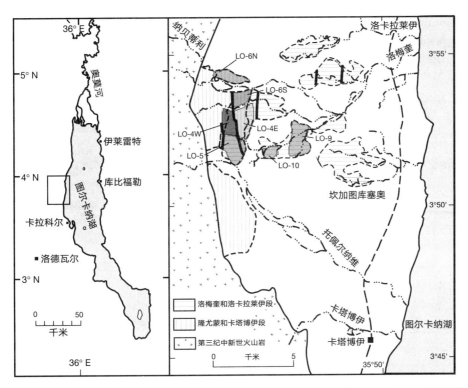

图 2. 该地图表示了在上洛梅奎段搜集到的化石的位置和简化的地质情况。卡塔博伊和洛梅奎段间的界线是图卢博尔凝灰岩的基部，用一条穿过 LO-4E 和 LO-4W 的虚线表示。断层用粗线表示；小断层都省略掉了。LO-4E 和 LO-4W 用不同的明暗度表示以相互区分。

颅骨的描述与比较

KNM-WT 40000 颅骨的总大小处于南方古猿阿法种和南方古猿非洲种的范围内。其保存下来了两个主要部分：脑颅，具有眼窝上缘和侧缘、缺少大部分颅底；面部，缺少前臼齿、前部牙齿齿冠和右门齿根。颅顶大部分由于死后板障扩张和来自下后方向的压缩而严重变形了（图 1a 和 1b）。保存较好的面部区域显示出鼻区有些侧斜，右犬齿向前移位以及齿槽和颧骨突有些扩张（图 1c～1d），但是还可以根据它们可靠地估计其形态。

只有右 M^2 的齿冠保存下来的部分足以用来可靠地比较牙齿的尺寸。它非常小，比其他早期的古人类的已知范围还要小（图 3a）。同样地，KNM-WT 38350 的 M^1 的齿冠的估值（表 1）与南方古猿湖畔种、南方古猿阿法种和能人的最小值一致，而比其他的非洲早期古人类的尺寸范围要小[5-7]。两个标本的臼齿釉质厚度都与南方古猿湖畔种和南方古猿阿法种的相当。CT 扫描表明 KNM-WT 40000 的 P^3 和 P^4 都有一个舌侧的齿根和两个完全分离的颊侧齿根。这种形态被认为是人科动物的祖先种才

to be the ancestral hominoid condition[8], is commonly found in *Paranthropus*, but is variable among species of *Australopithecus*. The P[3] of KNM-WT 38350 has three well-separated roots (Fig. 1e). Its P[4] seems to be two-rooted, but the deeply grooved buccal root may split more apically. Relative to M[2] crown size, the canine roots of KNM-WT 40000 are smaller in cross-section at the alveolar margin than in *Ardipithecus ramidus* and *A. anamensis*, similar in size to *A. afarensis*, *A. africanus* and *H. habilis*, and larger than in *P. boisei*. Exposed surfaces and CT scans demonstrate that the I[1] and I[2] roots in KNM-WT 40000 are straight and similar in size. At the level of the alveolar margin the cross-sectional area of the I[2] root is about 90% of that of the I[1] root, whereas this is typically 50–70% in other known hominid taxa.

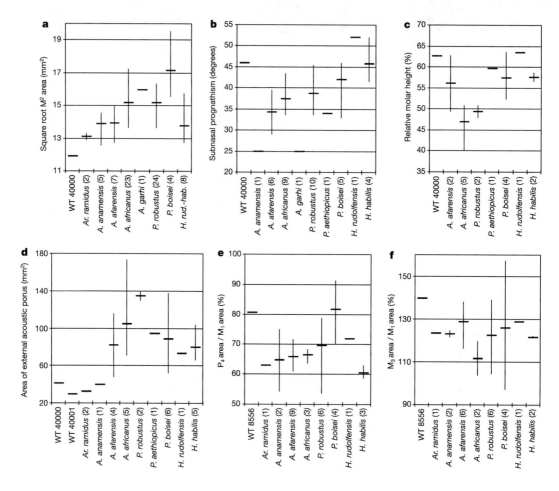

Fig. 3. Mean and range of characters of specified hominins. **a**, Square root of M[2] crown area (buccolingual × mesiolingual diameters). **b**, Angle of subnasal prognathism (nasospinale–prosthion to postcanine alveolar plane). **c**, Malar height[8] relative to orbitoalveolar height (orbitale to alveolar margin aligned with malar surface). **d**, Area of the external acoustic porus (π × long axis × short axis). **e**, Crown area of P[4] relative to that of M[1] × 100. **f**, Crown area of M[3] relative to that of M[1] × 100. All measurements are taken from originals, directly or as given in refs 8, 9, 14, 18, 37, 41–44, except for some South African crania taken from casts. Numbers in parentheses indicate sample size.

具有的特征[8]，通常是在傍人属中发现的，但是在南方古猿的物种中具有较多变异。KNM-WT 38350 的 P^3 具有三个完全分开的齿根（图 1e）。其 P^4 看上去似乎具有两个齿根，但是有较深槽的颊侧齿根可能在靠近根尖部分开。相对于 M^2 的齿冠尺寸而言，KNM-WT 40000 的犬齿根在齿槽边缘的横截面处比地猿原始种和南方古猿湖畔种的要小，而与南方古猿阿法种、南方古猿非洲种和能人的尺寸相近，比傍人鲍氏种的要大。暴露出来的表面和 CT 扫描证实 KNM-WT 40000 的 I^1 和 I^2 的齿根很直，大小相似。在齿槽边缘水平面上，I^2 齿根的横截面面积大约是 I^1 齿根的 90%，而这一数值在其他已知的人科动物分类单元中多为 50% ~ 70%。

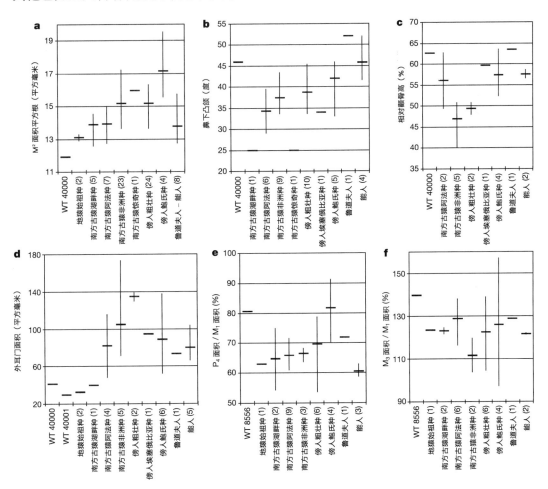

图 3. 特定古人类特征的平均值和范围。**a**，M^2 齿冠面积（颊舌侧 × 近中舌侧直径）的平方根。**b**，鼻下凸颌的角度（鼻棘点–齿槽中点与犬齿后部牙齿牙槽牙平面之间）。**c**，相对于眶齿槽高度（眶最下点到与颧骨表面在一条直线上的齿槽边缘的距离）的颧骨高度[8]。**d**，外耳门的面积（π × 长轴 × 短轴）。**e**，P$_4$ 的齿冠面积与 M$_1$ 的齿冠面积之比 × 100。**f**，M$_3$ 的齿冠面积与 M$_1$ 的齿冠面积之比 × 100。所有测量值都是直接从原始标本测量得到或者来自参考文献 8、9、14、18、37、41~44 中给出的数据，除了有些南非颅骨的尺寸是根据模型测量的。圆括号中的数字代表标本尺寸。

The incisor alveoli of KNM-WT 40000 are aligned coronally, just anterior to the bicanine line, and the overlying nasoalveolar clivus is flat both sagittally and transversely. There is no canine jugum visible on the preserved left side, reflecting the modest size of the canine root. At 32 mm (chord distance nasospinale to prosthion) the clivus is among the longest of all early hominins. Subnasal prognathism is moderate, expressed by a more vertically orientated clivus than in nearly all specimens of *Australopithecus* and *Paranthropus* (Fig. 3b). The nasal aperture lies in the same coronal plane as the nasoalveolar clivus and there are no anterior facial pillars (Fig. 1a–c). The nasal aperture is small and narrow, in contrast to the large, wide aperture in *A. africanus* and *P. robustus*. The midface of KNM-WT 40000 is dominated by the tall malar region (Fig. 3c) with a low and curved zygomaticoalveolar crest. At a level just below the nasal bones the transverse facial contour is flat (Fig. 1b). In both KNM-WT 40000 and KNM-WT 38350 the anterior surface of the zygomatic process of the maxilla is positioned between P^3 and P^4 (Fig. 1a, d, e), as is commonly seen in *Paranthropus*, but more anteriorly than in most *Australopithecus* specimens[9] or in *H. habilis*. The supraorbital region is *Australopithecus*-like, lacking both a frontal trigon as seen in *Paranthropus*, and a supratoral sulcus as seen in *H. habilis* (but not *H. rudolfensis*). Relative postorbital constriction (frontofacial index) of KNM-WT 40000 is similar to that in *Australopithecus*, *H. rudolfensis* and *H. habilis*, and less than in *P. boisei* (estimated frontofacial index[9] = 70). Its temporal lines converging on the frontal squama have a posteromedial course throughout (Fig. 1b). Around bregma the midline morphology is not well preserved, but the posterior half of the parietals show double, slightly raised temporal lines about 6 mm apart. These contribute posteriorly to indistinct compound temporal/nuchal lines. The original shape of the severely distorted mastoids cannot be reconstructed, but other parts of the left temporal are well preserved. The tubular tympanic lacks a petrous crest and forms a narrow external acoustic meatus with a small aperture. This combination constitutes the primitive hominin morphology, also seen in *Ar. ramidus* and *A. anamensis* (Fig. 3d). The mandibular fossa resembles that of specimens of *A. afarensis* and *A. africanus*. It is moderately deep, and the articular eminence, missing its lateral margin, is cylindrical with a moderately convex sagittal profile. The preserved posterior half of the foramen magnum suggests that it was probably oval in shape, rather than the heart shape seen in *P. boisei* and probably *P. aethiopicus*. Regarding the endocranial aspect, the reasonably well preserved occipital surface lacks any indication of the occipital/marginal venous sinus system characteristic of *A. afarensis*, *P. boisei* and *P. robustus*. Bilateral sulci suggest that the transverse/sigmoid sinus system was well developed. Endocranial capacity is difficult to estimate because of the distorted vault. However, comparing hominin glabella–opisthion arc lengths[8] with that of KNM-WT 40000 (259 mm; an estimate inflated by diploic expansion) suggests a value in the range of *Australopithecus* or *Paranthropus*.

The sex of KNM-WT 40000 is difficult to infer. Interpretation of the canine root size proves inconclusive without a suitable comparative context. The small M^2 crown size could suggest that the specimen is female. However, the close proximity and slightly raised aspect of the temporal lines on the posterior half of the parietals is not seen in known female hominin crania, including the *Paranthropus* specimens KNM-ER 732, KNM-ER 407 and DNH7, and suggests that KNM-WT 40000 could be male.

KNM-WT 40000 的门齿齿槽呈冠状排列，刚刚位于双犬齿线的前面，上面的鼻齿槽斜坡在矢状方向和横向上都是平的。在保存下来的左侧没有看到犬齿轭，这反映了犬齿齿根具有最适度的大小。斜坡的长度（从鼻棘点到牙槽中点的弦距离）是 32 毫米，这是所有早期古人类中最长的之一。鼻下凸颌适中，具有比几乎所有南方古猿和傍人属标本都更垂直的斜坡（图 3b）。鼻孔与鼻齿槽斜坡位于同一冠状面上，没有前面柱（图 1a ~ 1c）。鼻孔小而窄，与南方古猿非洲种和傍人粗壮种的大而宽的鼻孔形成对比。KNM-WT 40000 的中面部主要有一个高高的颧骨区（图 3c），颧骨区有一个低而弯曲的颧骨齿槽脊。在刚刚位于鼻骨之下的水平面上，横向的面部轮廓是平的（图 1b）。在 KNM-WT 40000 和 KNM-WT 38350 中，上颌骨颧突的前表面都位于 P^3 和 P^4 之间（图 1a、1d 和 1e），这在傍人中是很常见的，但是比大部分南方古猿标本[9] 或能人中的位置要更靠前一些。眶上区域与南方古猿相似，既缺少傍人中所见的额三角，也没有能人（并非鲁道夫人）中所见到的圆枕上沟。KNM-WT 40000 的相对眶后缩狭（额面指数）与南方古猿、鲁道夫人和能人的都相似，而比傍人鲍氏种的（额面指数估计值[9] = 70）要小。其收敛于额鳞部的颞线有一个贯穿后中的路径（图 1b）。前囟周围，中线的形态没有很好地保存下来，但是顶骨的后半部分显示出大约相隔 6 毫米远的、双重的、稍微抬高的颞线。再向后形成了不明确的复合颞/项线。严重变形的乳突的原始形状无法重建出来了，但是左颞骨的其余部分保存完好。管状鼓室缺少岩脊，形成了一个具有小孔的狭窄的外耳道。这种组合构成了原始的古人类形态，在地猿原始种和南方古猿湖畔种中也见到过（图 3d）。其下颌窝与南方古猿阿法种和南方古猿非洲种标本的很像。其深度适中；关节隆起呈圆柱形，侧边缘丢失，有一个凸度适中的矢状面。保存下来的枕骨大孔的后半部分表明，其形状可能是卵圆形，而非傍人鲍氏种中见到（傍人埃塞俄比亚种中也可能见到）的心形。至于颅内方面，保存相当完好的枕骨表面缺少任何南方古猿阿法种、傍人鲍氏种和傍人粗壮种特征性的枕窦/边缘窦系统。双侧沟表明横窦/乙状窦系统很发达。颅容量很难估计，因为颅顶已经变形了。但是，将古人类眉间-枕后点弧长[8] 与 KNM-WT 40000 的（259 毫米；由板障扩张获得的估计值）比较，得到的数值处于南方古猿或傍人的范围内。

KNM-WT 40000 的性别很难推断。对于犬齿齿根尺寸的说明证明是不确定的，没有适当的可用于比较的方面。小的 M^2 齿冠尺寸能够表明该标本是雌性。但是，在顶骨后半部分颞线极为接近并且较高，这在已知的雌性古人类颅骨（包括傍人标本 KNM-ER 732、KNM-ER 407 和 DNH7）中并没有见到，这一特点表明 KNM-WT 40000 可能是雄性。

With incisor alveoli close to the bicanine line and anteriorly positioned zygomatic processes, the face of KNM-WT 40000 resembles the flat, orthognathic-looking faces of both *Paranthropus* and *H. rudolfensis* cranium KNM-ER 1470. However, KNM-WT 40000 lacks most of the derived features that characterize *Paranthropus* (Table 2), and its facial architecture differs from the latter in much the same way as has been described for KNM-ER 1470 (refs 8, 10). Facial flatness in *Paranthropus* results from the forward position of the anteroinferiorly sloping malar region, whose main facial surface approximates the plane of the nasal aperture, but whose orientation contrasts with the more horizontally inclined nasoalveolar gutter[11]. In KNM-WT 40000 and KNM-ER 1470, it is the flat and orthognathic nasoalveolar clivus that aligns with the plane of the nasal aperture, whereas the anteriorly set, tall malar region is more vertically orientated. KNM-WT 40000 lacks the derived short nasal bones and everted lateral nasal margin of KNM-ER 1470, and is less orthognathic in the midfacial region than this specimen; however, on balance this is the hominin face that KNM-WT 40000 most closely resembles.

Additional Material

The right maxilla fragment KNM-WT 38343A preserves three well-separated P[3] roots, and its damaged canine seems low-crowned when compared with *A. afarensis* canines of similar size and degree of wear. The right temporal bone KNM-WT 40001 lacks the squama and petrous apex, but is otherwise well preserved (Fig. 1f, g). It shows a combination of characters not seen in any other hominin specimen. The projecting mastoid process is rounded, with an anteriorly positioned tip. It has a well-developed digastric fossa in the form of a deep, narrow groove that runs posterolaterally from the stylomastoid foramen, fully demarcating the mastoid process from the adjacent nuchal plane. The tympanic element is long, inferosuperiorly shallow and lacks a petrous crest. The external acoustic porus is the smallest of any known hominin temporal bone (Fig. 3d). The articular eminence is as broad mediolaterally (38 mm) as in *P. aethiopicus* and *P. boisei*, and similar to the largest found in *A. afarensis*. Compared with KNM-WT 40000 the eminence is relatively flat sagittally, and the mandibular fossa is shallow.

The partial mandibles KNM-WT 8556 and KNM-WT 16006 have been assigned to *A. afarensis*[3]. However, KNM-WT 8556 shows a more derived morphology than this species by having a flat, more horizontal post-incisive plane, a more superiorly positioned genioglossal pit, a molarized lower fourth premolar (P_4) and a large M_3 (ref. 3). Indeed, relative to its *Australopithecus*-sized M_1 (refs 5, 6, 12, 13), the P_4 and M_3 crowns of KNM-WT 8556 are enlarged to an extent only seen in *P. boisei* (Fig. 3e, f). All unworn molars in the Lomekwi sample are characterized by low occlusal relief and numerous secondary fissures. Most of the lower molars, including the KNM-WT 16006 M_3, have a well-developed protostylid, a feature that is usually absent in *A. afarensis*, but common in *A. africanus*[14]. The two I[2]s are lower crowned than in *A. afarensis*, *A. africanus*[14] and *P. robustus*[14]. Inability to distinguish between first and second molars makes meaningful intertaxon comparisons of these elements difficult.

KNM-WT 40000 的门齿槽靠近双犬齿线，颧突前置，面部与傍人和鲁道夫人颅骨 KNM-ER 1470 的扁平而正颌的面部很像。然而，KNM-WT 40000 缺少大部分表征傍人的衍生特征(表2)，其面部结构与后者不同，而与 KNM-ER 1470 描述过的情况几乎一样(参考文献8和10)。傍人的面部扁平的特点是由于向前下方倾斜的颧骨区的位置向前，其主面部表面接近鼻孔平面，但是其方向与更加水平倾斜的鼻齿槽沟形成对比[11]。在 KNM-WT 40000 和 KNM-ER 1470 中，扁平而正颌的鼻齿槽斜坡与鼻孔平面走向一致，而位于前面的高的颧骨区更加垂直。KNM-WT 40000 缺少 KNM-ER 1470 所具有的衍生的短小的鼻骨和外翻的侧鼻缘，中面部的正颌性比该标本差；但是，总的说来，这是与 KNM-WT 40000 最为相像的古人类面部。

其 他 材 料

右上颌骨碎片 KNM-WT 38343A 保存下来了三个完全分离的 P^3 齿根，当与南方古猿阿法种的具有相似大小和磨损程度的犬齿比较时，发现其损坏的犬齿似乎是低齿冠的。右颞骨 KNM-WT 40001 缺少鳞部和岩部尖，但是同样保存状况很好(图1f和1g)。其显示出在其他古人类标本中所没有看到过的特征的组合。突出的乳突呈圆形，有一个位于前面的尖端。它有一个发达的二腹肌窝，呈深而狭窄的沟状，从茎乳孔向后外侧延伸，完全将乳突与相邻的项面分隔开来。鼓室长，上下方向浅，缺少岩脊。外耳门是现在已知的古人类颞骨中最小的(图3d)。关节隆起在中外侧像傍人埃塞俄比亚种和傍人鲍氏种的一样宽阔(38毫米)，与南方古猿阿法种中发现的最大值很相似。与 KNM-WT 40000 相比，该隆起在矢状方向上相对平一些，下颌窝很浅。

部分下颌骨 KNM-WT 8556 和 KNM-WT 16006 都被划分到了南方古猿阿法种中[3]。但是，KNM-WT 8556 显示出比该物种更加衍生的形态，具有扁平而更加水平的门齿后平面、位置靠上的颏舌肌窝、臼齿化的下第四前白齿(P_4)和一个大的 M_3(参考文献3)。实际上，相对于其尺寸与南方古猿相当的 M_1(参考文献5、6、12和13)，KNM-WT 8556 的 P_4 和 M_3 的齿冠扩大到了只有在傍人鲍氏种中才见到过的程度(图3e和3f)。洛梅奎标本中所有没磨损的臼齿都具有的特征是低的咬合面脊纹和许多次级沟纹。大部分下白齿，包括 KNM-WT 16006 的 M_3，都具有发达的原副尖，这是南方古猿阿法种中通常不具备而南方古猿非洲种中很常见的特征[14]。两颗 I^2 的齿冠都比南方古猿阿法种、南方古猿非洲种[14]和傍人粗壮种[14]的低。无法区分第一和第二白齿，这使得很难对这些部位在分类单元之间进行有意义的比较。

Taxonomic Discussion

The hominin specimens recovered from the Kataboi and lower Lomekwi Members show a suite of features that distinguishes them from established hominin taxa, including the only contemporaneous eastern African species, *A. afarensis*. Compared with the latter, the morphology of *K. platyops* is more derived facially, and more primitive in its small external acoustic porus and the absence of an occipital/marginal sinus system. These finds not only provide evidence for a taxonomically more diverse middle Pliocene hominin record, but also show that a more orthognathic facial morphology emerged significantly earlier in hominin evolutionary history than previously documented. This early faciodental diversity concerns morphologies that functionally are most closely associated with mastication. It suggests a diet-driven adaptive radiation among hominins in this time interval, which perhaps had its origins considerably earlier. Furthermore, the presence in *K. platyops* of an anteriorly positioned zygomatic process in combination with a small M^1 and M^2 indicates that such characters are more independent than is suggested by developmental and functional models that link such facial morphology in *Paranthropus* with postcanine megadontia[11,15].

At present it is unclear whether the Lomekwi hominin fossils sample multiple species. Apart from the paratype maxilla KNM-WT 38350 with its small molar size and anteriorly positioned zygomatic process, the other specimens cannot be positively associated morphologically with the *K. platyops* holotype. These are therefore not included in the paratype series, and are left unassigned until further evidence emerges. Differences between the tympanic and mandibular fossa morphologies of the KNM-WT 40000 and KNM-WT 40001 temporal bones can perhaps be accommodated within a single species, but their shared primitive characters do not necessarily imply conspecificity. Affiliation of the KNM-WT 8556 mandible with the *K. platyops* types is not contradicted by its molarized P_4, which is consistent with an anteriorly positioned zygomatic process. However, its M_1 is larger than would be inferred from the smaller upper molars of the types, and with a 177 mm^2 crown area it is also larger than any in the combined sample of ten isolated M_1s and M_2s (139–172 mm^2). One isolated $M^{1/2}$ (KNM-WT 38362) is significantly larger than the molars of the *K. platyops* types, whereas another (KNM-WT 38337) is similar in size to the holotype's M^2.

The marked differences of the KNM-WT 40000 cranium from established hominin taxa, both with respect to individual features and their unique combination, fully justify its status as a separate species. It is worth noting that comparisons with *Australopithecus bahrelghazali* cannot be made directly, because this species was named on the basis of the limited evidence provided by an anterior mandible fragment[16]. Specific distinction of *A. bahrelghazali* from *A. afarensis* has yet to be confirmed[17], and Lomekwi specimens differ from *A. bahrelghazali* in symphyseal morphology and incisor crown height.

The generic attribution of KNM-WT 40000 is a more complex issue, in the absence of

分类学讨论

从卡塔博伊和下洛梅奎段发掘到的古人类标本显示出一套与已确定的古人类分类单元不同的特征，与唯一的同时代的东非物种——南方古猿阿法种也不相同。与后者相比，肯尼亚扁脸人的面部形态更具有衍生性，其较小的外耳门和枕窦/边缘窦系统的缺失则更加原始。这些发现不仅为分类学上更加多样的中上新世古人类记录提供了证据，而且也表明，更加正颌的面部形态在古人类演化史上出现的时间比之前文件证明的早得多。这一早期的面牙多样性涉及那些功能上与咀嚼关系最密切的形态。它表明在这一时间段的古人类中存在着饮食驱动的适应性辐射，其起源的时间可能更早。此外，肯尼亚扁脸人位置靠前的颧突与小型的 M^1 和 M^2 组合暗示了这些特征更具有独立性，而非发育和功能学模型所倡导的将傍人中的这种面部形态与较大的犬齿后齿联系起来[11,15]。

现在还不清楚洛梅奎古人类化石标本是否包含多个物种。除了副模标本上颌骨 KNM-WT 38350 及其小型臼齿尺寸和前置的颧突外，不能肯定其余标本在形态上与肯尼亚扁脸人正模标本有关。因此这些不被包括在副模标本系列中，直到出现进一步的证据时，才能对这些标本的归属进行判定。KNM-WT 40000 与 KNM-WT 40001 颞骨的鼓室和下颌窝形态之间的差异或许可以被归结为同一个物种内的差异，但是它们共有的原始特征未必意味着同种性。KNM-WT 8556 下颌骨归属于肯尼亚扁脸人类型，与其臼齿化的 P_4 并不冲突，这与前置的颧突是一致的。然而，其 M_1 比根据该类型较小的上臼齿推断出来的要大，齿冠面积达 177 平方毫米，比十颗游离的 M_1 和 M_2 的所有标本中的任何一个（139~172 平方毫米）都大。一颗游离的 $M^{1/2}$ （KNM-WT 38362）比肯尼亚扁脸人类型的臼齿要大得多，但是另一个标本（KNM-WT 38337）则与正模标本的 M^2 的大小相似。

KNM-WT 40000 颅骨与已确定的古人类分类单元在个体特征与其特异的结合方面的明显差异，都充分地说明了将其当成一个单独的物种的合理性。值得说明的是，不能直接与南方古猿羚羊河种进行比较，因为该物种是根据一块前下颌骨碎片提供的有限证据而命名的[16]。南方古猿羚羊河种与南方古猿阿法种的特异差别还没有得到证实[17]，洛梅奎标本与南方古猿羚羊河种在下颌联合部的形态和门齿齿冠高度方面有差异。

KNM-WT 40000 属一级别的划分归类是一个更加复杂的问题，对于该属分类的

consensus over the definition of the genus category[4]. The specimen lacks almost all of the derived features of *Paranthropus* (Table 2), and there are no grounds for assigning it to this genus unless it can be shown to represent a stem species. However, the fact that the facial morphology of KNM-WT 40000 is derived in a markedly different way renders this implausible. As KNM-WT 40000 does not show the derived features associated with *Homo*[4] (excluding *H. rudolfensis* and *H. habilis*) or the strongly primitive morphology of *Ardipithecus*[18], the only other available genus is *Australopithecus*. We agree with the taxonomically conservative, grade-sensitive approach to hominin classification that for the moment accepts *Australopithecus* as a paraphyletic genus in which are clustered stem species sharing a suite of key primitive features, such as a small brain, strong subnasal prognathism, and relatively large postcanine teeth. However, with its derived face and small molar size, KNM-WT 40000 stands apart from species assigned to *Australopithecus* on this basis. All it has in common with such species is its small brain size and a few other primitive characters in the nasal, supraorbital and temporal regions. Therefore, there is no firm basis for linking KNM-WT 40000 specifically with *Australopithecus*, and the inclusion of such a derived but early form could well render this genus polyphyletic. In a classification in which *Australopithecus* also includes the "robust" taxa and perhaps even species traditionally known as "early *Homo*"[4], this genus subsumes several widely divergent craniofacial morphologies. It could thus be argued that the inclusion of KNM-WT 40000 in *Australopithecus* would merely add yet another hominin species with a derived face. This amounts to defining *Australopithecus* by a single criterion, those hominin species not attributable to *Ardipithecus* or *Homo*, which in our view constitutes an undesirable approach to classification. Thus, given that KNM-WT 40000 cannot be grouped sensibly with any of the established hominin genera, and that it shows a unique pattern of facial and dental morphology that probably reflects a distinct dietary adaptive zone, we assign this specimen to the new genus *Kenyanthropus*.

Despite being separated by about 1.5 Myr, KNM-WT 40000 is very similar in its facial architecture to KNM-ER 1470, the lectotype of *H. rudolfensis*. The main differences amount to the more primitive nasal and neurocranial morphology of KNM-WT 40000. This raises the possibility that there is a close phylogenetic relationship between the two taxa, and affects our interpretation of *H. rudolfensis*. The transfer of this species to *Australopithecus* has been recommended[4,19], but *Kenyanthropus* may be a more appropriate genus. The identification of *K. platyops* has a number of additional implications. As a species contemporary with *A. afarensis* that is more primitive in some of its morphology, *K. platyops* weakens the case for *A. afarensis* being the sister taxon of all later hominins, and thus its proposed transfer to *Praeanthropus*[1,20]. Furthermore, the morphology of *K. platyops* raises questions about the polarity of characters used in analyses of hominin phylogeny. An example is the species' small molar size, which, although probably a derived feature, might also imply that the larger postcanine dentition of *A. afarensis* or *A. anamensis* does not represent the primitive hominin condition. Finally, the occurrence of at least one additional hominin species in the middle Pliocene of eastern Africa means that the affiliation of fragmentary specimens can now be reassessed. For example, the attribution of the 3.3 Myr old KNM-ER 2602 cranial fragment to *A. afarensis*[21] has been questioned[8],

定义还没有达成一致[4]。标本缺少几乎所有的傍人的衍生特征（表2），除非可以表明其代表了主干物种，否则没有依据将其划分到该属。但是，KNM-WT 40000 的面部形态的衍生方式如此迥异，使其不可能代表主干物种。鉴于 KNM-WT 40000 并没有显示出与人属（除了鲁道夫人和能人）相关的衍生特征[4]或者地猿的非常原始的形态[18]，其他唯一一个可考虑的属就是南方古猿了。我们同意分类学上保守的、级别敏感的古人类的分类方式，该方式目前将南方古猿认为是一种并系属，在该属中聚集了共有一套关键的原始特征的主干物种，这里所说的关键的原始特征包括较小的脑、强壮的鼻下凸颌和相对大的犬齿后部牙齿等。然而，KNM-WT 40000 具有衍生的面部和小的臼齿，这些将其与划分为南方古猿的物种分开了。它与该类物种共有的全部特征就是其小的脑和鼻骨、眶上区和颞区的一些其他原始特征。因此，没有确定的依据将 KNM-WT 40000 特异性地与南方古猿联系起来，将这样一个衍生但尚属早期的形式包括进来可以很好地使这个属复杂化。在一种分类中，南方古猿也包括"粗壮"的分类单元，甚至可能包括传统意义上称为"早期人属"的物种[4]，该属包含几种差别很大的颅面形态。因此可以这样说，将 KNM-WT 40000 囊括进南方古猿仅会增加另一个具有衍生面部的古人类物种。这等同于将南方古猿用唯一的标准定义，即那些不能被归入地猿或人属的古人类物种，依我们看来，这形成了一个不理想的分类方式。因此，鉴于 KNM-WT 40000 不能明智地与任何已确定的古人类属聚类到一起，以及鉴于其表明了一种独特的面部和牙齿形态模式，该模式可能反映了一个与众不同的饮食适应区，我们将该标本划分到一个新属中，即肯尼亚人。

尽管分隔约 150 万年，但是 KNM-WT 40000 在面部结构上与鲁道夫人的选模标本 KNM-ER1470 很相似。主要差异包括 KNM-WT 40000 更加原始的鼻骨和脑颅形态。这提出了一种可能性，即这两种分类单元之间存在密切的系统发育关系，并且影响了我们对鲁道夫人的解释。已经有人建议将鲁道夫人归入南方古猿[4,19]，但是肯尼亚人可能是一个更适合的属。对肯尼亚扁脸人的鉴定具有许多额外的含义。作为与某些形态更加原始的南方古猿阿法种同时代的一个物种，肯尼亚扁脸人削弱了南方古猿阿法种作为所有后来的古人类的姐妹分类单元的理由，以及将其归入 *Praeanthropus* 的建议[1,20]。此外，肯尼亚扁脸人的形态引发了关于分析古人类系统发育时使用的特征的极性的问题。一个例子是该物种的小型臼齿尺寸，这尽管可能是一种衍生性状，但是也可能意味着南方古猿阿法种或南方古猿湖畔种的较大犬齿后部牙齿并不代表原始的古人类状态。最后，在东非的中上新世时期至少出现了另外一个古人类物种，这意味着现在可以重新估计碎片标本的归属了。例如，将 330 万年前的 KNM-ER 2602 颅骨碎片划分为南方古猿阿法种[21]已经被质疑了[8]，现在

and evaluating its affinities with *K. platyops* is now timely.

Geological Context and Dating

KNM-WT 40000 was collected near the contact of the Nachukui Formation with Miocene volcanic rocks in the northern tributary of Lomekwi (Nabetili). It is situated 12 m above the Lokochot Tuff, and 8 m below the β-Tulu Bor Tuff (Fig. 4). Along Nabetili, the Lokochot Tuff is pinkish-grey and contains much clay and volcanic detritus. It is overlain by a volcanic pebble conglomerate, followed by a pale brown quartz-rich fine sandstone that includes a burrowed fine-sandstone marker bed 10–15 cm thick. The Lokochot Tuff is replaced by a thick volcanic clast conglomerate in the central part of Lomekwi. The contact between the fine sandstone and the overlying dark mudstone can be traced from Nabetili to the hominin locality. Locally the mudstone contains volcanic pebbles at the base, and it has thin pebble conglomerate lenses in the upper part at the hominin locality, and also contains $CaCO_3$ concretions. The hominin specimen and other vertebrate fossils derive from this mudstone. Overlying the dark mudstone at the hominin site is a brown mudstone (8 m) that directly underlies the β-Tulu Bor Tuff.

New $^{40}Ar–^{39}Ar$ determinations on alkali feldspars from pumice clasts in the Moiti Tuff and the Topernawi Tuff, stratigraphically beneath the Lokochot Tuff, were instrumental in re-investigating the lower portion of this section. The new results yield a mean age for the Topernawi Tuff of 3.96 ± 0.03 Myr; this is marginally older than the pooled age for the Moiti Tuff of 3.94 ± 0.03 Myr. Previous investigations[22,23] placed the Topernawi Tuff above the Moiti Tuff, mainly on the basis of the K/Ar ages on alkali feldspar from pumice clasts in the Topernawi Tuff (3.78, 3.71, 3.76 and 3.97 Myr, all ± 0.04 Myr)[23]. The older determination (3.97 Myr) was thought to result from contamination by detrital feldspar. South of Topernawi, however, the Topernawi Tuff has now been shown to underlie the Moiti Tuff, and to be in turn underlain by a tephra informally termed the "Nabwal tuff", previously thought to be a Moiti Tuff correlative. The correct sequence is shown in Fig. 4, and the new $^{40}Ar–^{39}Ar$ age data on the Moiti Tuff and Topernawi Tuff are provided as Supplementary Information.

正是对其与肯尼亚扁脸人的亲缘关系进行评估的时候。

地质学环境与定年

KNM-WT 40000 是在洛梅奎（纳贝蒂利）北部支流的中新世火山岩与纳丘库伊组接壤部位附近搜集到的。它位于洛科乔特凝灰岩之上 12 米，β 图卢博尔凝灰岩之下 8 米处（图 4）。沿着纳贝蒂利，洛科乔特凝灰岩呈略带粉红的灰色，包含大量黏土和火山碎屑。其上覆盖着火山卵石砾岩，紧接着是淡褐色的富含石英的细砂岩，其中包括挖掘出来的 10~15 厘米厚的细砂岩标准层。洛科乔特凝灰岩在洛梅奎中部被一层厚的火山碎屑砾岩取代了。在细砂岩和覆盖在上面的深色泥岩之间的衔接部分可以从纳贝蒂利一直追踪到古人类遗址处。在当地，泥岩在基部包含火山卵石，在古人类遗址的上半部分有薄层的卵石砾岩透镜体，还含有 $CaCO_3$ 结核。古人类标本和其他脊椎动物化石都是从这层泥岩中发现的。覆盖在古人类遗址深色泥岩上面的是一种褐色的泥岩（8 米），位于 β 图卢博尔凝灰岩的紧下面。

根据在地层学上位于洛科乔特凝灰岩下面的莫伊蒂凝灰岩和托佩尔纳维凝灰岩中的浮岩碎屑的碱性长石确定的 $^{40}Ar–^{39}Ar$ 新数据，对于重新调查该剖面的下半部分是有帮助的。这些新结果产生的托佩尔纳维凝灰岩的平均年代是 396 万 ±3 万年；这比莫伊蒂凝灰岩的并合年代 394 万 ±3 万年古老的程度有限。之前的调查[22,23] 将托佩尔纳维凝灰岩置于莫伊蒂凝灰岩之上，主要的依据是从托佩尔纳维凝灰岩中的浮岩碎屑得到的碱性长石的 K/Ar 年代（378 万、371 万、376 万和 397 万年，都是 ±4 万年）[23]。确定的年代中较为古老的（397 万年）被认为是长石碎屑的污染造成的。但是，在托佩尔纳维南部，现在表明托佩尔纳维凝灰岩是位于莫伊蒂凝灰岩之下的，而反过来，一种非正式称为"纳布瓦尔凝灰岩"的火山碎屑又位于它的下面，后者之前被认为是莫伊蒂凝灰岩的相关物。正确的地层顺序见图 4 所示，莫伊蒂凝灰岩和托佩尔纳维凝灰岩的新的 $^{40}Ar–^{39}Ar$ 年代数据在补充信息中给出。

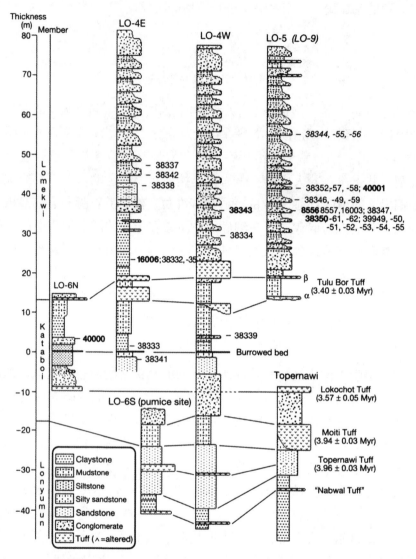

Fig. 4. Stratigraphic sections and placement of hominin specimens at sites in upper part of the Lomekwi drainage, west of Lake Turkana, northern Kenya. Specimen numbers are given without the prefix KNM-WT, and those in bold are discussed in the text. Placement of specimens is relative to the nearest marker bed in the section. Italicized numbers show the relative placement of specimens at LO-9 on section LO-5. The burrowed bed, a useful local marker, is used as stratigraphic datum (0 m). Representative ^{40}Ar–^{39}Ar analytical data on the Moiti Tuff and on the Topernawi Tuff are given as Supplementary Information. The date for the Tulu Bor Tuff is taken as the age of the Sidi Hakoma Tuff at Hadar[25], which is consistent with the age of the Toroto Tuff (3.32 ± 0.03 Myr)[45] that overlies the Tulu Bor at Koobi Fora. The age of the Lokochot Tuff is assigned from its placement at the Gilbert/Gauss Chron boundary[24,46]. The tuff formerly thought to be the Moiti Tuff at Lomekwi[22] has been informally called the "Nabwal tuff"[47]. Ages on the Tulu Bor and the Lokochot Tuffs are consistent with orbitally tuned ages of correlative ash layers in Ocean Drilling Program Core 722A in the Arabian Sea[48].

Linear interpolation between the Lokochot Tuff (3.57 Myr old)[24] and Tulu Bor Tuff (3.40 Myr)[25] yields an age of 3.5 Myr for KNM-WT 40000, and 3.53 Myr for the burrowed bed.

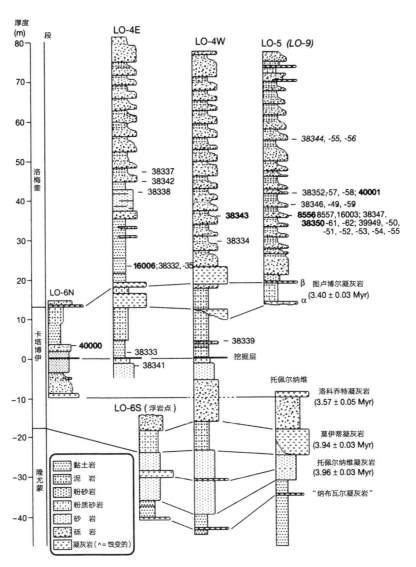

图 4. 肯尼亚北部图尔卡纳湖西侧的洛梅奎流域上游区的地层剖面和发现古人类标本的位置。标本号缺省了前缀 KNM-WT，黑体表示的是在正文中讨论的标本。标本的位置是与剖面中最近的标准层对应的。斜体数字表示 LO-9 处的标本在剖面 LO-5 的相对位置。挖掘层是一个有用的局部标志，被用来当作地层基准面 (0 米)。对莫伊蒂凝灰岩和托佩尔纳维凝灰岩进行的代表性的 ^{40}Ar–^{39}Ar 分析数据作为补充信息给出。图卢博尔凝灰岩的年代根据哈达尔的西迪哈克玛凝灰岩的年代得来[25]，这与覆盖在库比福勒的图卢博尔之上的托罗托凝灰岩的年代 (332 万年 ± 3 万年)[45] 一致。洛科乔特凝灰岩的年代根据其在吉尔伯特/高斯年代界线处的位置来判定[24,46]。在洛梅奎处之前认为属于莫伊蒂的凝灰岩[22] 被非正式地称作"纳布瓦尔凝灰岩"[47]。图卢博尔和洛科乔特凝灰岩的年代与阿拉伯海[48] 的海洋钻探计划岩芯 722A 中的相关灰层的轨道调谐年代一致。

　　洛科乔特凝灰岩 (357 万年)[24] 与图卢博尔凝灰岩 (340 万年)[25] 之间的线性插入得到 KNM-WT 40000 的年代为 350 万年，挖掘层的年代为 353 万年。紧邻挖掘层之下的

KNM-WT 38341 from immediately below the burrowed bed has an age near 3.53 Myr. KNM-WT 38333 and 38339, from between the burrowed bed and the α-Tulu Bor Tuff lie between 3.4 and 3.5 Myr. Other specimens from LO-4, LO-5, and LO-6 lie 16–24 m above the β-Tulu Bor Tuff, with ages near 3.3 Myr. Assuming linear sedimentation between the Tulu Bor Tuff and the Lokalalei Tuff (2.5 Myr)[23], specimens from LO-9 are around 3.2 Myr. The probable error on these age estimates is less than 0.10 Myr.

Palaeogeographically, the mudstone that contained KNM-WT 40000 at LO-6N was deposited along the northern margin of a shallow lake that extended to Kataboi and beyond[26,27]. Laterally discontinuous volcanic pebble conglomerates within the mudstone record small streams draining from hills to the west. Carbonate concretions at the hominin level are probably pedogenic, and indicate regional conditions with net evaporative loss. Other specimens between the burrowed bed and the Tulu Bor Tuff were also preserved in lake-margin environments, as is the case for KNM-WT 38341 that was collected below the burrowed bed. At LO-5, and in the upper part of LO-4E, strata were laid down by ephemeral streams draining the basin margin, principally the ancestral Topernawi, which deposited gravels in broad, shallow channels, and finer grained materials in interfluves. Specimens preserved in floodplain deposits of the ancestral Omo River that occupied the axial portion of the basin include those at LO-9, those less than 6 m above the Tulu Bor Tuff at LO-4E, and KNM-WT 38338. Thus, there is evidence for hominins occupying floodplains of major rivers, alluvial fans, and lake-margin environments 3.0–3.5 Myr ago. There is reasonable evidence that water sources were available to these hominins in channels of the ephemeral streams, and also possibly as seeps or springs farther out into the basin.

Palaeoecology and Fauna

Faunal assemblages from Lomekwi sites LO-4, LO-5, LO-6 and LO-9 indicate palaeoenvironments that were relatively well watered and well vegetated. The relative proportions of the bovids in the early collections from these sites indicate a mosaic of habitats, but with predominantly woodland and forest-edge species dominating[22]. Comparisons of the Lomekwi faunal assemblages with those from the few known hominin sites of similar age, Laetoli in Tanzania, Hadar in Ethiopia and Bahr el Ghazal in Chad, are of interest in view of the different hominin taxa represented. Hadar and Bahr el Ghazal, like Lomekwi, represent lakeshore or river floodplain palaeoenvironments[28,29], whereas Laetoli was not located near a water source; no aquatic taxa nor terrestrial mammals indicative of swamp or grassy wetlands were recovered[30]. The faunal assemblages of all four sites indicate a mosaic of habitats that seems to have included open grasslands and more wooded or forested environments[22,28,29,31,32]; the assemblages differ primarily in the indication of the nature of the dominant vegetation cover.

Although the mammalian faunal assemblage from Lomekwi is more similar to that from Hadar than to that from Laetoli, some mammalian species represented are different.

KNM-WT 38341 的年代接近 353 万年。从挖掘层与 α 图卢博尔凝灰岩之间发掘到的 KNM-WT 38333 和 38339 的年代介于 340 万年到 350 万年之间。从 LO-4、LO-5 和 LO-6 挖掘出来的其他标本位于 β 图卢博尔凝灰岩之上 16~24 米处，年代接近 330 万年。假定图卢博尔凝灰岩与洛卡拉莱伊凝灰岩(250 万年)[23]之间存在线性沉积作用，那么从 LO-9 挖掘出来的标本的年代就约为 320 万年。这些年代估计值的可能误差小于 10 万年。

从古地理学角度看，在 LO-6N 处包含 KNM-WT 40000 的泥岩是沿着一个浅湖的北部边缘沉积下来的，这个湖一直延伸到卡塔博伊甚至更远处[26,27]。在泥岩中的侧面的不连续的火山卵石砾岩记录下了从山丘流到西部的小溪流。在古人类地层中的碳酸盐结核可能是成土的，暗示着具有净蒸发损失的区域性情况。位于挖掘层与图卢博尔凝灰岩之间的其他标本也在湖缘环境中保存下来了，正如 KNM-WT 38341 的情况一样，该标本是在挖掘层之下的地层中搜集到的。在 LO-5，以及 LO-4E 的上半部分，地层由从盆地边缘流出的季节河(主要是原始的托佩尔纳维)沉积而成，它们将砾石在宽阔而浅的河道中沉积下来，细粒的材料在江河分水区沉积下来。原始的奥莫河占据了盆地的轴向部分，在其河漫滩沉积物中保存的标本包括在 LO-9 的那些标本，在 LO-4E 的图卢博尔凝灰岩之上不到 6 米处的标本，以及 KNM-WT 38338。因此，有证据表明，300 万 ~ 350 万年前，有古人类居住在大河的河漫滩、冲积扇以及湖缘环境中。有可靠证据表明，这些古人类可以得到的水源来自季节河的河道，可能还有深入盆地内部的渗流或泉水。

古生态学与动物群

来自洛梅奎遗址 LO-4、LO-5、LO-6 和 LO-9 的动物群组合暗示当时的古环境是相对水源充足、植物茂盛的。从这些遗址搜集到的早期标本中牛科动物的相对比例暗示着生境的镶嵌性，但林地和森林边缘物种占主导地位[22]。将洛梅奎动物群组合与已知的少数相似年代的古人类遗址(坦桑尼亚的莱托里、埃塞俄比亚的哈达尔和乍得的加扎勒河省)发现的动物群组合相比，从所代表的不同古人类分类单元的视点来看很有趣。哈达尔和加扎勒河省像洛梅奎一样，代表着湖岸或河流河漫滩的古环境[28,29]，而莱托里并不位于水源附近；没有发现暗示沼泽或多草湿地存在的水生分类单元或陆地哺乳动物[30]。所有四处遗址的动物群组合都暗示着生境的镶嵌性，它似乎包括开放的草地和更加多树木或多森林的环境[22,28,29,31,32]；这些动物群组合的主要区别在于优势植被的性质的含义不同。

尽管洛梅奎的哺乳动物群组合与哈达尔的相似性胜过其与莱托里的，但是有些代表性的哺乳动物物种是不同的。在洛梅奎，布鲁姆狮尾狒很常见，并且是主要的

At Lomekwi, *Theropithecus brumpti* is common and is the dominant cercopithecid, as it is elsewhere in the Turkana Basin at this time. This species is generally considered to indicate more forested or closed woodland habitats. In the Hadar Formation, *Theropithecus darti* is the common *Theropithecus* species and is associated with lower occurrences of the water-dependent reduncines and higher occurrences of alcelaphines and/or *Aepyceros*, which indicates drier woodlands and grasslands[33]. Differences in the representation of other common species at the two sites that are less obviously linked to habitat include *Kolpochoerus limnetes*, *Tragelaphus nakuae* and *Aepyceros shungurensis* at Lomekwi, as opposed to *K. afarensis*, *T. kyaloae* and an undescribed species of *Aepyceros* at Hadar (K. Reed, personal communication). The general indication is that the palaeoenvironment at Lomekwi may have been somewhat more vegetated and perhaps wetter than that persisting through much of the Hadar Formation. At both sites more detailed analyses will be essential to further develop an understanding of how subtle temporal changes in the faunal assemblages relate to hominin occurrences.

Note added in proof: If the hominin status of the recently published Lukeino craniodental specimens[49] is confirmed, this would support the suggestion that small molar size is the primitive rather than the derived hominin condition.

(**410**, 433-440; 2001)

Meave G. Leakey[*], Fred Spoor[†], Frank H. Brown[‡], Patrick N. Gathogo[‡], Christopher Kiarie[*], Louise N. Leakey[*] & Ian McDougall[§]

[*] Division of Palaeontology, National Museums of Kenya, P.O. Box 40658, Nairobi, Kenya
[†] Department of Anatomy & Developmental Biology, University College London, WC1E 6JJ, UK
[‡] Department of Geology & Geophysics, University of Utah, Salt Lake City, Utah 84112, USA
[§] Research School of Earth Sciences, The Australian National University, Canberra ACT 0200, Australia

Received 31 January; accepted 16 February 2001.

References:

1. Strait, D. S., Grine, F. E. & Moniz, M. A. A reappraisal of early hominid phylogeny. *J. Hum. Evol.* **32**, 17-82 (1997).

2. Ward, C., Leakey, M. & Walker, A. The new hominid species *Australopithecus anamensis*. *Evol. Anthrop.* **7**, 197-205 (1999).

3. Brown, B., Brown, F. & Walker, A. New hominids from the Lake Turkana Basin, Kenya. *J. Hum. Evol.* (in the press).

4. Wood, B. A. & Collard, M. C. The human genus. *Science* **284**, 65-71 (1999).

5. Leakey, M. G., Feibel, C. S., McDougall, I. & Walker, A. C. New four-million-year-old hominid species from Kanapoi and Allia Bay, Kenya. *Nature* **376**, 565-571 (1995).

6. Leakey, M. G., Feibel, C. S., McDougall, I., Ward, C. & Walker, A. New specimens and confirmation of an early age for *Australopithecus anamensis*. *Nature* **393**, 62-66 (1998).

7. Kimbel, W. H., Johanson, D. C. & Rak, Y. Systematic assessment of a maxilla of *Homo* from Hadar, Ethiopia. *Am. J. Phys. Anthrop.* **103**, 235-262 (1997).

8. Wood, B. A. *Koobi Fora Research Project* Vol. 4 (Clarendon Press, Oxford, 1991).

9. Lockwood, C. A. & Tobias, P. V. A large male hominin cranium from Sterkfontein, South Africa, and the status of *Australopithecus africanus*. *J. Hum. Evol.* **36**, 637-685 (1999).

10. Bilsborough, A. & Wood, B. A. Cranial morphometry of early hominids: facial region. *Am. J. Phys. Anthrop.* **76**, 61-86 (1988).

11. Rak, Y. *The Australopithecine Face* (Academic, New York, 1983).

12. Lockwood, C. A., Kimbel, W. H. & Johanson, D. C. Temporal trends and metric variation in the mandibles and dentition of *Australopithecus afarensis*. *J. Hum. Evol.* **39**, 23-55 (2000).

13. Moggi-Cecchi, J., Tobias, P. V. & Beynon, A. D. The mixed dentition and associated skull fragments of a juvenile fossil hominid from Sterkfontein, South Africa. *Am. J.*

猴科动物，就像当时在图尔卡纳盆地的其他地方一样。该物种通常被认为暗示着更加森林化或者更加封闭的林地生境。在哈达尔组，达提狮尾狒是常见的狮尾狒属的物种，与较少出现的依赖水的苇羚科和经常出现的狷羚科和（或）高角羚属相关，暗示着更加干旱的林地和草地[33]。在这两处遗址，与生境关联不那么明显的其他常见物种的差别包括洛梅奎的 *Kolpochoerus limnetes*、*Tragelaphus nakuae* 和 *Aepyceros shungurensis*，而哈达尔的 *K. afarensis*、*T. kyaloae* 和一种未描述的高角羚属物种（里得，个人交流）与之刚好形成对照。通常认为，与哈达尔组大部分地区长期存在的古环境相比，洛梅奎的古环境可能植被更加丰富，或许也更加潮湿。在这两处遗址，需要进行更加详细的分析以进一步理解动物群组合在时间上的微妙变化如何与古人类出现相关。

　　附加说明：如果最近发表的卢凯伊诺颅牙标本[49]的古人类地位得到了证实，那么将支持如下建议，即小型的臼齿尺寸是原始的，而非衍生的古人类特征。

　　　　　　　　　　　　　　　　　　　　　　　（刘皓芳 翻译；赵凌霞 审稿）

   Phys. Anthrop. **106**, 425-465 (1998).

14. Robinson, J. T. The dentition of the Australopithecinae. *Transv. Mus. Mem.* **9**, 1-179 (1956).

15. McCollum, M. A. The robust australopithecine face: a morphogenetic perspective. *Science* **284**, 301-305 (1999).

16. Brunet, M. *et al. Australopithecus bahrelghazali*, une nouvelle espèce d'Hominidé ancien de la région de Koro Toro (Tchad). *C.R. Acad. Sci. Ser. IIa* **322**, 907-913 (1996).

17. White, T. D., Suwa, G., Simpson, S. & Asfaw, B. Jaws and teeth of *A. afarensis* from Maka, Middle Awash, Ethiopia. *Am. J. Phys. Anthrop.* **111**, 45-68 (2000).

18. White, T. D., Suwa, G. & Asfaw, B. *Australopithecus ramidus*, a new species of early hominid from Aramis, Ethiopia. *Nature* **371**, 306-312 (1994).

19. Wood, B. & Collard, M. The changing face of genus *Homo. Evol. Anthrop.* **8**, 195-207 (2000).

20. Harrison, T. in *Species, Species Concepts, and Primate Evolution* (eds Kimbel, W. H. & Martin, L. B.) 345-371 (Plenum, New York, 1993).

21. Kimbel, W. H. Identification of a partial cranium of *Australopithecus afarensis* from the Koobi Fora Formation. *J. Hum. Evol.* **17**, 647-656 (1988).

22. Harris, J. M., Brown, F. & Leakey, M. G. Stratigraphy and paleontology of Pliocene and Pleistocene localities west of Lake Turkana, Kenya. *Cont. Sci. Nat. Hist. Mus. Los Angeles* **399**, 1-128 (1988).

23. Feibel, C. S., Brown, F. H. & McDougall, I. Stratigraphic context of fossil hominids from the Omo Group deposits, northern Turkana Basin, Kenya and Ethiopia. *Am. J. Phys. Anthrop.* **78**, 595-622 (1989).

24. McDougall, I., Brown, F. H., Cerling, T. E. & Hillhouse, J. W. A reappraisal of the geomagnetic polarity time scale to 4 Ma using data from the Turkana Basin, East Africa. *Geophys. Res. Lett.* **19**, 2349-2352 (1992).

25. Walter, R. C. & Aronson, J. L. Age and source of the Sidi Hakoma Tuff, Hadar Formation, Ethiopia. *J. Hum. Evol.* **25**, 229-240 (1993).

26. Brown, F. H. & Feibel, C. S. in *Koobi Fora Research Project*, Vol. 3, *Stratigraphy, artiodactyls and paleoenvironments* (ed. Harris, J. M.) 1-30 (Clarendon, Oxford, 1991).

27. Feibel, C. S., Harris, J. M. & Brown, F. H. in *Koobi Fora Research Project*, Vol. 3, *Stratigraphy, artiodactyls and paleoenvironments* (ed. Harris, J. M.) 321-346 (Clarendon, Oxford, 1991).

28. Johanson, D. C., Taieb, M. & Coppens, Y. Pliocene hominids from the Hadar Formation, Ethiopia (1973-1977): stratigraphic, chronologic and paleoenvironmental contexts, with notes on hominid morphology and systematics. *Am. J. Phys. Anthrop.* **57**, 373-402 (1982).

29. Brunet, M. *et al.* The first australopithecine 2,500 kilometres west of the Rift Valley (Chad). *Nature* **378**, 273-240 (1995).

30. Leakey, M. D. & Harris, J. M. *Laetoli, a Pliocene Site in Northern Tanzania* (Clarendon, Oxford, 1987).

31. Harris, J. M. in *Laetoli, a Pliocene Site in Northern Tanzania* (eds Leakey, M. D. & Harris, J. M.) 524-531 (Clarendon, Oxford, 1987).

32. Kimbel, W. H. *et al.* Late Pliocene *Homo* and Oldowan tools from the Hadar Formation (Kadar Hadar Member), Ethiopia. *J. Hum. Evol.* **31**, 549-561 (1996).

33. Eck, G. G. in *Theropithecus, the Rise and Fall of a Primate Genus* (ed. Jablonski, N. G.) 15-83 (Cambridge Univ. Press, 1993).

34. White, T. D. New fossil hominids from Laetolil, Tanzania. *Am. J. Phys. Anthrop.* **46**, 197-230 (1977).

35. Wood, B. Origin and evolution of the genus *Homo. Nature* **355**, 783-790 (1992).

36. Suwa, G. *et al.* The first skull of *Australopithecus boisei. Nature* **389**, 489-492 (1997).

37. Keyser, A. W. The Drimolen skull: the most complete australopithecine cranium and mandible to date. *S. Afr. J. Sci.* **96**, 189-193 (2000).

38. Kimbel, W. H., White, T. D. & Johanson, D. C. Cranial morphology of *Australopithecus afarensis*: a comparative study based on a composite reconstruction of the adult skull. *Am. J. Phys. Anthrop.* **64**, 337-388 (1984).

39. Walker, A., Leakey, R. E., Harris, J. H. & Brown, F. H. 2.5-Myr *Australopithecus boisei* from west of Lake Turkana, Kenya. *Nature* **322**, 517-522 (1986).

40. Kimbel, W. H., White, T. D. & Johanson, D. C. in *Evolutionary History of the "Robust" Australopithecines* (ed. Grine, F. E.) 259-268 (Aldine de Gruyter, New York, 1988).

41. Asfaw, B. *et al. Australopithecus garhi*: A new species of early hominid from Ethiopia. *Science* **284**, 629-635 (1999).

42. Grine, F. E. & Strait, D. S. New hominid fossils from Member 1 "Hanging Remnant" Swartkrans Formation, South Africa. *J. Hum. Evol.* **26**, 57-75 (1994).

43. Johanson, D. C., White, T. D. & Coppens, Y. Dental remains from the Hadar Formation, Ethiopia: 1974-1977 collections. *Am. J. Phys. Anthrop.* **57**, 545-603 (1982).

44. Tobias, P. V. in *Olduvai Gorge* Vol. 4 (Cambridge Univ. Press, 1991).

45. McDougall, I. K-Ar and $^{40}Ar/^{39}Ar$ dating of the hominid-bearing Pliocene-Pleistocene sequence at Koobi Fora, Lake Turkana, northern Kenya. *Geol. Soc. Am. Bull.* **96**, 159-175 (1985).

46. Brown, F. H., Shuey, R. T. & Croes, M. K. Magnetostratigraphy of the Shungura and Usno Formations, southwestern Ethiopia: new data and comprehensive reanalysis. *Geophys. J. R. Astron. Soc.* **54**, 519-538 (1978).

47. Haileab, B. *Geochemistry, Geochronology and Tephrostratigraphy of Tephra from the Turkana Basin, Southern Ethiopia and Northern Kenya*. Thesis, Univ. Utah (1995).

48. deMenocal, P. B. & Brown, F. H. in *Hominin Evolution and Climatic Change in Europe* Vol. 1 (eds Agustí, J., Rook, L. & Andrews, P.) 23-54 (Cambridge Univ. Press, 1999).

49. Senut, B. *et al.* First hominid from the Miocene (Lukeino Formation, Kenya). *C.R. Acad. Sci. Paris* **332**, 137-144 (2001).

Supplementary information is available on *Nature*'s World-Wide Web site (http://www.nature.com) or as paper copy from the London editorial office of *Nature*.

Acknowledgements. We thank the Government of Kenya for permission to carry out this research and the National Museums of Kenya for logistical support. The National Geographic Society funded the field work and some laboratory studies. Neutron irradiations were facilitated by the Australian Institute of Nuclear Science and Engineering and the Australian Nuclear Science and Technology Organisation. We also thank the Ethiopian Ministry of Information and Culture, the National

Museum of Ethiopia, B. Asfaw, Y. Bayene, C. Howell, D. Johanson, W. Kimbel, G. Suwa and T. White for permission to make comparisons with the early Ethiopian hominins and numerous people including N. Adamali, B. Asfaw, C. Dean, C. Feibel, A. Griffiths, W. Kimbel, R. Kruszynski, K. Kupczik, R. Leakey, D. Lieberman, J. Moore, K. Patel, D. Plummer, K. Reed, B. Sokhi, M. Tighe, T. White, and B. Wood for their help. Caltex (Kenya) provided fuel for the field expeditions, and R. Leakey allowed us the use of his aeroplane. The field expedition members included U. Bwana, S. Crispin, G. Ekalale, M. Eragae, J. Erus, J. Ferraro, J. Kaatho, N. Kaling, P. Kapoko, R. Lorinyok, J. Lorot, S. Hagemann, B. Malika, W. Mangao, S. Muge, P. Mulinge , D. Mutinda, K. Muthyoka, N. Mutiwa, W. Mutiwa, B. Onyango, E. Weston and J. Wynn. A. Ibui, F. Kyalo, F. Kirera, N. Malit, E. Mbua, M. Muungu, J. Ndunda, S. Ngui and A. Mwai provided curatorial assistance. The Leakey Foundation awarded a grant to F.B.

Correspondence and requests for materials should be addressed to M.G.L. (e-mail: meave@swiftkenya.com).

Observation of High-energy Neutrinos Using Čerenkov Detectors Embedded Deep in Antarctic Ice

E. Andrés *et al.*

Editor's Note

Neutrinos are chargeless and almost massless elementary particles. They interact with other matter only through the weak force, and therefore can penetrate large amounts of matter—neutrinos from the Sun are constantly streaming through the Earth. This weak interaction makes them difficult to detect. Here an international team of scientists report a proof-of-concept experiment called AMANDA, in which they set detectors in clear ice in the Antarctic and looked for the characteristic flashes of light emitted when a particle called a muon is created by interaction of a neutrino with an atomic nucleus. A subsequent project using a much larger volume of ice as the detector, called IceCube, is now operating near the South Pole.

Neutrinos are elementary particles that carry no electric charge and have little mass. As they interact only weakly with other particles, they can penetrate enormous amounts of matter, and therefore have the potential to directly convey astrophysical information from the edge of the Universe and from deep inside the most cataclysmic high-energy regions[1]. The neutrino's great penetrating power, however, also makes this particle difficult to detect. Underground detectors have observed low-energy neutrinos from the Sun and a nearby supernova[2], as well as neutrinos generated in the Earth's atmosphere. But the very low fluxes of high-energy neutrinos from cosmic sources can be observed only by much larger, expandable detectors in, for example, deep water[3,4] or ice[5]. Here we report the detection of upwardly propagating atmospheric neutrinos by the ice-based Antarctic muon and neutrino detector array (AMANDA). These results establish a technology with which to build a kilometre-scale neutrino observatory necessary for astrophysical observations[1].

HIGH-ENERGY neutrinos must be generated in the same astrophysical sources that produce high-energy cosmic rays[1]. These sources are a matter of speculation, but are thought to reside in shocked or violent environments such as are found in supernova remnants, active galactic nuclei, and gamma-ray bursters. The interaction of any high-energy proton or nucleus with matter or radiation in the source will produce neutrinos, some of which will have line-of-sight trajectories to Earth. AMANDA detects neutrinos with energies above a few tens of GeV by observing the Čerenkov radiation from muons that are produced in neutrino–nucleon interactions in the ice surrounding the detector

借助南极冰下深处的切伦科夫
探测器观测高能中微子

安德烈斯等

编者按

中微子是无电荷且几乎无质量的基本粒子。它们只能通过弱力与其他物质相互作用，因此可以穿透大量物质——来自太阳的中微子就不断地流经地球。这种弱相互作用使中微子难以被检测。本文中，一个国际科学家团队报告了名为 AMANDA 的概念验证型实验，他们将探测器设置在南极透明的冰层中，并寻找由中微子与原子核相互作用而产生的被称为 μ 子的粒子发出的特征性闪光。随后的冰立方 (IceCube) 项目使用更大体量的冰作为探测器，目前已在南极附近运行。

中微子是一种不携带电荷且质量极小的基本粒子。由于这类粒子与其他粒子的相互作用很弱，能够穿透大量的物质，因此很可能可以直接传递来自宇宙边缘以及大部分高能激变区域深处的天体物理信息[1]。然而，中微子极强的穿透力也使得对该粒子的探测变得非常困难。在以前的研究中，研究人员通过地下探测器已经探测到来自太阳和附近超新星的低能中微子[2]，以及地球大气中产生的中微子。但是，来自宇宙源的高能中微子通量很低，只能通过更大的、可扩展的探测器才能观测到，比如放置于深水[3,4]或冰[5]中的探测器。本文将报道通过基于冰的南极 μ 子和中微子探测器阵列 (AMANDA) 对向上传播的大气中微子的探测。这些结果将确立一种技术，可以用于建设千米尺度的中微子天文台，从而用于天体物理的观测[1]。

发射高能宇宙射线的天体物理源也必然产生高能中微子[1]。这些源目前尚属推测，但是一般认为存在于极端剧烈环境中，比如超新星遗迹、活动星系核和伽玛射线暴。这些源中，任何高能质子或原子核与物质或辐射相互作用，都会产生中微子，其中一些将具有指向地球的视线径迹。中微子与探测器周围冰层或底下岩床中的核子相互作用可以产生 μ 子。通过探测 μ 子发出的切伦科夫辐射，AMANDA 可以探测到能量高于数十 GeV 的中微子[6]。图 1 给出了我们用于探测切伦科夫辐射的光

or in the bedrock below[6]. This Čerenkov light is detected by an array of photomultiplier tubes (Fig. 1), which are buried deep in the ice in order to minimize the downward flux of muons produced in cosmic-ray interactions in the atmosphere. These muons constitute the main background for AMANDA. To ensure that the detected muons are produced by a neutrino, we use the Earth as a filter and look for upwardly propagating muons that perforce must have been produced by a neutrino that passed through the Earth. From the relative arrival times of the Čerenkov photons, measured with a precision of a few nanoseconds, we can reconstruct the track of the muon. The direction of the neutrino and muon are collinear within an angle $\theta_{\nu-\mu} \approx 1.5/\sqrt{E_\nu}$ degrees, where E_ν is measured in TeV, thus enabling us to search for point sources of high energy neutrinos.

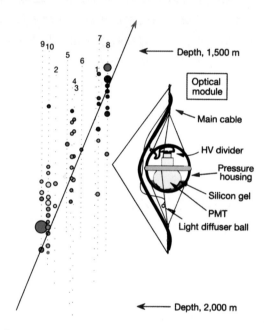

Fig. 1. The AMANDA-B10 detector and a schematic diagram of an optical module. Each dot represents an optical module. The modules are separated by 20 m on the inner strings (1 to 4), and by 10 m on the outer strings (5 to 10). The coloured circles show pulses from the photomultipliers for a particular event; the sizes of the circles indicate the amplitudes of the pulses and the colours correspond to the time of a photon's arrival. Earlier times are in red and later ones in blue. The arrow indicates the reconstructed track of the upwardly propagating muon.

Upwardly propagating atmospheric neutrinos are a well understood source that can be used to verify the detection technique. The results reported here are from analyses of experimental data acquired in 138 days of net operating time during the Antarctic winter of 1997. At that time the detector consisted of 302 optical modules deployed on ten strings at depths of between 1,500 m and 2,000 m (Fig. 1). The instrumented volume is a cylinder of approximately 120 m in diameter and 500 m in height. An optical module consists of an 8-inch photomultiplier tube housed in a glass pressure vessel. A cable provides the high voltage and transmits the anode current signals to the data acquisition electronics at the surface. Figure 1 also shows a representative event that has satisfied the selection criteria for an upwardly moving muon. The effective detection area for muons varies from about

电倍增管阵列，它们被埋设于冰层深处，以最大限度地减小大气中宇宙射线相互作用产生的 μ 子的向下通量。这些大气 μ 子构成了 AMANDA 的主要背景信号。为了确保探测到的 μ 子确实产生于中微子，我们将整个地球作为一个大过滤器，探测那些一定是由穿过地球的中微子产生的向上传播的 μ 子。探测器探测到的切伦科夫光子的相对到达时间的精度在纳秒量级，我们通过记录这些到达时间可以重建 μ 子的径迹。中微子与 μ 子的方向在夹角 $\theta_{\nu-\mu} \approx 1.5$ 度$/(E_\nu)^{1/2}$ 内共线，其中 E_ν 单位为 TeV，这可以使我们寻找到高能中微子点源。

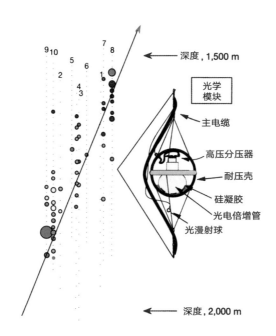

图 1. AMANDA-B10 探测器及其光学模块原理图。图中每个小点代表一个光学模块。在内部串，光学模块之间的距离为 20 m(1 到 4)；在外部串，模块之间的距离为 10 m(5 到 10)。不同颜色的圆圈表示某个事例中光电倍增管产生的脉冲，其中圆圈的尺寸大小代表该脉冲的幅度，颜色对应光子到达的时间。较早的时间用红色表示，较晚的时间用蓝色表示。图中箭头表示斜向上运动的 μ 子的重建径迹。

向上传播的大气中微子作为源已经为人们所了解，因此可以用大气中微子验证我们的探测技术。我们分析了在 1997 年南极冬季获取的 138 天净运转时间的实验数据，并将结果在本文中予以报告。当时，实验采用的探测器由 302 个光学模块构成，这些模块埋设为 10 串，深度在 1,500 m 到 2,000 m 之间（图 1）。整个探测器设备是一个圆柱体，其直径大约为 120 m，高度约为 500 m。每个光学模块包括一个封装于玻璃质压力容器中的 8 英寸光电倍增管。一根电缆用于提供光电倍增管所需的高电压，并将阳极电流信号传输到位于地面的数据采集电子设备。图 1 中也给出了一个满足向上运动的 μ 子选择标准的典型事件。根据 μ 子能量的不同，探测系统的有效

3,000 m^2 at 100 GeV to about 4×10^4 m^2 for the higher-energy muons ($\geqslant$ 100 TeV) that would be produced by neutrinos coming from, for example, the same cosmic sources that produce gamma-ray bursts[7].

Because a knowledge of the optical properties of the ice is essential for track reconstruction, these have been studied extensively[8,9]. The absorption length of blue and ultraviolet light (the relevant wavelengths for our purposes) varies between 85 m and 225 m, depending on depth. The effective scattering length, which combines the mean free path λ with the average scattering angle θ through $\lambda/(1 - \langle \cos\theta \rangle)$, varies from 15 m to 40 m. In order to reconstruct the muon tracks we use a maximum-likelihood method, which incorporates the scattering and absorption of photons as determined from calibration measurements. A bayesian formulation of the likelihood takes into account the much larger rate of downward muons relative to the upward signal and is particularly effective in decreasing the chance for a downward muon to be mis-reconstructed as upward. (See refs 6 and 10–13 for more information on optical properties of ice, calibration, and analysis techniques.)

Certain types of events that might appear to be upwardly propagating muons must be considered and eliminated. Rare cases, such as muons that undergo catastrophic energy loss through bremsstrahlung, or that are coincident with other muons, must be investigated. To this end, a series of requirements or quality criteria, based on the characteristic time and spatial pattern of photons associated with a muon track and the response of our detector, are applied to all events that, in a first assessment, appear to be upwardly moving muons. For example, an event that has a large number of optical modules hit by prompt (that is, unscattered) photons, has a high quality. By making these requirements (or "cuts") increasingly selective, we eliminate correspondingly more of the background of false upward events while still retaining a significant fraction of the true upwardly moving muons. Because there is a large space within which the parameters defining these cuts can be optimized, two different and independent analyses of the same set of data have been undertaken. These analyses yielded comparable numbers of upwardly propagating muons (153 in analysis A, 188 in analysis B). Comparison of these results with their respective Monte Carlo simulations shows that they are consistent with each other in terms of the number of events, the number of events in common and, as discussed below, the expected properties of atmospheric neutrinos.

In Fig. 2a, the number of experimental events is compared to simulations of background and signal as a function of the (identical) quality requirements placed on the three types of events: experimental data, simulated upwardly moving muons from atmospheric neutrinos, and a simulated background of downwardly moving cosmic-ray muons. For simplicity in presentation, the levels of the individual cuts have been combined into a single parameter representing the overall event quality. Figure 2b shows ratios of the quantities plotted above. As the quality level is increased, the ratios of simulated background to experimental data, and of experimental data to simulated signal, both continue their rapid decrease, the former toward zero and the latter toward 0.7. Over the same range, the ratio of experimental data to the simulated sum of background and signal remains nearly constant. We conclude that

探测面积范围从 3,000 m² 到 40,000 m²，对应从 100 GeV 到高能 μ 子（≥ 100 TeV）的情形，产生后者的中微子应该来自宇宙源，比如同样产生伽玛射线暴的宇宙源[7]。

为了重建中微子的径迹，首先要弄清冰的光学特性。人们对此已进行了广泛的研究[8,9]。根据冰层深度的不同，蓝光和紫外光（我们所用的相关波长）的吸收长度在 85 m 到 225 m 之间变化。由平均自由程 λ 和平均散射角 θ 构成的有效散射长度 λ/(1 − ⟨cosθ⟩) 为 15 ~ 40 m。我们采用最大似然方法重建 μ 子径迹，考虑了通过刻度测量得到的光子的散射和吸收。考虑到向下 μ 子的事例率远远大于向上的信号，对似然值的贝叶斯分析可以有效减小将向下 μ 子错误重建为向上 μ 子的可能性。（参考文献 6 以及 10 ~ 13 介绍了更多关于冰的光学特性、刻度以及分析技术。）

在探测过程中，可能存在某些看似是向上传播的 μ 子的伪事件，对这种情况必须予以考虑并剔除。同时，也必须要考虑那些罕见的情况，比如 μ 子经过韧致辐射损失了大部分能量，或者与其他 μ 子同时出现的情况。为此，一系列基于 μ 子径迹预期产生的光子时间和空间特征图样以及基于探测器的响应的要求或者说质量标准在首次评估中用于筛选那些看似是向上运动的 μ 子事件。比如，瞬发（即未被散射）光子触发大量光学模块，该事件就具有较高的品质。通过不断提高选择的标准（或者说"判选"），我们可以相应地剔除更多的向上 μ 子伪事件背景，同时仍然保留大部分真正向上运动的 μ 子。由于构成筛选标准的各参数具有较大的优化空间，我们对相同的实验数据进行了两种不同且互相独立的分析，分别称为 A 分析和 B 分析。A、B 两种分析结果给出的向上运动的 μ 子数目相当，分别为 153 和 188。这两个结果分别与其相应的蒙特卡罗模拟进行比较，发现二者在事件数目、共同事件数目以及如下所述的大气中微子预期特性方面彼此一致。

如图 2a 所示，我们将实验事件的数量与背景和信号的模拟结果作为（相同）品质判据的函数进行了对比，对以下三类事件——实验数据、模拟的大气中微子产生的向上运动的 μ 子、模拟的向下运动的宇宙射线 μ 子背景——进行了分析。为了简化演示，多个筛选标准合并成了一个代表整体事件品质的参数。图 2b 显示了图 2a 绘制的数量间的比值。随着品质水平的增加，模拟背景与实验数据的比值以及实验数据与模拟信号的比值，同时迅速减小，其中前者趋于 0 而后者趋于 0.7。在上述范围内，实验数据与另两者（模拟的背景和信号）之和的比值几乎保持不变。因此，我们得出以下结论：品质筛选标准可以剔除实验数据中误重建的向下运动的 μ

the quality requirements have reduced the presence of wrongly reconstructed downward muons in the experimental data to a negligible fraction of the signal and that the experimental data behave in the same way as the simulated atmospheric neutrino signal for events that pass the stringent cuts. The estimated uncertainty on the number of events predicted by the signal Monte Carlo simulation, which includes uncertainties in the high-energy atmospheric neutrino flux, the *in situ* sensitivity of the optical modules, and the precise optical properties of the ice, is +40%/−50%. The observed ratio of experiment to simulation (0.7) and the expectation (1.0) therefore agree within the estimated uncertainties.

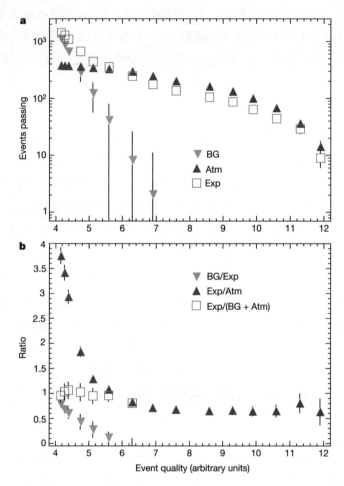

Fig. 2. Experimental data confront expectations. The numbers of reconstructed upwardly moving muon events for the experimental data (Exp) from analysis A are compared to simulations of background cosmic-ray muons (BG) and simulations of atmospheric neutrinos (Atm) as a function of "event quality", a variable indicating the combined severity of the cuts designed to enhance the signal. The comparison begins at a quality level of 4. Cuts were made on a number of parameters including the reconstructed zenith angle (> 100 degrees), maximum likelihood of the reconstruction, topological distributions of the detected photons, and the number of optical modules recording unscattered photons. The optimum levels of the cuts were determined by comparing the relative rejection rates for Monte Carlo simulated neutrino events and background events. **b**, Ratios of the quantities shown in **a**.

子，直至与信号相比达到可以忽略的水平；经过严格筛选后，实验数据与模拟大气中微子信号的行为规律相同。蒙特卡罗模拟预测的信号事件数的估算不确定度为+40%/−50%，该不确定度包括了高能大气中微子通量、光学模块的实地灵敏度以及冰层准确光学特性的不确定度。因此，在上述估算不确定度内，观察到的实验和模拟比值（0.7）与期望值（1.0）相符合。

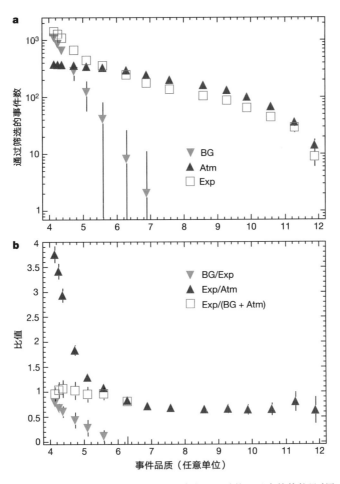

图 2. 实验数据与期望值比较。A 分析中重建实验数据中向上运动的 μ 子事件的数目（用 Exp 表示），与模拟宇宙射线 μ 子背景（BG）以及模拟大气中微子（Atm），作为"事件品质"的函数进行了比较。其中"事件品质"这一变量表示用于增强信号的若干筛选标准的组合的严格程度。图中对比起始于品质水平 4。筛选标准由若干参量构成，这些参量包括重建天顶角（＞100 度）、重建径迹的最大似然值、探测光子的拓扑分布以及记录未散射光子的光学模块数目。通过比较蒙特卡罗模拟中微子事件和背景事件的相对排斥率，可以确定筛选标准的最佳水平。b 图给出了 a 图中数量的比值。

The shape of the zenith-angle distribution of the 188 events from analysis B is compared to a simulation of the atmospheric neutrino signal in Fig. 3, where the absolute Monte Carlo rate has been normalized to the experimental rate. The variation of the measured rate with zenith angle is reproduced by the simulation to within the statistical uncertainty. We note that the tall geometry of the detector favours the more vertical muons. The arrival directions of the upwardly moving muons observed in both analyses are shown in Fig. 4. A statistical analysis indicates no evidence for point sources in these samples. The agreement between experiment and simulation of atmospheric neutrino signal, as demonstrated in Figs 2 and 3, taken together with comparisons for a number of other variables (to be published elsewhere) leads us to conclude that the upcoming muon events observed by AMANDA are produced mainly by atmospheric neutrinos with energies of about 50 GeV to a few TeV. The background in this event sample is estimated to be $15 \pm 7\%$ events, and is due to misreconstructions.

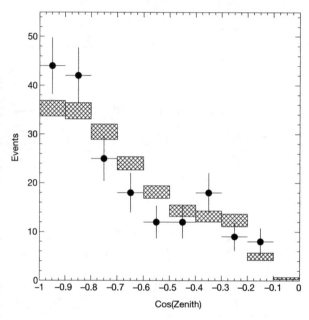

Fig. 3. Reconstructed zenith angle distribution. The data points are experimental data (from analysis B) and the shaded boxes are a simulation of atmospheric neutrino events, the widths of the boxes indicating the error bars. The overall normalization of the simulation has been adjusted to fit the data. The possible effects of neutrino oscillations on the flux and its zenith angle dependence estimated from the Super-Kamiokande measurements[20], are expected to be small in our energy range.

From the consistency of the selected event sample with muons generated by atmospheric neutrinos, and in particular the absence of an excess of high-energy events with a large number of optical modules that had been hit, we can determine an upper limit on a diffuse extraterrestrial neutrino flux. Assuming a hard E^{-2} spectrum characteristic of shockwave acceleration, we expect to reach a sensitivity of order $dN/dE_v = 10^{-6} E_v^{-2}$ cm^{-2} s^{-1} sr^{-1} GeV^{-1}. This value is low enough to be in the range where a number of models[14-19] predict fluxes, a few of which are larger[15,16]. Most recent estimates are smaller[17-19]. The present level of sensitivity

图 3 给出了 B 分析中 188 个事件的天顶角分布与模拟大气中微子信号的对比，其中蒙特卡罗绝对事例率根据实验值进行了归一化。在统计不确定度之内，测量事例率随天顶角的变化与模拟结果一致。我们注意到，探测器较高的几何尺寸会更利于探测到垂直方向的 μ 子。在 A 和 B 两种分析中，观测到的向上运动的 μ 子的到达方向如图 4 所示。统计分析表明，这些样本尚不足以提供点源存在的证据。图 2 与图 3 已经证明实验与大气中微子信号模拟的一致性；同时对比其他变量（具体研究结果即将发表）的话，我们得出以下结论：AMANDA 观测到的上行 μ 子事件主要是由那些能量介于 50 GeV 到若干 TeV 之间的大气中微子产生。事件样本中的背景事件估计为 15%±7%，这是由于错误的径迹重建造成的。

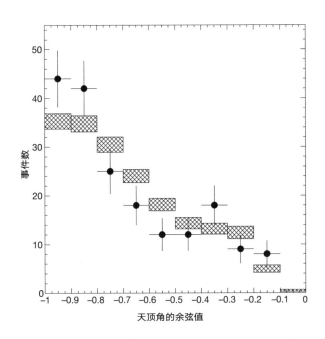

图 3. 重建径迹的天顶角分布。图中数据点是取自分析 B 的实验数据，阴影矩形框表示大气中微子事件的模拟，其中矩形框的宽度代表误差棒。模拟的整体归一化已经经过调整以拟合数据。中微子振荡对通量及通量的天顶角依赖的可能影响可以通过超级神冈实验的测量估算出来[20]，在我们研究的能量范围内，该影响预计很小。

根据筛选的事件样本与大气中微子产生的 μ 子的一致性，特别是触发大量光学模块的高能事件不存在超出的情况下，我们可以确定地外弥散中微子通量的上限。假设激波加速具有 E^{-2} 形式的硬谱特征，那么我们预计灵敏度的量级将达到 $\mathrm{d}N/\mathrm{d}E_v = 10^{-6} E_v^{-2} \ \mathrm{cm}^{-2} \cdot \mathrm{s}^{-1} \cdot \mathrm{sr}^{-1} \cdot \mathrm{GeV}^{-1}$。该值很小，位于很多模型预测的通量范围之内[14-19]，其中一些通量会更大些[15,16]。不过近期的估算值大多数相对更小[17-19]。我们设想可

and the prospects for improving it through longer exposure times and better determination of muon energy illustrate the ability of large-area detectors to test theoretical models that assume the hadronic origin of TeV photons from active galaxies—models which would be difficult to confirm or exclude without the ability to observe high-energy neutrinos.

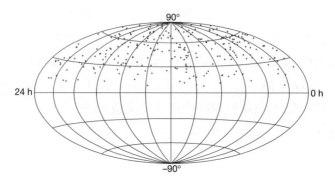

Fig. 4. Distribution in declination and right ascension of the upwardly propagating events on the sky. The 263 events shown here are taken from the upward muons contained in both analysis A and analysis B. The median difference between the true and the reconstructed muon angles is about 3 to 4 degrees.

Searches for neutrinos from gamma-ray bursts, for magnetic monopoles, supernova collapses and for a cold dark matter signal from the centre of the Earth are also in progress and, with only 138 days of data, yield limits comparable to or better than those from smaller underground neutrino detectors that have operated for a much longer period (see refs 10–13).

From 1997 to 1999 an additional nine strings were added in a concentric cylinder around AMANDA-B10. This larger detector, called AMANDA-II, consists of 677 optical modules and has an improved acceptance for muons over a larger angular interval. Data are being taken now with the larger array. Yet the fluxes of very high energy neutrinos predicted by theoretical models[14] or derived from the observed flux of ultra high energy cosmic rays[19] are sufficiently low that a neutrino detector having an effective area up to a square kilometre is required for their observation and study[1,14]. Plans are therefore being made for a much larger detector, IceCube, consisting of 4,800 photomultipliers to be deployed on 80 strings. This proposed neutrino telescope would have an effective area of about 1 km^2, an energy threshold near 100 GeV and a pointing accuracy for muons of better than one degree for high-energy events. In conclusion, the observation of neutrinos by a neutrino telescope deep in the Antarctic ice cap, a goal that was once thought difficult if not impossible, represents an important step toward establishing the field of high-energy neutrino astronomy first envisioned over 40 years ago.

(**410**, 441-443; 2001)

以通过更长时间的曝光以及更好地确定 μ 子的能量来改善灵敏度，目前的灵敏度水平和这些设想说明了大面积探测器检验理论模型的能力。这些理论模型假设活动星系发出的 TeV 能量的光子来源于强子，如果没有观测高能中微子能力，证实或排除这些模型将会非常困难。

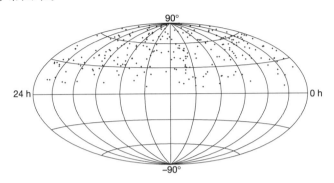

图 4. 向上传播的 μ 子事件在天空中赤纬和赤经上的分布。图中所示的 263 个事件取自 A 和 B 两个分析中包含的共同的上行 μ 子事件。μ 子的真实角度和重建角度之差的中值约为 3 ～ 4 度。

寻找来自伽玛射线暴的中微子，寻找磁单极子和超新星坍缩，寻找来自地球中心的冷暗物质，这些工作也在进行之中。而且，仅使用 138 天的数据，就可给出与那些已长时间运行的小型地下中微子探测器相媲美或更好的上限（见参考文献 10 ～ 13）。

从 1997 年到 1999 年，在 AMANDA-B10 周围的同心圆柱体中又添加了 9 串探测器。这个更大的探测器被称为 AMANDA-II，由 677 个光学模块构成，在更大的角度范围内改善了 μ 子的接收度。目前，这个更巨大的阵列正在采集数据。但是无论是理论模型预言[14]还是由超高能宇宙射线通量观测值推导出的[19]高能中微子通量都是非常低的，以至于为了观测研究这些中微子，探测器的有效面积需要达到一平方千米[1,14]。所以，目前正在计划研制更大的探测器，该探测器被称为冰立方（IceCube），由分布在 80 根串上的 4,800 个光电倍增管构成。这一计划中的中微子探测器将具有 1 km² 大小的有效面积，其探测能量阈值接近 100 GeV，对于高能事件的 μ 子的指向精度优于 1 度。总而言之，通过南极冰盖下的中微子望远镜观测到中微子意味着人们朝 40 多年前就试图建立的高能中微子天文学研究领域迈出了重要的一步，尽管这一设想一度被认为是极难甚至是无法实现的目标。

（金世超 翻译；曹俊 审稿）

E. Andrés[*], P. Askebjer[†], X. Bai[‡], G. Barouch[*], S. W. Barwick[§], R. C. Bay[∥], K.-H. Becker[¶], L. Bergström[†], D. Bertrand[#], D. Bierenbaum[§], A. Biron[⋆], J. Booth[§], O. Botner[**], A. Bouchta[⋆], M. M. Boyce[*], S. Carius[††], A. Chen[*], D. Chirkin[∥¶], J. Conrad[**], J. Cooley[*], C. G. S. Costa[#], D. F. Cowen[‡‡], J. Dailing[§], E. Dalberg[†], T. DeYoung[*], P. Desiati[†], J.-P. Dewulf[#], P. Doksus[*], J. Edsjö[†], P. Ekström[†], B. Erlandsson[†], T. Feser[§§], M. Gaug[⋆], A. Goldschmidt[∥∥], A. Goobar[†], L. Gray[*], H. Haase[⋆], A. Hallgren[**], F. Halzen[*], K. Hanson[‡‡], R. Hardtke[*], Y. D. He[∥], M. Hellwig[§§], H. Heukenkamp[⋆], G. C. Hill[*], P. O. Hulth[†], S. Hundertmark[§], J. Jacobsen[∥∥], V. Kandhadai[*], A. Karle[*], J. Kim[§], B. Koci[*], L. Köpke[§§], M. Kowalski[⋆], H. Leich[⋆], M. Leuthold[⋆], P. Lindahl[††], I. Liubarsky[∥], P. Loaiza[**], D. M. Lowder[∥], J. Ludvig[∥∥], J. Madsen[*], P. Marciniewski[**], H. S. Matis[∥∥], A. Mihalyi[‡‡], T. Mikolajski[⋆], T. C. Miller[‡], Y. Minaeva[†], P. Miočinović[∥], P. C. Mock[§], R. Morse[*], T. Neunhöffer[§§], F. M. Newcomer[‡‡], P. Niessen[⋆], D. R. Nygren[∥∥], H. Ögelman[*], C. Pérez de los Heros[**], R. Porrata[§], P. B. Price[∥], K. Rawlins[*], C. Reed[§], W. Rhode[¶], A. Richards[∥], S. Richter[⋆], J. Rodríguez Martino[†], P. Romenesko[*], D. Ross[§], H. Rubinstein[†], H.-G. Sander[§§], T. Scheider[§§], T. Schmidt[⋆], D. Schneider[*], E. Schneider[§], R. Schwarz[*], A. Silvestri[¶⋆], M. Solarz[∥], G. M. Spiczak[‡], C. Spiering[⋆], N. Starinsky[*], D. Steele[*], P. Steffen[⋆], R. G. Stokstad[∥∥], O. Streicher[⋆], Q. Sun[†], I. Taboada[‡‡], L. Thollander[†], T. Thon[⋆], S. Tilav[⋆], N. Usechak[§], M. Vander Donckt[#], C. Walck[†], C. Weinheimer[§§], C. H. Wiebusch[⋆], R. Wischnewski[⋆], H. Wissing[⋆], K. Woschnagg[∥], W. Wu[§], G. Yodh[§] & S. Young[§]

[*] Department of Physics, University of Wisconsin, Wisconsin, Madison 53706, USA

[†] Fysikum, Stockholm University, S-11385 Stockholm, Sweden

[‡] Bartol Research Institute, University of Delaware, Newark, Delaware 19716, USA

[§] Department of Physics and Astronomy, University of California, Irvine, California 92697, USA

[∥] Department of Physics, University of California, Berkeley, California 94720, USA

[¶] Fachbereich 8 Physik, BUGH Wuppertal, D-42097 Wuppertal, Germany

[#] Brussels Free University, Science Faculty CP230, Boulevard du Triomphe, B-1050 Brussels, Belgium

[⋆] DESY-Zeuthen, D-15735 Zeuthen, Germany

[**] Department of Radiation Sciences, Uppsala University, S-75121 Uppsala, Sweden

[††] Department of Technology, Kalmar University, S-39129 Kalmar, Sweden

[‡‡] Department of Physics and Astronomy, University of Pennsylvania, Philadelphia, Pennsylvania 19104, USA

[§§] Institute of Physics, University of Mainz, Staudinger Weg 7, D-55099 Mainz, Germany

[∥∥] Institute for Nuclear and Particle Astrophysics, Lawrence Berkeley National Laboratory, Berkeley, California 94720, USA

Received 15 September 2000; accepted 25 January 2001.

References:

1. Gaisser, T. K., Halzen, F. & Stanev, T. Particle physics with high-energy neutrinos. *Phys. Rep.* **258**, 173-236 (1995).

2. Totsuka, Y. Neutrino astronomy. *Rep. Prog. Phys.* **55**, 377-430 (1992).

3. Roberts, A. The birth of high-energy neutrino astronomy: A personal history of the DUMAND project. *Rev. Mod. Phys.* **64**, 259-312 (1992).

4. Balkanov, V. A. *et al.* An upper limit on the diffuse flux of high energy neutrinos obtained with the Baikal detector NT-96. *Astropart. Phys.* **14**, 61-67 (2000).

5. Lowder, D. M. *et al.* Observation of muons using the polar ice as a Čerenkov detector. *Nature* **353**, 331-333 (1991).

6. Andres, E. *et al.* The AMANDA neutrino telescope: Principle of operation and first results. *Astropart. Phys.* **13**, 1-20 (2000).

7. Waxman, E. & Bahcall, J. N. High-energy neutrinos from cosmological gamma ray burst fireballs. *Phys. Rev. Lett.* **78**, 2992-2295 (1997).

8. Askjeber, P. *et al.* Optical properties of the south pole ice at depths between 0.8 and 1 km. *Science* **267**, 1147-1150 (1995).

9. Price, B. P. Implications of optical properties of ocean, lake, and ice for ultrahigh-energy neutrino detection. *Appl. Opt.* **36**, 1965-1975 (1997).

10. Wischnewski, R. *et al.* in *Proc. 26th Int. Cosmic Ray Conf., Salt Lake City* Vol. 2 (eds Kieda, D., Salamon, M. & Dingus, B.) 229-232 (1999).

11. Dalberg, E. *et al.* in *Proc. 26th Int. Cosmic Ray Conf., Salt Lake City* Vol. 2 (eds Kieda, D., Salamon, M. & Dingus, B.) 348-351 (1999).

12. Bay, R. *et al.* in *Proc. 26th Int. Cosmic Ray Conf., Salt Lake City* Vol. 2 (eds Kieda, D., Salamon, M. & Dingus, B.) 225-228 (1999).

13. Niessen, P. *et al.* in *Proc. 26th Int. Cosmic Ray Conf., Salt Lake City* Vol. 2 (eds Kieda, D., Salamon, M. & Dingus, B.) 344-347 (1999).

14. Learned, J. G. & Mannheim, K. High-energy neutrino astrophysics. *Ann. Rev. Nucl. Sci.* **50**, 679-749 (2000).

15. Szabo, A. P. & Protheroe, R. J. Implications of particle acceleration in active galactic nuclei for cosmic rays and high-energy neutrino astronomy. *Astropart. Phys.* **2**, 375-392 (1994).

16. Stecker, F. W. & Salamon, M. H. High-energy neutrinos from quasars. *Space Sci. Rev.* **75**, 341-355 (1996).

17. Nellen, L., Mannheim, K. & Biermann, P. L. Neutrino production through hadronic cascades in AGN accretion disks. *Phys. Rev. D* **47**, 5270-5274 (1993).

18. Mannheim, K., Protheroe, R. J. & Rachen, J. P. Cosmic ray bound for models of extragalactic neutrino production. *Phys. Rev. D* **63**, 023003-1–023003-16 (2001).

19. Waxman, E. & Bahcall, J. N. High energy neutrinos from astrophysical sources. An upper bound. *Phys. Rev. D* **59**, 023002-1–023002-8 (1999).

20. Fukuda, Y. *et al.* (Super-Kamiokande Collaboration) Measurement of the flux and zenith-angle distribution of upward through-going muons by Super-Kamiokande. *Phys. Rev. Lett.* **82**, 2644-2648 (1999).

Acknowledgements. This research was supported by the following agencies: US National Science Foundation, Office of Polar Programs; US National Science Foundation, Physics Division; University of Wisconsin Alumni Research Foundation; US Department of Energy; Swedish Natural Science Research Council; Swedish Polar Research Secretariat; Knut and Allice Wallenberg Foundation, Sweden; German Ministry for Education and Research; US National Energy Research Scientific Computing Center (supported by the Office of Energy Research of the US Department of Energy); UC-Irvine AENEAS Supercomputer Facility; Deutsche Forschungsgemeinschaft (DFG). D.F.C. acknowledges the support of the NSF CAREER programme and C.P.d.l.H. acknowledges support from the European Union 4th Framework of Training and Mobility of Researchers.

Correspondence and requests for materials should be addressed to F.H. (e-mail: halzen@pheno.physics.wisc.edu).

Linkage Disequilibrium in the Human Genome

D. E. Reich *et al.*

Editor's Note

By this time, it was known that single nucleotide polymorphisms (SNPs)—the single-letter DNA changes thought to underlie disease susceptibility and individual variation—"travel" together, with one SNP carrying information about its SNP neighbours. So armed with the recent SNP map, geneticist Eric Lander and colleagues set out to see if the method of linkage disequilibrium (LD), where blocks of SNPs are correlated back to an ancestral chromosome, can be used to map disease-causing genes. This study shows, at least in the northern European population tested, that using LD to map disease-related genes might be easier and more practical than had been thought. The results also shed light on human history, suggesting this particular LD pattern might reflect the relatively recent migration of anatomically modern humans.

With the availability of a dense genome-wide map of single nucleotide polymorphisms (SNPs)[1], a central issue in human genetics is whether it is now possible to use linkage disequilibrium (LD) to map genes that cause disease. LD refers to correlations among neighbouring alleles, reflecting "haplotypes" descended from single, ancestral chromosomes. The size of LD blocks has been the subject of considerable debate. Computer simulations[2] and empirical data[3] have suggested that LD extends only a few kilobases (kb) around common SNPs, whereas other data have suggested that it can extend much further, in some cases greater than 100 kb[4-6]. It has been difficult to obtain a systematic picture of LD because past studies have been based on only a few (1–3) loci and different populations. Here, we report a large-scale experiment using a uniform protocol to examine 19 randomly selected genomic regions. LD in a United States population of north-European descent typically extends 60 kb from common alleles, implying that LD mapping is likely to be practical in this population. By contrast, LD in a Nigerian population extends markedly less far. The results illuminate human history, suggesting that LD in northern Europeans is shaped by a marked demographic event about 27,000–53,000 years ago.

To characterize LD systematically around genes, each of the 19 regions that we studied was anchored at a "core" SNP in the coding region of a gene. The core SNP was chosen from a database of more than 3,000 coding SNPs that had been identified by screening in a multi-ethnic panel (see Methods), subject to two requirements. First, "finished" genomic sequence was available for 160 kb in at least one direction from the core SNP; second, the frequency of the minor (less common) allele was at least 35% in the multi-ethnic panel.

人类基因组中的连锁不平衡

赖希等

编者按

如今众所周知，单核苷酸多态性(SNP)——被认为影响疾病的易感性和个体差异性的 DNA 单个碱基变化——是结伴"旅行"的，即一个 SNP 会携带关于其邻位的信息。因此，结合目前的 SNP 图谱，遗传学家埃瑞克·兰德及其同事着手研究连锁不平衡 (LD，即 SNP 的板块都是和一个祖先染色体相关联的)的方法能否被用于发现致病基因。本研究显示，至少在北欧人群的测试中，利用 LD 来寻找致病基因也许比预想的要更简单更实际。该结果还可用于阐释人类历史，提示这种独特的 LD 模式可能反映出解剖学上现代人类的迁徙。

随着高密度全基因组单核苷酸多态性(SNP)图谱的获得 [1]，人类遗传学的一个核心问题就是现在是否有可能利用连锁不平衡(LD)绘制出致病基因图谱。LD 指的是相邻等位基因之间的相互联系，表现为来自单一祖先染色体的"单体型"。LD 区域的大小一直是争议的焦点。计算机模拟 [2] 和经验性数据 [3] 显示 LD 位于常见 SNP 周围，大约只延伸数千个碱基(kb)，而其他数据显示其范围更广，在某些情况下可能大于 100 kb[4-6]。一直以来，LD 的系统性图谱难以获得，因为过去的研究都仅仅基于几个(1~3)位点以及不同的人群。本文我们报告一个采用了统一规范，针对 19 个随机选取的基因组区域的大规模实验。结果发现拥有北欧血统的美国人群其 LD 一般从常见等位位点延伸 60 kb，说明在这一群体中进行 LD 图谱的绘制是可行的。相反，在一个尼日利亚人群中，LD 区域的缩短十分明显。这些结果勾画出人类演化历史，提示大约 27,000~53,000 年前的某个重大的人口统计学事件导致了北欧人群中 LD 的形成。

为了系统性地鉴定基因附近的 LD，在我们研究的 19 个区域中，每个区域都在其编码区内锚定了一个"核心"SNP。核心 SNP 选自一个含有超过 3,000 个编码 SNP 的数据库，这些 SNP 从多民族人群筛查中选定(见方法)，要求满足以下两个条件。其一，从核心 SNP 起始，至少一个方向上 160 kb 范围内可获得完整基因组序列；其二，次等位基因位点(频率较低的那个位点)在这个多族群数据库中的频率至少为 35%。

We focused on high-frequency SNPs for several reasons. First, they tend to be of high frequency in all populations[7], facilitating cross-population comparisons. Second, LD around common alleles represents a "worst case" scenario: LD around rare alleles is expected to extend further because such alleles are generally young[8] and there has been less historical opportunity for recombination to break down ancestral haplotypes[2]. Third, LD around common alleles can be measured with a modest sample size of 80–100 chromosomes to a precision within 10–20% of the asymptotic limit (see Methods). Last, LD around common alleles will probably be particularly relevant to the search for genes predisposing to common disease[9].

To identify SNPs at various distances from the core SNP, we resequenced subregions of around 2 kb centred at distances 0, 5, 10, 20, 40, 80 and 160 kb in one direction from the core SNP using 44 unrelated individuals from Utah. Altogether, we screened 251,310 bp (see Methods) and found an average heterozygosity of $\pi = 0.00070$, consistent with past studies[1]. A total of 272 "high frequency" polymorphisms were identified (Table 1).

Table 1. Distribution of regions and SNPs within regions

Gene identification*	Chromosome	Local recombination rate (cM Mb⁻¹)†	Span of region in physical map (cR)‡	Number of high-frequency polymorphisms at distances (kb) from core SNPs§						
				1	5	10	20	40	80	160
BMP8‖	1	1.4	60.88–61.04	1	1	1	2	2	1	3
ACVR2B‖	3	1.1	37.27–37.43	2	1	2	3	0	3	1
TGFBI	5	1.4	152.16–152.00	1	3	1	0	1	7	0
DDR1‖	6	–	46.046–45.89	1	4	1	4	2	2	0
GTF2H4	6	–	46.059–46.22	3	3	3	6	2	0	0
COL11A2‖	6	–	48.27–48.43	1	0	1	2	3	1	1
LAMB1‖	7	2.3	106.58–106.42	0	0	5	3	3	2	4
WASL	7	0.5	122.99–122.83	2	1	0	4	2	2	1
SLC6A12	12	3.3	3.62–3.78	2	8	2	1	6	0	0
KCNA1	12	3.3	8.50–8.66	1	1	5	2	1	1	0
SLC2A3	12	2.1	16.33–16.49	1	2	1	1	5	0	2
ARHGDIB‖	12	1.2	21.84–22	3	9	1	2	2	1	2
PCI‖	14	4.3	98.41–98.57	0	7	0	0	4	14	1
PRKCBI	16	1.0	32.50–32.66	0	2	3	0	4	3	0
NFI	17	1.0	38.10–38.26	1	0	2	2	5	1	0
SCYA2‖	17	3.0	40.21–40.37	0	5	1	1	3	1	1
PAI2	18	2.7	64.17–64.01	2	0	2	1	5	2	10
IL17R‖	22	5.9	14.48–14.32	1	1	2	1	2	0	4
HCF2‖	22	2.0	17.91–17.75	1	1	2	2	1	0	3

* Abbreviations from LocusLink (www.ncbi.nlm.nih.gov/LocusLink/list.cgi).

我们着眼于高频率的 SNP 有以下几点原因。首先，它们倾向于在所有人群中高频出现 [7]，这有利于跨人群的比对。其次，常见等位位点附近的 LD 实际代表了一种"最差"的情况：而罕见位点附近的 LD 能够延伸更长，因为这些等位基因通常出现得比较晚 [8]，历史上发生重组而破坏原始单体型的概率也更低一些 [2]。第三，常见等位基因位点附近的 LD 可以通过对 80～100 个染色体这样一个适中的样本数量测量获得，精度在渐进极限的 10%～20% 之间（见方法）。最后，常见等位基因位点附近的 LD 很可能与寻找常见疾病易感基因更为相关 [9]。

为了确定距离核心 SNP 不同位置的 SNP，我们选取同一方向上距离核心 SNP 0、5、10、20、40、80 和 160 kb 的位置作为中心，对中心附近 2 kb 的子区域进行重测序，样本来自犹他州 44 个不相关的个体。总体来说，我们对 251,310 个碱基对（见方法）进行了上述筛查，发现平均杂合度 $\pi = 0.00070$，与既往的研究一致 [1]。总共检测到 272 个"高频"多态性位点（表 1）。

表 1. 区域分布以及区域内的 SNP

基因代码 *	染色体	局部重组率 (cM · Mb⁻¹)†	物理图谱中该区域的范围 (cR)‡	离核心 SNP 不同距离(kb)的高频多态位点数目§						
				1	5	10	20	40	80	160
BMP8‖	1	1.4	60.88～61.04	1	1	1	2	2	1	3
ACVR2B‖	3	1.1	37.27～37.43	2	1	2	3	0	3	1
TGFBI	5	1.4	152.16～152.00	1	3	1	0	1	7	0
DDR1‖	6	–	46.046～45.89	1	4	1	4	2	2	0
GTF2H4	6	–	46.059～46.22	3	3	3	6	2	0	0
COL11A2‖	6	–	48.27～48.43	1	0	1	2	3	1	1
LAMB1‖	7	2.3	106.58～106.42	0	0	5	3	3	2	4
WASL	7	0.5	122.99～122.83	2	1	0	4	2	2	0
SLC6A12	12	3.3	3.62～3.78	2	8	2	1	6	0	0
KCNA1	12	3.3	8.50～8.66	1	1	5	2	1	1	0
SLC2A3	12	2.1	16.33～16.49	1	2	1	1	5	0	2
ARHGDIB‖	12	1.2	21.84～22	3	9	1	2	2	1	2
PCI‖	14	4.3	98.41～98.57	0	7	0	0	4	14	1
PRKCBI	16	1.0	32.50～32.66	0	2	3	0	4	3	0
NFI	17	1.0	38.10～38.26	1	0	2	2	5	1	0
SCYA2‖	17	3.0	40.21～40.37	0	5	1	1	3	1	1
PAI2	18	2.7	64.17～64.01	2	0	2	1	5	2	10
IL17R‖	22	5.9	14.48～14.32	1	1	2	1	2	0	4
HCF2‖	22	2.0	17.91～17.75	1	1	2	2	1	0	3

* 缩写来自 LocusLink（www.ncbi.nlm.nih.gov/LocusLink/list.cgi）。

† For three regions the genetic and physical maps were inconsistent and no estimates were made.

‡ Span of region within a radiation hybrid map (http://www.ncbi.nlm.nih.gov/genome/seq/HsHome.shtml): 1 Mb ≈ 1 centirad. Position of the core SNP is listed first.

§ Number of SNPs discovered with at least 15 copies of the minor allele (successfully genotyped in at least 32 individuals) and in Hardy-Weinberg equilibrium using a significance criterion of $P < 0.02$.

‖ The ten regions selected for follow-up genotyping.

We measured LD between two SNPs using the classical statistic D′ (see Methods)[10]. D′ has the same range of values regardless of the frequencies of the SNPs compared[11]. Its sign (positive or negative) depends on the arbitrary choice of the alleles paired at the two loci. We chose the pair of SNPs that caused D′ > 0 in Utah so that, in comparisons with other populations, the sign of D′ indicates whether the same or opposite allelic association is present. In a large sample, |D′| of 1 indicates complete LD; 0 corresponds to no LD. The degree of LD needed for effective mapping depends on the details of a particular study[2]. A useful measure is the "half-length" of LD (the distance at which the average |D′| drops below 0.5).

Comparing the 19 randomly selected regions, LD has a half-length of about 60 kb (Fig. 1). Significant P-values for LD occur in greater than 50% of cases at distances of ≤ 80 kb. LD therefore extends much further than a previous prediction[2], and our data indicate that, in general, blocks of LD are large. Although the average extent is large, there is great variation in LD across the genomic regions (see also ref. 12), which is apparent in the different rates at which LD declines around the core SNP (Fig. 2). For example, |D′| > 0.5 for at least 155 kb around the *WASL* gene, but for less than 6 kb around *PCI* (Fig. 2). The variability across different genomic regions within the same population sample provides a context for explaining why past empirical studies, each based on one to three regions[3-5], have produced such different results. Large variations in LD are expected because of stochastic factors, such as different gene histories across loci[13]. Differences in recombination rates among regions can also affect the extent of LD. We observe a significant and important correlation ($P < 0.005$) between LD and the estimated local recombination rate (Fig. 2, inset).

Another feature of the data is that, near the range of distances at which LD drops off, there is often considerable variability in |D′| values at neighbouring SNPs (for example, around *IL17R*, *SCYA2*, *TGFBI* and *BMP8* in Fig. 2). Such a wide scatter of LD, even for markers close to each other, has been noted before[14], and is due to the underlying haplotypic structure of LD. SNPs marking sections of chromosome with short extents of correlation are likely to display much lower |D′| values than SNPs marking long haplotypes. In regions of high haplotype diversity, several SNPs may have to be genotyped to have a good chance of tagging most haplotypes. LD-based gene mapping may therefore require clusters of closely spaced SNPs to have maximal power.

† 其中三个区域的基因图谱和物理图谱不一致，因此没有进行估算。

‡ 放射性杂交图谱中的区域范围（www.ncbi.nlm.nih.gov/genome/seq/HsHome.shtml）：1Mb≈1 厘拉德。核心 SNP 的位置列在前。

§ 在至少 15 个次等位位点拷贝（在至少 32 个个体中成功确定基因型）中发现的 SNP 数量，并且在哈迪–温伯格平衡中显著性标准 $P < 0.02$。

‖ 这十个区域用于后续的基因分型。

我们使用经典的 D′ 值量化两个 SNP 之间的 LD（见方法）[10]。不论相比较的 SNP 出现频率如何，D′ 都具有相同的取值范围 [11]。其符号（正或者负）取决于两位点间等位位点对的随机选择。我们选择了在犹他州人群中 D′ > 0 的 SNP 对，这样在和其他人群进行比较时，D′ 的符号表示是否存在相同或者相反的等位相关性。在大样本中，|D′| 为 1 表示完全的 LD，0 表示不存在 LD。有效的图谱绘制所需 LD 的程度需要根据具体研究细节确定 [2]。一种可行的度量方式是利用 LD 的"半长"（即平均 |D′| 降至 0.5 以下的距离）。

对比这 19 个随机选取的区域，LD 的半长度大约是 60 kb（图 1）。在距离 ≤80 kb 的情况下，超过 50% 的样本 LD 的 P 值具有显著性。因此 LD 延伸的距离比先前的预计要长得多 [2]，而且我们的数据显示，总体上 LD 区域都相当大。尽管平均尺度很大，整个基因组区域中 LD 的变化也很大（也可见参考文献 12），但很明显在核心 SNP 附近 LD 衰减的速率不同（图 2）。比如，WASL 基因附近，|D′| > 0.5 的区域至少有 155 kb，但是对于 PCI 基因附近则小于 6 kb（图 2）。同一人群中这种不同基因组区域间的变异，可以用来解释为什么过去的经验性研究尽管只是基于 1 到 3 个区域 [3-5]，却产生如此多不同的结果。因为存在随机因素，比如不同位点间的基因历史不同 [13]，LD 的这种巨大变异是可以预期的。区域间重组率的不同也能影响到 LD 的程度。我们观察到 LD 和预测的局部重组率之间存在显著且重要的相关性（$P < 0.005$）（图 2，右下角插图）。

这些数据的另一个特征是在 LD 急剧下降的区间附近，相邻 SNP 间的 |D′| 值有较大波动（例如，图 2 中 IL17R、SCYA2、TGFBI 和 BMP8 附近）。LD 这种即使在相邻位点也产生大范围波动以前也曾发现过 [14]，是由于 LD 潜在的单体型结构引起的。标记染色体区域的 SNP 如果相关性延伸较短，相比于长单体型的 SNP 可能显示出更低的 |D′| 值。在具有高度单体型变化的区域，可能需要多个 SNP 被分型确定才有可能代表大多数的单体型。因此，绘制基于 LD 的基因图谱需要位置关系紧密的 SNP 集合才能达到最大效果。

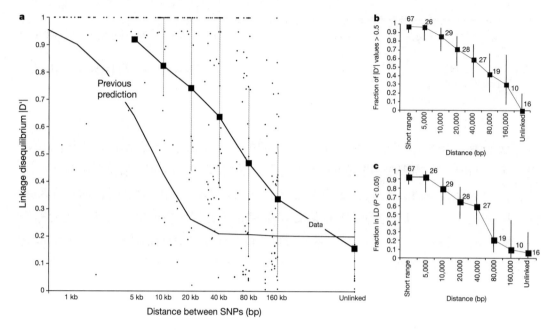

Fig. 1. LD versus physical distance between SNPs. For each distance from the core SNP (Table 1), we chose the SNP with the largest number of copies of the minor allele for comparison to SNPs at other distances. At a given distance, all comparisons are independent. **a**, Average |D′| values for each distance separation ("Data"; dotted lines indicate the 25th and 75th percentiles), compared with a prediction[2] based on simulations (see Methods). |D′| values for shorter physical distances were calculated by looking within contiguously sequenced stretches of DNA containing at least two SNPs, and picking the two with the most minor alleles. Unlinked marker comparisons are obtained by comparing SNPs in the 40-kb bin in each row of Table 1 to those in the next row. **b**, **c**, Fraction of |D′| values greater than 0.5 (**b**) and proportion of significant ($P < 0.05$) associations (**c**) between two SNPs separated by a given distance (as assessed by a likelihood ratio test[10]). Bars indicate 95% central confidence intervals. The number of data points used to make the calculations are shown.

Why does LD extend so far? LD around an allele arises because of selection or population history—a small population size, genetic drift or population mixture—and decays owing to recombination, which breaks down ancestral haplotypes[15]. The extent of LD decreases in proportion to the number of generations since the LD-generating event. The simplest explanation for the observed long-range LD is that the population under study experienced an extreme founder effect or bottleneck: a period when the population was so small that a few ancestral haplotypes gave rise to most of the haplotypes that exist today. Our simulations show that a severe bottleneck (inbreeding coefficient $F \geqslant 0.2$) occurring 800–1,600 generations ago (about 27,000–53,000 years ago assuming 25 years per generation) could have generated the LD observed (Fig. 3). In principle, long-range LD could also be generated by population mixture[16], but the degree of LD is much greater than would arise from the mixing of even extremely differentiated populations. An alternative explanation for the observed long-range LD is that the recombination rates in the regions studied might be markedly less than the genome-wide average. This could happen if recombination occurred primarily in well-separated hotspots and our regions fell between them (Fig. 3). However, under this hypothesis, the regions would be expected to show long-range LD in all populations, and this pattern is not observed (see below).

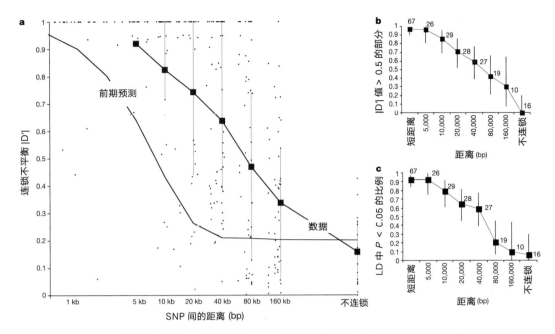

图 1. LD 与 SNP 间的物理距离。对于距离核心 SNP 的不同距离（表 1），我们选择具有次等位基因位点中最大拷贝数目的 SNP 用于与其他距离的 SNP 进行对比。在距离给定的前提下，所有的对比都是独立的。**a**，每个距离间隔的平均 |D'| 值（"数据"；虚线表示 25 和 75 百分位），与模拟推算得出的预期值[2]进行比较（见方法）。更短物理距离的 |D'| 值是通过找出含有至少两个 SNP 的连续 DNA 序列，并且挑选最罕见的两个等位位点计算获得的。不连锁的标志是通过对照表 1 每行中每 40 kb 分箱的 SNP 与其下一行进行比较获得的。**b** 和 **c**，在一定距离（通过似然比检验的估算获得[10]）的两个 SNP 之间，|D'| 值超过 0.5(**b**) 和具有显著性关联（$P < 0.05$）(**c**) 的部分。条形表示 95% 的置信区间。图中显示了计算所用的数据点的个数。

　　LD 为什么一直在延伸？等位基因附近 LD 的出现是由于自然选择或者种群历史——一个小规模种群，遗传漂变或者种群混合——由于发生了打破祖先单体型的重组而衰减导致的[15]。LD 长度的减小与 LD 产生事件之后世代的数目成比例。对较长 LD 的最简单解释就是我们所研究的人群经历了极端的建立者效应或者瓶颈效应：有一段时间种群太小以至于一些祖先单体型衍生出目前存在的大部分单体型。我们的模拟显示大约 800 ~ 1,600 代以前（大约 27,000 ~ 53,000 年前，假设每代 25 年）发生了一次严重的瓶颈效应（近交相关系数 $F \geqslant 0.2$），由此可能产生了我们观察到的 LD（图 3）。原则上说，长 LD 还能够通过种群混合产生[16]，但 LD 的程度要远远超过极端分化的种群混合而产生的 LD。另一种对长 LD 的解释就是所研究区域的重组率显著低于基因组的平均水平。如果重组主要发生在分离热点区且我们的研究区域刚好落入其中，这种现象可能出现（图 3）。但是，根据这个假设，这些区域应该在所有的人群中都显示长 LD，而事实观察到的并非如此（见下文）。

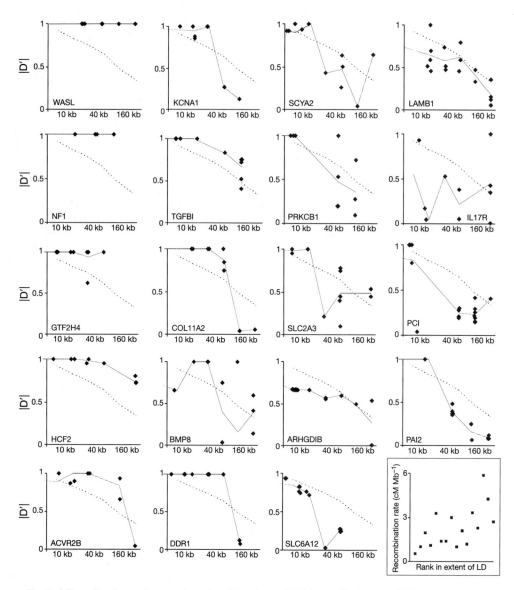

Fig. 2. LD profiles for each genomic region. The chosen SNP is usually the core SNP itself, unless the core SNP could not be readily genotyped or another SNP with more minor alleles had been identified within 1 kb. In both cases, we substituted the closest high-frequency SNP. The chosen SNP is compared with every other high-frequency SNP in the same genomic region. Solid line indicates average |D'| values for each distance for which SNPs were available; for comparison, the dashed line indicates the consensus LD curve from Fig. 1. The extent of LD was calculated by performing a least-squares linear regression to the average |D'| values at each distance from the chosen SNP; more sloped lines indicate less LD. The regions are ordered according to the extent of LD (most extensive LD, top left; least extensive LD, bottom right). Inset (bottom right) shows the rank of each region in terms of LD extent versus the estimated recombination rate per unit of physical distance. For each 160-kb region of interest, we looked for the closest pair of flanking genetic markers from the Marshfield map[34] subject to the condition that they were separated by a non-zero genetic distance on the map. We divided genetic map distance by the physical map distance on the basis of the available draft genome sequence[35]. We analysed the 16 regions for which the genetic and physical map orderings of markers were locally consistent (one-sided Spearman rank correlation, $P < 0.005$).

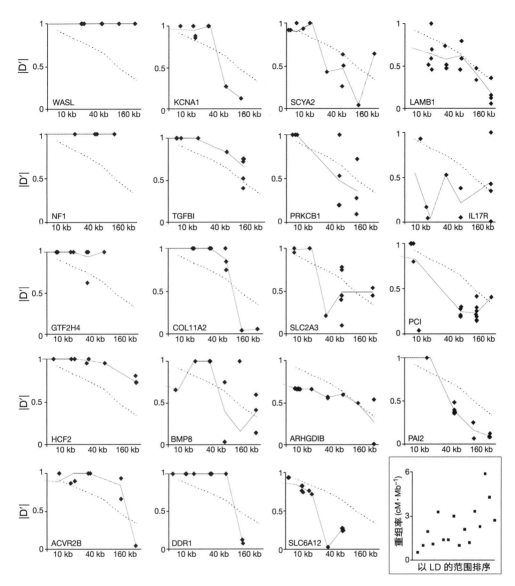

图 2. 每个基因组区域的 LD 概况。选择的 SNP 通常是核心 SNP 本身，除非核心 SNP 不能准确分型或者在 1 kb 范围内发现具有更小的次等位基因频率的 SNP。在这两种情况下，我们都替换成最近的高频率 SNP。所选取的 SNP 与同一个基因组区域的每个其他高频率 SNP 进行对比。实线表示每个距离上可获得 SNP 的平均 |D′| 值；作为对照，虚线表示从图 1 中获得的一致的 LD 曲线。LD 长度的计算是通过对每个距离上选定 SNP 的平均 |D′| 值进行最小二乘线性回归拟合；斜率越大表示 LD 越小。不同区域根据 LD 的长度进行排列（最长的位于左上，最小的位于右下）。插图（右下角）横纵轴分别代表依照 LD 范围大小将各个区域排序而成的各个序列与每一单位物理距离所估算的重组率。对于每个感兴趣的 160 kb 区域，我们根据它们在图谱上被一个非零遗传距离分隔的条件，从马什菲尔德图谱 [34] 中寻找最近的两侧基因标志。根据可获得的基因草图序列 [35]，我们用物理图谱距离来划分基因图谱的距离。我们分析了基因和物理图谱标志顺序局部一致的 16 个区域（单侧斯皮尔曼等级相关性，$P < 0.005$）。

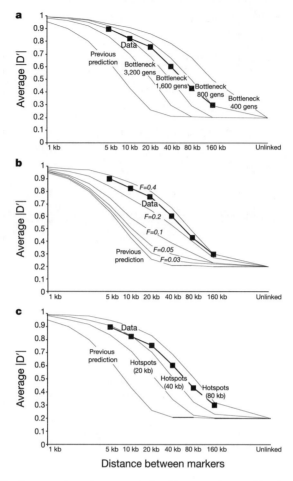

Fig. 3. Effect on LD of assumptions about population history and recombination. **a**, The effect of a population bottleneck instantaneously reducing the population to a constant size of 50 individuals for 40 generations ($F = 0.4$) and occurring 400, 800, 1,600 or 3,200 generations ago. **b**, The effect of bottleneck intensity (F) for a bottleneck that occurred 800 generations ago. **c**, The effect of variation in recombination rate, assuming that all recombination in the genome occurs at hotspots randomly distributed according to a Poisson process with an average density of one every 20, 40 and 80 kb, and with a genome-wide average rate of 1.3 centiMorgans per megabase per generation[35]. Results are compared to a no-hotspot model for the same historical hypothesis. ("Data" refers to the mean $|D'|$ values at each physical distance separation, obtained from Fig. 1a).

To confirm our findings of long-range LD and to investigate the reasons for its occurrence, we next examined a representative subset of SNPs in two additional samples. We first studied another north-European sample (48 southern Swedes) and found LD in a nearly identical pattern to that observed in Utah, both in terms of the overall magnitude of LD and the particular alleles that were associated (indicated by the sign of D′) (Fig. 4). The similarity in LD patterns may be due to the same historical event, which occurred deep in European prehistory before the separation of the ancestors of these two groups. This suggests that the long-range LD pattern is general in northern Europeans[3,17].

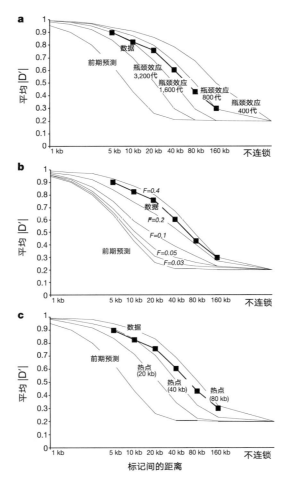

图 3. 关于种群历史和重组的假设对 LD 的影响。**a**，按照 400，800，1,600 或者 3,200 代以前发生的将人群数量即刻减少到 50 个恒定个体并持续 40 代（$F = 0.4$）的人口瓶颈效应影响。**b**，一个发生于 800 代前的瓶颈效应强度（F）的影响。**c**，变异对重组率的影响，假设基因组中所有的重组发生在热点上且服从泊松过程，每 20、40 和 80 kb 的平均密度为 1，并且整个基因组范围内的平均速率是 1.3 厘摩每百万碱基每代[35]。所有结果均和相同的历史事件假设但没有热点的模型进行比较。（"数据"指的是图 1a 中得到的每个物理距离分隔的 |D′| 平均值）。

为了证实我们发现的长 LD 并且研究其出现的原因，接下来我们在另外两个不同样本中检查了一个有代表性的 SNP 子集。首先我们研究了另一个北欧人群样本（48 名南瑞典人）并且发现无论是总体的 LD 大小还是相关的特定等位基因（由 D′ 结果推测）都和犹他州样本发现的模式几乎一致（图 4）。LD 整体模式的相似性可能是由于群体经历了相同的历史事件。该事件可能发生于史前的欧洲，彼时两个群体的祖先尚未分离。这提示在北欧人群中长 LD 的模式普遍存在[3,17]。

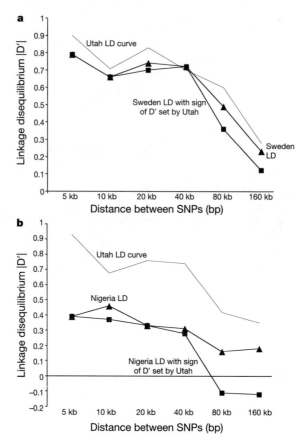

Fig. 4. LD curve for Swedish and Yoruban samples. To minimize ascertainment bias, data are only shown for marker comparisons involving the core SNP. Alleles are paired such that D' > 0 in the Utah population. D' > 0 in the other populations indicates the same direction of allelic association and D' < 0 indicates the opposite association. **a**, In Sweden, average D' is nearly identical to the average |D'| values up to 40-kb distances, and the overall curve has a similar shape to that of the Utah population (thin line in **a** and **b**). **b**, LD extends less far in the Yoruban sample, with most of the long-range LD coming from a single region, *HCF2*. Even at 5 kb, the average values of |D'| and D' diverge substantially. To make the comparisons between populations appropriate, the Utah LD curves are calculated solely on the basis of SNPs that had been successfully genotyped and met the minimum frequency criterion in both populations (Swedish and Yoruban).

We next studied 96 Yorubans (from Nigeria), believed to share common ancestry with northern Europeans about 100,000 years ago[18]. At short distances, the Nigerian and European-derived populations typically show the same allelic combinations (Fig. 4): D' has the same sign and a similar magnitude, indicating a common LD-generating event tracing far back in human history. However, the half-length of LD seems to extend less than 5 kb (Fig. 4) in the Yorubans. Markedly shorter range LD in sub-Saharan Africa has also been observed in several studies of single regions[19,20] (although two other studies did not show a clear trend[21,22]). Our results indicate that the pattern of shorter LD in sub-Saharan African populations may be general.

Notably, LD in the Nigerians is largely a subset of what is seen in the northern Europeans.

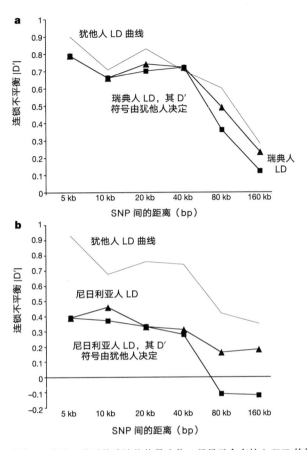

图 4. 瑞典和约鲁巴人的 LD 曲线。为了使确认偏倚最小化，只显示含有核心 SNP 的标志对比数据。等位基因位点进行配对使犹他州人群中 D′ > 0。其他人群中 D′ > 0 说明具有相同方向的等位位点具有相关性，而 D′ < 0 则说明具有相反的关系。**a**，在瑞典人中，平均 D′ 与长达 40 kb 距离的平均 |D′| 值几乎一致，总体曲线的形状也与犹他州人群类似（**a** 和 **b** 中的细线）。**b**，约鲁巴样本中 LD 的长度短一些，其中最长的 LD 来自 *HCF2* 一个单独的区域。即便只有 5 kb，平均 | D′| 值和 D′ 值最终还是发生了分离。为了在两个人群间进行恰当的比较，犹他州人群的 LD 曲线完全基于成功分型并且符合两个人群（瑞典人和约鲁巴人）的最低频率标准的 SNP 进行计算。

　　随后我们研究了 96 名约鲁巴人（来自尼日利亚），他们被认为在 100,000 年前与北欧人拥有相同的祖先[18]。在短的 LD 中，来自尼日利亚和欧洲的群体都显示出相同的等位基因位点重组（图 4）：D′ 有相同的迹象和类似的大小，说明在人类历史中经历了共同的 LD 产生事件。但是，约鲁巴人中 LD 的半长似乎少于 5 kb（图 4）。数个针对单个区域的研究也发现撒哈拉沙漠以南的非洲人 LD 范围明显缩短[19,20]（尽管另外两个研究没有显示明显的趋势[21,22]）。我们的结果表明撒哈拉沙漠以南非洲人群中更短的 LD 模式可能是普遍存在的。

　　显然，尼日利亚人的 LD 在很大程度上是观察到的北欧人 LD 的子集。约鲁巴

The Yoruban haplotypes are generally contained within the longer Utah haplotypes, and there is little Yoruban-specific LD (85% of observations of substantial LD ($|D'| > 0.5$) in Yorubans are also substantial and of the same sign in Utah). The vast difference in the extent of LD between populations points to differences owing to population history, probably a bottleneck or founder effect that occurred among the ancestors of north Europeans after the divergence from the ancestors of the Nigerians. The short extent of LD in Nigerians is more consistent with the predictions of a computer simulation study assuming a simple model of population expansion[2].

Could the apparent differences in the extent of LD among populations be due to "ascertainment bias" in the identification of the SNPs? The core SNPs are probably not subject to bias because they were identified in a multi-ethnic population. The neighbouring SNPs were identified in the Utah population and subsequently studied in the other populations, and thus they may be susceptible to ascertainment bias. However, we selected only SNPs with high frequency in Utah and most of these satisfied the high-frequency criterion for use in the other populations (87% in Sweden and 71% in Nigeria). Thus, the inferences about LD are not likely to be much different from what would have been obtained had we used SNPs ascertained in the Yoruban sample. Moreover, the cross-population comparisons (Fig. 4) minimize ascertainment bias because they involve only the core SNP, and because they calculate LD in each population using only the SNPs present in both.

What was the nature of the population event that created the long-range LD? The event could be specific to northern Europe, which was substantially depopulated during the Last Glacial Maximum (30,000–15,000 years ago), and subsequently recolonized by a small number of founders[23,24]. Alternatively, the long-range LD could be due to a severe bottleneck that occurred during the founding of Europe or during the dispersal of anatomically modern humans from Africa[19,20,25,26] (the proposed "Out of Africa" event) as recently as 50,000 years ago. Under the first hypothesis, the strong LD at distances ≥ 40 kb would be absent in populations not descended from northern Europeans. Under the second hypothesis, the same pattern of long-range LD could be observed in a variety of non-African populations. Regardless of the timing and context of the bottleneck, the severity of the event (in terms of inbreeding) can be assessed from our data. To have a strong effect on LD, a substantial proportion of the modern population would have to be derived from a population that had experienced an event leading to an inbreeding coefficient of at least $F = 0.2$ (Fig. 3). This corresponds to an effective population size (typically less than the true population size[15]) of 50 individuals for 20 generations; 1,000 individuals for 400 generations; or any other combination with the same ratio.

Our results have implications for disease gene mapping, suggesting a possible two-tiered strategy for using LD. The presence of large blocks of LD in northern European populations suggests that genome-wide LD mapping is likely to be possible with existing SNP resources[1]. Although the large blocks should make initial localization easier, they may also limit the resolution of mapping to blocks of DNA in the range of 100 kb[27]. Populations

单体型基本上都包含在更长的犹他州单体型中，几乎没有约鲁巴人独有的 LD(85% 约鲁巴人常见的 LD(| D′| > 0.5)在犹他州人中也很常见，而且具有相同的标志)。不同人群间 LD 长度的明显差别归因于人群历史的差异，北欧人祖先与尼日利亚人祖先分离之后可能发生了瓶颈效应或者建立者效应。尼日利亚人中的短 LD 与基于人群扩张简单模型假设的计算机模拟研究预测结果一致 [2]。

这种人群中 LD 的长度有明显差别的现象是否是由于 SNP 鉴定过程中存在的 "确认偏倚" 导致的？核心 SNP 不太可能受偏倚的影响，因为它们是在多种族人群中鉴定出来的。在犹他州人群中鉴定出的相邻 SNP，随后在其他人群中进行了研究，因此它们有可能受到确认偏倚的影响。但是，我们在犹他州人群中仅仅选择了高频出现的 SNP，其大部分都满足了可以用于其他人群的高频出现的标准(瑞典人群有 87%，尼日利亚人群有 71%)。因而我们现在推断出的 LD 与这些在约鲁巴人群样本中确定出的 SNP 所具有的 LD 不会有太大的差别。此外，跨人群比较使确认偏倚最小化(图 4)，因为这种比较仅仅使用了核心 SNP，还因为仅仅使用了共有的 SNP 来计算每个人群的 LD。

那么产生长 LD 的人群事件的本质是什么？这些事件可能是北欧人特有的，即在末次冰盛期(30,000 ~ 15,000 年前)人口的大量减少，随后少量的建立者重新定居 [23,24]。另一种可能是，这种长 LD 是由于在欧洲的建立过程中或是 50,000 年前解剖学上的现代人从非洲分离 [19,20,25,26](所谓"走出非洲"事件)所产生的严重瓶颈效应。根据第一个假设，长度 ≥ 40 kb 的大型 LD 应该不存在于非北欧后代人群中。根据第二个假设，这类长 LD 应该可以出现于各种非非洲人群中。不考虑这种瓶颈效应的时间和背景，从我们的数据可以推断出该事件的严重程度(以近亲繁殖的程度表示)。如果对 LD 产生巨大影响，很大部分的现代人类将是来自于经历过能导致近交系数至少是 $F = 0.2$ 的严重事件的群体(图 3)。这就相当于一个有效群体(通常小于真实的人口 [15])有 50 个个体，经历 20 代繁衍，或 1,000 个个体，经历 400 代繁衍，或是任何相同比例的组合。

我们的结论对疾病的基因图谱绘制具有重要意义，给出了在两个层次上利用 LD 的可能策略。北欧人群中存在大的 LD 板块，说明在现有 SNP 资源下获得基因组水平的图谱是可能实现的 [1]。尽管大的 LD 板块可能会使初始的定位变得容易，但是这同时也限制了将图谱的 DNA 解析为 100 kb 大小的分辨率 [27]。含有更小 LD

with much smaller blocks of LD (for example, Yorubans) may allow fine-structure mapping to identify the specific nucleotide substitution responsible for a phenotype[12]. Our study also has implications for LD as a tool to study population history[19-22]. Simultaneous assessment of LD at multiple regions of the genome provides an approach for studying history with potentially greater sensitivity to certain aspects of history than traditional methods based on properties of a single locus.

Methods

Core SNPs were identified by screening more than 3,000 genes in a multi-ethnic panel of 15 European Americans, 10 African Americans, and 7 East Asians (see ref. 28 for details; a full description of this database will be presented elsewhere). DNA used for sequencing was obtained from the Coriell Cell Repositories. Identification numbers for these Utah samples from the CEPH mapping panel were NA12344, 06995, 06997, 07013, 12335, 06990, 10848, 07038, 06987, 10846, 10847, 07029, 07019, 07048, 06991, 10851, 07349, 07348, 10857, 10852, 10858, 10859, 10854, 10856, 10855, 12386, 12456, 10860, 10861, 10863, 10830, 10831, 10835, 10834, 10837, 10836, 10838, 10839, 10841, 10840, 10842, 10843, 10845 and 10844. We did follow-up genotyping in 48 Swedes (healthy individuals from a case/control study of adult-onset diabetes) and in 96 Yoruban males from Nigeria (healthy individuals from a case/control study of hypertension).

SNPs were discovered by DNA sequencing[28] in the 44 individuals from Utah; we sequenced about 2 kb centred at each distance from the core SNP ≥ 5 kb, with about 1 kb sequenced around the core SNP itself. When no polymorphism of sufficiently high frequency was found, a nearby subregion of about two further kilobases was resequenced; this occurred in only 18% of the cases. Polymorphisms were identified and genotypes were scored automatically using Polyphred[29] and checked manually by at least two different scorers. SNPs in Hardy-Weinberg disequilibrium or showing evidence of breakdown of LD over short physical distances (< 2.5 kb) were triple-checked. Of the 275 high-frequency SNPs (that is, SNPs with at least 15 observed copies of the minor allele), three were discarded because of a Hardy-Weinberg P value of < 0.02; one of the SNPs used in the analysis had a nominally significant P value ($P < 0.05$). To assess the accuracy of scoring, we rescored 26 randomly chosen high-frequency SNPs; only seven discrepancies were found among 1,144 genotypes. For cases in which follow-up genotyping was done, the discrepancy rate was 47 out of 1,484 (3%) between genotypes obtained by both methods.

Genotyping of SNPs was performed by single-base extension followed by mass spectroscopy (Sequenom)[30], fluorescence polarization (LJL Biosystems)[31] or detection on a sequencing gel (Applied Biosystems) . For the ten regions selected for follow-up genotyping (Table 1), we chose at most one "representative" SNP at each distance from the core SNP (each column in Table 1) according to the criterion that it had the highest number of minor alleles of all SNPs at that distance from the core SNP. For other populations, only those SNPs that, when genotyped, had a minimum number of minor alleles were included in studies of LD. For Yorubans, the cutoff was 25 alleles (76% of SNPs met the criterion); for Swedes, the cutoff was 15 alleles (89%). The fact that most of the SNPs we studied in Utah are also present in high frequency in these other populations indicates that the

板块的人群(比如约鲁巴人)则能够进行精细结构的图谱绘制,便于鉴定出造成某种表型的特异性核苷酸替换 [12]。我们的研究同时也显示 LD 可以作为研究人群历史的工具 [19-22]。同时评估基因组中多个区域的 LD 可以作为研究人群历史的方法,相对于基于单个位点性质的传统方法来说在一些方面具有更高的敏感度。

方　　法

核心 SNP 是通过对一个多民族人群中超过 3,000 个基因进行筛查确定的,其中包括 15 个欧裔美国人、10 个非裔美国人和 7 个东亚人(详见参考文献 28,此数据库的详细描述将另文表述)。用于测序的 DNA 来自科里尔细胞库。CEPH 图谱标记中这些犹他州人群样本的编号是 NA12344、06995、06997、07013、12335、06990、10848、07038、06987、10846、10847、07029、07019、07048、06991、10851、07349、07348、10857、10852、10858、10859、10854、10856、10855、12386、12456、10860、10861、10863、10830、10831、10835、10834、10837、10836、10838、10839、10841、10840、10842、10843、10845 和 10844。我们对另外 48 个瑞典人(一项成人糖尿病的病例 / 对照研究中的健康个体)和 96 个尼日利亚约鲁巴男性(一项高血压的病例 / 对照研究中的健康个体)进行了基因分型。

SNP 是从 44 个犹他州人群个体中用 DNA 测序 [28] 方法获得的。我们测定了每个距离核心 SNP ≥ 5 kb 处周边 2 kb 的序列,同时也对距离核心 SNP 周边 1 kb 区域进行了测序。如果没有足够高频率的基因多态位点,则将周边测序范围再扩大 2 kb;这种情况出现的比例是 18%。多态位点发现以后,就用 Polyphred 软件 [29] 自动进行基因型的评分,并且至少两个不同的计分人进行手工复查。哈迪–温伯格不平衡中的 SNP 或者在短物理距离(< 2.5 kb)显示 LD 断开证据的 SNP 需要进行三重检查。在这 275 个高频 SNP 中(即 SNP 含有至少 15 个次等位基因位点的可见拷贝),有 3 个由于哈迪–温伯格 P 值 < 0.02 而被舍弃;用于分析的其中一个 SNP 只是在名义上有显著性 P 值(P < 0.05)。为了评分的准确性,我们重新将 26 个随机选取的高频 SNP 进行了评分;在 1,144 个基因型中只发现了 7 处差异。对于进行基因分型的样本,两种方法获得的基因型差异率是 47/1,484(3%)。

SNP 的基因分型使用的是单碱基延伸法,随后使用质谱(Sequenom)[30]、荧光偏振(LJL Biosystems)[31] 或者测序胶(Applied Biosystems)[32] 检测。对于选定进行后续分型的 10 个区域而言(表 1),我们以距离核心 SNP 一定距离的所有 SNP 中出现最多的次等位位点作为标准,选择一个最具代表性的 SNP(表 1 中的每一栏)。对于其他人群,只有那些分型后具有最低数量要求的次等位基因位点才纳入 LD 的研究。约鲁巴人的阈值是 25 个等位基因位点(76% 的 SNP 满足标准);瑞典人的阈值是 15 个等位位点(89% 的 SNP 满足标准)。我们在犹他州人群中研究的 SNP 大多数也高频存在于其他人群的事实表明,LD 的评估不太可能受

assessment of LD is not likely to be subject to large ascertainment bias.

Heterozygosity[15] (π) was calculated as the average of $2jk/n(n-1)$ for all base pairs screened, with j and k equal to the number of copies of the minor and major alleles, respectively ($n=j+k$). A base was considered screened if it had Phred quality scores[29] of ≥ 15 in ≥ 10 individuals. D' values between markers with alleles A/a and B/b (allele frequencies, c_A, c_a, c_B and c_b; haplotype frequencies, c_{AB}, c_{Ab}, c_{aB} and c_{ab}) were obtained by dividing $c_{AB}-c_Ac_B$ by its maximum possible value: $\min(c_Ac_b$, $c_ac_B)$ if $D>0$ and $\min(c_Ac_B, c_ac_b)$ otherwise. An implementation of the expectation maximization algorithm was used to infer haplotype frequencies for pairs of SNPs both for actual and simulated data[33]. A likelihood ratio test was used to assess significance of associations between pairs of SNPs[10].

Computer simulations were based on a model related to that in ref. 2, assuming a population that was constant at an effective size of 10,000 individuals until 5,000 generations ago, when it expanded instantaneously to a size of 100,000,000 (an arbitrary value). This model captures many of the features of more complicated growth, as the effect of population growth on LD is not dependent on the precise details of population growth or the final population size when the growth is moderately fast[2]. Bottlenecks were modelled as described, with the population crashing to a constant size for a fixed number of generations before re-expanding. (The effect of a bottleneck on LD depends primarily on the F-value, the inbreeding coefficient, which is defined as the probability that two alleles randomly picked from the population after the bottleneck derive from the same ancestral allele just before the bottleneck.) Coalescent simulations were used to generate gene genealogies under these models for markers separated by a specified recombination distance (see ref. 13 for a more detailed description of the theory behind these simulations). Simulations were run 2,000 times with sample-size distributions mimicking our data. SNPs were generated by distributing mutations on the simulated gene genealogies at a mutation rate of 6×10^{-5} per generation, under an "infinite alleles"mutation model. The mutation rate was chosen such that the probability of high frequency SNPs in a 2-kb stretch of DNA sequenced in 44 samples (for the model of a simple expansion 5,000 generations ago) was similar to what we observed (about 70%). We also tested mutation rates ten times higher and found that inferences about LD were essentially unchanged.

An extreme hypothesis of population mixture and its effect on LD were assessed in a simulated, mixed population of European Americans and sub-Saharan Africans. For the first simulated mixture, we constructed a mixture of 22 Yorubans and 22 samples from Utah, and used data from the ten core SNPs genotyped in both populations. For the second simulation, we used 26 SNPs from a previous study[7] that had been found to have a minor allele frequency of at least 15% in either African Americans or European Americans; we chose at most one SNP per gene, picking the first listed SNP (in Table A1 of ref. 7) that met our minimum frequency criterion. We found much stronger LD even at 40 kb (56% with $|D'| > 0.5$) (Fig. 1) in our actual data than in the simulated, admixed populations. For the 45 possible pairwise comparisons of the ten core SNPs in the simulated mix of Yoruban and Utah samples, no values of $|D'| > 0.5$ were observed. For the 325 possible pairwise comparisons from the second study, only 11% showed $|D'| > 0.5$. This suggests that admixture probably did not generate the strong signal of LD at long physical distances seen in Utah.

(**411**, 199-204; 2001)

到大量的确认偏倚的影响。

对于所有筛选的碱基对而言，杂合度[15](π)等于 $2jk/n(n-1)$ 的均值，其中 j 和 k 分别相当于次等位位点和主等位位点的拷贝数，$n=j+k$。如果一个碱基在 $\geqslant 10$ 个个体中的 Phred 质量评分[29] $\geqslant 15$，则被认为通过筛选。等位基因 A/a 和 B/b（等位基因频率 c_A、c_a、c_b 和 c_b；单体型频率 c_{AB}、c_{Ab}、c_{aB} 和 c_{ab}）间标志的 D′ 值的计算是将 $c_{AB} - c_A c_B$ 除以其最大可能值：如果 $D > 0$，取 $c_A c_b$ 和 $c_a c_B$ 的较小值，反之则是 $c_A c_B$ 和 $c_a c_b$ 的较小值。最大期望值算法用来推断实际和模拟数据中成对 SNP 的单体型频率[33]。似然比检验用来评价成对 SNP 之间相关性的显著性[10]。

计算机模拟是基于参考文献 2 中的一个相关模型，模型假设一个种群一直保持稳定在有效群体数量为 10,000 个个体直到 5,000 代，突然扩张到 100,000,000（一个随意假设的值）个个体。在增长速度比较快的情况下，人群增长对 LD 的影响不取决于其增长的准确细节或最终大小[2]，因此该模型具有更复杂增长模式的许多特点。瓶颈效应的模型如前述进行设置，在重新扩张前人群的数量在固定的代数中坍缩至常数值。（瓶颈效应对 LD 的影响主要取决于 F 值，即近交系数，其定义是瓶颈效应后从群体中随机选取的两个等位基因位点来源于瓶颈效应之前同一个祖先位点的可能性。）联合模拟用来生成这些模型设定下的基因谱系以获得特定的重组距离分隔的标志物（有关模拟的详细理论陈述详见参考文献 13）。在样本量的分布类似于我们数据的条件下进行了 2,000 次模拟。SNP 是在突变率为每代 6×10^{-5}，"无限等位位点"突变模型下，通过将突变分布在模拟的基因谱系中产生。突变率的设定是为了保证 44 个样本中所测序的 2 kb 序列中出现高频 SNP 的可能性（对于 5,000 代之前简单扩张的模型）与我们所观察到的类似（大约 70%）。我们也检验了突变率增高至 10 倍以后的情况，发现关于 LD 相关结论没有改变。

对于人群混合的极端假设及其对 LD 的影响是在模拟的、混合了欧裔美国人和南撒哈拉非洲人中进行评估的。对于第一组模拟的混合，我们构建了 22 个约鲁巴人和 22 个犹他州人的混合样本，并且使用两个人群中都进行了分型的 10 个核心 SNP 的数据。对于第二个模拟，我们使用先前研究[7]的 26 个 SNP，已发现无论在非裔美国人还是欧裔美国人中，出现次等位基因位点的频率都不小于 15%；每个基因根据第一个满足我们最小频率标准的 SNP 最多选择一个位点（见参考文献 7 表 A1）。我们发现即便是 40 kb 处，实际数据比模拟的混合人群中的 LD 程度更强（56% 的 |D′| > 0.5）（图 1）。在模拟混合的约鲁巴人和犹他州人样本中 10 个核心 SNP 的 45 种设对比较中，没有看到 |D′| > 0.5 的值。在第二个研究的 325 个可设对比较中，只有 11% 出现 |D′| > 0.5。这说明人群混合可能无法形成像犹他州人群中所见到的那种在长物理距离上产生 LD 的强信号。

（毛晨晖 翻译；曾长青 审稿）

David E. Reich[*], **Michele Cargill**[*†], **Stacey Bolk**[*], **James Ireland**[*], **Pardis C. Sabeti**[‡], **Daniel J. Richter**[*], **Thomas Lavery**[*], **Rose Kouyoumjian**[*], **Shelli F. Farhadian**[*], **Ryk Ward**[‡] & **Eric S. Lander**[*§]

[*] Whitehead Institute/MIT Center for Genome Research, Nine Cambridge Center, Cambridge, Massachusetts 02142, USA

[‡] Institute of Biological Anthropology, University of Oxford, Oxford OX2 6QS, UK

[§] Department of Biology, MIT, Cambridge, Massachusetts 02139, USA

[†] Present address: Celera Genomics, 45 West Gude Drive, Rockville, Maryland 20850, USA.

Received 11 December 2000; accepted 13 March 2001.

References:

1. Sachidanandam, R. *et al.* A map of human genome sequence variation containing 1.42 million single nucleotide polymorphisms. *Nature* **409**, 928-933 (2001).

2. Kruglyak, L. Prospects for whole-genome linkage disequilibrium mapping of common disease genes. *Nature Genet.* **22**, 139-144 (1999).

3. Dunning, A. M. *et al.* The extent of linkage disequilibrium in four populations with distinct demographic histories. *Am. J. Hum. Genet.* **67**, 1544-1554 (2000).

4. Abecasis, G. R. *et al.* Extent and distribution of linkage disequilibrium in three genomic regions. *Am. J. Hum. Genet.* **68**, 191-197 (2001).

5. Taillon-Miller, P. *et al.* Juxtaposed regions of extensive and minimal linkage disequilibrium in human Xq25 and Xq28. *Nature Genet.* **25**, 324-328 (2000).

6. Collins, A., Lonjou, C. & Morton, N. E. Genetic epidemiology of single-nucleotide polymorphisms. *Proc. Natl Acad. Sci. USA* **96**, 15173-15177 (1999).

7. Goddard, K. A. B., Hopkins, P. J., Hall, J. M. & Witte, J. S. Linkage disequilibrium and allele-frequency distributions for 114 single-nucleotide polymorphisms in five populations. *Am. J. Hum. Genet.* **66**, 216-234 (2000).

8. Watterson, G. A. & Guess, H. A. Is the most frequent allele the oldest? *Theor. Pop. Biol.* **11**, 141-160 (1977).

9. Lander, E. S. The new genomics: global views of biology. *Science* **274**, 536-539 (1996).

10. Schneider, S., Kueffler, J. M., Roessli, D. & Excoffier, L. Arlequin (ver. 2.0): A software for population genetic data analysis (Genetics and Biometry Laboratory, Univ. Geneva, Switzerland, 2000).

11. Lewontin, R. C. On measures of gametic disequilibrium. *Genetics* **120**, 849-852 (1988).

12. Jorde, L. B. Linkage disequilibrium and the search for complex disease genes. *Genome Res.* **10**, 1435-1444 (2000).

13. Hudson, R. R. in *Oxford Surveys in Evolutionary Biology* (eds Futuyma, D. J. & Antonovics, J.) 1-44 (Oxford Univ. Press, Oxford, 1990).

14. Clark, A. G. *et al.* Haplotype structure and population genetic inferences from nucleotide-sequence variation in human lipoprotein lipase. *Am. J. Hum. Genet.* **63**, 595-612 (1998).

15. Hartl, D. L. & Clark, A. G. *Principles of Population Genetics* (Sinauer, Massachusetts, 1997).

16. Chakraborty, R. & Weiss, K. M. Admixture as a tool for finding linked genes and detecting that difference from allelic association between loci. *Proc. Natl Acad. Sci. USA* **85**, 9119-9123 (1988).

17. Eaves, I. A. *et al.* The genetically isolated populations of Finland and Sardinia may not be a panacea for linkage disequilibrium mapping of common disease genes. *Nature Genet.* **25**, 320-322 (2000).

18. Goldstein, D. B., Ruiz Linares, A., Cavalli-Sforza, L. L. & Feldman, M. W. Genetic absolute dating based on microsatellites and the origin of modern humans. *Proc. Natl Acad. Sci. USA* **92**, 6723-6727 (1995).

19. Tishkoff, S. A. *et al.* Global patterns of linkage disequilibrium at the CD4 locus and modern human origins. *Science* **271**, 1380-1387 (1996).

20. Tishkoff, S. A. *et al.* Short tandem-repeat polymorphism/*Alu* haplotype variation at the *PLAT* locus: Implications for modern human origins. *Am. J. Hum. Genet.* **67**, 901-925 (2000).

21. Kidd, J. R. *et al.* Haplotypes and linkage disequilibrium at the phenylalanine hydroxylase locus, *PAH*, in a global representation of populations. *Am. J. Hum. Genet.* **66**, 1882-1899 (2000).

22. Mateu, E. *et al.* Worldwide genetic analysis of the *CFTR* region. *Am. J. Hum. Genet.* **68**, 103-117 (2001).

23. Housley, R. A., Gamble, C. S., Street, M. & Pettitt, P. Radiocarbon evidence for the Late glacial human recolonisation of northern Europe. *Proc. Prehist. Soc.* **63**, 25-54 (1994).

24. Richards, M. *et al.* Tracing European founder lineages in the Near Eastern mtDNA pool. *Am. J. Hum. Genet.* **67**, 1251-1276 (2000).

25. Reich, D. E. & Goldstein, D. B. Genetic evidence for a Paleolithic human population expansion in Africa. *Proc. Natl Acad. Sci. USA* **95**, 8119-8123 (1998).

26. Ingman, M., Kaessmann, H., Pääbo, S. & Gyllensten, U. Mitochondrial genome variation and the origin of modern humans. *Nature* **408**, 708-713 (2000).

27. Altshuler, D., Daly, M. & Kruglyak, L. Guilt by association. *Nature Genet.* **26**, 135-137 (2000).

28. Cargill, M. *et al.* Characterization of single-nucleotide polymorphisms in coding regions of human genes. *Nature Genet.* **22**, 231-238 (1999).

29. Nickerson, D. B., Tobe, V. O. & Taylor, S. L. PolyPhred: automating the detection and genotyping of single nucleotide substitutions using fluorescence-based sequencing. *Nucleic Acids Res.* **25**, 2745-2751 (1997).

30. Ross, P. Hall, L., Smirnov, I. & Haff, L. High level multiplex genotyping by MALDI-TOF mass spectroscopy. *Nature Biotech.* **16**, 1347-1351 (1998).

31. Chen, X., Levine, L. & Kwok, P. Y. Fluorescence polarization in homogenous nucleic acid analysis. *Genome Res.* **9**, 492-498 (1999).

32. Lindblad-Toh, K. *et al.* Large-scale discovery and genotyping of single-nucleotide polymorphisms in the mouse. *Nature Genet.* **24**, 381-386 (2000).

33. Excoffier, L. & Slatkin, M. Maximum-likelihood estimation of molecular haplotype frequencies in a diploid population. *Mol. Biol. Evol.* **12**, 921-927 (1995).

34. Broman, K. W., Murray, J. C. Sheffield, V. C., White, R. L. & Weber, J. L. Comprehensive human genetic maps: individual and sex-specific variation in recombination. *Am. J. Hum. Genet.* **63**, 861-689 (1998).

35. Lander, E. S. *et al.* Initial sequencing and analysis of the human genome. *Nature* **409**, 860-921.

Acknowledgements. We thank L. Groop for the Swedish samples; R. Cooper and C. Rotimi for the Yoruban samples; and D. Altshuler, M. Daly, D. Goldstein, J. Hirschhorn, C. Lindgren and S. Schaffner for discussions. This work was supported in part by grants from the National Institutes of Health, Affymetrix, Millennium Pharmaceuticals, and Bristol-Myers Squibb Company, and by a National Defense Science and Engineering Fellowship to D.E.R.

Correspondence and requests for materials should be addressed to D.E.R. (e-mail: reich@genome.wi.mit.edu) or E.S.L. (e-mail: lander@genome.wi.mit.edu).

Rapid X-ray Flaring from the Direction of the Supermassive Black Hole at the Galactic Centre

F. K. Baganoff *et al.*

Editor's Note

Although it is generally accepted that there is a black hole with a mass of several million solar masses at the centre of our Milky Way galaxy, there is less light (especially X-rays) coming from the gas swirling into the black hole than standard theory predicts. This led to some uncertainty about what was powering the source of radio emission identified with the Galactic Centre, called Sagittarius A*. Here Frederick Baganoff, lead scientist for one of the instruments on board the Chandra space-based X-ray observatory, and his colleagues report rapid X-ray flares from the Galactic Centre, which imply that the X-rays must be generated within a tiny volume of space. This is consistent with its origin in gas that is falling into the supermassive black hole.

The nuclei of most galaxies are now believed to harbour supermassive black holes[1]. The motions of stars in the central few light years of our Milky Way Galaxy indicate the presence of a dark object with a mass of about 2.6×10^6 solar masses (refs 2, 3). This object is spatially coincident with the compact radio source Sagittarius A* (Sgr A*) at the dynamical centre of the Galaxy, and the radio emission is thought to be powered by the gravitational potential energy released by matter as it accretes onto a supermassive black hole[4,5]. Sgr A* is, however, much fainter than expected at all wavelengths, especially in X-rays, which has cast some doubt on this model. The first strong evidence for X-ray emission was found only recently[6]. Here we report the discovery of rapid X-ray flaring from the direction of Sgr A*, which, together with the previously reported steady X-ray emission, provides compelling evidence that the emission is coming from the accretion of gas onto a supermassive black hole at the Galactic Centre.

OUR view of Sgr A* in the optical and ultraviolet wavebands is blocked by the large visual extinction, $A_V \approx 30$ magnitudes[7], caused by dust and gas along the line of sight. Sgr A* has not been detected in the infrared owing to its faintness and to the bright infrared background from stars and clouds of dust[8]. We thus need to detect X-rays from Sgr A* in order to constrain the spectrum at energies above the radio-to-submillimetre band and to test whether gas is accreting onto a supermassive black hole (see above).

We first observed the Galactic Centre on 21 September 1999 with the imaging array of the Advanced CCD (charge-coupled device) Imaging Spectrometer (ACIS-I) aboard the

银河系中心超大质量黑洞方向上的快速 X 射线耀发

巴加诺夫等

编者按

尽管人们普遍都接受了在我们的银河系中心存在着一个几百万太阳质量的黑洞这一观点，但是气体旋转着被黑洞吸积所释放的光（尤其是 X 射线）比标准理论所预测的要少。这使得被证认来自于银河系中心人马座 A* 的射电辐射的来源究竟是什么有了一些不确定性。在这篇文章中，弗雷德里克·巴加诺夫（钱德拉空基 X 射线大文台上装载的一架仪器的首席科学家）和他的同事们报道了来自银河系中心的快速 X 射线耀发，这也意味着这些 X 射线肯定是来自于一个很小的空间体积内。这个结论与它是来自于正在被超大质量黑洞吸积的气体的理论相符合。

当前，人们普遍相信大部分星系的中心存在超大质量黑洞[1]。而我们银河系中非常靠近中心（距离约几光年以内）的恒星运动表明银河系中心存在一个质量大约为 $2.6×10^6$ 太阳质量的暗天体（参考文献 2 和 3）。该天体在空间分布上正好与位于银河系动力学中心的致密射电源人马座 A*(Sgr A*) 一致，其射电辐射则被认为来源于一个超大质量黑洞吸积物质所释放的引力势能[4,5]。然而，Sgr A* 在各个波段，特别是在 X 射线波段，要暗于预期，这使得人们对上述射电辐射来源产生了一些疑问。直到最近，人们才首次发现了该天体存在 X 射线辐射的强有力证据[6]。在本工作中，我们报告在 Sgr A* 方向发现了快速 X 射线耀发。连同之前发现的稳定 X 射线辐射，这一发现为 Sgr A* 致密射电源的射电辐射来源于银河系中心超大质量黑洞的气体吸积过程提供了极具说服力的证据。

我们在光学和紫外波段对 Sgr A* 的观测受限于视线方向上尘埃和气体产生的显著消光，A_V 约为 30 星等[7]。在红外波段，Sgr A* 没有被探测到，其原因一方面在于它本身暗弱，另一方面在于恒星以及尘埃云的红外背景明亮[8]。因此，为了限制 Sgr A* 的能量在射电到亚毫米波段以上的光谱，并验证气体是否吸积到超大质量黑洞上（见上），我们需要对 Sgr A* 的 X 射线进行探测。

1999 年 9 月 21 日，我们利用安装在钱德拉 X 射线天文台的高级 CCD（电荷耦合器件）成像光谱仪的成像阵列（ACIS-I）[9] 对银河系中心进行了首次观测，发现了

Chandra X-ray Observatory[9] and discovered an X-ray source coincident within $0.35'' \pm 0.26''$ (1σ) of the radio source[6]. The luminosity in 1999 was very weak—$L_X \approx 2 \times 10^{33}$ erg s^{-1} in the 2–10 keV band—after correction for the inferred neutral hydrogen absorption column $\mathcal{N}_H \approx 1 \times 10^{23}$ cm^{-2}. This is far fainter than previous X-ray observatories could detect[6].

We observed the Galactic Centre a second time with Chandra ACIS-I from 26 October 2000 22:29 until 27 October 2000 08:19 (UT), during which time we saw a source at the position of Sgr A* brighten dramatically for a period of approximately 10,000 s (10 ks). Figure 1 shows surface plots for both epochs of the 2–8 keV counts integrated over time from a $20'' \times 20''$ region centred on the radio position of Sgr A*. The modest peak of the integrated counts at Sgr A* in 1999 increased by a factor of about 7 in 2000, despite the 12% shorter exposure. The peak integrated counts of the fainter features in the field show no evidence of strong variability, demonstrating that the flaring at Sgr A* is intrinsic to the source.

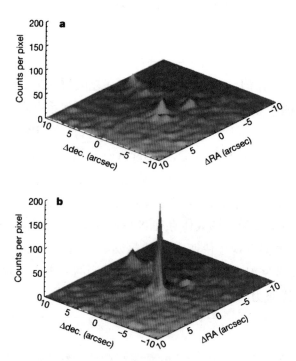

Fig. 1. Surface plots of the 2–8 keV counts within a $20'' \times 20''$ field centred on Sgr A* at two epochs. The data were taken with Chandra ACIS-I on (**a**) 21 September 1999 and (**b**) 26–27 October 2000. The effective exposure times were 40.3 ks and 35.4 ks, respectively. The spatial resolution is 0.5″ per pixel. An angle of 1″ on the sky subtends a projected distance of about 0.04 pc at the galactocentric distance of 8.0 kpc (ref. 30). The peak integrated counts per pixel at the position of Sgr A* increased by a factor of 7 from the first epoch to the second, despite the slightly smaller exposure time (~12%) in the second epoch. The low-level peak a few arcseconds to the southwest of Sgr A* is the infrared source IRS 13, and the ridge of emission to the northwest is from a string of unresolved point sources. The fainter features in the field are reasonably consistent between the two epochs, considering the limited Poisson statistics and the fact that these stellar sources may themselves be variable; this consistency shows that the strong variations at Sgr A* are intrinsic to the source.

与射电源位置在 $0.35'' \pm 0.26''$ (1σ) 以内相吻合的一个 X 射线源[6]。根据推断的柱密度为 $N_H \approx 1 \times 10^{23}$ cm^{-2} 的中性氢吸收进行修正之后，该 X 射线源的光度是非常低的 ($2 \sim 10$ keV 波段内，光度 $L_X \approx 2 \times 10^{33}$ erg $\cdot$ s^{-1})，远比之前各 X 射线天文台的探测极限暗弱[6]。

2000 年 10 月 26 日 22:29 到 2000 年 10 月 27 日 08:19 (UT) 这段时间里，我们利用钱德拉 ACIS-I 对银河系中心进行了第二次观测。这段时间中，我们看到在 Sgr A* 的位置上有一个在大约 10,000 s (10 ks) 的时段内剧烈变亮的源。图 1 显示了以 Sgr A* 射电源为中心的一个 $20'' \times 20''$ 区域内两次观测对时间积分的 $2 \sim 8$ keV 光子计数曲面分布图。尽管曝光时间少了 12%，但是 1999 年的计数曲面分布图上 Sgr A* 位置处的不太大的峰值在 2000 年增加到了大约 7 倍。视场中其他较暗的峰没有明显的光变，表明 Sgr A* 的这个耀发来自其自身的内禀光变。

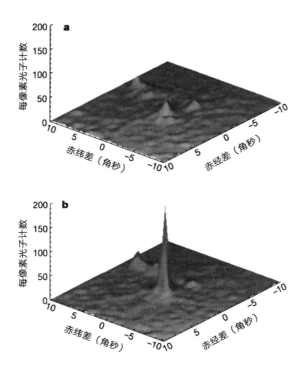

图 1. 两个不同时期以 Sgr A* 为中心的 $20'' \times 20''$ 的天区中 $2 \sim 8$ keV 波段光子计数的曲面分布图。图中所用数据来源于钱德拉 ACIS-I 于 (a) 1999 年 9 月 21 日和 (b) 2000 年 10 月 26 ~ 27 日的观测数据。有效曝光时间分别为 40.3 ks 和 35.4 ks。空间分辨率为每像素 $0.5''$。在距离银心 8.0 kpc 远的位置处 (参考文献 30)，天空中 $1''$ 的角度约对应 0.04 pc 的投影距离。Sgr A* 位置处每像素的峰值积分光子计数从第一个时期到第二个时期增加到了 7 倍，尽管第二个时期的曝光时间略短 (约少 12%)。在 Sgr A* 西南方向外几个角秒处的较低水平的峰值对应的是红外源 IRS 13，而西北方向出现的脊状辐射来源于一串未分辨的点源。考虑到受限的泊松统计，以及这些恒星级天体本身的光度也可能发生变化，两个时期中视场内较暗的特征大致是一致的；这种一致性反映出 Sgr A* 位置处的强烈光变来自于源的内禀原因。

Figure 2 shows light curves of the photon arrival times from the direction of Sgr A* during the observation in 2000. Figure 2a and b shows hard-band (4.5–8 keV) and soft-band (2–4.5 keV) light curves constructed from counts within an angular radius of 1.5″ of Sgr A*. Both bands exhibit roughly constant, low-level emission for the first 14 ks or so, followed by a 6-ks period of enhanced emission beginning with a 500-s event (4.4σ significance using 150-s bins). At 20 ks, flare(s) of large relative amplitude occur, lasting about 10 ks, and finally the emission drops back to the low state for the remaining 6 ks or so. About 26 ks into the observation, the hard-band light curve drops abruptly by a factor of 5 within a span of around 600 s and then partially recovers within a period of 1.2 ks. The soft-band light curve shows a similar feature, but it appears to lag the hard-band event by a few hundred seconds and is less sharply defined.

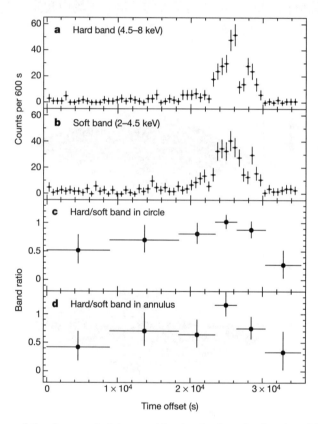

Fig. 2. Light curves of the photon arrival times and band ratios from the direction of Sgr A* on 26–27 October 2000. **a**, Hard-band (4.5–8 keV) counts. **b**, Soft-band (2–4.5 keV) counts. **c**, Hard/soft band ratio within a circle of radius 1.5″. **d**, Hard/soft band ratio within an annulus of inner and outer radii 0.5″ and 2.5″, respectively. The x axis shows the time offset from the start of the observation at 26 October 2000 22:29 (UT). The data are shown with 1σ error bars. The single 2.8-hour period of flaring activity which we have detected so far during a total of 21 hours of observations yields a (poorly determined) duty cycle of approximately 1/8. Our continuing observations of this source with Chandra will permit us to refine this value. During the quiescent intervals at the beginning and end of this observation, the mean count rate in the 2–8-keV band was $(6.4 \pm 0.6) \times 10^{-3}$ counts s^{-1}, consistent with the count rate we measured in

图 2 显示了 2000 年观测期间 Sgr A* 方向上关于光子到达时间的光变曲线。图 2a 和 2b 分别显示了根据 Sgr A* 的 1.5″ 角半径范围内的光子计数构建得到的硬 X 射线波段 (4.5～8 keV) 和软 X 射线波段 (2～4.5 keV) 的光变曲线。两个波段在最初的大概 14 ks 内都展现出大致稳定的低水平辐射，随后以一个 500 s 的事件作为开始呈现出一个时段为 6 ks 的辐射增强 (以 150 s 分区间，其显著性为 4.4σ)。在 20 ks 的时候，发生相对大幅度的耀发，持续大约 10 ks。在剩余的约 6 ks 里，整体辐射回落到之前较低的水平。在观测到大约 26 ks 的时候，硬 X 射线波段的光变曲线在约 600 s 的时间间隔内突然下降为原来的五分之一，然后在 1.2 ks 的时段内得到了部分的恢复。软 X 射线波段显示了类似的特征，但是似乎要比硬 X 射线波段的事件滞后几百秒，且特征没有那么分明。

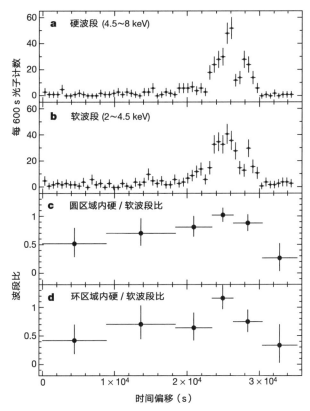

图 2. 2000 年 10 月 26～27 日 Sgr A* 方向上的光子到达时间的光变曲线以及波段比。**a**，硬 X 射线波段 (4.5～8 keV) 光子计数。**b**，软 X 射线波段 (2～4.5 keV) 光子计数。**c**，半径为 1.5″ 的圆区域内硬/软 X 射线波段比。**d**，环区域 (内环半径 0.5″，外环半径 2.5″) 内硬/软 X 射线波段比。x 轴表示从 2000 年 10 月 26 日 22:29(UT) 起算的时间偏移。数据显示了 1σ 的误差棒。目前我们在整个 21 小时的观测中仅探测到一次 2.8 小时的耀发活动，占空比约为 1/8(结果仍不确定)。我们利用钱德拉对这个源进行后续观测将能够修改进这一数值。在本次观测开始和结束的平静状态下，2～8 keV 波段的平均光子计数率为 (6.4±0.6) × 10⁻³ 个/秒，这个结果与我们在 1999 年测量的计数率 (5.4±0.4) × 10⁻³ 个/秒相一致。在提

1999, which was $(5.4 \pm 0.4) \times 10^{-3}$ counts s^{-1}. The detected count rate at the peak of the flare was 0.16 ± 0.01 counts s^{-1} within an extraction circle of radius 1.5″, but this is about 30% less than the true incident count rate owing to pile-up of X-rays in the detector during the 3.2-s integration time for each CCD (charge-coupled device) read out.

The band-ratio time series in Fig. 2c, defined as the ratio of hard-band counts to soft-band counts, suggests that the spectrum "hardened" (that is, became flatter and extended more strongly to higher energies) during the flare. The difference between the band ratio measured at the peak of the flare and the average of the band ratios during the quiescent periods at the beginning and end of the observation is 0.63 ± 0.21 (that is, 3σ). The peak-flare band ratio in Fig. 2c is affected to some extent by the effects of pile-up (see Fig. 2 legend), which would tend to harden the spectrum; however, the band ratios in Fig. 2d, which were computed using the non-piled-up data extracted from the wings of the point spread function, also show evidence for spectral hardening with 2.7σ significance. The sizes of dust-scattering haloes in the Galactic Centre are typically greater than 1′ (ref. 10), so dust-scattered X-rays from the source contribute a negligible fraction of the emission within the source extraction region that we used; hence dust-scattered X-rays cannot account for the spectral variations. We therefore conclude that the spectral hardening during the flare is likely to be real.

The quiescent-state spectra in 1999 and 2000 and the peak flaring-state spectrum in 2000 are shown in Fig. 3. We fitted each spectrum individually using a single power-law model with corrections for the effects of photoelectric absorption and dust scattering[10]. The best-fit values and 90% confidence limits for the parameters of each fit are presented in the first three lines of Table 1. The column densities for the three spectra are consistent, within the uncertainties, as are the photon indices of both quiescent-state spectra. Next, we fit a double power-law model to the three spectra simultaneously, using a single photon index for both quiescent spectra, a second photon index for the flaring spectrum, and a single column density for all three spectra. The best-fit models for each spectrum from the simultaneous fits are shown as solid lines in Fig. 3; the parameter values are given in the last line of Table 1. Using these values, we derive an absorption-corrected 2–10 keV luminosity of $L_X = (2.2^{+0.4}_{-0.3}) \times 10^{33}$ erg s^{-1} for the quiescent-state emission and $L_X = (1.0 \pm 0.1) \times 10^{35}$ erg s^{-1} for the peak of the flaring-state emission, or around 45 times the quiescent-state luminosity. We note that previous X-ray observatories did not have the sensitivity to detect such a short-duration, low-luminosity flare in the Galactic Centre[6]. The best-fit photon index $\Gamma_f = 1.3^{+0.5}_{-0.6}$ $(N(E) \propto E^{-\Gamma})$ for the flaring-state spectrum is slightly flatter than, but consistent with, systems thought to contain supermassive black holes[11]. Here E is the energy of a photon in keV, $N(E)$ is the number of photons with energies in the differential interval E to $E+dE$, and Γ is the power-law index of the photon number distribution.

取数据的半径为 1.5″ 的圆区域里，耀发峰值处探测的计数率为 0.16±0.01 个/秒。因为在每个 CCD(电荷耦合器件)读出时有 3.2 s 的积分时间，X 射线会在探测器中有堆积效应，因此这个数值比真实的入射光子计数率低大约 30%。

　　图 2c 显示的波段比(其定义为硬 X 射线波段光子计数与软 X 射线波段光子计数之比)随时间的一连串变化表明在耀发的过程中光谱有"硬化"的现象(即光谱变得更平且更强地延伸到高能端)。耀发时峰值处与观测初期以及末期平静时期的波段比的差别为 0.63±0.21(即 3σ)。图 2c 中的耀发峰的波段比在一定程度上受堆积效应的影响(见图 2 图注)，该效应倾向于使光谱硬化；然而，图 2d 中的波段比利用从点扩散函数的两翼提取的无堆积效应的数据计算得到，它也在 2.7σ 的显著性上显示光谱硬化的证据。银河系中心尘埃散射晕的尺度通常认为大于 1′(参考文献 10)，因此，这个源被尘埃散射的 X 射线对我们用来提取数据的区域内辐射的贡献可以忽略；所以尘埃散射的 X 射线不足以解释所观测到的光谱变化。我们因而得到这样的结论：耀发过程中的光谱硬化很可能是真实的。

　　图 3 中显示了 1999 年和 2000 年平静状态下的光谱以及 2000 年峰值耀发状态下的光谱。我们利用单一幂律模型对这些光谱逐个进行了拟合，其中对光电吸收和尘埃散射的效应做了修正[10]。表 1 的前三行分别列举了每个拟合的最佳拟合值及其 90% 置信限。三条光谱的柱密度在不确定度范围内是一致的，同时两条平静状态下光谱的光子幂指数也是一致的。随后，我们同时对三条光谱进行双幂律模型拟合，过程中对两条平静状态下的光谱采用一个统一的光子幂指数，对耀发光谱采用另一个光子幂指数，而对三条光谱采用统一的柱密度。图 3 中的实线显示了同时拟合情况下每条谱线的最佳拟合模型；表 1 的最后一行列举了对应的参数值。利用这些结果，我们得到了这个源在 2～10 keV 波段做了吸收修正后的光度：平静状态下 $L_X = (2.2^{+0.4}_{-0.3}) \times 10^{33}$ erg·s^{-1}，耀发状态下的峰值 $L_X = (1.0 \pm 0.1) \times 10^{35}$ erg·s^{-1}(大约是平静状态下的 45 倍)。我们注意到过去的 X 射线天文台的灵敏度不够高，无法探测到银河系中心这样持续时间短、光度低的耀发信号[6]。对于耀发状态下的光谱，其光子幂指数的最佳拟合结果为 $\Gamma_f = 1.3^{+0.5}_{-0.6}$ ($N(E) \propto E^{-\Gamma}$)，这个结果比被认为包含超大质量黑洞的系统的理论预言[11]稍平，但仍相互一致。这里，E 是以 keV 为单位的光子能量，$N(E)$ 是能量在 E 到 $E+dE$ 微分区间内的光子数目，Γ 是光子数分布的幂律指数。

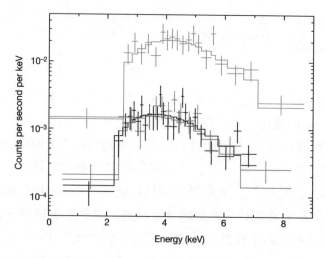

Fig. 3. X-ray spectra of the Chandra source at the position of Sgr A*. The data are shown as crosses with vertical bars indicating the 1σ errors in the count rate and horizontal bars the energy range of each bin. The events have been grouped to yield 10 counts per bin. The counts in the 1999 (black) and 2000 (red) quiescent-state spectra were extracted using a source radius of 1.5″. A non-piled-up, peak flaring-state spectrum (green) was extracted from the wings of the point spread function, using an annulus with inner and outer radii of 0.5″ and 2.5″, during the time interval 23.7–26.3 ks after the start of the 2000 observation. The solid lines are the best-fit models for each spectrum, obtained by fitting an absorbed, dust-scattered, double power-law model to the three spectra simultaneously. The best-fit values and 90% confidence intervals for the model parameters are given in the last line of Table 1. The spectra are well-fitted using a single photon index for both quiescent spectra, a second photon index for the flaring spectrum, and a single column density for all three spectra. The column density from the simultaneous fits corresponds to a visual extinction $A_V = 29.6^{+5.0}_{-6.1}$ magnitudes, which agrees well with infrared-derived estimates of $A_V \approx 30$ magnitudes[7]; we thus find no evidence in our X-ray data of excess gas and dust localized around the supermassive black hole. This places an important constraint on the maximum contribution to the infrared spectrum of Sgr A* produced by local dust reprocessing of higher energy photons from the accretion flow. The spectral models used in ref. 6 did not account for dust scattering; hence a higher column density was needed to reproduce the low-energy cut-off via photoelectric absorption alone. We note that there is no sign of an iron $K\alpha$ emission line in the flaring-state spectrum.

Table 1. Spectral fits

Spectrum	$\mathcal{N}_H$	Γ_q	Γ_f	χ^2/ν
1999 quiescent	$5.8^{+1.3}_{-1.4}$	$2.5^{+0.8}_{-0.7}$	...	19/22
2000 quiescent	$5.0^{+1.9}_{-2.3}$	$1.8^{+0.7}_{-0.9}$	...	7.6/12
2000 flaring	$4.6^{+2.0}_{-1.6}$	...	$1.0^{+0.8}_{-0.7}$	12/17
All	$5.3^{+0.9}_{-1.1}$	$2.2^{+0.5}_{-0.7}$	$1.3^{+0.5}_{-0.6}$	45/55

Best-fit parameter values and 90% confidence intervals for power-law models, corrected for photoelectric absorption and dust scattering[10]. $\mathcal{N}_H$ is the neutral hydrogen absorption column in units of 10^{22} H atoms cm^{-2}. Γ_q and Γ_f are the photon-number indices of the quiescent-state and peak flaring-state spectra ($\mathcal{N}(E) \propto E^{-\Gamma}$). χ^2 is the value of the fit statistic for the best-fit model, and ν is the number of degrees of freedom in the fit. The parameter values for the spectrum marked "All" were derived by fitting an absorbed, dust-scattered, double power-law model to the three spectra simultaneously (see text and Fig. 3).

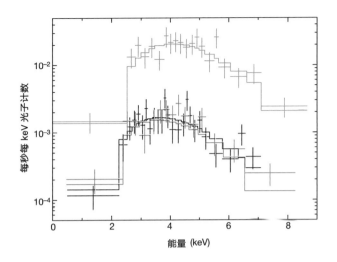

图 3. Sgr A* 位置处钱德拉源的 X 射线光谱。数据用交叉十字表示，其中纵向棒代表了光子计数率的 1σ 的误差，而横向棒代表了每个区间的能量范围。这些事件被分组以保证每个区间有 10 个计数。我们在源附近半径为 1.5″ 的范围内从 1999 年（黑色）和 2000 年（红色）的平静状态下的光谱中提取了光子计数，并于 2000 年观测开始的 23.7 ~ 26.3 ks 的时间间隔内使用一个环形范围（内环半径 0.5″、外环半径 2.5″）从点扩散函数的两翼中提取了耀发状态下无堆积效应的峰值光谱（绿色）。实线是每条光谱的模型最佳合合结果，这是基于一个考虑吸收和尘埃散射的双幂律模型对三条光谱同时进行拟合得到的。模型参数的最佳拟合结果以及 90% 置信区间列在了表 1 的最后一行。在对两条平静状态下的光谱采用统一的光子幂指数，对耀发光谱采用另一个光子幂指数并对三条光谱采用统一的柱密度的情况下，我们可以很好地拟合这些光谱。同时拟合所得到的柱密度对应目视波段消光 $A_v = 29.6^{+3.0}_{-3.0}$ 星等，这个结果与从红外数据推断的 $A_v \approx 30$ 星等[7]相吻合；因此我们在 X 射线数据中并没有发现超大质量黑洞附近存在额外气体和尘埃的证据。局域尘埃会吸收吸积流中更高能量的光子再发出红外辐射，这里得到的结果对这个过程为 Sgr A* 红外光谱所提供的最大贡献给出非常重要的限制。文献 6 中使用的光谱模型没有考虑尘埃散射；因此若只用光电吸收来再现低能端的截断，就需要更高的柱密度。我们注意到，在耀发状态下的光谱中没有铁的 Kα 发射线的迹象。

表 1. 谱线拟合

光谱	N_H	Γ_q	Γ_f	x^2/ν
1999 年的平静状态	$5.8^{+1.3}_{-1.4}$	$2.5^{+0.8}_{-0.7}$	...	19/22
2000 年的平静状态	$5.0^{+1.9}_{-2.3}$	$1.8^{+0.7}_{-0.9}$	...	7.6/12
2000 年的闪耀状态	$4.6^{+2.0}_{-1.6}$	...	$1.0^{+0.8}_{-0.7}$	12/17
总体	$5.3^{+0.9}_{-1.1}$	$2.2^{+0.5}_{-0.7}$	$1.3^{+0.5}_{-0.6}$	45/55

利用考虑了光电吸收和尘埃散射修正[10]的幂律模型进行拟合的最佳参数值以及 90% 置信区间。N_H 为中性氢吸收柱密度，单位是 10^{22} 个氢原子/cm^2。Γ_q 和 Γ_f 分别为平静状态和峰值耀发状态下光谱的光子数指数（$N(E) \propto E^{-\Gamma}$）。x^2 是最佳拟合模型的拟合统计值，而 ν 是拟合中自由度的个数。标记为"总体"的光谱的参数值是基于一个考虑吸收和尘埃散射的双幂律模型对三条光谱同时进行拟合得到的（详情见正文和图 3）。

If we view the outburst as a single event, the rise/fall timescales of a few hundred seconds and the 10-ks duration are consistent with the light-crossing and dynamical timescales for the inner part (at less than 10 Schwarzschild radii; $R_S \equiv 2GM/c^2$) of the accretion flow around a black hole of 2.6×10^6 solar masses ; here R_S is the radius of the black-hole event horizon (that is, the boundary at which the escape velocity equals the speed of light, c), G is the gravitational constant and M is the mass of the black hole. Although we cannot strictly rule out an unrelated contaminating source as the origin of the flare (for example, an X-ray binary, for which little is known about such short-timescale, low-luminosity events as we have detected; W. Lewin, personal communication), this explanation seems unlikely, as the characteristic angular scales of the young and old stellar clusters around Sgr A* are 5–20″ (ref. 7), whereas the flaring source lies within 1/3″ of the radio position. These clusters contain up to a million solar masses of stars and stellar remnants[2]; hence it is rather improbable that there would be only one very unusual stellar X-ray source in the image and that it would be fortuitously superposed on Sgr A*. Furthermore, it is not clear that X-ray binaries can be easily formed or long endure near Sgr A*, given the high velocity dispersion and high spatial density of the stars in its deep gravitational potential well[6,12].

Strong, variable X-ray emission is a characteristic property of active galactic nuclei; factors of 2–3 variations on timescales ranging from minutes to years are typical for radio-quiet active galactic nuclei[11]. Moderate- to high-luminosity active galactic nuclei (that is, Seyfert galaxies and quasars) show a general trend of increasing variability with decreasing luminosity[13]. However, this trend does not extend to low-luminosity active galactic nuclei (LLAGN), which show little or no significant variability on timescales less than a day[14]. Assuming the X-ray flare is from Sgr A*, it is remarkable that this source— generally thought to be the nearest and least luminous example of accretion onto a central supermassive black hole—has shown a factor of 45 variation that is an order of magnitude more rapid than the fastest observed variation of similar relative amplitude by a radio-quiet active galactic nucleus of any luminosity class[15]. We note that flares of similar luminosity would be undetectable by Chandra in the nucleus of even the nearest spiral galaxy, M31. LLAGN emit $L_X \gtrsim 10^{38}$ erg s^{-1} (ref. 14), so it should be kept in mind that the astrophysics of accretion onto even the LLAGN may differ substantially from that of Sgr A*. This makes Sgr A* a useful source for testing the theory of accretion onto supermassive black holes in galactic nuclei.

The faintness of Sgr A* at all wavelengths requires that the supermassive black hole be in an extremely quiet phase, either because the accretion rate is very low, or because the accretion flow is radiatively inefficient, or both[5]. A variety of theoretical scenarios, usually based on advective accretion models[16-19], jet–disk models[20], or Bondi–Hoyle models[21,22], have developed this idea. An important prediction of the advective accretion models is that the X-ray spectrum in the Chandra energy band should be dominated by thermal bremsstrahlung emission from hot gas in the outer regions of the accretion flow ($R \gtrsim 10^3 R_S$), but a region this large could not produce the rapid, large-relative-amplitude variations we have seen. Thus, the properties of the X-ray flare are inconsistent with the advective accretion flow models. The low luminosity and short timescales of this event are also inconsistent with

794

如果我们将爆发看作是一次单一事件，那么一个几百秒的增强/减弱的时标以及 10 ks 的持续时间与一个 2.6×10^6 太阳质量的黑洞周围的吸积流内部区域（小于 10 倍施瓦西半径；$R_S \equiv 2GM/c^2$）的光线穿越以及动力学时标相符合；这里 R_S 是黑洞事件视界的半径（即逃逸速度等于光速 c 的边界），G 是引力常数，而 M 是黑洞质量。尽管我们不能严格地排除此次耀发起源于一个无关的污染源（比如一个 X 射线双星，对于探测到它的这类短时标、低光度的事件，我们知之甚少；与卢因的个人交流），但是这个解释的可能性很低，因为 Sgr A* 周围年轻和年老星团的特征角尺度大约为 $5''\sim20''$（参考文献 7），而耀发源位于射电源位置附近 $1/3''$ 以内。这些星团包含多达百万太阳质量的恒星以及恒星残迹[2]，因此，在图像中仅出现一个非常特殊的恒星级 X 射线源，且其很偶然地与 Sgr A* 的位置重合，这种概率是相当低的。此外，在 Sgr A* 深引力势阱中[6,12]，恒星拥有高速度弥散以及高空间密度，在这种情况下，X 射线双星能否在 Sgr A* 附近顺利形成或者长时间存活，目前并不清楚。

强且变化的 X 射线辐射是活动星系核的一个特征属性；而射电宁静活动星系核一般会在从分钟到年的时标上产生 $2\sim3$ 倍的光变[11]。对于中到高光度活动星系核（即赛弗特星系和类星体）而言，它们大体的趋势是光度越低，光变越大[13]。然而，这种趋势并不能用于描述低光度活动星系核（LLAGN），因为它们在短于一天的时标下几乎没有显著的光变[14]。假设 X 射线耀发来源于 Sgr A*，那么值得注意的是，这个源（一般被认为是最近的而且是最暗的中心超大质量黑洞吸积的例子）展现了一个 45 倍的快速光变，这比拥有类似相对幅度的任何光度级的射电宁静活动星系核观测到的最快光变[15]还要快一个量级。我们注意到，即使是最近的漩涡星系 M31 的核存在类似光度的耀发，钱德拉望远镜也无法探测到。LLAGN 的 X 射线光度为 $L_X \gtrsim 10^{38}$ $erg \cdot s^{-1}$（参考文献 14），即使是吸积到 LLAGN 上的天体物理过程也可能与吸积到 Sgr A* 上的大不相同，这一点我们也需要牢记在心。因此，Sgr A* 成为检验星系核超大质量黑洞吸积理论的有用的源。

在各个波段 Sgr A* 都非常的暗，这就需要超大质量黑洞处在一个极端宁静的状态，这或者是因为吸积率非常低，或者是因为吸积流的辐射效率低下，又或者两个原因都有[5]。基于径移吸积模型[16-19]、喷流–盘模型[20]或是邦迪–霍伊尔模型[21,22]等，大量理论图景都提出了这一想法。径移吸积模型的一个重要的预言是钱德拉能量波段的 X 射线光谱应当由吸积流外围区域（$R \gtrsim 10^3 R_S$）高温气体的热轫致辐射所主导，但是这样大的区域无法产生我们所看到的快速、相对大幅度的光变。因此，X 射线耀发的性质与径移吸积流模型不相符。这一事件的低光度、短时标特征，也与

tidal disruption of a star by a central supermassive black hole[23].

In all models, the radio-to-submillimetre spectrum of Sgr A* is cyclo-synchrotron emission from a combination of sub-relativistic and relativistic electrons (and perhaps positrons) spiralling around magnetic field lines either in a jet or in a static region within the inner $10R_S$ of the accretion flow. The electron Lorentz factor inferred from the radio spectrum of Sgr A* is $\gamma_e \approx 10^2$. If the X-ray flare were produced via direct synchrotron emission, then the emitting electrons would need $\gamma_e \gtrsim 10^5$. For the 10–100 G magnetic field strengths predicted by the models, the cooling time of the particles would be ~1–100 s. Thus, the approximately 10-ks duration of the flare would require repeated injection of energy to the electrons. On the other hand, if the X-rays were produced via up-scattering of the submillimetre photons off the relativistic electrons, a process called synchrotron self-Comptonization (SSC), then $\gamma_e \approx 10^2-10^3$ would be required, and the cooling time would be of the order of hours, which is consistent with the duration of the flare. The rapid turn-off of the X-ray emission might then be attributed to the dilution of both photon and electron densities in an expanding plasma.

The X-ray spectra of radio-quiet quasars and active galactic nuclei are thought to be produced by thermal Comptonization of infrared-to-ultraviolet seed photons from a cold, optically thick, geometrically thin accretion disk by hot electrons in a patchy corona above the disk[13,24]. The X-ray spectra of these sources generally "soften" (that is, become steeper and extend less strongly to higher energies) as they brighten[15]. In contrast, the extremely low luminosity of Sgr A* precludes the presence of a standard, optically thick accretion disk[5]; hence, the dominant source of seed photons would be the millimetre-to-submillimetre synchrotron photons.

The energy released by an instability in the mass accretion rate or by a magnetic reconnection event near the black hole would shock-accelerate the electrons, causing the synchrotron spectrum to intensify and to extend farther into the submillimetre band. Consequently, the Compton up-scattered X-ray emission would harden as the X-ray intensity increased, exactly as observed. We note that the millimetre-band spectrum of Sgr A* has been observed to harden during one three-week flare[25] and one three-day flare[26], as would be required by the current SSC models for Sgr A* (refs 20, 22).

To test the SSC models, we measured the flux density of Sgr A* at a wavelength of 3 mm with the Millimeter Array at the Owens Valley Radio Observatory, simultaneous with part of the 2000 Chandra measurement. Unfortunately, the available observing window (20:10–02:30 UT) preceded the X-ray flare (04:03–06:50 UT) by a few hours. The observed flux density of Sgr A* was 2.05 ± 0.3 Jy, consistent with previously reported measures[27,28]. Recently, a 106-day quasi-periodicity has been reported in the centimetre band from an analysis of 20 years of data taken with the Very Large Array (VLA)[29]. A weekly VLA monitoring program detected a 30% increase in the radio flux density of Sgr A* beginning around 24 October 2000 and peaking on 5 November 2000. This increase was seen at 2 cm, 1.3 cm, and 7 mm (R. McGary, J.-H. Zhao, W. M. Goss and G. C. Bower, personal communication).

中心超大质量黑洞潮汐瓦解恒星[23]不相符。

在所有模型中，Sgr A* 从射电到亚毫米波段的光谱来源于亚相对论性和相对论性电子（或许还有正电子）组合沿喷流中或者吸积流 $10R_s$ 以内稳定区域中的磁场线旋进所产生的回旋–同步加速辐射。从 Sgr A* 的射电谱推断，其电子的洛伦兹因子 γ_e 约为 10^2。如果 X 射线耀发是通过直接的同步加速辐射产生的，那么发出辐射的电子的洛伦兹因子需要达到 $\gamma_e \gtrsim 10^5$。对于模型所预言的 $10 \sim 100$ G 的磁场强度，粒子的冷却时间大约为 $1 \sim 100$ s。因此只有向电子重复注入能量，才能维持大约 10 ks 的耀发。另一方面，如果 X 射线是通过相对论性电子对亚毫米波光子的向上散射（此过程称为同步加速辐射的自康普顿效应（SSC））产生的，那么就需要 $\gamma_e \approx 10^2 \sim 10^3$，而冷却时间将是小时量级，这与耀发所持续的时间相符。X 射线辐射的快速减弱则可以归因于膨胀等离子体中光子和电子密度的稀释。

射电宁静类星体与活动星系核的 X 射线光谱被认为是由来自冷的、光学厚的、几何薄的吸积盘的从红外到紫外的种子光子被盘上方零散的冕中的热电子热康普顿化产生的[13,24]。这些源的 X 射线谱通常随着亮度的增强会出现"软化"的现象（即光谱变得更陡，更少能延伸到较高能量）[15]。相比之下，Sgr A* 的极低的光度排除了一个标准的、光学厚的吸积盘存在[5]的可能性；因此，种子光子的主要来源将是从毫米波到亚毫米波的同步加速辐射光子。

黑洞附近质量吸积率的不稳定性或者磁重联事件释放的能量将通过激波的方式加速电子，导致同步加速辐射谱增强并进一步延伸到亚毫米波段。因此，正如观测到的，随着 X 射线辐射强度的增加，康普顿向上散射造成的 X 射线辐射将会硬化。我们注意到已经观测到 Sgr A* 毫米波段的光谱在一次历时三周[25]和一次历时三天[26]的耀发过程中发生了硬化，而这正是目前 Sgr A* 的 SSC 模型（参考文献 20 和 22）所要求的。

为了检验 SSC 模型，我们利用欧文斯谷射电天文台的毫米波阵列以及部分钱德拉 2000 年的观测数据对 Sgr A* 在波长为 3 mm 处的流量密度进行了测量。不巧的是，可用的观测时段（20:10 至 02:30 UT）比 X 射线耀发事件（04:03 至 06:50 UT）早了几个小时。观测到 Sgr A* 的流量密度为 2.05 ± 0.3 Jy，这与以前公布的结果[27,28]是一致的。最近报道，甚大阵（VLA）20 年数据[29]分析得到了 Sgr A* 在厘米波段的一个 106 天的准周期性。一个 VLA 的每周监控项目探测到 Sgr A* 的射电流量密度大约从 2000 年 10 月 24 日起开始了一次 30% 的增强，在 2000 年 11 月 5 日达到峰值。这次增强在 2 cm、1.3 cm 以及 7 mm 的波长上看到了（与麦加里、赵军辉、戈斯和

The timing of the X-ray flare and the rise in the radio flux density of Sgr A* suggests that there is a connection between the two events, providing additional indirect support for the association of the X-ray flare with Sgr A* and further strengthening the case that it was produced via either the SSC or direct synchrotron processes. Definitive evidence for these ideas will require detection of correlated variations in the radio-to-submillimetre and X-ray wavebands through future coordinated monitoring projects.

(**413**, 45-48; 2001)

F. K. Baganoff[*], M. W. Bautz[*], W. N. Brandt[†], G. Chartas[†], E. D. Feigelson[†], G. P. Garmire[†], Y. Maeda[††], M. Morris[§], G. R. Ricker[*], L. K. Townsley[†] & F. Walter[‖]

[*] Center for Space Research, Massachusetts Institute of Technology, Cambridge, Massachusetts 02139-4307, USA

[†] Department of Astronomy and Astrophysics, Pennsylvania State University, University Park, Pennsylvania 16802-6305, USA

[‡] Institute of Space and Astronautical Science, 3-1-1 Yoshinodai, Sagamihara, 229-8501, Japan

[§] Department of Physics and Astronomy, University of California at Los Angeles, Los Angeles, California 90095-1562, USA

[‖] Department of Astronomy, California Institute of Technology, Pasadena, California 91125, USA

Received 27 April; accepted 2 August 2001.

References:

1. Richstone, D. *et al.* Supermassive black holes and the evolution of galaxies. *Nature* **395** (suppl. on optical astronomy) A14-A19 (1998).

2. Genzel, R., Pichon, C., Eckart, A., Gerhard, O. E. & Ott, T. Stellar dynamics in the Galactic Centre: proper motions and anisotropy. *Mon. Not. R. Astron. Soc.* **317**, 348-374 (2000).

3. Ghez, A. M., Morris, M., Becklin, E. E., Tanner, A. & Kremenek, T. The accelerations of stars orbiting the Milky Way's central black hole. *Nature* **407**, 349-351 (2000).

4. Lynden-Bell, D. & Rees, M. J. On quasars, dust and the Galactic Centre. *Mon. Not. R. Astron. Soc.* **152**,461-475 (1971).

5. Melia, F. & Falcke, H. The supermassive black hole at the Galactic Center. *Annu. Rev. Astron. Astrophys.* **39**, 309-352 (2001).

6. Baganoff, F. K. *et al.* Chandra X-ray spectroscopic imaging of Sgr A* and the central parsec of the Galaxy. *Astrophys. J.* (submitted); also preprint astro-ph/0102151 at ⟨xxx.lanl.gov⟩ (2001).

7. Morris, M. & Serabyn, E. The galactic center environment. *Annu. Rev. Astron. Astrophys.* **34**, 645-702 (1996).

8. Menten, K. M., Reid, M. J., Eckart, A. & Genzel, R. The position of Sagittarius A*: accurate alignment of the radio and infrared reference frames at the Galactic Center. *Astrophys. J.* **475**, L111-L114 (1997).

9. Weisskopf, M. C., O'Dell, S. L. & van Speybroeck, L. P. Advanced X-Ray Astrophysics Facility (AXAF). *Proc. SPIE* **2805**, 2-7 (1996).

10. Predehl, P. & Schmitt, J. H. M. M. X-raying the interstellar medium: ROSAT observations of dust scattering halos. *Astron. Astrophys.* **293**, 889-905 (1995).

11. Mushotzky, R. F., Done, C. & Pounds, K. A. X-ray spectra and time variability of active galactic nuclei. *Annu. Rev. Astron. Astrophys.* **31**, 717-761 (1993).

12. Davies, M. B., Blackwell, R., Bailey, V. C. & Sigurdsson, S. The destructive effects of binary encounters on red giants in the Galactic Centre. *Mon. Not. R. Astron. Soc.* **301**, 745-753 (1998).

13. Nandra, K., George, I. M., Mushotzky, R. F., Turner, T. J. & Yaqoob, T. ASCA observations of Seyfert 1 galaxies. I. Data analysis, imaging, and timing. *Astrophys. J.* **476**, 70-82 (1997).

14. Ptak, A., Yaqoob, T., Mushotzky, R., Serlemitsos, P. & Griffiths, R. X-ray variability as a probe of advection-dominated accretion in low-luminosity active galactic nuclei. *Astrophys. J.* **501**, L37-L40 (1998).

15. Ulrich, M.-H., Maraschi, L. & Urry, C. M. Variability of active galactic nuclei. *Annu. Rev. Astron. Astrophys.* **35**, 445-502 (1997).

16. Narayan, R., Mahadevan, R., Grindlay, J. E., Popham, R. G. & Gammie, C. Advection-dominated accretion model of Sagittarius A*: evidence for a black hole at the Galactic center. *Astrophys. J.* **492**, 554-568 (1998).

17. Quataert, E. & Narayan, R. Spectral models of advection-dominated accretion flows with winds. *Astrophys. J.* **520**, 298-315 (1999).

18. Ball, G. H., Narayan, R. & Quataert, E. Spectral models of convection-dominated accretion flows. *Astrophys. J.* **552**, 221-226 (2001).

19. Blandford, R. D. & Begelman, M. C. On the fate of gas accreting at a low rate on to a black hole. *Mon. Not. R. Astron. Soc.* **303**, L1-L5 (1999).

20. Falcke, H. & Markoff, S. The jet model for Sgr A*: radio and X-ray spectrum. *Astron. Astrophys.* **362**, 113-118 (2000).

21. Melia, F. An accreting black hole model for Sagittarius A*. II: A detailed study. *Astrophys. J.* **426**, 577-585 (1994).

22. Melia, F., Liu, S. & Coker, R. Polarized millimeter and submillimeter emission from Sagittarius A* at the Galactic center. *Astrophys. J.* **545**, L117-L120 (2000).

23. Rees, M. J. Tidal disruption of stars by black holes of 10^6–10^8 solar masses in nearby galaxies. *Nature* **333**, 523-528 (1988).

鲍尔的个人交流）。X 射线耀发的时间与 Sgr A* 射电波段流量密度的增强的时间相一致，这意味着两个事件之间存在联系，这为本次 X 射线耀发与 Sgr A* 之间的关联提供了额外的间接支持，并且进一步强化了该事件起源于 SSC 或者直接的同步加速辐射过程的情形。要获得这些想法的决定性证据，我们还需要通过未来的协调监测项目探测射电到亚毫米波段与 X 射线波段之间的相关光变。

（刘项琨 翻译；陈阳 审稿）

24. Haardt, F., Maraschi, L. & Ghisellini, G. X-ray variability and correlations in the two-phase disk-corona model for Seyfert galaxies. *Astrophys. J.* **476**, 620-631 (1997).

25. Tsuboi, M., Miyazaki, A. & Tsutsumi, T. in *The Central Parsecs of the Galaxy* (ed. Falcke, H. *et al.*) Vol. 186, 105-112 (ASP Conf. Ser., Astronomical Society of the Pacific, San Francisco, 1999).

26. Wright, M. C. H. & Backer, D. C. Flux density of Sagittarius A at λ = 3 millimeters. *Astrophys. J.* **417**, 560-564 (1993).

27. Serabyn, E. *et al.* High frequency measurements of the spectrum of SGR A*. *Astrophys. J.* **490**, L77-L81 (1997).

28. Falcke, H. *et al.* The simultaneous spectrum of Sagittarius A* from 20 centimeters to 1 millimeter and the nature of the millimeter excess. *Astrophys. J.* **499**, 731-734 (1998).

29. Zhao, J.-H., Bower, G. C. & Goss, W. M. Radio variability of Sagittarius A*—a 106 day cycle. *Astrophys. J.* **547**, L29-L32 (2001).

30. Reid, M. J. The distance to the center of the Galaxy. *Annu. Rev. Astron. Astrophys.* **31**, 345-372 (1993).

Acknowledgements. We thank M. Begelman for useful comments. This work has been supported by a grant from NASA.

Correspondence and requests for materials should be addressed to F.K.B.(e-mail: fkb@space.mit.edu).

Appendix: Index by Subject
附录：学科分类目录

Physics
物理学

Chemistry
化学

Biology
生物学

Astronomy
天文学

Geoscience
地球科学